ASPECTS OF TODAY´S COSMOLOGY

Edited by **Antonio Alfonso-Faus**

Aspects of Today´s Cosmology
Edited by Antonio Alfonso-Faus

Published by InTech

Janeza Trdine 9, 51000 Rijeka, Croatia

Publishing Process Manager Martina Blecic

Technical Editor Teodora Smiljanic

Cover Designer InTech Design Team

Additional hard copies can be obtained from orders@intechopen.com

Aspects of Today´s Cosmology, Edited by Antonio Alfonso-Faus
p. cm.
ISBN 978-953-307-626-3

For EU product safety concerns:
IN TECH d.o.o., Prolaz Marije Krucifikse Kozulić 3, 51000 Rijeka, Croatia,
info@intechopen.com or visit our website at intechopen.com.

Meet the editor

Professor Antonio Alfonso-Faus holds a Diploma in Space Science (Physics and Dynamics) from the University College, London (U.K. 1965), a M.Sci. and a Ph.D. degree in Physics and Mathematics from the University of Minnesota (U.S.A., 1968), and a Ph.D. degree in Aeronautical Engineering from the Technical University of Madrid (T.U.M., Spain, 1969). He was a Research Student, and a Research Fellow, in the field of Space Research (ESRO in Europe and NASA in the U.S.A.). He held the chair of mechanics, as Full Professor (E.U.I.T. Aeronáutica, Madrid, Spain, 1986). At present he is an Emeritus Professor at the Technical University of Madrid (Spain, 2010). His research work includes: the interplanetary magnetic field, the interaction of the solar wind with the Moon, the Moon's wake observed by the satellite Explorer 35, the angular momentum of the sun, the rotation of the planets, the structure and evolution of the universe and quantum gravity as related to entropy-information theory.

Contents

Preface

We have a history of cosmology, as a science, that goes back to about 100 years. It was Albert Einstein's general relativity, published at the initial decades of the last century, the starting point for applying the scientific method to the knowledge of our universe. The first model was built by Einstein applying his field equations to cosmology. It was a static model, a constant size universe. Soon it was realized that the universe should be expanding: the Hubble's red shift law from distant galaxies had this interpretation. And reversing the time, going back into the past, it was clear that close to the beginning the universe should have been very small, and in a state of very high density and temperature. Putting things forward in time, we get the idea of an initial explosion: the so-called big-bang. But this intuitive idea had many problems built in. One of them was how to explain the present size of the visible universe: about 10^{28}cms. To arrive at such a large size things in the past needed to have been going much faster than today. And this is where the idea of an initial INFLATION (by Guth and Linde), a very rapid exponential expansion, came into the picture. To be validated it had to predict observable properties. One of them was the flatness of the universe, flatness to a high degree because of the rapid exponential expansion that irons the initial curvature of space-time. And this is what is observed.

The initial, and very rapid, exponential expansion had to be quickly stopped by some braking agent: the attractive gravitational force seems to be a good candidate. But the action of this attractive force could not completely stop the expansion of the universe. There is no evidence of any shrinking of the universe in the past: it has always been expanding. And we know that gravity has always been present. Today we know that the present state of the universe is that of an accelerated expansion. And there is evidence of zero acceleration at about half way back in time from today. Apparently gravity was able to cancel the inertial acceleration left after inflation. But it did not reverse the expansion. After this zero point for acceleration the universe went on expanding more and more, and today this expansion is observed to be accelerating. Even more intriguing is the fact that, extrapolating this acceleration to the future, the universe will probably disaggregate to infinite in a rather short time, considered in terms of cosmological scales. It is obvious that a pushing expanding force of some kind is still present. One conclusion is that gravity is not able to reverse the expansion of the universe: there is another agent present, stronger than gravity, which probably will "soon" produce a doomsday for our universe. Certainly gravity must have been operating always, since the very beginning, and most probable will always be there. But there is something very important and more powerful that overcomes the attractive gravitational force. It was Einstein again the first cosmologist that realized this, and he added in his equation the so called lambda term to push and overcome gravity. Initially he just wanted to equilibrate gravity and to get a static universe. But the Hubble findings, and the same Einstein´s cosmological equations, soon inclined the scientific cosmological community to accept the idea that the universe had to be expanding.

This book presents some aspects of the picture presented above. Some scientists approach here the subject from different points of view. The book presents then a versatile picture: it is the result of the work of many scientists in the field of cosmology, in accordance with their expertise and particular interests. It is a collection of different aspects produced by important scientists in the field of cosmology. It is a bit representative of the odyssey that we have in cosmology, following the effort to understand our universe. And it has challenging subjects, like the possible doomsday that is pending confirmation from the expected experimental data to be obtained within the next decade.

Each chapter of the book has its particular value: comprehensive reviews, (inflation by Prof. Sergio del Campo), new approaches to cosmology (Prof. Tartaglia, Dr. Olmo, Prof. Vallée, Profs. Fahr and Skaliwska), dark matter and dark energy (Drs. Disney, Capistrano and Maia, Abada and Nasri), new cosmological models (Prof. Khadekar et al., Dr. Socorro et al., Prof. Vigoureux, Prof. Bali Raj), more mathematical approaches (Drs. Cho and Hong, Dr. Zecca, Dr. Cabo, Dr. Ortzas and Smith, Dr. Mena et al., Drs. Li, Xiao and Zhu), and my own contribution to the possible finite lifetime of the universe.

It has been an honor to me to have had the opportunity to read these papers. I want to thank all the authors for their contribution to the science of cosmology.

Let everybody meet the challenges of the future, trying to find the right answers to them.

Dr. Antonio Alfonso-Faus
Emeritus Professor (UPM)
Spain

Part 1

Inflation

1

Warm Inflationary Universe Models

Sergio del Campo
Pontificia Universidad Catolica de Valparaiso,
Instituto de Fisica, Curauma, Valparaiso.
Chile

1. Introduction

The most appealing cosmological model to date is the standard hot big-bang scenario. This model rests on the assumption of the cosmological principle that the universe is both homogeneous and isotropic at large scale (Peebles, 1991; 1993; 1994; Weinberg, 2008).
Even though this model could explain observational facts such that the approximately 3-K microwave background radiation (Penzias & Wilson, 1965), the primordial abundances of the light elements[1] (Alpher et al., 1948; Gamow, 1946), the Hubble expansion (Hubble, 1929; Hubble & Humason, 1931) and the present acceleration (Perlmutter et al., 1999; Riess et al., 1998), it presents some shortcomings ("puzzles") when this is traced back to very early times in the evolution of the universe. Among them we distinguish the horizon, the flatness, and the monopole problems. In dealing with these "puzzles", the standard big-bang model demands an unacceptable amount of fine-tuning concerning the initial conditions for the universe.
Inflation has been proposed as a good approach for solving most of the cosmological "puzzles" (Guth, 1981)[2]. The essential feature of any inflationary universe model proposed so far is the rapid (accelerated) but finite period of expansion that the universe underwent at very early times in its evolution.
This brief accelerated expansion serves, apart of solving most of the cosmological problems mentioned previously, to produce the seeds that, in the course of the subsequent eras of radiation and matter dominance, developed into the cosmic structures (galaxies and clusters thereof) that we observe today. In fact, the present popularity of the inflationary scenario is entirely due to its ability to generate a spectrum of density perturbations which lead to structure formation in the universe. In essence, the conclusion that all the observations of microwave background anisotropies performed so far support inflation, rests on the consistency of the anisotropies with an almost Harrison-Zel'dovich power spectrum predicted by most of the inflationary universe scenarios (Peiris et al., 2003).
The different inflationary universe model could be classified depending how the scale factor, $a(t)$, evolves with the cosmological time, t. One of the first models considered that the scale factor follows a de Sitter law of expansion, i.e. $a(t) \sim \exp Ht$, with H the Hubble "parameter". Examples of these models are "old inflaton" (Guth, 1981), "new inflation" (Albrecht & Steinhardt, 1982; Linde, 1982), "chaotic inflation" (Linde, 1983; 1986), and some corrections to this model (Cárdenas et al., 2003). Also, were described models in which the scale factor follows a power law, i.e. $a(t) \sim t^n$, with $n > 1$ (Lucchin & Matarrese, 1985). Models that present this sort of behavior are "extended inflation"

---

[1] For an historical review on this point, see the Alpher & Herman's article (Alpher & Herman, 1988).
[2] A complete description of inflationary scenarios can be found in the book by Linde (Linde, 1990a).

(La & Steinhardt, 1989) and its applications (Barrow & Maeda, 1990; Campuzano et al., 2006; del Campo & Vilenkin, 1989; del Campo & Herrera, 2003; 2005), "chaotic extended inflation" (Linde, 1990b), "hyperextended inflation" (Steinhardt & Accetta, 1990), which corresponds to a generalization of the extended models. Various studied of this sort of scenario have been presented in the literature (del Campo, 1991; De Felice & Trodden, 2004; Liddle & Wands, 1992). Also, there exist a particular scenario of "intermediate inflation" (Barrow, 1990; Barrow & Saich, 1990) in which the scale factor evolves as $a(t) \sim \exp At^f$, where A is constant and f is a free parameter which ranges $0 < f < 1$. In this sort of scenario, the expansion of the universe is slower than standard de Sitter inflation, but faster than power law inflation. The main motivation to study this latter kind of model becomes from string/M theory. This theory suggests that in order to have a ghost-free action high order curvature invariant corrections to the Einstein-Hilbert action must be proportional to the Gauss-Bonnet (GB) term (Boulware & Deser, 1985; 1986). This kind of theory has been applied to the study of accelerated cosmological solutions (Nojiri et al., 2005). In particular, very recently, it has been found that (Sanyal, 2007) for a dark energy model the GB interaction in four dimensions with a dynamical dilatonic scalar field coupling leads to a solution of the form a $a(t) = a_0 \exp At^f$. One of the problems that arises in these kind of models is due to the characteristic of the scalar inflaton potential, $V(\phi)$, that it does not present a minimum. The usual mechanism introduced to bring inflation to an end becomes useless. In fact, the standard mechanism is described by the stage of oscillations of the scalar field which is an essential part of the so-called reheating mechanism, where most of the matter and radiation of the universe was created, via the decay of the inflaton field, while the temperature grows in many orders of magnitude. It is at this point where the big bang universe is recovered. Here, the reheating temperature, the temperature associated to the temperature of the universe when the big bang model begins, is of particular interest. In this epoch the radiation domination begins, where there exist a number of particles of different kinds. In order to bring the intermediate inflationary period to an end it is introduced a special mechanisms of reheating via the introduction of a new scalar field, the so called curvaton field (del Campo & Herrera, 2007a; Lyth & Wands, 2002; Mollerach, 1990).

Another possible way of schematizing inflationary models is the classification scheme in term of *large-field*, *small-field* and *hybrid* models (Lyth & Riotto, 1999). In the case of large-field inflation, (where the inflaton potential, $V(\phi)$, satisfies the inequalities $V'' > 0$ and $(logV)'' < 0$, with the primes denoting the derivatives with respect to the inflaton field) we have that the scalar inflaton potential is usually taken to be a polynomial, $V(\phi) = \lambda^4(\phi/\phi_c)^n$, where λ^4 represent the vacuum energy density during inflation, ϕ_0 represents the change of the inflaton field during inflation and n is a real number, or exponential, such that $V(\phi) = \lambda^4 \exp(\phi/\phi_c)$. A typical example of this kind of model is chaotic inflation (Linde, 1983; 1986). The most appealing property that these sort of models have is they do not need special initial conditions for inflation to start (the start fine-tuning). Of course this fine-tuning has nothing to do with the fine-tuning needed during the evolution of inflation (the dynamic fine-tuning). Also, these models are interesting for their simplicity. They predict a significant amount of tensor perturbations due to the scalar inflaton field gets across the trans-Planckian distance during inflation (Lyth, 1997) (a fact that should be checked by astronomical observations). However, due to the inflaton crosses the trans-Planckian boarder, there appear some problems when one wants to calculate the trans-Planckian expectation value of the inflaton field.

There exist other type of inflationary models that do not need trans-Planckian expectation values of the inflaton field. These kind of models are part of the so-called small-field (they characterize by $V'' < 0$ and $(logV)'' < 0$). They have been discussed in the context of D-brane inflation (Baumann et al., 2007) in the supersymmetric standard model(Allahverdi et al., 2006) and in supergravity (Lalak & Turzynski, 2008). In each of these cases some fine-tuning of the effective inflaton potential is required (see Ref. (Linde & Westphal, 2008) for recent treatment of these issues).

The third category of inflationary universe models are called hybrid inflation(in this case the inflaton potential satisfies $V'' > 0$ and $(logV)'' > 0$) (Linde, 1991; 1994). Here, are introduced two scalar fields: one of the fields is the inflaton field, ϕ, which is responsible for the slow-roll period of inflation, the other one, χ takes care of the end of inflation. In this process, inflation ends abruptly and is followed by a regime during which topological defects (like global string (Shafi & Vilenkin, 1984; Vilenkin & Everett, 1982)) could be produced. Perhaps, these topological defect might play an interesting role in giving an appropriated expression for density perturbation which is important for understanding the large scale structure in galaxy formation (Vilenkin & Shellard, 2000). One of the problems that confront hybrid inflation is related with the fine tuning needed at the beginning of inflation (only a small fraction of possible initial conditions give rise to successful inflation). This problem is solved if it is considered nonrenomalizable coupling between the two scalar fields ϕ and χ. Also, it was found that hybrid inflation is not compatible with the supersymmetric standard models. Here it is found that the gravitinos are overproduced by the inflaton decay (Kawasaki et al., 2006a;b) and thus, in this context hybrid inflation is disfavored. The solution of this problem needs to take some fine tuning.

Beside of the possible classification of the different inflationary universe scenarios presented above we may add, in general term, that there are two main competing scenarios in regard to the *slow roll* inflation: The standard inflationary model is divided into two regimes: the slow roll and reheating epochs. In the slow roll period the universe inflates and all interactions between the inflaton scalar field and any other field are typically neglected. Subsequently, a reheating period is invoked to end the brief acceleration. After reheating, the universe is filled with relativistic fluid and thus the universe is connected with the radiation big bang phase.

Warm inflation is an alternative mechanism for having successful inflation. As is well known, warm inflation[3] - as opposed to the conventional "cool" inflation (Kolb & Tuner, 1990; Liddle & Lyth, 2000) - has the attractive feature of not necessitating a reheating phase at the end of the accelerated expansion thanks to the decay of the inflaton into radiation and particles during the slow roll (Berera, 1995; 1997; Berera & Fang, 1995; del Campo et al., 2008). Thus, the temperature of the Universe does not drop dramatically and the Universe can smoothly proceed into the decelerated, radiation-dominated era essential for a successful big bang nucleosynthesis (Peebles, 1993). This scenario has further advantages, namely: (i) the slow-roll condition $\dot{\phi}^2 \ll V(\phi)$ can be satisfied for steeper potentials, (ii) the density perturbations originated by thermal fluctuations may be larger than those of quantum origin (Berera, 2000; Gupta et al, 2002; Taylor & Berera, 2000), (iii) it may provide a very interesting mechanism for baryogenesis (Brandenberger & Yamaguchi, 2003) and (iv) it may also be considered as a model, which comes from an effective high dimensional theory. Different applications of warm inflation have been presented in the literature (Cid et al., 2007; del Campo & Herrera, 2007b; 2008; Herrera et al., 2006).

Apart of the advantage described above, warm inflation was criticized on the basis that the inflaton cannot decay during the slow roll (Yokoyama & Linde, 1999). However, in recent years, it has been demonstrated that the inflaton can indeed decay during the slow-roll phase - see (Bastero-Gil & Berera, 2005; Berera & Ramos, 2005a; Hall & Moss, 2005) and references therein - whereby it now rests on solid theoretical grounds.

We should mention that in warm inflation, dissipative effects are important during inflation, so that radiation production occurs concurrently with the accelerating expansion. The

[3] For a nice review on warm inflationary scenarios see the article (Berera et al., 2009).

dissipating effect arises from a friction term which describes the processes of the scalar field dissipating into a thermal bath via its interaction with other fields. In fact, we may say that the decay of the scalar field is described by means of an interaction Lagrangian. For instance, the authors of (Berera & Ramos, 2003; 2005b; Hall et al., 2004a) take the interaction terms of the form $\frac{1}{2}\lambda^2\phi^2\chi^2$ and $g\chi\overline{\psi}\psi$ where the inflationary period presents a two-stage decay chain $\phi \to \chi \to \psi$. In this case, they reported that the damping term Γ becomes $\lambda^3 g^2\phi/256\pi^2$.
Also, warm inflation shows how thermal fluctuations during inflation may play a dominant role in producing the initial perturbations. In such models, the density fluctuations arise from thermal rather than quantum fluctuations (Berera, 2000; Berera & Fang, 1995; Hall et al., 2004b; Moss, 1985). These fluctuations have their origin in the hot radiation and influence the inflaton through a friction term in the equation of motion of the inflaton scalar field (Berera, 1996; del Campo et al., 2007c). Among the most attractive features of these models, warm inflation ends when the universe heats up to become radiation dominated; at this epoch the universe stops inflating and smoothly enters a radiation dominated big bang phase (Berera, 1995; 1997). The matter components of the universe are created by the decay of either the remaining inflationary field or the dominant radiation fluid.
In this chapter we present the warm inflationary universe scenarios in some detail. The chapter will develop recent advances on this area of continuous research, and their possible implications in the near future, specially, those related with the confrontations with new astrophysical observations, which will put strong constraints on these kind of inflationary universe models. In order to do this, our guideline has been to concentrate on resent results that seem likely still to be of general concern to those researchers that show interest in this subject. Here, we pretend to indulge in recollections of different works on this area of research that have been put forward in the literature. In this way, the intention of this chapter is to make these developments accessible to someone who is interested in understanding how the warm inflationary universe models works. Throughout this chapter we use units in which $c = \hbar = k_B = 1$.

2. Warm inflation at work

We start by considering a spatially flat Friedmann-Robertson-Walker (FRW) universe filled with a self-interacting inflaton scalar field ϕ, of energy density, $\rho_\phi = \frac{1}{2}\dot{\phi}^2 + V(\phi)$ (with $V(\phi) = V$ the scalar potential), and a radiation energy density, ρ_γ.
The corresponding Friedmann equation reads

$$3H^2 = \kappa\left(\rho_\phi + \rho_\gamma\right). \tag{2.1}$$

Here, the constant κ is given by $\kappa = 8\pi G = 8\pi/m_P^2$, with m_P the Planck mass.
The dynamics of the cosmological model, for ρ_ϕ and ρ_γ in the warm inflationary scenario is described by the equations

$$\dot{\rho}_\phi + 3H\left(\rho_\phi + P_\phi\right) = -\Gamma\dot{\phi}^2, \tag{2.2}$$

and

$$\dot{\rho}_\gamma + 4H\rho_\gamma = \Gamma\dot{\phi}^2, \tag{2.3}$$

where $P_\phi = \frac{1}{2}\dot{\phi}^2 - V$ and Γ represents the dissipation coefficient and it is responsible of the decay of the scalar field into radiation during the inflationary era. Γ can be assumed to be a constant or a function of the scaler field ϕ, or the temperature T, or both (Berera, 1995; 1997). On the other hand, Γ must satisfy $\Gamma > 0$ in agreement with the Second Law of Thermodynamics. Dots mean derivatives with respect to the cosmological time.

During the inflationary epoch the energy density associated to the scalar field dominates over the energy density associated to the radiation field (Berera, 2000; Hall et al., 2004b; Moss, 1985) i.e. $\rho_\phi > \rho_\gamma$, the Friedmann equation (2.1) reduces to

$$H^2 \approx \frac{\kappa}{3}\rho_\phi, \tag{2.4}$$

and from Eqs. (2.2) and (2.4), we can write

$$\dot{\phi}^2 = -\frac{2\,\dot{H}}{\kappa\,(1+Q)}, \tag{2.5}$$

where Q is the rate defined as

$$Q = \frac{\Gamma}{3H}. \tag{2.6}$$

For the strong (weak) dissipation regime, we have $Q \gg 1$ ($Q \ll 1$).
We also consider that during warm inflation the radiation production is quasi-stable (Berera, 2000; Hall et al., 2004b; Moss, 1985), i.e. $\dot{\rho}_\gamma \ll 4H\rho_\gamma$ and $\dot{\rho}_\gamma \ll \Gamma\dot{\phi}^2$. From Eq.(2.3) we obtained that the energy density of the radiation field becomes

$$\rho_\gamma = \frac{\Gamma\dot{\phi}^2}{4H} = -\frac{\Gamma\,\dot{H}}{2\kappa\,H\,(1+Q)}, \tag{2.7}$$

which could be written as $\rho_\gamma = C_\gamma\,T^4$, where $C_\gamma = \pi^2\,g_*/30$ and g_* is the number of relativistic degrees of freedom. Here T is the temperature of the thermal bath.
From Eqs.(2.5) and (2.7) we get that

$$T = \left[-\frac{\Gamma\,\dot{H}}{2\kappa\;C_\gamma H\,(1+Q)}\right]^{1/4}. \tag{2.8}$$

From first principles in quantum field theory the dissipation coefficient Γ is computed for models in cases of low-temperature regimes (Moss & Xiong, 2006) (see also Ref. Berera & Ramos (2001)). Here, was developed the dissipation coefficients in supersymmetric models which have an inflaton together with multiplets of heavy and light fields. In this approach, it was used an interacting supersymmetric theory, which has three superfields Φ, X and Y with a superpotential, $W = \frac{1}{\sqrt{2}}g\Phi X^2 - \frac{1}{\sqrt{2}}hXY^2$. The scalar components of the superfields are ϕ, χ and y respectively [4]. In the low -temperature regime, i.e. where their masses satisfy $m_\chi, m_\psi > T > H$, the dissipation coefficient, when χ and y are singlets, becomes (Moss & Xiong, 2006)

$$\Gamma \simeq 0.64\,g^2\,h^4\left(\frac{g\,\phi}{m_\chi}\right)^4\frac{T^3}{m_\chi^2}. \tag{2.9}$$

This latter equation can be rewritten as

$$\Gamma \simeq C_\phi\,\frac{T^3}{\phi^2}, \tag{2.10}$$

[4] This potential could be easily modified to produce Hybrid inflation (Moss & Xiong, 2006).

where $C_\phi = 0.64\, h^4\, \mathcal{N}$. Here $\mathcal{N} = \mathcal{N}_\chi \mathcal{N}_{decay}^2$, where $\mathcal{N}_\chi$ is the multiplicity of the X superfield and $\mathcal{N}_{decay}$ is the number of decay channels available in X's decay (Bueno Sanchez et al., 2008; Moss & Xiong, 2006).
From Eq.(2.8) the above equation becomes

$$\Gamma^{1/4}\,(1+Q)^{3/4} \simeq \left[\frac{-2\,\dot{H}}{9\,\kappa\, C_\gamma\, H}\right]^{3/4} \frac{C_\phi}{\phi^2}, \tag{2.11}$$

which determines the dissipation coefficient in the strong (or weak) dissipative regime in terms of scalar field ϕ and the parameters of the model.
In general the scalar potential can be obtained from Eqs.(2.1) and (2.7)

$$V(\phi) = \frac{1}{\kappa}\left[3\,H^2 + \frac{\dot{H}}{(1+Q)}\left(1+\frac{3}{2}\,Q\right)\right], \tag{2.12}$$

which could be expressed explicitly in terms of the scalar field, ϕ, by using Eqs.(2.5) and (2.11), in the weak (or strong) dissipative regime.

3. The inclusion of viscous pressure

Usually, for the sake of simplicity, in studying the dynamics of warm inflation the particles created in the decay of the inflaton are treated as radiation thereby ignoring altogether the existence of particles with mass in the fluid thus generated. However, the very existence of these particles necessarily alters the dynamics as they modify the fluid pressure in two important ways: (i) its hydrodynamic, equilibrium, pressure is no longer $p_\gamma = \rho_\gamma/3$, with ρ_γ the energy density of the radiation fluid, but the slightly more general expression $p = (\gamma - 1)\rho$ where the adiabatic index, γ, is bounded by $1 \leq \gamma \leq 2$. (ii) It naturally arises a non-equilibrium, viscous, pressure Π, via two different mechanisms: (a) the inter-particle interactions (Huang, 1987), and (b) the decay of particles within the fluid (Zeldovich, 1970).
Concerning the latter mechanism, it is well known that the decay of particles within a fluid can be formally described by a bulk viscous pressure, Π. This is so because the decay is an entropy-producing scalar phenomenon linked to the spontaneous widening of the phase space and the bulk viscous pressure is also an scalar entropy-producing agent. In the case of warm inflation it has been proposed that the inflaton can excite a heavy field and trigger the decay of the latter into light fields (Berera & Ramos, 2003; 2005a).
Recently, a detailed analysis of the dynamics of warm inflation with viscous pressure showed that when $\Pi \neq 0$ the inflationary region takes a larger portion of the phase space associated to the autonomous system of differential equations than otherwise (Mimoso et al., 2006). It then follows that the viscous pressure facilitates inflation and lends support to the warm inflationary scenario.
For the viscous pressure we shall assume the usual fluid dynamics expression $\Pi = -3\zeta H$ (Huang, 1987), where ζ denotes the phenomenological coefficient of bulk viscosity and H the Hubble function. This coefficient is a positive-definite quantity (a restriction imposed by the second law of thermodynamics) and in general it is expected to depend on the energy density of the fluid. We shall resort to the WMAP data to restrict the aforesaid coefficient. In this case Eq.(2.3) becomes

$$\dot{\rho} + 3H(\rho + p + \Pi) = \dot{\rho} + 3H(\gamma\rho + \Pi) = \Gamma\dot{\phi}^2. \tag{3.1}$$

In this section we shall restrict our analysis to the strong (or high) dissipation regime, i.e., $Q \gg 1$. The reason for this limitation is the following. Outside this regime radiation and particles produced both by the decay of the inflaton and the decay of the heavy fields will be much dispersed by the inflationary expansion, whence they will have little chance to interact and give rise to a non-negligible bulk viscosity. Likewise, because a much lower number of heavy fields will be excited the number of decays of heavy fields into lighter ones will diminish accordingly. (The weak dissipation regime ($R \leq 1$) has been considered by Berera and Fang (Berera & Fang, 1995) and Moss (Moss, 1985). Further, if R is not big, the fluid will be largely diluted and the mean free path of the particles will become comparable or even larger than the Hubble horizon. Hence, the regime will no longer be hydrodynamic but Knudsen's and the hydrodynamic expression $\Pi = -3\zeta H$ we are using for the viscous pressure will become invalid.

3.1 Scalar and tensor perturbations in presence of viscosity

We introduce the dimensionless slow-roll parameters ε and η (Kolb & Tuner, 1990; Linde, 1990b; Lyth, 2000), as a function of the inflaton scalar potential, $V(\phi)$ and its two first derivatives, $V_{,\phi} = dV(\phi)/d\phi$ and $V_{,\phi\phi} = d^2V(\phi)/d\phi^2$,

$$\varepsilon \equiv -\frac{\dot{H}}{H^2} = \frac{1}{2(1+Q)}\left[\frac{V_{,\phi}}{V}\right]^2, \tag{3.2}$$

and

$$\eta \equiv -\frac{\ddot{H}}{H\dot{H}} \simeq \frac{1}{(1+Q)}\left[\frac{V_{,\phi\phi}}{V} - \frac{1}{2}\left(\frac{V_{,\phi}}{V}\right)^2\right]. \tag{3.3}$$

In order to find scalar (density) and tensor (gravitational) perturbations we take the perturbed FRW metric in the longitudinal gauge which is given by

$$ds^2 = (1+2\Phi)\,dt^2 - a(t)^2(1-2\Psi)\,\delta_{ij}\,dx^i dx^j, \tag{3.4}$$

where the functions $\Phi = \Phi(t,\mathbf{x})$ and $\Psi = \Psi(t,\mathbf{x})$ denote the gauge-invariant variables of Bardeen (Bardeen, 1980). Introducing the Fourier components $e^{i\mathbf{kx}}$, with k the wave number, the following set of equations, in the momentum space, follow from the perturbed Einstein field equations -to simplify the writing we omit the subscript k-

$$\Phi = \Psi, \tag{3.5}$$

$$\dot{\Phi} + H\Phi = \frac{1}{2}\left[-\frac{(\gamma\rho+\Pi)\,a\,v}{k} + \dot{\phi}\,\delta\phi\right], \tag{3.6}$$

$$(\delta\phi)\ddot{} + [3H+\Gamma]\,(\delta\phi)\dot{} + \left[\frac{k^2}{a^2} + V_{,\phi\phi} + \dot{\phi}\Gamma_{,\phi}\right]\delta\phi \;= 4\dot{\phi}\,\dot{\Phi} - \left[\dot{\phi}\,\Gamma + 2V_{,\phi}\right]\Phi, \tag{3.7}$$

$$(\delta\rho)\dot{} + 3\gamma H\delta\rho + ka(\gamma\rho+\Pi)v + 3(\gamma\rho+\Pi)\dot{\Phi} - \dot{\phi}^2\Gamma_{,\phi}\delta\phi - \Gamma\dot{\phi}[2(\delta\phi)\dot{} + \dot{\phi}\Phi] = 0, \tag{3.8}$$

and

$$\dot{v} + 4Hv + \frac{k}{a}\left[\Phi + \frac{\delta p}{(\rho+p)} + \frac{\Gamma\dot{\phi}}{(\rho+p)}\delta\phi\right] = 0, \tag{3.9}$$

where

$$\delta p = (\gamma - 1)\delta\rho + \delta\Pi, \qquad \delta\Pi = \Pi\left[\frac{\zeta_{,\rho}}{\zeta}\,\delta\rho + \Phi + \frac{\dot{\Phi}}{H}\right], \tag{3.10}$$

and the quantity v arises upon splitting the velocity field as $\delta u_j = -\frac{iak_j}{k}\, v\, e^{i\mathbf{kx}}$ $(j = 1,2,3)$ (Bardeen, 1980).
Since the inflaton and the matter-radiation fluid interact with each other isocurvature (i.e., entropy) perturbations emerge alongside the adiabatic ones. This occurs because warm inflation can be understood as an inflationary model with two basics fields (Oliveira, 2002; Starobinski & Yokoyama, 1995; Starobinski & Tsujikawa, 2001). In this context, dissipative effects themselves can produce a variety of spectral ranging from red to blue (Berera, 2000; Hall et al., 2004a; Oliveira, 2002), thus producing the running blue to red spectral suggested by WMAP data (Hinshaw et al., 2009; Komatsu et al., 2009; 2011; Larson et al., 2011; Spergel et al., 2007).
When looking for non-decreasing adiabatic and isocurvature modes on large scales, $k \ll aH$ (which depend only weakly on time), it is permissible to neglect $\dot{\Phi}$ and those terms with two-times derivatives. This together with the slow-roll approximation, the above equations simplify enough so we can find solutions in such a way that expressions for the corresponding scalar and tensor perturbations could be written down.
Here, the density perturbation becomes given by the expression[5]

$$\delta_H^2 \approx \frac{2}{25\,\pi^2}\exp[-2\widetilde{\Im}(\phi)]\left[\frac{T_r}{\widetilde{\varepsilon}\,Q^{1/2}\,V^{3/2}}\right], \tag{3.11}$$

where $\widetilde{\varepsilon} \approx \frac{1}{2Q}\left[\frac{V_{,\phi}}{V}\right]^2$ denotes the dimensionless slow-roll parameter in the high dissipation phase, i.e. $\widetilde{\varepsilon} = \varepsilon(Q \gg 1)$, T_r stands for the temperature of the thermal bath and the function $\widetilde{\Im}(\phi)$ result to be

$$\widetilde{\Im}(\phi) = -\int\left\{\frac{\Gamma_{,\phi}}{\Gamma} + \frac{3}{8\,G(\phi)}\left[1 - \left((\gamma-1) + \Pi\,\frac{\zeta_{,\rho}}{\zeta}\right)\frac{\Gamma_{,\phi}V_{,\phi}}{3\gamma\Gamma\,H}\right](\ln(V))_{,\phi}\right\}d\phi. \tag{3.12}$$

The scalar spectral index n_s is defined by

$$n_s - 1 = \frac{d\ln\,\delta_H^2}{d\ln k}, \tag{3.13}$$

which, upon using Eqs.(3.11) and (3.13), results to be given by

$$n_s \approx 1 - \left[\widetilde{\varepsilon} + 2\,\widetilde{\eta} + \left(\frac{2\widetilde{\varepsilon}}{Q}\right)^{1/2}\left[2\widetilde{\Im}_{,\phi} - \frac{Q_{,\phi}}{2R}\right]\right], \tag{3.14}$$

where

$$\widetilde{\eta} \approx \frac{1}{Q}\left[\frac{V_{,\phi\phi}}{V} - \frac{1}{2}\left(\frac{V_{,\phi}}{V}\right)^2\right] \tag{3.15}$$

[5] See Ref. (del Campo et al., 2007c) for details.

stands for the second slow-roll parameter, η, when $Q \gg 1$.

One interesting feature of the seven-year data gathered by the WMAP experiment is a significant running in the scalar spectral index $dn_s/d\ln k = \alpha_s$ (Komatsu et al., 2011). Dissipative effects can lead to a rich variety of spectral from red to blue (Berera, 2000; Hall et al., 2004a; Oliveira, 2002). From Eq.(3.14) it is seen that in our model the running of the scalar spectral index is given by

$$\alpha_s \simeq -\sqrt{\frac{2\,\widetilde{\varepsilon}}{Q}}\,[\widetilde{\varepsilon}_{,\phi} + 2\widetilde{\eta}_{,\phi}] - \frac{\widetilde{\varepsilon}}{Q}\left[\left(\frac{\widetilde{\varepsilon}_{,\phi}}{\widetilde{\varepsilon}} - \frac{Q_{,\phi}}{Q}\right)\left(2\widetilde{\Im}_{,\phi} - \frac{Q_{,\phi}}{2Q}\right) + \left(4\widetilde{\Im}_{,\phi\phi} - (\ln(Q))_{,\phi\phi}\right)\right]. \quad (3.16)$$

In models with only scalar fluctuations, the marginalized value of the derivative of the spectral index can be approximated by $dn_s/d\ln k = \alpha_s \sim -0.05$ for WMAP only (Spergel et al., 2007). In including the SN "Constitution" sample[6] of type Ia supernovae (Hicken et al., 2009), which presents a proof for the current acceleration of the universe, and the Baryonic Acoustic Oscillations (BAOs), which are the sound oscillations of the primeval baryon-photon fluid prior to the recombination epoch[7] (Eisenstein et al., 1998), WMAP-7 presented the range $-0.065 < \alpha_s < 0.010$ (Komatsu et al., 2011; Larson et al., 2011) for the running scalar spectral index α_s.

With regard to the generation of tensor perturbations during inflation gives rise to stimulated emission in the thermal background of gravitational waves (Bhattacharya et al., 2006). As a consequence, an extra temperature dependent factor, $\coth(k/2T)$, where, k and T stand for the wave number and the temperature, respectively, enters the spectrum, $A_g^2 \propto k^{n_g}$. Thus it now reads,

$$A_g^2 = 2\left(\frac{H}{2\pi}\right)^2 \coth\left[\frac{k}{2T}\right] \simeq \frac{V}{6\,\pi^2}\coth\left[\frac{k}{2T}\right], \quad (3.17)$$

the spectral index being

$$n_g = \frac{d}{d\,\ln k}\ln\left[\frac{A_g^2}{\coth[k/2T]}\right] = -2\,\varepsilon, \quad (3.18)$$

where we have used Eq.(3.2).

A quantity of prime interest is the tensor-scalar ratio, defined as $R(k_0) = \left.\left(\frac{A_g^2}{P_{\mathcal{R}}}\right)\right|_{k=k_0}$ where $P_{\mathcal{R}} \equiv 25\delta_H^2/4$ and k_0 is known as the pivot point. Its expression in the high dissipation limit, $R \gg 1$, follows from using Eqs. (3.11) and (3.17),

$$R(k_0) = \left.\left(\frac{A_g^2}{P_{\mathcal{R}}}\right)\right|_{k=k_0} = \frac{2}{3}\left.\left[\left(\frac{\widetilde{\varepsilon}\,r^{1/2}\,V^{5/2}}{T_r}\right)\exp[2\,\widetilde{\Im}(\phi)]\coth\left(\frac{k}{2T}\right)\right]\right|_{k=k_0}. \quad (3.19)$$

[6] This corresponds to an extension of the "Union" sample (Kowalski et al., 2008).

[7] Quite recently, the size of the BAO peak was detected in the large-scale correlation function clustering of approximately 44,000 luminous red galaxies from the Sloan Digital Sky Survey (SDSS) (Eisenstein et al., 2005)

In the case in which we consider a chaotic scalar potential, i.e. $V(\phi) = \frac{1}{2} m^2 \phi^2$, where $m > 0$ is a free parameter, and (as mentioned above) we restrict ourselves to study the high dissipation regime ($Q \gg 1$).
From Eq.(3.11), the scalar power spectrum results to be

$$P_{\mathcal{R}}(k_0) \approx \frac{1}{2\pi^2} \left[8\gamma\Gamma_0 V(\phi_0)^{1/2} + 2\sqrt{3}m^2(1-2\gamma) \right. \\ \left. +3\sqrt{3}\zeta_0\Gamma_0(2-3\gamma)\right]^{3/2} \left[\frac{\Gamma_0^{1/2}\, T_r}{3^{1/4} m^2\, V(\phi_0)^{3/4}}\right], \tag{3.20}$$

Likewise, Eq.(3.19) provides us with the tensor-scalar ratio

$$R(k_0) \approx \frac{2}{3} \left[8\gamma\Gamma_0 V(\phi_0)^{1/2} + 2\sqrt{3}m^2(1-2\gamma) \right. \\ \left. +3\sqrt{3}\zeta_0\Gamma_0(2-3\gamma)\right]^{-3/2} \left[\frac{3^{1/4} m^2 V(\phi_0)^{7/4}}{\Gamma_0^{1/2} T_r}\right] \coth\left(\frac{k}{2T}\right), \tag{3.21}$$

where $V(\phi_0)$ and ϕ_0 stand for the potential and the scalar field, respectively, when the perturbation, of scale $k_0 = 0.002\text{Mpc}^{-1}$, was leaving the horizon.
By resorting to the WMAP three-year data, $P_{\mathcal{R}}(k_0) \simeq 2.3 \times 10^{-9}$ and $R(k_0) = 0.095$, and choosing the parameters $\gamma = 1.5$, $m = 10^{-6}\,\mathrm{m}_P$, $T \simeq T_r \simeq 0.24 \times 10^{16}$ GeV and $k_0 = 0.002\,\text{Mpc}^{-1}$, it follows from Eqs. (3.20) and (3.21) that $V(\phi_0) \simeq 1.5 \times 10^{-11}\,\mathrm{m}_P^4$ and $\zeta_0 \simeq 3 \times 10^{-6}\,\mathrm{m}_P^3$. When the scale k_0 was leaving the horizon the inflaton decay rate Γ_0 is seen to be of the order of $10^{-3}\,\mathrm{m}_P$. Thus Eq. (3.16) tells us that one must augment ζ_0 by two orders of magnitude to have a running spectral index α_s close to the observed value (Spergel et al., 2007).
While cool inflation typically predicts a nearly vanishing bispectrum, and hence a small (just a few per cent) deviation from Gaussianity in density fluctuations -see e.g. (Gangui et al., 1994)-, warm inflation clearly predicts a non-vanishing bispectrum. The latter effect arises from the non-linear coupling between the the fluctuations of the inflaton and those of the radiation. This can produce a moderate non-Gaussianity (Gupta, 2006; Gupta et al, 2002) or even a stronger one -likely to be detected by the PLANCK satellite (Ade et al., 2011; PLANCK Collaboration, 2009)- if the aforesaid nonlinear coupling is extended to subhorizon scales (Moss & Xiong, 2007). Because Π implies an additional coupling between the radiation and density fluctuations it is to be expected that non-Gaussianity will be further enhanced. Perhaps, this could serve to observationally constrain Π by future experiments.
Thus, our model presents two interesting features: (i) Related to the fact that the dissipative effects plays a crucial role in producing the entropy mode, they can themselves produce a rich variety of spectral ranging from red to blue. The possibility of a spectrum which does run so is particularly interesting because it is not commonly seen in inflationary models which typically predict red spectral. (ii) The viscous pressure may tell us about how the matter-radiation component behaves during warm inflation. Specifically, it will be very interesting to know how the viscosity contributes to the large scale structure of the Universe. In this respect, we anticipate that the PLANCK mission (Ade et al., 2011; PLANCK Collaboration, 2009) will significantly enhance our understanding of the large scale structure by providing us with high quality measurements of the fundamental power spectrum over an larger wavelength range than the WMAP experiment.

3.2 Viscosity and the stability of warm inflation

Any inflationary model -whether "cold" or "warm"- must fulfill the requirement of stability[8]; that is to say, its inflationary solutions ought to be attractors in the solution space of the relevant cosmological solutions. It means, in practice, that the scalar field, ϕ, must approach an asymptotic attractor characterized by $\dot{\phi} \simeq -\frac{\partial V}{\partial \phi}(3H)^{-1}$ in cold inflation, and $\dot{\phi} \simeq -\frac{(\partial V/\partial\phi)}{3H(1+Q)}$ in warm inflation (see e.g. Liddle et al. (1994); Salopek & Bond (1990)). This ensures that the system will stay sufficiently near to the slow-roll solution for many Hubble times. Here V denotes the scalar field potential and H the Hubble expansion rate.

In the case of warm inflation the conditions for stability have been considered by de Oliveira and Ramos (Oliveira & Ramos, 1998) and, recently, more fully by Moss and Xiong (Moss & Xiong, 2008) who allowed the scalar potential and the damping rate to depend not only on the inflaton field but on the temperature of the radiation gas as well. This automatically introduces two further slow-roll parameters and renders the conditions for a successful warm inflationary scenario even less restrictive.

Here, we want to study the stability of warm inflationary solutions by considering the presence of massive particles and fields in the radiation fluid as well as the existence of a the viscous pressure, Π, associated to the resulting mixture of heavy and light particles.

The corresponding field equations are those described previously, but now we will take both the scalar potential and the damping rate as a function of the temperature, i.e. $V = V(\phi, T)$ and $\Gamma = \Gamma(\phi, T)$.

The total pressure becomes

$$p = \frac{1}{2}\dot{\phi}^2 - V(\phi, T) + (\gamma - 1)Ts + \Pi, \tag{3.22}$$

where we have included the entropy density, s, that follows from the thermodynamical relation $s = -\partial f/\partial T \simeq -V_{,T}$, when the Helmholtz free-energy, $f = (1/2)\dot{\phi}^2 + V(\phi, T) + \rho_\gamma - Ts$, is dominated by the scalar potential.

The conservation of the stress-energy can be expressed as

$$T\dot{s} + 3H(\gamma Ts + \Pi) = \Gamma\dot{\phi}^2 . \tag{3.23}$$

Making $u = \dot{\phi}$, the slow roll equations take the form

$$u = \frac{-V_{,\phi}}{3H(1+Q)}, \qquad Ts = \frac{Qu^2 + 3H\zeta}{\gamma}, \qquad 3H^2 = V(\phi, T) . \tag{3.24}$$

To find the conditions for the validity of the slow roll approximation, we perform a linear stability analysis to see whether the system remains close to the slow roll solution for many Hubble times. In cold inflationary scenario, the slow roll equation is of first order in the time derivative. Choosing the inflaton field as independent variable, the conservation equations (2.1) and (3.23) can be written as first order equations in the derivative with respect to ϕ, indicated by a prime,

$$x' = F(x) , \tag{3.25}$$

[8] For more details on this subsection see Ref. (del Campo et al., 2010)

where

$$x = \begin{pmatrix} u \\ s \end{pmatrix}. \tag{3.26}$$

Thus, the system (2.1), (3.23) becomes

$$u' = -3H - \Gamma - V_{,\phi} u^{-1}, \tag{3.27}$$

$$s' = -3H\gamma s u^{-1} - 3H\Pi (Tu)^{-1} + T^{-1}\Gamma u. \tag{3.28}$$

Here the Hubble rate and entropy density are determined by (2.2) and $s \simeq -V_{,T}$, respectively. Taking a background $\overline{x}$ which satisfies the slow roll equations (3.24), the linearized perturbations satisfy

$$\delta x' = M(\overline{x})\delta x - \overline{x}', \tag{3.29}$$

where

$$M = \begin{pmatrix} A & B \\ C & D \end{pmatrix}, \tag{3.30}$$

is the matrix of first derivatives of F evaluated at the slow roll solution. Linear stability demands that its determinant be positive and its trace negative.

The matrix elements read,

$$A = \frac{H}{u}\left\{-3(1+Q) - \frac{\epsilon}{(1+Q)^2}\right\}, \tag{3.31}$$

$$B = \frac{H}{s}\left\{-cQ - \frac{Q}{(1+Q)^2}\epsilon + b(1+Q)\right\}, \tag{3.32}$$

$$C = \gamma\frac{Hs}{u^2}\left(6 - \frac{\epsilon}{(1+Q)^2}\right)\left\{1 + \frac{\Pi}{\gamma^2\rho_\gamma}\left(\frac{6(1+Q)^2 - 2\epsilon}{6(1+Q)^2 - \epsilon}\right)\right\}, \tag{3.33}$$

$$D = \gamma\frac{H}{u}\left(c - 4 - \frac{Q\epsilon}{\gamma^2(1+Q)^2}\right) + \frac{H\Pi}{u\gamma\rho_\gamma}\left\{c - \frac{Q\epsilon}{\gamma^2(1+Q)^2} + \frac{3\Pi}{2\gamma^2\rho_\gamma}\right\}. \tag{3.34}$$

In the strong regime ($Q \gg 1$), the determinant and trace of M assume the comparatively simple expressions

$$\det M = \frac{3\gamma Q H^2}{u^2}\left(4 - 2b + c + (c-2b)\frac{\Pi}{\gamma^2\rho_\gamma} - \frac{3}{2}\frac{\Pi^2}{\gamma^4\rho_\gamma^2}\right), \tag{3.35}$$

and

$$\mathrm{tr} M = \frac{H}{u}\left\{-3Q + \gamma(c-4) + \frac{\Pi}{2\gamma^3\rho_\gamma}\left(2\gamma^2 c + 3\frac{\Pi}{\rho_\gamma}\right)\right\}. \tag{3.36}$$

Sufficient conditions for stability are that M varies slowly and that

$$|c| \leq \frac{4 - 3\sigma^2/2}{1+\sigma} - 2b, \qquad b \geq 0, \tag{3.37}$$

where $\sigma \equiv \frac{\Pi}{\gamma^2\rho_\gamma}$. Upon these conditions the determinant results positive and the trace negative, implying stability of the corresponding solution. Expression (3.37.1) generalizes Eq. (27) of Moss and Xiong (Moss & Xiong, 2008).

Since the chosen background is not an exact solution of the complete set of equations, the forcing term in equation (3.29) depends on $\overline{x}'$, and will be valid only if $\overline{x}'$ is small. The size of $\overline{x}'$ depends on the quantities $\dot{u}/(Hu)$ and $\dot{s}/(Hs)$. From the time derivative of (3.24.3) we obtain

$$\frac{\dot{H}}{H^2} = -\frac{\epsilon}{1+Q}. \tag{3.38}$$

Combining this with the other slow-roll equations, (3.24.1) and (3.24.2), we get

$$\frac{\dot{u}}{Hu} = \frac{1}{\Delta}\left[-\frac{c[A(1+Q)-BQ]-4}{1+Q}\epsilon + \frac{4Q}{1+Q}\beta + (Ac-4)\eta - \frac{3(1+Q)c}{1-f}b\right], \tag{3.39}$$

and

$$\frac{\dot{s}}{Hs} = \frac{3}{\Delta}\left[\frac{A(3+Q)-B(1+Q)}{1+Q}\epsilon + \frac{Q-1}{1+Q}A\beta\right.$$

$$\left. -2A\eta - \frac{(1+Q)[Ac(Q-1)+Q+1]c}{(1-f)Q}b\right], \tag{3.40}$$

where

$$\Delta = 4(1+Q) + Ac(Q-1), \qquad A = \frac{\rho_\gamma + \gamma^{-1}\Pi}{\rho_\gamma - \kappa\Pi}, \qquad B = \frac{\Pi}{\rho_\gamma - \kappa\Pi}, \tag{3.41}$$

$$f = -\frac{3}{2}\frac{(1+Q)^2}{Q}\frac{\zeta}{\gamma H\epsilon}, \qquad \kappa = \rho_\gamma\frac{\zeta_{,\rho_\gamma}}{\zeta}. \tag{3.42}$$

Notice that when $\Pi \to 0$, one has that $A \to 1$, $B \to 0$, $f \to 0$, and therefore the equations (3.38)-(3.40) reduce to the corresponding expressions in Ref. Moss & Xiong (2008). Obviously, the value of the parameter κ in this limit depends on the specific expression of the viscosity coefficient, ζ; but it does not alter the value of B in the said limit. In this limit, the κ parameter could take any value depending of the model. Its value does not affect the $\Pi \to 0$ limit.
The thermal fluctuations produce a power spectrum of scalar density fluctuations of the form (Moss & Xiong, 2008)

$$\mathcal{P}_s = \frac{\sqrt{\pi}}{2}\frac{H^3 T}{u^2}\sqrt{1+Q}. \tag{3.43}$$

Note that the power spectrum of fluctuations in inflationary models where the friction coefficient depends also on the temperature, i.e., $\Gamma = \Gamma(\phi, T)$, was considered recently in Ref. Graham & Moss (2009).
We calculate the spectral index by means of

$$n_s - 1 = \frac{\dot{\mathcal{P}}_s}{H\mathcal{P}_s}. \tag{3.44}$$

By virtue of the equations (3.38)-(3.40), we obtain

$$n_s - 1 = \frac{p_1\epsilon + p_2\beta + p_3\eta + p_4 b}{\Delta}, \tag{3.45}$$

where the p_i coefficients are given by

$$p_1 = -\frac{10(2+Q) - A(3+5c+Q) + B(1+Q+(5c/2)Q)}{1+Q}, \tag{3.46}$$

$$p_2 = \frac{A(Q-1) - 10Q}{1+Q}, \tag{3.47}$$

$$p_3 = \frac{8(1+Q) - A(2+2c+2Q+3cQ)}{1+Q}, \tag{3.48}$$

$$p_4 = \frac{3(1+Q)[1+(1+5c/2)Q]}{(1-f)Q}. \tag{3.49}$$

For $Q \gg 1$, and assuming c of order unity, the p_i coefficients reduce to

$$p_1 = -10 + A - B(1+5c/2);\ p_2 = A - 10;\ p_3 = 8 - A(2+3c);\ p_4 = \frac{3Q(1+5c/2)}{(1-f)}, \tag{3.50}$$

and $\Delta = Q(4+Ac)$. Therefore (3.45) becomes

$$n_s - 1 = -\frac{10 - A + B(1+5c/2)}{(4+Ac)Q}\epsilon - \frac{10-A}{(4+Ac)Q}\beta + \frac{8 - A(2+3c)}{(4+Ac)Q}\eta + \frac{3(1+5c/2)}{(4+Ac)(1-f)}b. \tag{3.51}$$

The tensor modes happen to be the same as in the cold inflationary models (Moss & Xiong, 2008), i.e.,

$$\mathcal{P}_T = H^2, \tag{3.52}$$

and the corresponding spectral index is

$$n_T - 1 = -\frac{2}{1+Q}\epsilon. \tag{3.53}$$

With the help of of (3.52), (3.43) and (3.24.1) the tensor-to-scalar amplitude ratio can be written as

$$r = \frac{2V_{,\phi}(\phi, T)}{9\sqrt{\pi}H^3 T(1+Q)^{5/2}}. \tag{3.54}$$

The recent WMAP seven-year results imply the upper-bound $r < 0.36$ (95% CL) (Larson et al., 2011) on the scalar-tensor ratio. Below, we shall make use of this bound to set constraints on the parameters of our models.
When applying the formalism of above to the specific case in which the thermodynamic potential is taken to be (Moss & Xiong, 2008)

$$V(\phi, T) = -\frac{\pi^2}{90}g_* T^4 - \frac{1}{12}m_\phi^2 T^2 + \frac{1}{2}m_\phi^2\phi^2, \tag{3.55}$$

where g_* is the effective number of thermal particles, and the damping coefficient may be written as

$$\Gamma(\phi, T) = \Gamma_0 \left(\frac{\phi}{\phi_0}\right)^m \left(\frac{T}{\tau_0}\right)^n, \tag{3.56}$$

with n and m real numbers and ϕ_0, τ_0, and Γ_0 some nonnegative constants. The damping term has a generic form given approximately by $\Gamma \sim g^4\phi^2\tau$, where g is the coupling constant (Hall et al., 2004a). From Ref. Hosoya & Sakagami (1984) the damping term, $\tau = \tau(\phi, T)$, is related to the relaxation time of the radiation and for the models with an intermediate particle decay, $\tau = \tau(\phi)$ is linked to the lifetime of the intermediate particle. Different choices of n and m have been adopted. For instance the case $n = m = 0$ was considered by Taylor and Berera (Taylor & Berera, 2000), whereas the choice $m = 2, n = -1$ corresponds to the damping term first calculated by Hosoya (Hosoya & Sakagami, 1984). This expression slightly differs from those in Hall et al. (2004a) and Zhang (2009), where a single index rather than two was considered.
As for the bulk viscosity coefficient we use the general expression

$$\zeta = \zeta_0 \rho_\gamma^\lambda, \tag{3.57}$$

where ζ_0 is a positive semi-definite constant and λ an integer that may take any of the two values: $\lambda = 1/2$, i.e., $\zeta \propto \rho_\gamma^{1/2}$ (Li et al., 2010) (see also Ref. Brevik & Gorbunova (2005)) and $\lambda = 1$, i.e., $\zeta \propto \rho_\gamma$ (del Campo et al., 2007c).

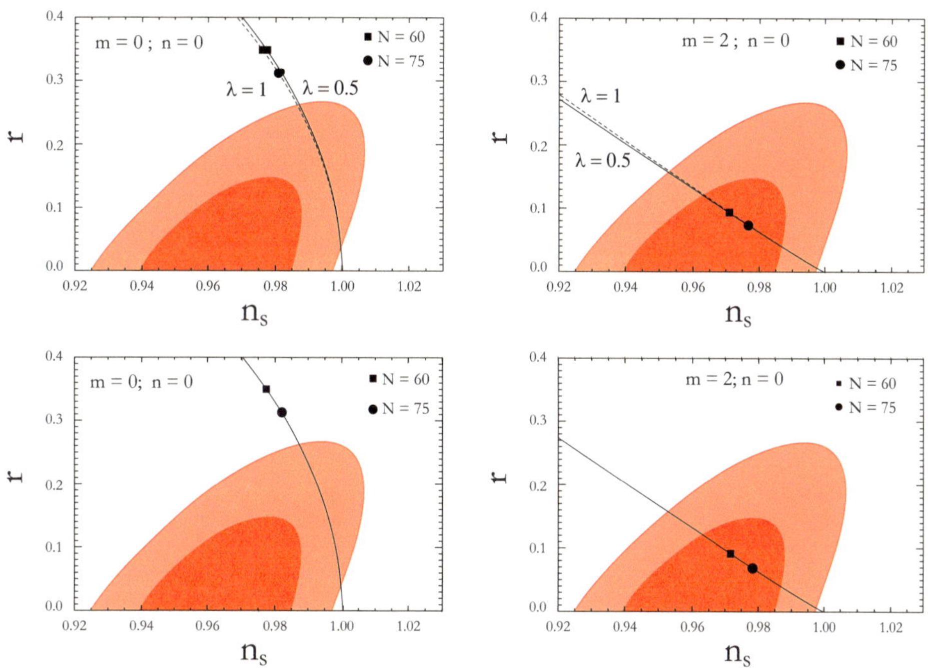

Fig. 1. Top row of panels: Plot of the tensor-scalar ratio r as a function of the spectral index n_s, for two values of the λ parameter in the case of example 1 (i.e., potential (3.55)). Bottom row: Same as the top row but assuming no viscosity ($\zeta_0 = 0$). In each panel the 68% and 95% confidence levels set by seven-year WMAP experiment are shown. The latter places severe limits on the tensor-scalar ratio (Larson et al., 2011).

Panel in Fig. 1	N	r	N	r
top left ($m = n = 0$)	60	0.351	75	0.314
top right ($m = 2, n = 0$)	60	0.094	75	0.074

Table 1. Results from first example with $\lambda = 1$ (The results for $\lambda = 1/2$ are very similar). Rows from top to bottom refers to panels of Fig. 1 from left to right.

Panel in Fig. 1	N	r	N	r
bottom left ($m = n = 0$)	60	0.350	75	0.318
bottom right ($m = 2, n = 0$)	60	0.094	75	0.074

Table 2. Results from first example with no viscosity, i.e., $\zeta_0 = 0$.

Figure 1 depicts the dependence of the tensor-scalar ratio, r, on the spectral index, n_s, for the model given by Eqs. (3.55), (3.56), and (3.57) when $\lambda = 0.5$ and when $\lambda = 1$. From Ref. (Larson et al., 2011), two-dimensional marginalized constraints (68% and 95% confidence levels) on inflationary parameters r and n_s, the spectral index of fluctuations, defined at $k_0 = 0.002$ Mpc^{-1}. The seven-year WMAP data (Larson et al., 2011) places stronger bounds on r than the five-year WMAP data (Hinshaw et al., 2009; Komatsu et al., 2009). In order to write down values that relate n_s and r, we used Eqs. (3.51) and (3.54), and the values $g_* = 100$, $\gamma = 1.5$, $\zeta_0 = (2/3)\zeta^{(1)}_{max}$, and $m_\phi = 0.75 \times 10^{-5}$, $T = 2.5 \times 10^{-6}$, $\Gamma_0 = 1.2 \times 10^{-6}$, $\tau_0 = 3.73 \times 10^{-5}$, $\phi_0 = 0.3$ for $m = 0, n = 0$; and $m_\phi = 2.5 \times 10^{-5}$, $T = 1.75 \times 10^{-6}$, $\Gamma_0 = 3.58 \times 10^{-6}$, $\tau_0 = 5.63 \times 10^{-5}$, $\phi_0 = 0.6$ for $m = 2, n = 0$, in Planck units (Hall et al., 2004a).
Figure1 suggests that the pair of indices ($m = 2$, $n = 0$), corresponding to the right panel, is preferred over the other pair of indices ($m = n = 0$), left panel. Likewise, it shows that there is little difference between choosing $\lambda = 1$ or $\lambda = 0.5$ as well as with the case of no viscosity, i.e., $\zeta_0 = 0$.
Table 2 indicates the value of the ratio r for $\lambda = 1$ and different choices of the pair of indices m and n when the number of e-folds is 60 and when it is 75. Very similar values (not shown) follow for $\lambda = 0.5$. All of them can be checked with the help of Eqs. (3.51) and (3.54).
A comparison of the results shown in both Tables indicates that only in the case of the pair ($m = n = 0$) with $N = 75$ (top and bottom left panels in Fig. 1) viscosity makes a non-negligible impact.

4. Warm inflation and non-Gaussianity

Due to the existence of a wide range of inflationary universe models it is important to discriminate between them. One of the features that can help us in this direction is the non-Gaussianity. In fact, non-Gaussian statistics (such that bispectrum) provides a powerful tool to observationally discriminate between different mechanisms for generating the curvature perturbation. But this feature not only well help us to discriminate between inflationary scenarios, but also, measurement (including an upper bound) of non-Gaussianity of primordial fluctuations is expected to have the potential to rule out many of inflationary models that have been put forward.
It has been notice that a single field, slow roll inflationary scenarios are known to produce negligible non-Gaussianity (Acquaviva et al., 2003; Maldacena, 2003), there exist now a variety of models available in the literature which may predict an observable signature. One important referent of this situation is warm inflation. The reason of this is due that warm inflation could be seen as a model which is analogous to a multi-field inflation scenario, which

is well know that can produce large non- Gaussianity which can be observed in the near future experiments such as PLANCK mission (Battefeld & Easther, 2007)
The constraint on the primordial non- Gaussianity is currently obtained from Cosmic Microwave Background measurements. WMAP sets the limit on the so-called local type of the primordial non-Gaussianity, which is parameterized by the constant dimensionless parameter f_{NL}. This parameter appears in the following expression

$$\Phi(\mathbf{x}) = \Phi_G(\mathbf{x}) + f_{NL}\left(\Phi_G^2(\mathbf{x}) - \langle\Phi_G^2(\mathbf{x})\rangle\right), \tag{4.1}$$

where Φ is Bardeeen's gauge-invariant potential, Φ_G is the Gaussian part of the potential and $\langle\ \rangle$ denotes the ensemble average. The ansatz (4.1) is known as the "local" form of non-Gaussianity[9].
The power spectrum $\mathcal{P}(k)$ of the Bardeen's gauge-invariant potential is defined by the two-point correlation function of the Fourier transform of the Bardeen's potential

$$\langle\Phi_G(\mathbf{k})\Phi_G(\mathbf{k}')\rangle = (2\pi)^3\delta^3\left(\mathbf{k}+\mathbf{k}'\right)\mathcal{P}(k), \tag{4.2}$$

where δ represents the Dirac's delta function. Similarly, The bispectrum $\mathcal{B}\left(k_1,k_2,k_3\right)$ becomes given by

$$\langle\Phi_G(\mathbf{k}_1)\Phi_G(\mathbf{k}_2)\Phi_G(\mathbf{k}_3)\rangle = (2\pi)^3\delta^3\left(\mathbf{k}_1+\mathbf{k}_2+\mathbf{k}_3\right)\mathcal{B}\left(k_1,k_2,k_3\right), \tag{4.3}$$

The δ^3 function in this last expression reflects translational invariance and ensures that $\mathcal{B}\left(k_1,k_2,k_3\right)$ depends on the three momenta in such a way that they form a triangle, i.e. $\mathbf{k}_1+\mathbf{k}_2+\mathbf{k}_3 = \mathbf{0}$. On the other hand, rotational invariance implies that the 3-spectrum function is symmetric in its arguments.
We should mentioned that the 3-point correlation function en general terms it has a very particular dependence on momenta. For instance, if it peaks when the three momenta are equal, then it is referred as *equilateral*. Now, if one of the three momenta is half of the other two, then this bispectrum is referred as *flattened*. Also, if one of the three momenta is much smaller than the other two, then we say that the bispectrum is *squeezed*. In general, the shape for the three-point spectrum could correspond to a superposition of two shapes, the *flattened* and the *equilateral* shapes, for instance (Senatore et al., 2010).
In general terms, the amount of non-Gaussianity in the bispectrum is expressed by the non-linear function f_{NL} which is given by

$$f_{NL}\left(k_1,k_2,k_3\right) = \frac{5}{6}\frac{\mathrm{B}\left(k_1,k_2,k_3\right)}{\mathcal{P}(k_1)\mathcal{P}(k_2)+\mathcal{P}(k_2)\mathcal{P}(k_3)+\mathcal{P}(k_3)\mathcal{P}(k_1)}, \tag{4.4}$$

where the numerical 5/6 factor is introduced for convenience when compared with the results of the cosmic microwave background radiation data (Komatsu & Spergel, 2001). Models in which the function f_{NL} results to be a constant are called *local* models. This kind of models arise naturally from the non-linear evolution of density perturbations on super-Hubble scales starting from Gaussian field fluctuations during the inflationary period. Other non-Gaussian models could give different expression for the bispectrum function, specially those expression which do not result from the inflationary evolution.

[9] This is not the only well-motivated form for a non-Gaussian curvature perturbation. It could be considered a non-Gaussian part of $\Phi(x)$ which need not be correlated with the gaussian part. For instance, consider a primordial curvature perturbation of the form $\Phi(\mathbf{x}) = \Phi_G(\mathbf{x}) + FNL[\Psi_G(\mathbf{x})]$, where FNL is some arbitrary nonlinear function and the field $\Psi_G(\mathbf{x})$ is a Gaussian field which is uncorrelated with $\Phi_G(x)$ (Barnaby, 2010).

The best observational limit on the non-gaussianity at present is from the WMAP seven-year data release (Komatsu et al., 2011), which gives $-10 < f_{NL} < 74$ with 95% confidence for a constant (or local) component, when combined with Large Scale Structure (LSS) data the bound becomes somewhat stronger $-1 < f_{NL}^{local} < 65$ (Slosar et al., 2008).
A description of non-Gaussianity for different models (those could have their genesis in inflationary universe models or any other different non-inflationary one) could be made by using the so called shape function (Fergusson & Shellard, 2009). This function becomes defined as

$$S\left(k_1,k_2,k_3\right)=\frac{1}{N}\left(k_1k_2k_3\right)^2 \mathrm{B}\left(k_1,k_2,k_3\right), \tag{4.5}$$

where N is a normalization factor, often taken to be $N = 1/f_{NL}$.
For instance, in the case of warm inflation it results to be

$$S^{Warm}\left(k_1,k_2,k_3\right)\propto\frac{3!}{\left(k_1k_2k_3\right)^3}\sum_{i\neq j=1}^{3}\left(k_ik_j\right)^2\left[k_i^2k_j^3-k_j^5+\sum_{l(\neq i\neq j)=1}^{3}k_l^5\right] \tag{4.6}$$

In the Fergusson and Shellard's paper (Fergusson & Shellard, 2009) it is described an improved methods for an efficient computation of the full CMB bispectrum for any general (nonseparable) primordial bispectrum, where was incorporated the flat sky approximation and a cubic interpolation. Following this approach, they have found a range for the non-linear parameter related to warm inflation

$$-107 < f_{NL}^{Warm} < 11. \tag{4.7}$$

Very recently it has been reported that for warm inflation in the strong regime the total bispectrum corresponds to a sum of two terms (Moss & Yeomans, 2011)

$$B=\frac{6}{5}f_{NL}^{local}\sum_{cycli}\mathcal{P}(k_1)\mathcal{P}(k_2)-\frac{6}{5}f_{NL}^{adv}\sum_{cycli}\left(k_1^{-2}+k_2^{-2}\right)\mathbf{k}_1\cdot\mathbf{k}_2\mathcal{P}(k_1)\mathcal{P}(k_2) \tag{4.8}$$

where f_{NL}^{Adv} represents the fluid's bulk motion (advection) terms. Here, in the case of equilateral triangles it is obtained that $f_{NL} = f_{NL}^{Local} + f_{NL}^{Adv}$.
It was found that the standard deviation of the parameter f_{NL}^{adv} is around 5 times larger than the standard deviation in the estimator f_{NL}^{Local}. For PLANCK (PLANCK Collaboration, 2009), the detection limit for f_{NL}^{Local} is expected to be around 5 - 10, depending on how successfully the backgrounds can be removed. This would imply that PLANCK would only be able to detect the presence of f_{NL}^{adv} if the value was at least 25. Certainly, the detection of the f_{NL}^{adv} contribution will demand an effort where new experiment of higher resolution need to be developed. This is an issue that has to be solved by implementing appropriated futures missions.

5. Comments and remarks

In this chapter we have considered a warm inflationary universe models. We have studied this scenario in which a viscous pressure is present in the matter-radiation fluid. We investigated the corresponding scalar and tensor perturbations. The contributions of the adiabatic and entropy modes were described explicitly. Specifically, a general relation for the density perturbations, Eq.(3.41), the tensor perturbations, Eq. (3.17), and the tensor-scalar ratio -as well as the dissipation parameter- are modified by a temperature dependent factor.

We have described various aspects of warm inflationary universe models when viscosity is taken into account. This feature is a very general characteristic in multiparticle and entropy producing systems and, in the context of warm inflation, it is of special significance when the rate of particle production and/or interaction is high. In this chapter we have focused on the strong regime described by the condition that $Q \gg 1$.
On the other hand, we have seen that one important fact of warm inflation in presence of viscosity is its stability. This feature becomes expressed by the inequalities given by (3.37). Upon these conditions the determinant (expressed by Eq. (3.35)) results positive and the trace (expressed by Eq. (3.36)) negative, implying stability of the corresponding solution.
The general expression for the spectral index, n_s, expressed by Eq. (3.45), depends explicitly on viscosity through the four p_i coefficients (see Eqs. (3.50)). The latter do not depend on the slow-roll parameters (ϵ, β, η, and b), as shown by equations (3.46)-(3.49).
In order to further ensure the stability of the warm viscous inflation, the slow-roll parameters must satisfy the following conditions

$$\epsilon \ll 1+Q, \quad |\beta| \ll 1+Q, \quad |\eta| \ll 1+Q,$$

as well as the condition on the slow-roll parameter that describes the temperature dependence of the potential, namely,

$$|b| \ll \frac{(1-f)Q}{1+Q}.$$

where f becomes given by $f \approx -\frac{3}{2}\frac{Q\zeta}{\gamma H\epsilon}$ in the strong regime.
These conditions give the necessary and sufficient condition for the existence of stable slow-roll solutions. Under these conditions, we got the same stability range obtained in the no-viscous case, so long as $\sigma = -8/3$. In this sense, the range of the slow-roll parameter c decreases when $-8/3 < \sigma < 0$, and increases when $\sigma < -8/3$.
To bring in some explicit results we have taken the constraint $n_s - r$ plane to first-order in the slow roll approximation. For the potential (Eq. (3.55)) we obtained that, when $\lambda = 0.5$ and $\lambda = 1$, the model is consistent with the WMAP seven year data for the pair of indices ($m = 2$, $n = 0$), see Fig. 1.
Note that in subsection 3.2 we did not address the case, in which we have that the coefficient of dissipation, Γ, does not depend on the inflaton field and the temperature. In this case a more detailed and laborious calculation for the density perturbation would be necessary in order to check the validity of expression (3.43). This is an issue that deserve further study.
The observational bound on the $|f_{NL}|$ parameter, which gives a limit on non-Gaussianity, comes from the WMAP seven-year data release (Komatsu et al., 2011), which combined with LSS data it becomes $-1 < f_{NL}^{local} < 65$. In this respect the PLANCK satellite observations have a predicted sensitivity limit of around $|f_{NL}| \sim 5$ (Komatsu & Spergel, 2001). The prediction of warm inflation lies well above the PLANCK threshold, with a specific angular dependence, should provide a means to test warm inflation observationally.
Finally, in general terms, when we count with a more precise set of data about the detection of deviations from a Gaussian distribution will allow us to check the predictions from warm inflationary universe scenario, or any another specific theoretical model.

6. Acknowledgments

Some of the perspectives presented here in this chapter incorporates the contributions I have developed with different colleagues. Particularly, I would like to thank Ramón Herrera and Diego Pavón, with whom I have been involved in the study of warm inflation. This work was funded by Comision Nacional de Ciencias y Tecnología through FONDECYT Grants 1110230, 1090613, 1080530 and by DI-PUCV Grant 123.710/2011.

7. References

Acquaviva, V.; Bartolo, N.; Matarrese, S. & Riotto A., (2003). Second-order cosmological perturbations from inflation. *Nucl. Phys.* B, 667, 119-148

Ade, P.A.R. et al., (2011). PLANCK early results: The power spectrum Of cosmic infrared background anisotropies. arXiv:1101.2028 [astro-ph]

Albrecht, A. & Steinhardt, P. J., (1982). Cosmology for grand unified theories with radiatively induced symmetry breaking. *Phys. Rev. Lett.*, 48, 1220-1223

Allahverdi, R.; Enqvist, K.; Garcia-Bellido, J. & Mazumdar, A., (2006). Gauge invariant MSSM inflaton. *Phys. Rev. Lett.* 97, 191304

Alpher, R.A.; Bethe, H. & Gamow, G. (1948). The origin of chemical elements. *Phys. Rev.*, 73, 803-804

Alpher, R.A. & Herman, H. (1988). Reflections on early work on "big bang" cosmology. *Phys. Today*, 41, (8) 24-34

Bardeen, J., (1980). Gauge invariant cosmological perturbations. *Phys. Rev.* D, 22, 1882-1905

Barnaby, N., (2010). Nongaussianity from particle production during inflation. *Adv. Astron.* 2010, 156180

Barrow, J.D., (1990). Graduated inflationary universes. *Phys. Lett.* B, 235, 40-43

Barrow, J.D. & Maeda K., (1990). Extended inflationary universes. *Nucl. Phys.* B, 341, 294-308

Barrow, J.D. & Saich, P., (1990). The behavior of intermediate inflationary universes. *Phys. Lett.* B, 249, 406-410

Bastero-Gil, M. & Berera, A., (2005). Determining the regimes of cold and warm inflation in the SUSY hybrid model. *Phys. Rev.* D, 71, 063515

Battefeld T. & Easther R., (2007). Non-Gaussianities in multi-field inflation. *JCAP*, 0703, 20

Baumann, D.; Dymarsky, A.; Klebanov, I. R.;McAllister L. & Steinhardt P.J., (2007). A delicate universe. *Phys. Rev. Lett.* 99, 141601

Berera, A., (1995). Warm inflation. *Phys. Rev. Lett.*, 75, 3218-3221

Berera, A., (1996). Thermal properties of an inflationary universe. *Phys. Rev.* D, 54, 2519-2534

Berera, A., (1997). Interpolating the stage of exponential expansion in the early universe: A possible alternative with no reheating. *Phys. Rev.* D, 55, 3346-3357

Berera, A., (2000). Warm inflation at arbitrary adiabaticity: A model, an existence proof for inflationary dynamics in quantum field theory. *Nuclear Physics* B, 585, 666-714

Berera, A.; Gleiser, M. & Ramos, R.O., (2009). Strong dissipative behavior in quantum field theory. *Phys. Rev.* D, 58, 123508

Berera, A. & Fang, L. Z., (1995). Thermally induced density perturbations in the inflation era. *Phys. Rev. Lett.*, 74, 1912-1915

Berera, A. & Ramos, O.R., (2001). The affinity for scalar fields to dissipate. *Phys. Rev.* D, 63, 103509

Berera, A. & Ramos, O.R., (2003). Construction of a robust warm inflation mechanism. *Phys. Lett.* B, 567, 294-304

Berera, A. & Ramos, O.R., (2005a). Dynamics of interacting scalar fields in expanding space-time. *Phys. Rev.* D, 71, 023513

Berera, A. & Ramos, O.R., (2005b). Absence of isentropic expansion in various inflation models. *Phys. Lett.* B, 607, 1-7

Berera, A.; Moss, I.G. & Ramos, O.R., (2009). Warm inflation and its microphysical basis. *Report. Prog. Phys.* 72, 026901

Bhattacharya, K.; Mohanty, S. & Nautiyal, A., (2006). Enhanced polarization of CMB from thermal gravitational waves. *Phys. Rev. Lett.* 97, 251301
Boulware, D. G. & Deser, S., (1985). String-generated gravity models. *Phys. Rev. Lett.*, 55, 2656-2660
Boulware, D. G. & Deser, S., (1986). Effective Gravity Theories With Dilatons. *Phys. Lett.* B, 175, 409-412
Brandenberger, R. H. & Yamaguchi, M., (2003). Spontaneous baryogenesis in warm inflation. *Phys. Rev.* D, 68, 023505
Brevik, I. & Gorbunova, O., (2005). Dark energy and viscous cosmology. *Gen. Rel. Grav.* 37, 2039
Bueno Sanchez, J.C.; Bastero-Gil, A.; Berera, A. & Dimopoulos, K., (2008). Warm hilltop inflation. *Phys. Rev.* D, 77, 123527
Campuzano C.; del Campo, S. & Herrera, R., (2006). Extended curvaton reheating in inflationary models. *JCAP*, 0606, 017
Cárdenas V.H.; del Campo, S. & Herrera, R., (2003). R^2 corrections to chaotic inflation. *Mod. Phys. Lett.* A, 18, 2039-2050
Cid M.A.; del Campo, S. & Herrera, R., (2007). Warm inflation on the brane. *JCAP*, 0710, 005
del Campo, S. & Vilenkin, A., (1989). Initial condition for extended inflation. *Phys. Rev.* D, (Rapid Communications), 40, 688-690
del Campo, S., (1991). Quantum cosmology for hyperextended inflation. *Phys. Lett.* B, 259, 34-37
del Campo, S. & Herrera, R., (2003). Extended open inflationary universes. *Phys. Rev.* D, 67, 063507
del Campo, S. & Herrera, R., (2005). Extended closed inflationary universes. *Class.Quant.Grav.* 22, 2687-2700
del Campo, S. & Herrera, R., (2007a). Curvaton field and intermediate inflationary universe model. *Phys. Rev.* D, 76, 103503
del Campo, S. & Herrera, R., (2007b). Warm inflation in the DGP brane-world model. *Phys.Lett.* B, 653, 122-128
del Campo, S.; Herrera, R. & Pavón, D., (2007c). Cosmological perturbations in warm inflationary models with viscous pressure. *Phys. Rev.* D, 75, 083518
del Campo, S. & Herrera, R., (2008). Warm-Chaplygin inflationary universe model. *Phys.Lett.* B, 665, 100-105
del Campo, S.; Herrera, R. & Saavedra, J., (2008). Tachyon warm inflationary universe model in the weak dissipative regime. *Eur. Phys. J.* C, 59, 913-916
del Campo, S.; Herrera, R.; Pavón, D. & Villanueva, J.R. , (2010). On the consistency of warm inflation in the presence of viscosity. *JCAP*, 08, 002
De Felice, A. & Trodden, M., (2004). Baryogenesis after hyperextended inflation. *Phys.Rev.* D, 72, 043512
Eisenstein, D.J.; Hu, W. & Tegmark, M., (1998). Cosmic complementarity: H(0) and Omega(m) from combining CMB experiments and redshift surveys. *Astrophys. J.*, 504, L57-L61
Eisenstein, D.J. et al., (2005). Detection of the baryon acoustic peak in the large-scale correlation function of SDSS luminous red galaxies. *Astrophys. J.* 633, 560-574
Fergusson, J.R. & Shellard, E.P.S., (2009). Shape of primordial non-Gaussianity and the CMB bispectrum. *Phys.Rev.* D, 80, 043510
Gamow, G., (1946). Expanding universe and the origin of elements. *Phys. Rev.*, 70, 572-573
Gangui, A.; Lucchin, F.; Matarrese S. & Mollerach, S., (1994). The Three point correlation function of the cosmic microwave background in inflationary models. *Astrophys. J.* 430, 447-457

Graham, C. & Moss, I.G., (2009), Density fluctuations from warm inflation. *JCAP*, 0907, 013
Guth, A., (1981). Inflationary universe: A possible solution to the horizon and flatness problems. *Phys. Rev.* D, 23, 347-356
Gupta, S., (2006). Dynamics and non-gaussianity in the weak-dissipative warm inflation scenario. *Phys. Rev.* D, 73, 083514
Gupta, S.; Berera, A.; Heavens A. F. & Matarrese S., (2002). Non-Gaussian signatures in the cosmic background radiation from warm inflation. *Phys. Rev.* D, 66, 043510
Hall, L.M.H. & Moss, I.G., (2005). Thermal effects on pure and hybrid inflation. *Phys. Rev.* D, 71, 023514
Hall, L.M.H.; Moss, I.G.& Berera, A., (2004a). Scalar perturbation spectra from warm inflation. *Phys. Rev.* D, 69, 083525
Hall, L.M.H.; Moss, I.G.& Berera, A., (2004b). Constraining warm inflation with the cosmic microwave background. *Phys. Lett.* B, 589, 1-6
Herrera, R.; del Campo, S. & Campuzano, C. (2006). Tachyon warm inflationary universe models. *JCAP*, 0610, 009
Hicken, M. et al., (2009), Improved Dark Energy Constraints from 100 New CfA Supernova Type Ia Light Curves. *Astrophys. J.*, 700, 1097-1140
Hinshaw, G. et al., (2009). Five-Year Wilkinson Microwave Anisotropy Probe (WMAP) Observations: Data processing, sky maps, and basic results. *Astrophys. J. Suppl.*, 180, 225-245
Hosoya, A. & Sakagami, M.-aki., (1984). Time development of Higgs field at finite temperature. *Phys. Rev.* D, 29, 2228-2239
Huang, K., (1987). *Stadistical mechanics*. Wiley, ISBN-13: 978-0471815181, New York, U.S.A.
Hubble, E., (1929). A relation between distance and radial velocity among extragalactic nebulae. *Proc. Natl. Acad. Sci.*, 15, (3), 168-173
Hubble, E. & Humason, M., (1931). The velocity-distance relation among extra-galactic nebulae. *Astrophys.J*, 74, 43-80
Kawasaki, M.; Takahashi, F. & Yanagida, T.T., (2006a). Gravitino overproduction in inflaton decay. *Phys. Lett.* B, 638, 8-12
Kawasaki, M.; Takahashi, F. & Yanagida, T.T., (2006b). The gravitino-overproduction problem in inflationary universe. *Phys. Rev.* D, 74, 043519
Kolb, E.W. & Tuner, M. S., (1990). *The Early Universe*. Addison- Wesley, Reading, ISBN-10: 0201626748, MA., U.S.A.
Komatsu, E. & Spergel, D.N., (2001). Acoustic signatures in the primary microwave background bispectrum. *Phys. Rev.* D, 63, 063002
Komatsu, E. et al., (2009). Five-Year Wilkinson Microwave Anisotropy Probe (WMAP) observations: Cosmological interpretation. *Astrophys. J. Suppl.*, 180, 330-376
Komatsu, E. et al., (2011). Seven-year Wilkinson Microwave Anisotropy Probe (WMAP) observations: Cosmological interpretation. *Astrophys. J. Suppl.*, 192, 18-75
Kowalski, M. et al., (2008), Improved cosmological constraints from new, old and combined supernova datasets. *Astrophys. J.*, 686, 749-778
La, D. & Steinhardt, P.J., (1989). Extended inflationary cosmology. *Phys. Rev. Lett.*, 62, 376-378
Lalak, Z. & Turzynski, K., (2008). Back-door fine-tuning in supersymmetric low scale inflation. *Phys. Lett.* B 659, 669-675
Larson, D. et al., (2011). Seven-Year Wilkinson Microwave Anisotropy Probe (WMAP) observations: Power spectra and WMAP-derived parameters. *Astrophys. J. Suppl.*, 192, 16-38
Li, W.; Ling, J.Y.; Wu, J.P. & Kuang, X.M., (2010). Thermal fluctuations in viscous cosmology. *Phys. Lett.* B, 687, 1-5

Liddle, A.R. & Wards, D., (1992). Hyperextended inflation: Dynamics and constraints. *Phys.Rev.* D, 45, 2665-2673
Liddle, A.R.; Parsons, P. & Barrow, J.D., (1994). Formalizing the slow roll approximation in inflation.*Phys. Rev.* D, 50, 7222-7232
Liddle, A.R. & Lyth, D., (2000). *Cosmological Inflation and Large Scale Structure*, Cambridge University Press, ISBN-13: 978-0521575980, Cambridge, England
Linde, A., (1982) A new inflationary universe scenario: A possible solution of the horizon, flatness, homogenfity, isotropy and primordial monopole problems. *Phys. Lett.* B, 108, 389-393
Linde, A., (1983). Chaotic inflation. *Phys. Lett.* B, 129, 177-181
Linde, A., (1986). Eternally existing selfreproducing chaotic inflationary universe. *Phys. Lett.* B, 75, 395-400
Linde, A., (1990a). *Particle Physics and Inflationary Cosmology*. Gordon and Breach, ISBN-13: 978-3718604890, New York, USA
Linde, A., (1990b). Extended chaotic inflation and spatial variations of the gravitational constant. *Phys. Lett.* B, 238, 160-165
Linde, A., (1991). Axions in inflationary cosmology. *Phys. Lett.* B, 259, 38-47
Linde, A., (1994). Hybrid inflation. *Phys. Rev.* D, 49, 748-754
Linde, A. & Westphal, A. (2008). Accidental inflation in string theory. *JCAP* 03, 005
Lucchin, F. & Matarrese, S., (1985). Power-law inflation. *Phys. Rev.* D, 32, 1316-1322
Lyth, D.H., (1997). What would we learn by detecting a gravitational wave signal in the cosmic microwave background anisotropy? *Phys. Rev. Lett.*,78, 1861-1863
Lyth, D.H., (2000). *Cosmological inflation and large scale structure*. Cambridge University Press, Cambridge, ISBN-13: 978-0521575980, England
Lyth, D.H. & Riotto, A., (1999). Particle physics models of inflation and the cosmological density perturbation. *Phys. Rept.*, 314, 1-146
Lyth, D.H. & Wands, D., (2002). Generating the curvature perturbation without an inflaton. *Phys. Lett.*, B, 524, 5-14
Maldacena, J. M., (2003). Non-Gaussian features of primordial fluctuations in single field inflationary models. *JHEP*, 0305, 013
Mimoso, J.P.; Nunes, A. & Pavón, D., (2006). Asymptotic behavior of the warm inflation scenario with viscous pressure. *Phys. Rev.* D, 73, 023502
Mollerach, S., (1990). Isocurvature baryon perturbations and inflation. *Phys. Rev.* D, 42, 313-325
Moss, I.G., (1985). Primordial inflation with spontaneous symmetry breaking. *Phys. Lett.* B 154, 120-124
Moss, I.G. & Xiong, C. (2006). Dissipation coefficients for supersymmetric inflatonary models. arXiv:0603266 [hep-ph]
Moss, I.G. & Xiong, Ch. (2007). Non-Gaussianity in fluctuations from warm inflation. *JCAP*, 0704, 007
Moss, I.G. & Xiong, Ch. (2008). On the consistency of warm inflation *JCAP*, 11, 023
Moss, I.G. & Yeomans, T. (2011). Non-Gaussianity in the strong regime of warm inflation. arXiv:1102.2833 [astro-ph.CO]
Nojiri, S.; Odintsov S.D. & Sasaki, M., (2005). Gauss-Bonnet dark energy. *Phys. Rev.* D, 71, 123509
Oliveira, H., (2002). Density perturbations in warm inflation and COBE normalization. *Phys. Lett.* B, 526, 1-8
Oliveira, H. & Ramos, R.O., (1998). Dynamical system analysis for inflation with dissipation.*Phys. Rev.* D, 57, 741-749

Peebles, P.J.E.; Schramm, D.N.; Turner, E.L. & Kron R.G., (1991). The case for the relativistic hot big bang cosmology. *Nature*, 352, 769 - 776.

Peebles, P.J.E., (1993). *Principles of Physical Cosmology*. Princeton University Press, ISBN-13: 978-0691019338, Princeton, New Jork, USA

Peebles, P.J.E.; Schramm, D.N.; Turner, E.L. & Kron R.G., (1994). The evolution of the universe. *Scientific American*, 271, 29-33

Peiris, H.V. et al., (2003). First-year Wilkinson Microwave Anisotropy probe (WMAP) observations: Implications for inflation. *Astrophys. J. Suppl.*, 148, 213-231

Penzias, A.A. & Wilson, R.W., (1965). A Measurement of excess antenna temperature at 4080-Mc/s.3pp. *Astrophys. J.*, 142, 419-421

Perlmutter, S. et al., (1999), Measurements of omega and lambda from 42 high-redshift supernova. *Astrophys. J.*, 517, 565-586

PLANCK Collaboration (2009). http://planck.caltech.edu/

Riess, A. et al. (1998). Observational evidence from superovae for an accelerating universe and a cosmological constant. *Astronom. J.*, 116, 1009-1038

Salopek, D.S. & Bond, J.R., (1990). Nonlinear evolution of long wavelength metric fluctuations in inflationary models. *Phys. Rev.* D, 42, 3936-3952

Sanyal, A.K., (2007). If Gauss-Bonnet interaction plays the role of dark energy. *Phys. Lett.* B, 645, 1-5

Senatore, L.; Smith, K.M. & Zaldarriaga, M., (2010). Non-Gaussianities in single field inflation and their optimal limits from the WMAP 5-year data. *JCAP*, 1001, 028

Shafi, Q. & Vilenkin. A., (1984). Spontaneously broken global symmetries and cosmology *Phys. Rev.* D, 29, 1870-1871

Slosar, A.; Hirata, C.; Seljac, U.; Ho, S. & Padmanabhan, N., (2008). Constraints on local primordial non-Gaussianity from large scale structure. *JCAP*, 0808, 031

Spergel, D.N. et al., (2007). Wilkinson Microwave Anisotropy probe (WMAP) three year results: Implications for cosmology. *Astrophys. J. Suppl.* 170, 377-408

Starobinski, A. & Yokoyama, J., (1995). *Proceedings of the Fourth Workshop on General Relativity and Gravitation*, edited by N. Nakao *et al.*, pp. 381, Kyoto University, Kyoto

Starobinsky, A. & Tsujikawa, S., (2001). Cosmological perturbations from multifield inflation in generalized Einstein theories. *Nucl. Phys.* B, 610, 383-410

Steinhardt, P.J. & Accetta, F.S., (1990). Hyperextended inflation. *Phys. Rev. Lett.*, 64, 2740-2743

Taylor A. N. & Berera A., (2000). Perturbation spectra in the warm inflationary scenario. *Phys. Rev.* D, 62, 083517

Vilenkin, A. & Everett, A.E., (1982). Cosmic strings and domain walls in models with Goldstone and pseudo-Goldstone bosons. *Phys. Rev. Lett.* 48, 1867-1870

Vilenkin, A. & Shellard E.P.S., (2000). *Cosmic strings and other topological defects*. Cambridge University Press, ISBN-13: 978-0521654760, Cambridge, England

Weinberg, S., (2008). *Cosmology*. Oxford University Press, ISBN-13: 978-0198526827, USA

Yokoyama, J. & Linde, A., (1999). Is warm inflation possible? *Phys. Rev.* D, 60, 083509

Zhang, Y., (2009). Warm inflation with a general form of the dissipative coefficient. *JCAP*, 0903, 023

Zeldovich, Ya. B., (1970). Particle production in cosmology. (In Russian). *Pisma Zh. Eksp. Teor. Fiz.* 12, 443-447 [*JETP lett.* 12, 307 (1970)]

Part 2

New Approaches to Cosmology

2

The Strained State Cosmology

Angelo Tartaglia
Politecnico di Torino
Italy

1. Introduction

When studying cosmology one is unavoidably faced with the problem of the relevance and meaning of the terms that are in use and any purely physical and mathematical discussion borders philosophy. In this respect we must move from the remark that any description of the cosmos needs the concepts of space and time. These two entities, so fundamental in physics, are indeed neither trivial nor obvious in any respect. Going back into the past to look for the thought of the first thinkers we see for instance that Aristotle could not accept the idea of an empty space, rejecting even space as something else from the extension of existing things. "Nothing" of course does not exist, so anything in between two objects has to be something: no void, no emptiness (Aristotle, 350 b.C.).
The situation with time is even worse. The ancient Grecian thought associated time with movement and with flow, however still in the antiquity but after a few centuries we find an interesting quote from St. Augustine which gives a vivid picture of the situation: "*What is time? If nobody asks me I know, however if I wish to answer anybody asking me, I don't know*" (Augustine, 398 a.D.). I do not want to enter philosophical issues but it is wise to be aware that such fundamental questions linger in the background of any scientific discussion on cosmology.
With the birth of modern physics the question regarding the nature of space and time was posed in different terms with respect to the past, but not really solved. Newton gave definitions attributing to space and time an absolute character: an immutable stage on which physical phenomena are played within an equally immutable regular flow setting the pace for all changes and movements (Newton, 1687). This simplified and solemn view was challenged at the end of the 19th century by the failure of the Galilean transformations to guarantee the invariance of Maxwell's equations. The ether affair and the Michelson-Morley null experiment gave their contribution and finally both space and time were revisited by Einstein in his brand new Special Relativity (SR) theory. In SR length and time measurements are both observer-dependent and a new absolute entity emerges: space-time. A full description of the properties of space-time required a few years and the work of a number of scientists, not only Einstein's. At the end the relation between space-time, on one side, and matter/energy, on the other, was cast into the world famous Einstein equations:

$$G_{\mu\nu} = \kappa T_{\mu\nu} \tag{1}$$

A problem still remained. It was and is with the nature of the left hand side of the equations. Usually space-time is thought of as a smart mathematical tool more than a physical entity,

even though it interacts with matter, as the equations say. This interpretation is not explicit and some doubts remain. On the physical nature of space-time I can report a quote from a speech of Einstein's pronounced in Leiden in the 20's of the past century (Einstein, 1920):
".... according to the general theory of relativity space is endowed with physical qualities; in this sense, therefore, there exists an ether. ... But this ether may not be thought of as endowed with the quality characteristic of ponderable media, as consisting of parts which may be tracked through time. ..."
Then space-time is real; Einstein's sentence was referred to the only space, but the implication is that the whole manifold has physical relevance even though it is not possible to treat it as matter.
That space-time is indeed something is clearly accepted by people who, since a long time and with poor results so far, are trying to quantize gravity. In these attempts space-time is often treated as a sort of field even though a subtle contradiction is implied. Fields need a background (space-time) to be described: what would the background of space-time be? Nobody has found a way out of this puzzle, at the moment.
I will not tackle directly the fundamental aspects of the problem; rather, I shall start from a simple remark. There is another branch of physics, classical physics, where a fully geometrical description is given: this is the theory of three-dimensional material continua and in particular the theory of elasticity. Even though at the beginning engineers and even physicists were not much attracted by that new mathematical language developed, at the end of the 19th century and first years of the 20th , by the Italian school (Ricci-Curbastro and Levi-Civita), after a while, thanks also to the onset of General Relativity, the whole machinery of tensor calculus was accepted. Today the elastic properties of continuous materials are currently accounted for and described in terms of tensors.
I shall elaborate on the correspondence between the general properties of space-time and the ones of ordinary material continua in order to work out a consistent description of the universe and its properties. As we shall see, the core of the theory expounded in the present chapter will be the presence in space-time of a strain energy that is the direct analogue of the elastic potential energy. The strain energy is associated with the curvature of space-time induced by the presence of matter/energy and/or by the presence of texture defects. This will be a classical approach to the other puzzling problem related with the vacuum energy. The idea of establishing a connection between a sort of rigidity of space-time and its vacuum energy is old (Sakharov, 1968), but usually implemented in terms of quantum physics and finally facing the problem of the huge mismatch between the values obtained from quantum computations and the value needed to account for the cosmological phenomena. Not all problems will be solved by this approach but many useful hints will be found.

2. Deformable continua

Let us start considering an N+n-dimensional space, where N and n are integers. We shall call this space *the embedding manifold* and we shall assume it is flat: the geometry in it is Euclidean. Let us cover the embedding manifold with some coordinates system that we denote with X^a (a runs from 1 to N+n).
Next we introduce two N-dimensional embedded spaces. The first will be our *reference manifold* and is assumed to be flat; the second embedded space will be the *natural manifold* and will be intrinsically curved (Eshelby, 1956). Each embedded manifold has its own coordinates; for them I use the symbols ξ^μ (reference manifold) and x^μ (natural manifold);

the μ index runs from 1 to N. In the embedding space the reference frame is expressed by n linear constraints:

$$F_i\left(X^1,....,X^{N+n}\right) = \text{constant} \tag{2}$$

Viceversa the natural frame is fixed by n generally non-linear constraints:

$$H_i\left(X^1,....,X^{N+n}\right) = \text{constant} \tag{3}$$

The index i runs from 1 to n. Eq.s (2) and (3) permit to express n of the embedding coordinates in terms of the other N on the two submanifolds. In practice the N coordinates defined on each submanifold will be functions of the N+n coordinates of the embedding space: $\xi^\mu = \xi^\mu\left(X^1,....,X^{N+n}\right)$ and $x^\mu = x^\mu\left(X^1,....,X^{N+n}\right)$. For obvious convenience n will be as small as possible, i.e. in most cases it will be n = 1; however for peculiar natural frames containing singularities one more dimension can be insufficient to give a flat embedding, so more will be required.

As an additional assumption, suppose, for the moment, that the natural manifold is sufficiently regular and all functional dependences are smooth and differentiable as many times as needed. As a consequence it will be possible to directly express the coordinates on the reference manifold as functions of those on the natural manifold and viceversa.

Once the above definitions and conditions have been declared we may establish a one to one correspondence between points located on the two embedded manifolds. This correspondence is embodied in an **u** vector field: each **u** vector goes from a point in the reference to a point in the natural manifold. The flatness of the embedding space permits a global definition of the vector field. The situation described so far is summarized in fig. 1. The vector **u** field is called the *displacement vector field*; whenever it is non-uniform we say that the natural manifold is distorted with respect to the reference one.

Considering pairs of arbitrarily near positions on both manifolds we may compare the corresponding line elements. Let us write

$$d\sigma^2 = \eta_{\mu\nu} d\xi^\mu d\xi^\nu \tag{4}$$

for the reference manifold. Due to the flatness condition it must also be

$$\eta_{\mu\nu} = \delta_{\alpha\beta} \frac{\partial y^\alpha}{\partial \xi^\mu} \frac{\partial y^\beta}{\partial \xi^\nu} \tag{5}$$

The y's are Cartesian coordinates and the metric tensor $\eta_{\mu\nu}$ corresponds to an Euclidean geometry in N dimensions.

For the natural manifold it will be

$$ds^2 = g_{\mu\nu} dx^\mu dx^\nu \tag{6}$$

Both line elements (5) and (6) can of course be expressed in the embedding space as

$$ds^2 = \delta_{ab} dX^a dX^b \tag{7}$$

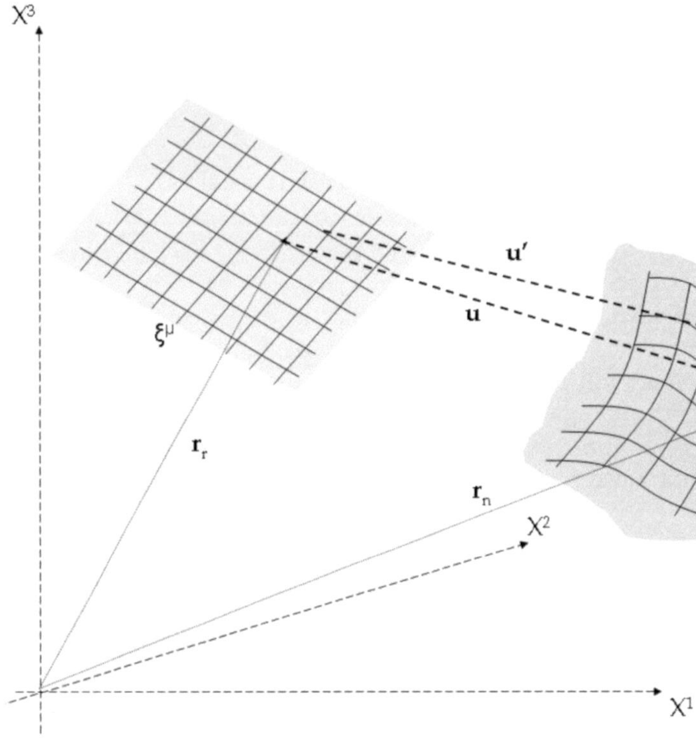

Fig. 1. The embedding space with the two embedded manifolds. The figure represents a three-dimensional embedding of two bidimensional manifolds, but the scheme can be applied to any number of dimensions.

where Cartesian coordinates are assumed, for simplicity; Latin indices from the first part of the alphabet (as a, b, c...) run from 1 to N+n. One goes from (7) to (5) or (6) applying respectively the constraints (2) and (3) and remarking that (see fig. 1) it is:

$$\mathbf{r}_n = \mathbf{r}_r + \mathbf{u} \tag{8}$$

Summing up and using (8) we see that the difference between (6) and (4) is:

$$ds^2 - d\sigma^2 = \delta_{\mu a}\frac{\partial u^a}{\partial x^\nu} + \delta_{\nu a}\frac{\partial u^a}{\partial x^\mu} + \delta_{ab}\frac{\partial u^a}{\partial x^\mu}\frac{\partial u^b}{\partial x^\nu} \tag{9}$$

The difference (9) has been written in terms of the coordinates on the natural manifold. Using on both sides the same coordinates, eq. (9), together with (4) and (6), leads to:

$$g_{\mu\nu} - \eta_{\mu\nu} = 2\varepsilon_{\mu\nu} \tag{10}$$

The elements $\varepsilon_{\mu\nu}$ belong to a rank 2 symmetric tensor in N dimensions: it is called the *strain tensor*.

So far the correspondences we have established may be though of as being purely formal, however if we consider a physical situation we may think of obtaining the natural manifold from the reference one by continuous deformation. In this case the displacement vector tells us from where to where a given point has been moved during the process and the differential part of the displacement does indeed represent the strain induced in the manifold.

2.1 Defects

The conceptual framework outlined in the previous section permits to introduce another important notion: the one of *defect* or *texture defect*.

Defects play an important role in the analysis of the properties of crystals or, in general, of material continua. A consistent description for them was worked out between the end of the 19th and the beginning of the 20th century (Volterra, 1904) and that is the picture I shall use in the following.

Consider the situation represented in fig. 2, whose general structure is the same as that of fig.1. We say we have a defect whenever a whole region $\mathcal{C}$ of the reference manifold corresponds to a point O (or a line or any other lower dimensional subset) of the natural manifold, while, for the rest, the correspondence remains one to one.

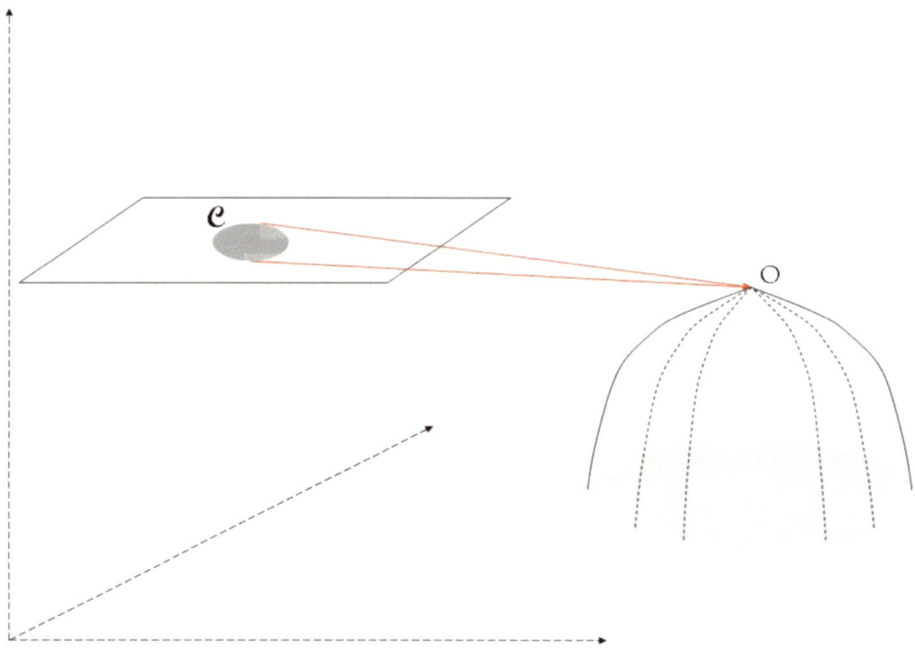

Fig. 2. Defects in continuous manifolds. Point O corresponds to a whole region $\mathcal{C}$ of the reference manifold. The natural manifold has non-zero strain.

The presence of a defect implies a non-zero strain tensor in the natural manifold and the strain is singular in correspondence of the defect. Defects also induce peculiar symmetries in the natural manifold: a pointlike defect induces a central (spherical) symmetry; a straight linear defect implies a cylindrical symmetry, etc. A whole classification of defects, on the basis of the corresponding symmetries, exists in terms of dislocations and disclinations. Volterra's classification has been extended to space-time by Puntigam and Soleng (Puntigam, 1997) who identified the 10 possible types of distortions existing in four dimensions; they wanted to apply the idea of topological defects to the study of cosmic strings. I will not enter into further details, since the general concepts are enough for the purpose of this chapter.

2.2 Elasticity

In physical terms, strain is not enough to account for what happens. We must say something about the causes of the distortion of the manifold and their interrelation with the effects. In other words, when we try to deform a material system (the reference manifold of our abstract representation) we expect it to react back to our action. In three dimensions the reaction is in term of stresses in the bulk of the material: strains are relative changes in the linear sizes; stresses are forces per unit surface and altogether they form the rank 2 symmetric stress tensor, $\sigma_{\mu\nu}$. Stresses and strains are mutually and causally connected to each other; in this connection consists the *elasticity* of the material. The simplest assumption we can make is that the relation between strain and stress is linear. Indeed if we exclude discontinuities in the behaviour of the continuum we are analyzing, linearity is in any case the lowest order approximation for the strain/stress functional dependence. Let us then limit our study to the linear elasticity case; its basic equation is Hooke's law, which, in tensor notation, is written:

$$\sigma_{\mu\nu} = C_{\mu\nu}{}^{\alpha\beta}\varepsilon_{\alpha\beta} \tag{11}$$

The $C_{\mu\nu}{}^{\alpha\beta}$'s are the elements of a rank 4 completely symmetric tensor, which we can call the elastic modulus tensor; it contains the properties of the material at the linear approximation level. Eq. (11) is a tensor equation so it is covariant and locally coinciding with its expression on the tangent space; this means that the upper or lower position of the indices is simply a matter of convenience in order to exploit Einstein's summation convention[1].

If we assume that our material continuum is locally isotropic, simple symmetry arguments tell us that the elastic modulus tensor only depends on two parameters, known as the Lamé coefficients, λ and μ, of the material. Explicitly one has:

$$C_{\alpha\beta\mu\nu} = \lambda\eta_{\alpha\beta}\eta_{\mu\nu} + \mu\left(\eta_{\alpha\mu}\eta_{\beta\nu} + \eta_{\alpha\nu}\eta_{\beta\mu}\right) \tag{12}$$

Eq. (12) is written for an arbitrary choice of the coordinates; using Cartesian coordinates the η's would be replaced by Kronecker δ's. Using (12) Hooke's law becomes:

$$\sigma_{\mu\nu} = \lambda\varepsilon^{\alpha}{}_{\alpha}\eta_{\mu\nu} + 2\mu\varepsilon_{\mu\nu} \tag{13}$$

2.2.1 Deformation energy

It is convenient to write down the elastic potential energy of the strained state, which is $W = \frac{1}{2}\sigma_{\mu\nu}\varepsilon^{\mu\nu}$. Using eq. (13) we obtain:

$$W = \frac{1}{2}\lambda\varepsilon^2 + \mu\varepsilon_{\mu\nu}\varepsilon^{\mu\nu} \tag{14}$$

[1] Some care will be required when treating a manifold with Lorentzian signature.

Now I have posed the trace of the strain tensor $\varepsilon^{\alpha}{}_{\alpha} = \varepsilon$ for short.
Eq. (14) could have been written also considering the lowest significant terms of the Helmholtz free energy F_H of the material, written in terms of strain. In fact F_H must contain only scalar quantities and, besides a constant, its lowest order is the second, because the thermodynamical equilibrium must correspond to a minimum (Landau, 1986). Eq. (14) contains the only two second order scalars that can be built from the strain tensor.

3. Space-time and the universe

The whole description of strained continua is molded on three-dimensional examples, but the treatment holds for any number of dimensions. Of course one needs to generalize the interpretation of such things as the stresses and the energy, but formulae and criteria remain valid. So let us apply the theory to four dimensions and the Lorentzian signature, i.e. to space-time, treated as a physical continuum endowed with properties analogous to the ones of ordinary elastic materials.
As a first step I will generalize the action integral of space time plus matter/energy. The generalization consists in that a strained state is associated with a potential like the one expressed in eq. (14). The additional term will appear in the Einstein-Hilbert action that becomes:

$$S = \int \left(R + \frac{1}{2}\lambda\varepsilon^2 + \mu\varepsilon_{\mu\nu}\varepsilon^{\mu\nu} + \mathcal{L}_{mat} \right)\sqrt{-g}\,d^4x \tag{15}$$

Now the scalar curvature R plays the role of dynamical term, since it contains the derivatives of the Lagrangian coordinates, i.e. the elements of the metric tensor; $\mathcal{L}_{mat}$ is the Lagrangian density of matter/energy. Eq. (15) is the starting point for what I shall call the Strained State Theory (SST), which in the following will be applied to the Strained State Cosmology (SSC).
From (15) we can also derive generalized Einstein equations. The new elastic potential terms contribute an additional stress/energy tensor in the final equations. We may treat the strain tensor in the same way as we do with matter fields, only remembering that it must satisfy the constraint represented by eq. (10). In particular the indices of the strain tensor are raised and lowered using the full metric tensor. On this footing we obtain the new generalized version of eq. (1) in the form:

$$G_{\mu\nu} = T_{(e)\mu\nu} + \kappa T_{\mu\nu} \tag{16}$$

In explicit form it is:

$$T_{(e)\mu\nu} = \lambda\varepsilon\varepsilon_{\mu\nu} + 2\mu\varepsilon_{\mu\nu} \tag{17}$$

The tensor $T_{(e)\mu\nu}$ actually belongs to space-time (it is in a sense a self-interaction energy) but works as an effective additional term on the side of the sources.

3.1 A Robertson-Walker universe

It is commonly assumed that the universe has a Robertson-Walker (RW) symmetry, i.e. it is homogeneous and isotropic in space (cosmological principle). This conviction is based both on a priori arguments and on the observation. On the theoretical side: why should a given position or direction in space be more important than another? So let us assume all positions and directions are equivalent. In the 19th and at the beginning of the 20th century, as well as later on, at the time of the Hoyle-Gold-Bondi steady state cosmology, this argument was assumed to hold also for time: why should any given moment be "special"? The homogeneity of time together with the homogeneity and isotropy of space forms the so called "perfect cosmological principle".

The four-dimensional homogeneity has however almost completely been abandoned on the basis of observation. Strictly speaking a stationary universe had already been challenged by the Olbers' paradox (1826): why is the sky dark at night? However the crucial data came from Hubble's work at the end of the 20's of the last century: the redshift of the light coming from other galaxies tells us that the universe is expanding. Today, after the publication of the observations by the groups led by Adam Riess (Riess, 1998) and Saul Perlmutter (Perlmutter, 1999), we even think that the expansion of the universe is accelerating.

As for the homogeneity and isotropy of space the observational evidence is not so stringent. It is evident that locally the universe is neither homogeneous nor isotropic. One has to go to a large enough scale to override local inhomogeneities and anisotropies; how large? Actually we see large voids in the universe, then huge filaments made of galaxies, so that the cosmological principle is assumed to hold at a scale of at least hundreds of megaparsecs (Mpc). However it is also true that we have knowledge only of the visible part of the universe; of the rest we cannot say almost anything or even nothing at all. In fact various anisotropic solutions for the Einstein equations applied to cosmology have been studied and the possibility that some "local" inhomogeneity is responsible for what has been interpreted as an accelerated expansion has also been considered (Biswas, 2010).

I will not discuss further these issues, but will stay with the standard cosmology and accept that the cosmological principle holds on the average. This assumption greatly simplifies the discussion of the global behaviour of the universe and is synthetically expressed by the Robertson-Walker symmetry.

A question is however legitimate now: why is the RW symmetry there? If you just add a uniformly distributed dust to an empty Minkowskian space-time you do not obtain, as an unique outcome, a RW universe. A homogeneous distribution of matter is gravitationally unstable; does this preserve isotropy and lead to a singularity in the past? Not necessarily.

If I adopt the viewpoint of the SSC, I may think that space-time *per se* (the natural manifold) has a built-in RW symmetry independently from the presence of matter; the latter simply responds to the symmetry, reinforcing it. The primordial symmetry is in turn explained assuming the presence of a spacelike defect (a Cosmic Defect) within the manifold. Of course you might ask why the defect should be there, however we know that going back along the chain of "why?"'s sooner or later we exit the domain of physics. We can only try and minimizing the number of independent assumptions and if possible look for physically consistent interpretations of their meaning.

The approach of the Strained State Cosmology is best visualized in fig. 3, where the embedding of a Robertson-Walker space-time in a three-dimensional flat manifold is shown. O is the defect responsible for the RW symmetry. For convenience in making the drawing, the example of a closed space has been represented. For an open space the original defect would be linear (a ridge) and space-like. All geodetic lines starting from the defect are time-like; τ is the cosmic time; space is any space-like intersection between the natural manifold and an open surface (for instance a hyperplane) in the embedding space. Successive intersections of the natural manifold, in correspondence of increasing values of the cosmic time, evidence what the typical 3+1 human view reads as an expanding universe.

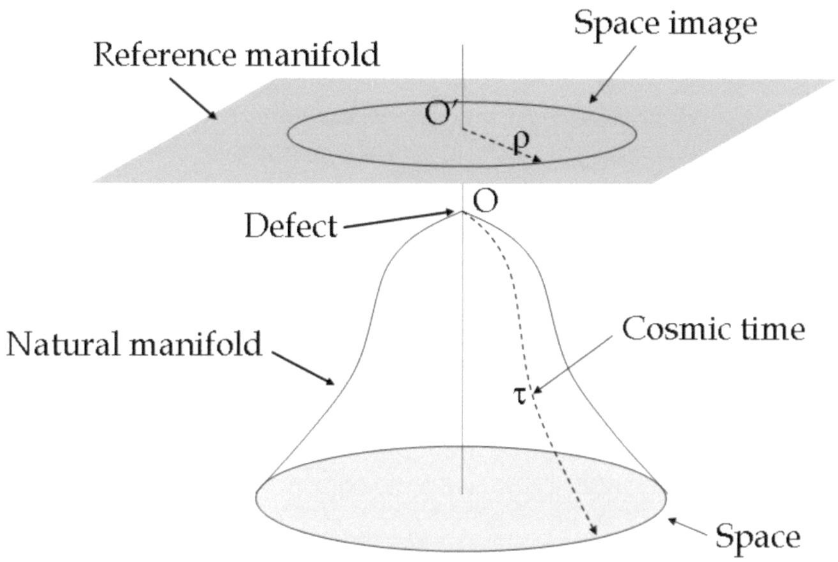

Fig. 3. Pictorial view of a Robertson-Walker universe embedded in a three-dimensional flat space. The picture corresponds to a closed universe.

The correspondence we establish between the reference and the natural manifold identifies an "image" of any given natural space in the reference. We must now write down and compare the corresponding line elements on the two manifolds. Due to the simple symmetry, the line element on the natural manifold is of course[2]:

$$ds^2 = d\tau^2 - a^2(\tau)\, dl^2 \tag{18}$$

The a function of the cosmic time is the *scale factor* of the universe; dl is the space length element.

As for the reference manifold you can in principle define the correspondence with the actual RW space-time in infinite different ways. Using the coordinates chosen for the natural manifold, you are left with four free functions for the choice of the coordinates on the reference, with the constraint that the reference has to be flat. In the specific case under consideration, however, the final symmetry reduces the free functions to only one and the reference line element is written:

$$d\sigma^2 = b^2(\tau)\, d\tau^2 + dl^2 \tag{19}$$

The function b of the cosmic time has been called gauge function in (Radicella, 2011) but this denomination is not entirely correct, since b does not correspond to a real freedom: since we assume that the deformation process is a real one, the way the correspondence between the

[2] Times are expressed as lengths.

unstrained and the strained manifold is established depends on the two Lamé coefficients of space-time, under the assumption of local isotropy.
From eq.s (18) and (19), using the definition (10), we easily obtain the non-zero elements of the strain tensor for a RW space-time:

$$\begin{cases} \varepsilon_{oo} = \dfrac{1-b^2}{2} \\ \varepsilon_{ii} = -\dfrac{1+a^2}{2} \end{cases} \tag{20}$$

Once we have the strain tensor, it is possible to deduce the potential term (14) in the action integral; indices are raised and lowered by means of the full RW metric tensor. It is:

$$W = \frac{\lambda}{8}\left(1 - b^2 + 3\frac{1+a^2}{a^2}\right)^2 + \frac{\mu}{4}\left[\left(1-b^2\right)^2 + 3\frac{\left(1+a^2\right)^2}{a^4}\right] \tag{21}$$

The other ingredients of the action integral, besides the matter/energy Lagrangian density, are:

$$R = -6\left(\frac{\ddot{a}}{a} + \frac{\dot{a}^2}{a^2}\right); \qquad \sqrt{-g} = a^3 \tag{22}$$

Dots denote derivatives with respect to time.
An expression for b^2 is immediately found imposing $dW/db = 0$ (i.e. extremizing the Lagrangian density with respect to the gauge function). Rejecting the inadmissible $b = 0$, the solution is:

$$b^2 = 2\frac{2\lambda + \mu}{\lambda + 2\mu} + \frac{3}{a^2}\frac{\lambda}{\lambda + 2\mu} \tag{23}$$

Given the solution (23) the only residual unknown is the scale factor a. Of course we should also specify the type of matter we consider. The simplest is to assume that matter/energy is made of dust plus radiation. Under these conditions, applying Hamilton's principle to the action integral (15) leads to:

$$H = \frac{\dot{a}}{a} = c\sqrt{\frac{3}{16}B\left(1 - \frac{(1+z)^2}{a_0^{\,2}}\right)^2 + \frac{\kappa}{6}(1+z)^3\left[\rho_{m0} + \rho_{r0}(1+z)\right]} \tag{24}$$

H is the Hubble parameter. The variable z is the redshift factor and use has been made of the relation $a(1+z)$ = constant = a_0; a_0 is the present value of the scale factor. ρ_{m0} and ρ_{r0} are the present values of the average matter and radiation densities in the universe; $\kappa = 16\pi G/c^2$ is the coupling constant between geometry and matter/energy. B combines the Lamé coefficients of space-time according to:

$$B = \frac{3}{2}\mu\frac{2\lambda + \mu}{\lambda + 2\mu} \tag{25}$$

The term proportional to B in the square root of eq. (24) is the contribution coming from the strain of the space-time; the rest is the standard cosmology of a RW universe filled up with dust and radiation.
The choice of the sign for the square root in (24) tells us whether the universe is expanding or contracting; the given behaviour is for ever. In the same time we see that the contribution from strain implies the onset of acceleration after an initial phase of deceleration. The dependence of the expansion rate on the scale factor is shown in fig. 4 in arbitrary units. At very early times ($z >> 1$) the strain contributes a radiation-like term boosting the expansion:

$$H_{z>>1} \cong cz^2 \sqrt{\frac{3}{16}\frac{B}{a_0^{\ 4}} + \frac{\kappa}{6}\rho_{r0}} \tag{26}$$

In late times ($z \to -1$) the Hubble parameter becomes constant: the expansion assumes an exponential trend at a rate depending only on B:

$$H_{z\to -1} \cong c\sqrt{\frac{3}{16}B}; \qquad a_\infty \approx e^{c\sqrt{\frac{3}{16}B}t} \tag{27}$$

We have so seen that the SSC is able to account for the accelerated expansion as being a consequence both of the presence of a cosmic defect (the Big Bang) and of the elastic properties of space-time.

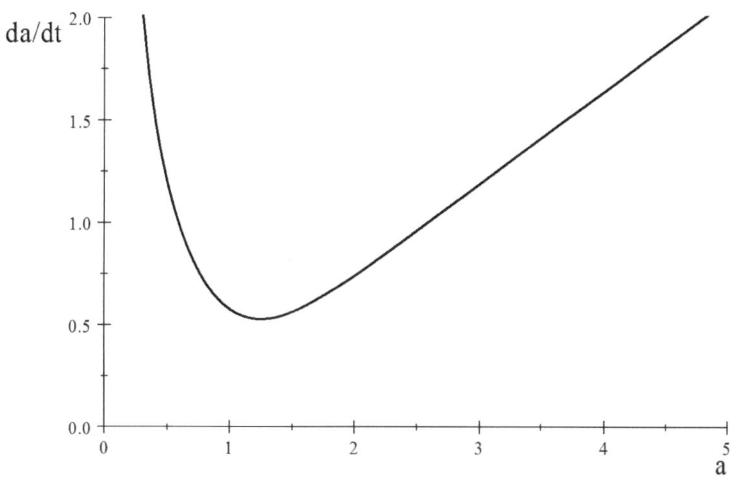

Fig. 4. Expansion rate of a RW universe according to the Strained State Theory. The graph is drawn giving arbitrary values to the parameters. The universe always expands; at the beginning the expansion decelerates, afterwards it accelerates.

What remains to be done is to find appropriate values for the parameters of the theory, which, at this stage, are B and a_0 besides ρ_{m0} and ρ_{r0}. This will be the subject of the next section.

4. Cosmological tests

In order to determine the optimal values for the parameters of the theory and to check its credibility we have considered four typical tests: the dependence of the luminosity of type Ia supernovae (SnIa) on the redshift; the Big Bang Nucleosynthesis (BBN); the acoustic horizon scale in the Cosmic Microwave Background; the Large Scale Structure (LSS) formation after the recombination era. The first test I have quoted is not in decreasing redshift order as the others are; the reason for privileging it is in that SnIa's have been the first evidence in favor of an accelerated expansion (Riess, 1998) (Perlmutter, 1999).

4.1 The luminosity curve of type Ia supernovae

Type Ia supernovae are thought to be the product of the implosion of a slowly rotating white dwarf star that accretes matter from a companion in a tightly bound binary system (Hillebrandt, 2000). These stars have masses that do not exceed the Chandrasekhar limit (Chandrasekhar, 1931), i.e roughly 1.38 solar masses. The mass limit and the implosion mechanism are such that the characteristic light curve of an SnIa is quite uniform and reproducible, so that this kind of objects can be used as *standard candles* for determining cosmic distances (Colgate, 1979).

In order to exploit the mentioned beautiful property of SnIa's we need the *luminosity distance* of the source which depends on the expansion mechanism of the universe. When expressed in terms of *distance modulus* and of the redshift parameter it is given by the formula (Weinberg, 1972):

$$m - M = 25 + 5\log_{10}\left[(1+z)\int_0^z \frac{dz'}{H(z')}\right] \tag{28}$$

M is the absolute magnitude of the source; m is the locally observed magnitude; H is the Hubble parameter and depends on the expansion model one uses. Formula (28) holds when distances are measured in Mpc.

When applying (28) to the luminosity data from SnIa's in the framework of the standard cosmology, one finds (Riess, 1998) (Perlmutter, 1999) that the sources appear to be dimmer than expected from the z value of the host galaxy. The immediate interpretation of this fact is that the expansion of the universe is indeed accelerated.

We applied the SST to try and fit the luminosity data from SnIa's using formulae (28) and (24) (Tartaglia, 2010). The experimental luminosities were from 307 SnIa's from the *Supernova Cosmology Project Union Survey* (Kowalski, 2008). The result is shown in fig. 5; the quality of the fit, if taken as the only test, is good. The free parameters of the theory, considering that for z values < 2 the radiation term is negligible, are three; the final reduced χ^2 is 1.017.

For comparison we use the Λ Cold Dark Matter (ΛCDM) scenario (Concordance Model), which is the simplest and most effective theory currently adopted in order to account for the properties of the universe. ΛCDM, when employed to fit the same data of SnIa's as above, gives χ^2 = 1.019. The problem with ΛCDM is that the physical nature of the cosmological constant Λ (or of the corresponding *dark energy*) remains a mystery.

For further analysis it is convenient to explicitly reproduce the χ^2 formula:

$$\chi^2_{SnIa} = \sum_i \left(\frac{d_i - d(z_i)}{\delta d_i}\right)^2 \tag{29}$$

The d_i's are the measured values of the distance modulus; $d(z_i)$ is the corresponding value given by the theory; δd_i are the variances of the experimental data; the sum is over the number of supernovae we use.
This first test is encouraging, but is not enough, so let us go on with more.

4.2 More tests

4.2.1 The abundance of primordial isotopes

The lightest elements up to lithium Li^7 (mentioning just the stable isotopes) formed after the baryogenesis phase, while the primordial plasma cooled and expanded (Big Bang Nucleosynthesis: BBN). The relative abundances of hydrogen, deuterium and helium that we find today as a residue of that time depend on the early expansion history, affecting both the temperature and the density of the plasma. Since the SST gives an additional contribution to the radiation density and pressure, as seen in formula (26), we do not expect it to influence the cross section of the nuclear reactions but the quantitative final result of BBN.

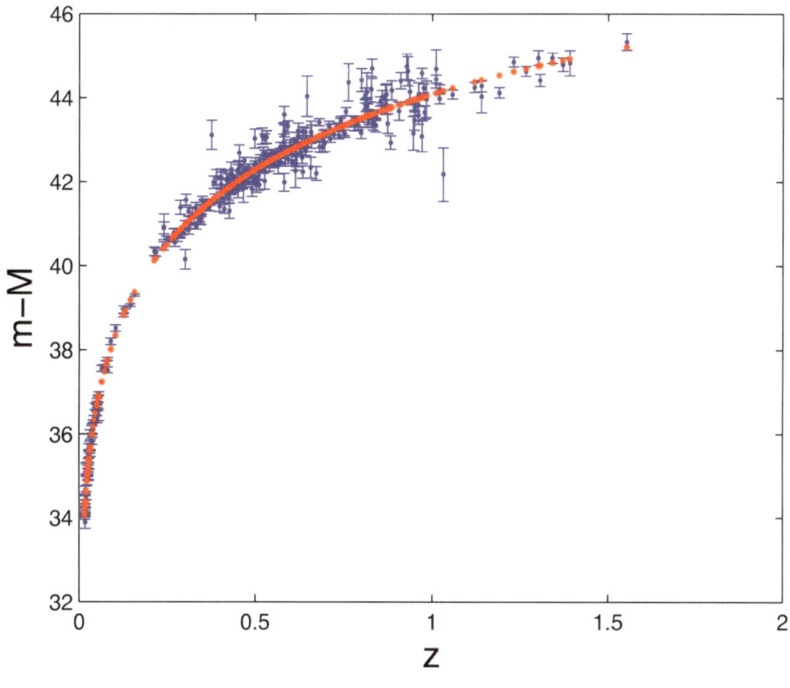

Fig. 5. Fit of the luminosity data from 307 Snia's obtained using the SST. The distance modulus is given as a function of the redshift parameter. The experimental data are shown with their error bars.

Let us recast (26) as:

$$H \cong cz^2 \sqrt{\frac{\kappa}{6}\left(1+\frac{B}{B_{a_0}}\right)\rho_{r0}} \tag{30}$$

where it is

$$B_{a_0} = \frac{8}{9} \kappa \rho_{r0} {a_0}^4 \tag{31}$$

The term in brackets in (30) acts as an effective boost factor for the radiation energy density $X_{boost} = 1 + B/B_{a0}$ enhancing the expansion rate. This fact would lead to an earlier freeze-out of the neutrons, then to a higher final abundance of He^4. Knowing the actual abundance of helium we can then put constraints on the value of the parameters of the SST. The primordial fraction of helium by mass, Y_p, is estimated using various methods and with good accuracy; see for instance (Izotov, 2010). We adopted a conservative attitude picking up the value $Y_p = 0.250 \pm 0.03$ (Iocco, 2009) obtained by an ample analysis of a number of different values in the literature. The ensuing constraint in the boost factor is $X_{boost} = 1.025 \pm 0.015$. Our final purpose is to perform a statistical analysis of the compatibility of SST with the data, so we work out the χ^2 constraint that follows from the quoted uncertainties:

$$\chi^2_{BBN} = \left(\frac{X_{boost} - 1.025}{0.015} \right) \tag{32}$$

4.2.2 Cosmic microwave background constraint

The analysis of the CMB spectrum is a complex task, but the scope of this discussion is limited to a compatibility check, so I shall pick out just one parameter whose value is affected both by the expansion factor at the matter/radiation equality time and by the history of the universe from the decoupling time to the present. The chosen parameter is the acoustic scale (Komatsu, 2011):

$$l_A = (1 + z_{LS}) \pi \frac{D_A(z_{LS})}{r_s(z_{LS})} \tag{33}$$

D_A is the angular diameter distance to the last scattering surface; r_s is the size of the sound horizon at recombination; $z_{LS} \sim 1090$ is the last scattering redshift. The mode of the expansion affects the position of the acoustic peaks which depends on the expansion factor at the equality scale a_e; in practice the position is influenced by the value of the boost factor for the radiation X_{boost}. The acoustic horizon formula will then be the same as for ΛCDM, but the equality scale factor is now boosted: $a_e = X_{boost}\rho_{r0}/\rho_{m0}$. As for the angular diameter distance, it depends on the total expansion history from the last scattering surface to present:

$$D_A(z_{LS}) = \frac{c}{(1 + z_{LS})} \int_0^{z_{LS}} \frac{dz}{H(z)} \tag{34}$$

The final value for l_A is not much sensitive to the choice of the cosmological model so we will make reference to the values obtained from WMAP-7 using ΛCDM (Komatsu, 2011). Our reference experimental (+ΛCDM) value is ${l_A}^{Obs} = 302.69 \pm 0.76 \pm 1.00$. The first uncertainty is the statistical error, the second is an estimate of the uncertainty connected with the choice of the model; the two uncertainties are mutually independent so they can be added in quadrature. Summing up we have the statistical constraint:

$$\chi^2_{CMB} = \left(\frac{l_A - 302.69}{1.26}\right)^2 \tag{35}$$

4.2.3 Large scale structure formation

If space-time is expanding in a radiation dominated universe matter density fluctuations cannot produce growing seeds for future structures. As we have seen, the presence of strain in early epochs effectively increases the radiation density, so retarding the onset of matter dominance. This is the reason why LSS poses further constraints on the SST. The effective boost, X_{boost}, affects the scale of the particle horizon at the equality epoch, $z_{eq} \cong 3150$ (Komatsu, 2011). On the other hand, the SST preserves the Newtonian limit of gravity even in presence of defects (Tartaglia, 2010), so that, in SSC, the growth of mass density perturbations is affected mainly through the modified expansion rate of the background. The horizon at the equality is imprinted in the matter transfer function. The constraint from LSS can be written as (Peacock, 1999):

$$(\Omega_{m0}h)_{apparent} = \frac{(\Omega_{m0}h)_{true}}{\sqrt{X_{boost}}} \tag{36}$$

Ω_{m0} is the mass density in units of the critical density $\rho_c = 3H_0^2/8\pi G$; H_0 is the Hubble constant; h is the Hubble constant in units of 100 km s^{-1}Mpc^{-1}.

According to the conclusions drawn from the analysis of the data from the 2dF Galaxy Redshift Survey (Cole, 2005) it is $(\Omega_{m0}h)_{apparent} = 0.168 \pm 0.016$. For consistency we make the same assumption as in ref. (Cole, 2005) on the index of the primordial power spectrum (n = 1). The related constraint on the cosmological parameters of the SSC is:

$$\chi_{LSS}{}^2 = \left(\frac{\left(\Omega_{m0}h / \sqrt{X_{boost}} - 0.168\right)}{0.016}\right)^2 \tag{37}$$

4.3 Global consistency

The various tests we have described in the previous sections must be satisfied together, so we must check for the global compatibility of the constraints when applied to SSC. The analysis has been made using standard Bayesian methods (Mackay, 2003). According to Bayes theorem the posterior probability p for a given parameter $\boldsymbol{P}$ given the data $\boldsymbol{d}$ is proportional to the product of the likelihood $\mathcal{L}$ of $\boldsymbol{P}$ times the prior probability for $\boldsymbol{P}$:

$$p(\mathbf{P}\,|\,\mathbf{d}) \propto \mathcal{L}(\mathbf{P}\,|\,\mathbf{d})\, p(\mathbf{P}) \tag{38}$$

The likelihood is expressed in terms of the total χ^2 as $\mathcal{L} \propto e^{-\chi^2/2}$ and the total χ^2 is in turn given by the sum of the independent values (29), (32), (35), (37):

$$\chi^2 = \chi^2_{SnIa} + \chi^2_{BBN} + \chi^2_{CMB} + \chi^2_{LSS} \tag{39}$$

For this analysis we use three parameters of the theory. The constraints we have considered do not require us to distinguish between baryonic and dark matter, so that we consider a single parameter density for the dustlike matter, ρ_{m0}. The strain related properties, in a RW symmetry, are accounted for by the B parameter. Finally, the present value of the scale factor is described in terms of B_{a0} (actually we shall use its inverse). A flat distribution for each parameter has been assumed. The relativistic energy density has been fixed at $\rho_{r0} \cong 7.8 \times 10^{-31}$ kg/m^3. The parameter space has been explored with Monte Carlo Markov chain methods (Lewis, 2002) running four chains, each one with 10^4 samples. Convergence criteria were safely satisfied, with the Gelman and Rubin ratio (Gelman, 1992) being ≤ 1.003 for each parameter. The final results are shown in fig. 6a,b,c.

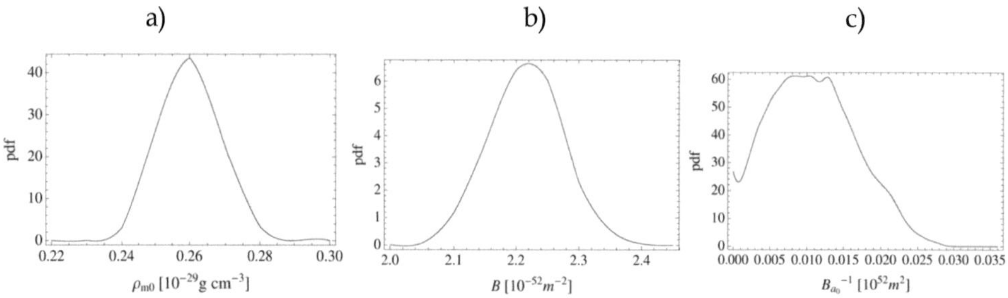

Fig. 6. Posterior probability density functions for the parameters of the SSC; the functions are normalized. Units are as in Table 1.

From the probability density functions we obtain the best estimates for the parameters. The corresponding amounts are listed in Table 1 where also the maximum likelihood values are reported in parentheses.

ρ_{m0} (10^{-26} kg×m^{-3})	B (10^{-52} m^{-2})	B_{a0}^{-1} (10^{52} m^2)
0.260 (0.258) ± 0.009	2.22 (2.22) ± 0.06	0.011 (0.009) ± 0.006

Table 1. Estimated values of the parameters. The numbers in brackets correspond to the maximum likelihood.

The estimated value for the present matter density, when expressed in terms of the critical density, becomes $\Omega_{m0} = 0.28 \pm 0.01$ which is consistent with the value commonly accepted for the sum of baryonic and dark matter.

4.3.1 Further compatibility checks

The theory, together with the values obtained in the previous section for the parameters, can be used to evaluate various cosmic quantities that can be verified with observation. For instance the calculated Hubble constant of SSC is $H_0 = 70.2 \pm 0.5$ km s^{-1}Mpc^{-1}, which compares well with $73 \pm 2 \pm 4$ km s^{-1}Mpc^{-1} obtained from high precision distance determination methods (Freedman, 2010). Another interesting quantity is the age of the universe; the SSC value is $T = 13.7$ Gy, fully compatible with the lowest limits obtained from the age of the oldest globular clusters and from radioactive dating.

5. Open problems and perspectives

The Strained State Theory applied to cosmology, at least in the case of a RW symmetry, performs well, as we have seen, however some aspects of the theory require further thought and clarification. Let us for instance consider a problem I have hardly touched in the previous sections: the signature of space-time. The logic of the method I have outlined here requires a totally undifferentiated flat manifold to start with. In other words the reference manifold should best be Euclidean. It is easy to verify however that the results concerning a RW universe can be obtained as well starting with a Minkowski reference manifold. The latter choice is in a sense friendlier because it has, from the start, the same signature as the final strained space-time which we want to describe. However we may ask where does the initial signature of a Minkowski space-time come from. Hopefully in the case of SSC the start can be Euclidean even if the final state has a Lorentzian signature. In the theory a cosmic defect is essential to define the global symmetry of the universe on a large scale and all timelike world lines stem out of that defect. Is the presence of a defect the condition for introducing the signature (in practice the light cones) in the natural manifold? The guess is that it is so, but the fact that the idea works in the case of the RW symmetry is not a proof, that should be given in general terms. In any case an important remark is that there must be no confusion between the reference manifold, which is Euclidean, and the local tangent space at any position in the natural manifold, which is instead Minkowskian.

The importance of the Cosmic Defect (CD) has been stressed more than once in this chapter. Are there other defects in the universe? The answer is in principle yes of course, but, if other defects exist, how and where do they show up? The CD is space-like and is the origin of the signature of space-time; if additional defects exist they could/should be time-like. A possibility is to have, for instance, a linear time-like defect; such defect would be surrounded, at any given moment, by a spherically symmetric space. If we think for instance to a big spherical cosmic void it could indeed be centered on a linear time-like defect. On the other side the present theory, for the essential, is not different from General Relativity: it is not locally distinguishable from GR, since the gravitational interaction is described in the same geometrical terms. The natural manifold admits locally a flat Minkowskian tangent space, just as in GR, and this means that the equivalence principle holds and also that the SST complies with the Newtonian limit. By the way the values obtained from the cosmological application and listed in Table 1 tell us that the scale at which deviations from the standard GR can be expected are very large, much wider than the solar system and even than a single galaxy. It is however true that the local spherical symmetry is also the typical Schwarzschild symmetry and there GR has a singular exact solution. Today black holes are well accepted and evidence for their existence, at least in the center of galaxies, is abundant. The conceptual problems posed by the singularity are bypassed by the cosmic censorship principle, so that people do not worry too much about them. Is there a connection between the black holes of GR and linear defects of the SST? The singularities of GR have to do with infinite matter densities; the defects of the SST are in the space-time as such and at most they influence the behaviour of surrounding matter. The singularity of a defect in a manifold is much friendlier than the singularities of GR. Are there horizons in SST too? All these open questions deserve further work and analysis. Remaining in the domain of defects, the properties of other symmetries need to be explored, first of all the screw symmetry which corresponds to the same symmetry as the one of the Kerr black holes.

Looking at the Lagrangian density contained in eq. (15) and in particular to the additional new elastic potential terms of eq. (14) we see that they look very much like the massive gravity Lagrangian density initially proposed by Fierz and Pauli (Fierz, 1939) (Dvali, 2008). This similarity is very strict when it is $\lambda = -2\mu$, however it must be kept in mind that the Fierz-Pauli Lagrangian was proposed in pursuit of a gravitational spin-2 field in a Minkowski background; furthermore the Fierz-Pauli Lagrangian is obtained by a

linearization process in which the deviation from the flat Minkowsky manifold is represented by a $h_{\mu\nu}$ tensor, whose elements are all small with respect to 1. When letting the mass of the graviton in the Fierz-Pauli theory go to zero, one is left with a linearized General Relativity, whose equations can be used both for the study of gravito-magnetic effects and for Gravitational Waves (GW). Fierz and Pauli's approach however has a problem: its limit for zero mass of the graviton does not smoothly reproduce the results of GR: it is the so called van Dam-Veltman-Zakharov (vDVZ) discontinuity (van Dam, 1970) (Zakharov, 1970). Furthermore a non-zero mass graviton implies the presence of a ghost when studying propagating modes. The debate on these problems and on massive gravity is open.
In any case we must remark that in the SST the strain tensor is not a perturbation of a flat Minkowski background, rather it expresses the difference (not necessarily small) with respect to an Euclidean reference, which is of course *not* the tangent space at any given event of the natural manifold. The behaviour of a strained space-time with respect to propagating perturbations, i.e. waves, must be studied, but we can expect it to be similar, even though not identical, with "massive gravity"; in particular we can expect subluminal waves and contributions to a cosmic thermal gravitational background according to some appropriate dispersion law.
As a last conceptual aspect to be considered with the SST I start from a simple remark. The classical theory of elasticity is the macroscopic manifestation of an underlying microscopic reality made of discrete particles with their interactions. Can we think the elasticity of space-time to have a similar origin? The idea, at first sight, seems reasonable, however the point is subtle. On one side, an underlying microscopic structure of space-time would bring us close to the attempts to quantize the space-time and gravity (and to their difficulties). On the other, we should face the problem I mentioned in the Introduction concerning the implicit request of a "background" (a super-space-time?) in which the microscopic structure of space-time would be located. Our current view of the universe, whether we are aware of it or not, is basically dualistic: on one side space-time with properties of its own; on the other side matter/energy described by quantum mechanics in terms of eigenstates and eigenvalues of quantum operators associated with physically meaningful parameters. The two sides of the duality resist against the attempts to reduce them to a single paradigm. Maybe this simply means that nobody has found the right way so far, but it could also be that they are mutually irreducible. If so the elasticity of the four-dimensional manifold could be a fundamental property of space-time and not the macroscopic approximation of some unknown microscopic structure.

6. Conclusion

In this chapter I have expounded a theory based on physical intuition, which extends to four dimensions what we already know in three when studying material continua. I have used concepts such as strain to describe the distortion induced in space-time either by the presence of matter/energy or by the presence of texture defects analogous to the ones we find in crystalline solids. The idea of an induced strain implies directly the existence of an analogue of the deformation energy. This distortion energy enters the Lagrangian of space-time as an additional potential and leads to a new dynamical history of the universe. The structure and fundaments of General Relativity are all preserved. As we have seen, the theory, when applied to a Robertson-Walker universe, passes various important consistency tests, while reproducing the luminosity/distance curve of type Ia supernovae (in practice it

accounts for the accelerated expansion). The values we found for the parameters of the theory tell us that locally it will be indistinguishable from GR, while producing emerging effects at cosmic scales. There are a number of developments to be pursued and difficulties to be discussed and overcome, but the way through seems not to be impassable.
Of course there are many theories that, in a way or another, account for the accelerated expansion while passing various cosmological consistency tests. First of all there is ΛCDM, which is reasonably simple and reasonably successful, though not exempt from drawbacks. How and why should we discard one theory and prefer another? Most often in cosmology new theories are introduced manipulating the Lagrangians or adding fields on heuristic bases; internal consistency is of course cared of, but physical intuition plays a minor role. Hundreds of papers appear every years discussing details of theories whose basic assumptions are motivated only by the final results one wants to obtain; the old Occam's razor (*entia non sunt multiplicanda praeter necessitatem*) is left behind and it is difficult, if not impossible, to think of crucial experiments that can discriminate among the theories. In this situation maybe the strategy of sticking as far as possible to what one already knows is sound and trying to build the least possible exotic physical scenario is advisable. This is the meaning of the Strained State Theory and of the Strained State Cosmology, which is not yet an accomplished paradigm, but aspires to become so. We have just started.

7. Acknowledgment

The present work could not have been carried to the level where it is now without the collaboration of a number of younger colleagues and students. I wish here to explicitly acknowledge the contributions by Ninfa Radicella and Mauro Sereno who helped me to clarify many aspects of the theory in fruitful discussions, besides co-authoring a couple of papers that have been used to support a good portion of the present chapter.

8. References

Aristotle of Stageira (350 b.C.), *Τὰ φυσικά*, book IV, 350 b.C.

Augustine of Hippo (398 A.D.), *Confessiones*, XI, 14 , 398 a.D.

Biswas, T. ; Notari, A. ; Valkenburg, W. (2010). Testing the void against cosmological data: fitting CMB, BAO, SN and H_0, *JCAP*, Vol. 11, p. 030.

Chandrasekhar, S. (1931). The Maximum Mass of Ideal White Dwarfs, *Ap. J.*, Vol. 74, p. 81.

Cole, S.; et al. (2005). The 2dF Galaxy Redshift Survey: Power-spectrum analysis of the final dataset and cosmological implications, *Mon. Not. R. Astron. Soc.*, Vol. 362, p. 505.

Colgate, S.A. (1979). Supernovae as a standard candle for cosmology, *Ap. J.*, Vol. 232, pp. 404-408

Dvali, G; Pujolàs, O.; Redi, M. (2008). Non-Pauli-Fierz Massive Gravitons, *Phys. Rev. Lett.* Vol. 101, p. 171303.

Einstein, A. (1920), in *The Collected Papers of Albert Einstein*, volume 7. Princeton University Press, Princeton, 2007.; Archives Online, http://alberteinstein.org/, Call. Nr.[1-41.00].

Eshelby, J. D. (1956). *Solid State Physics*, New York: Academic.

Fierz, M.; Pauli, W. (1939). On Relativistic Wave Equations for Particles of Arbitrary Spin in an Electromagnetic Field, *Proc. R. Soc. A*, Vol. 173, p. 211.

Freedman, W.L.; Madore, B. F. (2010). The Hubble Constant, *Annu. Rev. Astron. Astrophys.* Vol. 48, pp. 673-710.

Gelman, A. and Rubin, D. (1992). Inference from Iterative Simulation Using Multiple Sequences, *Stat. Sci.* Vol. 7, pp. 457-511.

Hillebrandt, W.; Niemeyer, J.C. (2000), Type Ia Supernova Explosion Models, *Annual Review of Astronomy and Astrophysics*, Vol. 38, pp. 191-230.

Izotov, Y.I.; Thuan, T.X. (2010). The Primordial Abundance of ^{4}He: Evidence for Non-standard Big Bang Nucleosynthesis, *Astrophys. J. Lett.*, Vol. 710, L67.

Komatsu, E; et al. (2011). Seven-year Wilkinson Microwave Anisotropy Probe (WMAP) Observations: Cosmological Interpretation, *Astrophysical J. Supplement Series*, Vol. 192, 18.

Kowalski, M. et al. (2008). Improved Cosmological Constraints from New, Old, and Combined Supernova Data Sets, *Astrophys. J.*, Vol. 686, pp. 749–778.

Landau, L. & Lifshitz, E. (1986). *Theory of Elasticity*, 3rd edn, Oxford: Pergamon 1986.

Lewis, A. & Bridle, S (2002). Cosmological parameters from CMB and other data: a Monte Carlo approach, *Phys. Rev. D*, Vol. 66, n. 10, pp. 103511-16.

Mackay, D.J.C. (2003). *Information Theory, Inference and Learning Algorithms*, Cambridge Press, October 2003

Newton, I. (1687), *Philosophiae Naturalis Principia Mathematica*

Peacock, J. A. (1999). *Cosmological Physics*, Cambridge University Press.

Perlmutter, S. *et al* (1999). Measurements of Omega and Lambda from 42 High-Redshift Supernovae, *Astrophys. J.*, Vol. 517, pp. 565–86.

Puntigam, R. A. & Soleng, H. H. (1997). Volterra Distortions, Spinning Strings, and Cosmic Defects, *Class. Quantum Grav.* Vol. 14, pp. 1129–49.

Radicella, N; Sereno, M. & Tartaglia, A. (2011), Cosmological constraints for the Cosmic Defect theory, *Int. J. Mod. Phys. D*, Vol. 20, n. 4, pp. 1039-1051.

Riess, A. G. *et al* (1998). Observational Evidence from Supernovae for an Accelerating Universe and a Cosmological Constant, *Astron. J.*, Vol. 116, pp. 1009–38.

Sakharov, A. D. (1968). *Sov. Phys. Dokl.*, Vol. 12, pp 1040.

Tartaglia, A.; Radicella, N. (2010). A tensor theory of spacetime as a strained material Continuum, *Class. Quantum Grav.*, Vol. 27, pp. 035001-035019.

van Dam, H.; Veltman, M. (1970). Massive and mass-less Yang-Mills and gravitational fields, *Nucl. Phys.* Vol. B 22, p. 397.

Volterra, V. (1904), *Ann. Sci. de l'École Normale Supérieure*, Vol. 24, pp. 401–517

Weinberg, S. (1972). *Gravitation and Cosmology: Principles and Applications of the General Theory of Relativity*, New York: Wiley.

Zakharov, V.I. (1970). *JETP Lett.*, Vol. 12, p. 312.

3

Introduction to Modified Gravity: From the Cosmic Speedup Problem to Quantum Gravity Phenomenology

Gonzalo J. Olmo
Departamento de Física Teórica and IFIC, Universidad de Valencia-CSIC. Burjassot, Valencia
Spain

1. Introduction

The reasons and motivations that lead to the consideration of alternatives to General Relativity are manifold and have changed over the years. Some theories are motivated by theoretical reasons while others are more phenomenological. One can thus find theories aimed at unifying different interactions, such as Kaluza-Klein theory (5-dimensional spacetime as a possible framework to unify gravitation and electromagnetism) or the very famous string theory (which should provide a unified explanation for *everything*, i.e., from particles to interactions); others appeared as spin-offs of string theory and are now seen as independent frameworks for testing some of its phenomenology, such is the case of the string-inspired "brane worlds" (which confine the standard model of elementary particles to a 4-dimensional brane within a larger bulk accessible to gravitational interactions); we also find modifications of GR needed to allow for its perturbative renormalization, or modifications aimed at avoiding the big bang singularity, effective actions related with non-perturbative quantization schemes, etcetera. All them are motivated by theoretical problems.

On the other hand, we find theories motivated by the need to find alternative explanations for the current cosmological model and astrophysical observations, which depict a Universe filled with some kind of *aether* or *dark energy* representing the main part of the energy budget of the Universe, followed by huge amounts of *unseen* matter which seems necessary to explain the anomalous rotation curves of galaxies, gravitational lensing, and the formation of structure via gravitational instability.

One of the goals of this chapter is to provide the reader with elementary concepts and tools that will allow him/her better understand different alternatives to GR recently considered in the literature in relation with the cosmic speedup problem and the phenomenology of quantum gravity during the very early universe. Since such theories are aimed at explaining certain observational facts, they must be able to account for the new effects they have been proposed for but also must be compatible with other observational and experimental constraints coming from other scenarios. The process of building and testing these theories is, in our opinion, a very productive theoretical exercise, since it allows us to give some freedom to our imagination but at the same time forces us to keep our feet on the ground.

Though there are no limits to imagination, experiments and observations should be used as a guide to build and put limitations on sensible theories. In fact, a careful theoretical interpretation of experiments can be an excellent guide to constrain the family of viable theories. In this sense, we believe it is extremely important to clearly understand the implications of the Einstein equivalence principle (EEP). We hope these notes manage to convey the idea that theorists should have a deep knowledge and clear understanding of the experiments related with gravitation.

We believe that $f(R)$ theories of gravity are a nice toy model to study a possible gravitational alternative to the *dark energy problem*. Their dynamics is relatively simple and they can be put into correspondence with scalar-tensor theories of gravity, which appear in many different contexts in gravitational physics, from extended inflation and extended quintessence models to Kaluza-Klein and string theory. On the other hand, $f(R)$ theories, in the Palatini version, also seem to have some relation with non-perturbative approaches to quantum gravity. Though such approaches have only been applied with certain confidence in highly symmetric scenarios (isotropic and anisotropic, homogeneous cosmologies) they indicate that the Big Bang singularity can be avoided quite generally without the need for extra degrees of freedom. Palatini $f(R)$ theories can also be designed to remove that singularity and reproduce the dynamical equations derived from isotropic models of Loop Quantum Cosmology. Extended Lagrangians of the form $f(R,Q)$, being Q the squared Ricci tensor, exhibit even richer phenomenology than Palatini $f(R)$ models. These are very interesting and promising aspects of these theories of gravity that are receiving increasing attention in the recent literature and that will be treated in detail in these lectures.

We begin with Newton's theory, the discovery of special relativity, and Nordström's scalar theories as a way to motivate the idea of gravitation as a curved space phenomenon. Once the foundations of gravitation have been settled, we shift our attention to the predictions of particular theories, paying special attention to $f(R)$ theories and some extensions of that family of theories. We show how the solar system dynamics can be used to reconstruct the form of the gravity Lagrangian and how modified gravity can be useful in modeling certain aspects of quantum gravity phenomenology.

2. From Newtonian physics to Einstein's gravity.

In his Principia Mathematica (1687) Newton introduced the fundamental three laws of classical mechanics:

- If no net force acts on a particle, then it is possible to select a set of reference frames (inertial frames), observed from which the particle moves without any change in velocity. This is the so called *Principle of Relativity* (PoR).
- From an inertial frame, the net force on a particle of mass m is $\vec{F} = m\vec{a}$.
- Whenever a particle A exerts a force on another particle B, B simultaneously exerts a force on A with the same magnitude in the opposite direction.

Using Newton's laws one could explain all kinds of motion. When a nonzero force acts on a body, it accelerates at a rate that depends on its *inertial* mass m_i. A given force will thus lead to different accelerations depending on the inertial mass of the body. In his Principia, Newton also found an explanation to Kepler's empirical laws of planetary motion: between any two bodies separated by a distance d, there exists a force called *gravity* given by $F_g = G\frac{m_1 m_2}{d^2}$. Here G is a constant, and m_1, m_2 represent the *gravitational* masses of those bodies. When one studies experimentally Newton's theory of gravity quickly realizes that there is a deep relation between the inertial and the gravitational mass of a body. It turns out that the acceleration a experienced by any two bodies on the surface of the Earth looks the same irrespective of the mass of those bodies. This suggests that inertial and gravitational mass have the same numerical values, $m_i = m_g$ (in general, they are proportional, being the proportionality constant the same for all bodies). This observation is known as *Newton's equivalence principle* or weak equivalence principle.

From Newton's laws it follows that Newtonian physics is based on the idea of *absolute space*, a background structure with respect to which accelerations can be effectively measured. However, the PoR implies that, unlike accelerations, absolute positions and velocities are not directly observable. This conclusion was challenged by some results published in 1865 by J.C. Maxwell. In Maxwell's work, the equations of the electric and magnetic field were improved by the addition of a new term (Maxwell's displacement current). The new equations predicted the existence of electromagnetic waves. The explicit appearance in those equations of a speed c was interpreted as the existence of a privileged reference frame, that of the *luminiferous aether*[1]. According to this, it could be possible to measure absolute velocities (at least with respect to the aether[2]).

This idea motivated the experiment carried out by Michelson and Morley in 1887 to measure the relative velocity of the Earth in its orbit around the sun with respect to the aether[3]. Despite the experimental limitations of the epoch, their experiment had enough precision to confirm that the speed of light is independent of the direction of the light ray and the position of the Earth in its orbit.

Motivated by this intriguing phenomenon, in 1892 Lorentz proposed that moving bodies contract in the direction of motion according to a specific set of transformations. In 1905 Einstein presented its celebrated theory of special relativity and derived the Lorentz transformations using the PoR and the observed constancy of the speed of light without assuming the presence of an aether. Therefore, though the principle of relative motion had been put into question by electromagnetism, it was salvaged by Einstein's reinterpretation[4].

As of that moment, it was understood that any good physical theory should be adapted to the new PoR. Fortunately, Minkowski (1907) realized that Lorentz transformations could be nicely interpreted in a four dimensional space-time (he thus invented the notion of spacetime as opposed to the well-known spatial geometry of the time). In this manner, a Lorentz-invariant theory should be constructed using geometrical invariants such as scalars and four-vectors, which represents a geometrical formulation of the PoR.

---

[1] The aether was supposed to have very special properties, such as a very high elasticity, and to exhibit no friction to the motion of bodies through it.

[2] The aether was assumed to be at rest because otherwise the light from distant stars would suffer distortions in their propagation due to local motions of this fluid.

[3] Note that the speed of sound is relative to the wind. Analogously, it was thought that the speed of light should be measured with respect to the aether. Due to the motion of the Earth, that speed should depend on the position of the Earth and the direction of the light ray. The interferometer was built on a rotating surface such that the full experiment could be rotated to observe periodic variations of the interference pattern.

[4] It is worth noting that Einstein's results did not rule out the aether, but they implied that its presence was irrelevant for the discussion of experiments.

2.1 A relativistic theory of gravity: Nordström's theory.

The acceptance of the new PoR led to the development of relativistic theories of gravity in which the gravitational field was represented by different types of fields, such as scalars (in analogy with Newtonian mechanics) or vectors (in analogy with Maxwell's electrodynamics). A natural proposal[5] in this sense consists on replacing the Newtonian equations by the following relativistic versions [Norton (1992)]

$$\nabla^2\phi = 4\pi G\rho \;\rightarrow\; \Box\phi = 4\pi G\rho \tag{1}$$

$$\frac{d\vec{v}}{dt} = -\vec{\nabla}\phi \;\rightarrow\; \frac{du_\mu}{d\tau} = -\partial_\mu\phi \tag{2}$$

This proposal, however, is unsatisfactory. From the assumed constancy of the speed of light, $\eta_{\mu\nu}u^\mu u^\nu = -c^2$, one finds that $u_\mu \frac{du^\mu}{d\tau} = 0$, which implies the unnatural restriction $u^\mu\partial_\mu\phi = \frac{d\phi}{d\tau} = 0$, i.e., the gravitational field should be constant along any observer's world line.
To overcome this drawback, Nordström proposed that the mass of a body in a gravitational field could vary with the gravitational potential [Nordström (1912)] . Nordström proposed a relativistic scalar theory of gravity in which the matter evolution equation (2) was modified to make it compatible with the constancy of the speed of light

$$F_\mu \equiv \frac{d(mu_\mu)}{d\tau} = -m\partial_\mu\phi \leftrightarrow m\frac{du_\mu}{d\tau} + u_\mu\frac{dm}{d\tau} = -m\partial_\mu\phi. \tag{3}$$

This equation implies that in a gravitational field m changes as $md\phi/d\tau = c^2dm/d\tau$, which leads to $m = m_0e^{\phi/c^2}$ and avoids the undesired restriction $d\phi/dt = 0$ of the theory presented before[6]. The matter evolution equation can thus be written as

$$\frac{du_\mu}{d\tau} = -\partial_\mu\phi - \frac{d\phi}{d\tau}u_\mu\,. \tag{4}$$

It is worth noting that this equation satisfies Newton's equivalence principle in the sense that the gravitational mass of a body is identified with its rest mass. Free fall, therefore, turns out to be independent of the rest mass of the body. However, Einstein's special theory of relativity had shown a deep relation between mass and energy that should be carefully addressed in the construction of any relativistic theory of gravity. The equation $E = mc^2$, where $m = \gamma m_0$ and $\gamma = 1/\sqrt{1-\vec{v}^2/c^2}$, states that kinetic energy increases the effective mass of a body, its inertia. Therefore, if inertial mass is the source of the gravitational field, a moving body could generate a stronger gravitational field than the same body at rest. By extension of this reasoning, one can conclude that bodies with different internal energies could fall differently in an external gravitational field. Einstein found this point disturbing and used it to criticize Nordström's theory. In addition, in this theory the gravitational potential ϕ of point particles goes to $-\infty$ at the location of the particle, thus implying that point particles are massless and, therefore, cannot exist. One is thus led to consider extended (or continuous) objects, which possess other types of inertia in the form of stresses that cannot be reduced to a mass. The source

[5] Another very natural proposal would be a relativistic theory of gravity inspired by Maxwell's electrodynamics, being $F_\mu \equiv mdu_\mu/d\tau = kG_{\mu\nu}u^\nu$ with $G_{\mu\nu} = -G_{\nu\mu}$. Such a proposal immediately implies that $F_\mu u^\mu = 0$ and is compatible with the constancy of c^2.

[6] Varying speed of light theories may also avoid the restriction $d\phi/d\tau = 0$, but such theories break the essence of special relativity by definition.

of the gravitational field, the right hand side of (1), should thus take into account also such stresses.
To overcome those problems and others concerning energy conservation pointed out by Einstein, Nordström proposed a second theory [Nordström (1913)]

$$\Box\phi = g(\phi)\nu \tag{5}$$

$$\mathcal{F}_\mu = -g(\phi)\nu\partial_\mu\phi \,. \tag{6}$$

where $\mathcal{F}$ represents the force per unit volume and $g(\phi)\nu$ is a density that represents the source of the gravitational field. To determine the functional form of $g(\phi)$ and find a natural correspondence between ν and the matter sources, Nordström proceeded as follows. Firstly, he defined the *gravitational mass* of a system using the right hand side of (5) and (6) as

$$M_g = \int d^3x g(\phi)\nu \,. \tag{7}$$

Then he assumed that the *inertial mass* of the system should be a Lorentz scalar made out of all the energy sources, which include the rest mass and stresses associated to the matter, the gravitational field, and the electromagnetic field. He thus proposed the following expression

$$m_i = -\frac{1}{c^2}\int d^3x[T_m + G_\phi + F_{em}] \,, \tag{8}$$

where the trace of the stress-energy tensor of the matter is represented by T_m, the trace of the electromagnetic field by F_{em} (which vanishes), and that of the gravitational field by G_m, being $G_{\mu\nu} = (2/\kappa^2)[\partial_\mu\phi\partial_\nu\phi - (1/2)\eta_{\mu\nu}(\partial_\lambda\phi\partial^\lambda\phi)]$ the stress-energy tensor of the (scalar) gravitational field.
To force the equivalence between inertial and gravitational mass in a system of particles immersed in an external gravitational field with potential ϕ_a, Nordström imposed that for such a system the following relation should hold

$$M_g = g(\phi_a)m_i \,. \tag{9}$$

Then he considered a stationary system on that gravitational field and showed that the contribution of the local gravitational field to the total inertia of the system was given by

$$-\frac{1}{c^2}\int d^3x G_\phi = -\frac{1}{c^2}\int d^3x(\phi - \phi_a)g(\phi)\nu \,. \tag{10}$$

Combining this expression with (9) and (8) one finds that

$$\int d^3x\left[T_m + g(\phi)\nu\left(\phi - \phi_a + \frac{c^2}{g(\phi_a)}\right)\right] = 0 \,. \tag{11}$$

Demanding proportionality between T_m and ν, one finds that $g(\phi) = C/(A+\phi)$. A natural gauge corresponds to $g(\phi) = -4\pi G/\phi$ because it allows to recover the Newtonian result $E_0 = mc^2 = M_g\phi_a$ that implies that the energy of a system with gravitational mass M_g in a field with potential ϕ_a is exactly $M_g\phi_a$. Therefore, from Nordström's second theory it follows that the inertial mass of a stationary system varies in proportion to the external potential whereas M_g remains constant, i.e., $m/\phi = constant$.

With the above results one finds that (5) and (6) turn into (from now on $\kappa^2 \equiv 8\pi G$)

$$\phi\Box\phi = -\frac{\kappa^2}{2}T_m \tag{12}$$

$$\frac{du_\mu}{d\tau} = -\partial_\mu \ln\phi - u_\mu \frac{d}{d\tau}\ln\phi\,. \tag{13}$$

Using these equations it is straightforward to verify that the total energy-momentum of the system is conserved, i.e., $\partial^\mu\left(T^\phi_{\mu\nu} + T^m_{\mu\nu}\right) = 0$, where one must take $T^m_{\mu\nu} = \rho\phi u_\mu u_\nu$ for pressureless matter because, as shown above, the inertial rest mass density of a system grows linearly with ϕ.

Nordström's second theory, therefore, represents a satisfactory example of relativistic theory of gravity in Minkowski space that satisfies the equivalence between inertial and gravitational mass and in which energy and momentum are conserved. Unfortunately, it does not predict any bending of light and also fails in other predictions that were important at the beginning of the twentieth century such as the perihelion shift of Mercury. Nonetheless, it admits a geometric interpretation that greatly simplifies its structure and puts forward the direction in which Einstein's work was progressing.

Considering a line element of the form $ds^2 = \phi^2(-dt^2 + d\vec{x}^2)$, Einstein and Fokker showed that the matter evolution equation (13) could be obtained by extremizing the path followed by a free particle in that geometry, i.e., by computing the variation $\delta\left(-mc^2\int ds\right) = 0$ [Einstein and Fokker (1914)] . This variation yields the geodesic equation[7]

$$\frac{d\tilde{u}^\mu}{d\tilde{\tau}} + \Gamma^\mu_{\alpha\beta}\tilde{u}^\alpha\tilde{u}^\beta = 0\,, \tag{14}$$

where $\Gamma^\mu_{\alpha\beta} = \partial_\alpha\phi\delta^\mu_\beta + \partial_\beta\phi\delta^\mu_\alpha - \eta^{\mu\rho}\partial_\rho\phi\eta_{\alpha\beta}$. The gravitational field equation also takes a very interesting form

$$R = 3\kappa^2\tilde{T}_m\,, \tag{15}$$

where $R = -(6/\phi^3)\eta^{\alpha\beta}\partial_\alpha\partial_\beta\phi$ and $\tilde{T}_m = T_m/\phi^4$ due to the conformal transformation that relates the background metric with the Minkowski metric. These last results represent generally covariant equations that establish a non-trivial relation between gravitation and geometry. Though this theory was eventually ruled out by observations, its potential impact on the eventual formal and conceptual formulation of Einstein's general theory of relativity must have been important.

2.2 To general relativity via general covariance

The Principle of Relativity together with Newton's ideas about the equivalence between inertial and gravitational mass led Einstein to develop what has come to be called the Einstein equivalence principle (EEP), which will be introduced later in detail. Einstein wanted to extend the principle of relativity not only to inertial observers (special relativity) but to all kinds of motion (hence the term general relativity). This motivated the search for generally

[7] To obtain (13) from the geodesic equation one should note that $d\tilde{\tau} = \phi d\tau$, $\tilde{u}_\mu = \phi u_\mu$, and that the indices in (13) are raised and lowered with $\eta_{\mu\nu}$.

covariant equations [8].
Though it is not difficult to realize that one can construct a fully covariant theory in Minkowski space, the consideration of arbitrary accelerated frames leads to the appearance of inertial or ficticious forces whose nature is difficult to interpret. This is due to the fact that Minkowski spacetime, like Newtonian space, is an absolute space. The possibility of writing the laws of physics in a coordinate (cartesian, polar,...) and frame (inertial, accelerated,...) invariant way, helped Einstein to realize that a local, homogeneous gravitational field is indistinguishable from a constant acceleration. This allowed him to introduce the concept of local inertial frame (LIF) and find a correspondence between gravitation and geometry, which led to a deep conceptual change: **there exists no absolute space**. This follows from the fact that, unlike other well-known forces, the local effects of gravity can always be eliminated by a suitable choice of coordinates (Einstein's elevator).
The forces of Newtonian mechanics, which were thought to be measured with respect to *absolute space*, were in fact being measured in an accelerated frame (static with respect to the Earth), which led to the appearance of the observed gravitational acceleration. According to Einstein, accelerations produced by interactions such as the electromagnetic field should be measured in LIFs. This means that they should be measured not with respect to absolute space but with respect to the local gravitational field (which defines LIFs). In other words, Einstein identified the Newtonian absolute space with the local gravitational field. Physical accelerations should, therefore, be measured in local inertial frames, where Minkowskian physics should be recovered. Gravitation, according to Einstein, was intrinsically different from the rest of interactions. It was a geometrical phenomenon.
The geometrical interpretation of gravitation implied that it should be described by a tensor field, the metric $g_{\mu\nu}$, which boils down to the Minkowski metric locally in appropriate coordinate systems (LIFs) or globally when gravitation is absent. This view made it natural to interpret the effects of a gravitational field on particles as geodesic motion. In the absence of non-gravitational interactions, particles should follow geodesics of the background metric, which are formally described by eq.(14) but with $\Gamma^{\mu}_{\alpha\beta}$, the so-called Levi-Civita connection, defined in terms of a symmetric metric tensor $g_{\mu\nu}$ as

$$\Gamma^{\mu}_{\alpha\beta} = \frac{g^{\mu\rho}}{2}\left[\partial_\alpha g_{\rho\beta} + \partial_\beta g_{\rho\alpha} - \partial_\rho g_{\alpha\beta}\right] . \tag{16}$$

To determine the dynamics of the metric tensor one needs at least ten independent equations, as many as independent components there are in $g_{\mu\nu}$. Since the source of the gravitational field must be related with the stress-energy tensor of matter and the dynamics of classical mechanics is generally governed by second-order equations, Einstein proposed the following set of tensorial equations

$$R_{\mu\nu} - \frac{1}{2}g_{\mu\nu}R = \kappa^2 T_{\mu\nu} , \tag{17}$$

where $R_{\mu\nu} \equiv R^{\rho}{}_{\mu\rho\nu}$ is the so-called Ricci tensor, $R = g^{\mu\nu}R_{\mu\nu}$ is the Ricci scalar, and $R^{\alpha}{}_{\beta\mu\nu} = \partial_\mu\Gamma^{\alpha}_{\nu\beta} - \partial_\nu\Gamma^{\alpha}_{\mu\beta} + \Gamma^{\alpha}_{\mu\lambda}\Gamma^{\lambda}_{\nu\beta} - \Gamma^{\alpha}_{\nu\lambda}\Gamma^{\lambda}_{\mu\beta}$ represents the components of the Riemann tensor, the field strength of the connection $\Gamma^{\alpha}_{\mu\beta}$, which here is defined as in (16).

[8] The idea of general covariance is nowadays naturally seen as a basic mathematical requirement in any theory based on the use of differential manifolds. In this sense, though general covariance forces the use of tensor calculus, it should be noted that it does not necessarily imply curved space-time. Note also that it is the connection, not the metric, the most important object in the construction of tensors.

Eq. (17) represents a system of non-linear, second-order partial differential equations for the ten independent components of the metric tensor. The conservation of energy and momentum is guaranteed independently for the left and the right hand sides of (17). The contraction[9] $\nabla^{\mu}(R_{\mu\nu} - \frac{1}{2}g_{\mu\nu}R) = 0$ follows from a geometrical identity, whereas $\nabla^{\mu}T_{\mu\nu} = 0$ follows if the Minkowski equations of motion for the matter fields are satisfied locally. The non-linearity of the equations manifests the fact that the energy stored in the gravitational field can source the gravitational field itself in a non-trivial way. Unlike Nordström's second theory, this set of tensorial equations imply that the gravitational field is sourced by the full stress-energy tensor, not just by its trace. This implies that electromagnetic fields, like any other matter sources, generate a non-zero Ricci tensor and, therefore, gravitate.

Einstein's theory was rapidly accepted despite its poor experimental verification. In fact, we had to wait until the 1960's to have the perihelion shift of Mercury and the deflection of light by the sun measured to within an accuracy of $\sim$ 1% and $\sim$ 50%, respectively. In 1959 Pound and Rebka were able to measure the gravitational redshift for the first time. Additionally, though Hubble's discoveries on the recession of distant galaxies had boosted Einstein's popularity, those observations were a mere qualitative verification of the effect and only recently has it been possible to contrast theory and observations with some confidence in the cosmological setting. It is therefore not surprising that between 1905 and 1960, there appeared at least 25 alternative relativistic theories of gravitation, where spacetime was flat and gravitation was a Lorentz-invariant field on that background. Though many researchers defended Einstein's idea of curved spacetime, others like Birkhoff did not [Birkhoff (1944)]:

> *The initial attempts to incorporate gravitational phenomena in flat space-time were not satisfactory. Einstein turned to the curved spacetime suggested by his principle of equivalence, and so constructed his general theory of relativity. The initial predictions, based on this celebrated theory of gravitation, were brilliantly confirmed. However, the theory has not led to any further applications and, because of its complicated mathematical character, seems to be essentially unworkable. Thus curved spacetime has come to be regarded by many as an auxiliary construct (Larmor) rather than as a physical reality.*

Such strong claims suggest that it was necessary a careful analysis of the foundations of Einstein's theory: is spacetime really curved or is gravitation a tensor-like interaction in a flat background? The next section is devoted to clarify these points and others that will help establish the foundations of gravitation theory.

2.3 The Einstein equivalence principle

The experimental facts that support the foundations of gravitation should never be underestimated since they provide a valuable guide in the construction of viable theories and in constraining the realm of speculation. In this sense, the experimental efforts carried out by Robert Dicke in the 1960's [Dicke (1964)] resulted in what has come to be called the Einstein equivalence principle (EEP) and constitute a fundamental pillar for gravitation theory. We will briefly review next the experimental evidence supporting it, and the way it enters in the construction of gravitation theories [Will (1993)]. The EEP states that [Will (2005)]

- Inertial and gravitational masses coincide (weak equivalence principle).
- The outcome of any non-gravitational experiment is independent of the velocity of the freely-falling reference frame in which it is performed (Local Lorentz Invariance).

[9] The differential operator ∇_{μ} represents a covariant derivative, which is the natural extension of the usual flat space derivative ∂_{μ} to spaces with non-trivial parallel transport.

- The outcome of any local non-gravitational experiment is independent of where and when in the universe it is performed (Local Position Invariance).

Let us briefly discuss the experimental evidence supporting the EEP.

2.3.1 Weak equivalence principle

A direct test of WEP is the comparison of the acceleration of two laboratory-sized bodies of different composition in an external gravitational field. If the principle were violated, then the accelerations of different bodies would differ. In Dicke's torsion balance experiment, for instance, the gravitational acceleration toward the sun of small gold and aluminum weights were compared and found to be equal with an accuracy of about a part in 10^{11}. One should note that gold and aluminum atoms have very different properties, which is important for testing how gravitation couples to different particles and interactions. For instance, the electrons in aluminum are non-relativistic whereas the k-shell electrons of gold have a 15% increase in their mass as a result of their relativistic velocities. The electromagnetic negative contribution to the binding energy of the nucleus varies as Z^2 and represents 0.5% of the total mass of a gold atom, whereas it is negligible in Al. Additionally, the virtual pair field, pion field, etcetera, around the gold nucleus would be expected to represent a far bigger contribution to the total energy than in aluminum. This makes it clear that a gold sphere possesses additional inertial contributions due to the electromagnetic, weak, and strong interactions that are not present (or are negligible) in the aluminum sphere. If any of those sources of inertia did not contribute by the same amount to the gravitational mass of the system, then gold and aluminum would fall with different accelerations.
The precision of Dicke's experiment was such that from it one can conclude, for instance, that positrons and other antiparticles fall down, not up [Dicke (1964)]. This is so because if the positrons in the pair field of the gold atom were to tend to fall up, not down, there would be an anomalous weight of the atom substantially greater for large atomic number than small.

2.3.2 Tests of local Lorentz invariance

The existence of a preferred reference frame breaking the local isotropy of space would imply a dependence of the speed of light on the direction of propagation. This would cause shifts in the energy levels of atoms and nuclei that depend on the orientation of the quantization axis of the state relative to our universal velocity vector, and on the quantum numbers of the state. This idea was tested by Hughes (1960) and Drever (1961), who examined the $J = 3/2$ ground state of the ^7Li nucleus in an external magnetic field. If the Michelson-Morley experiment had found $\delta \equiv c^{-2} - 1 \approx 10^{-3}$, the Hughes-Drever experiment set the limit to $\delta \approx 10^{-15}$. More recent experiments using laser-cooled trapped atoms and ions have reached $\delta \approx 10^{-17}$.
Currently, new ideas coming from quantum gravity (with a minimal length scale), braneworld scenarios, and models of string theory have motivated new ways to test Lorentz invariance by considering Lorentz-violating parameters in extensions of the standard model and also some astrophysical tests. So far, however, no compelling evidence for a violation of Lorentz invariance has been found.

2.3.3 Tests of local position invariance

Local position invariance can be tested by gravitational redshift experiments, which test the existence of spatial dependence on the outcome of local non-gravitational experiments, and by measurements of the fundamental non-gravitational constants that test for temporal

dependence. Gravitational redshift experiments usually measure the frequency shift $Z = \Delta\nu/\nu = -\Delta\lambda/\lambda$ between two identical frequency standards (clocks) placed at rest at different heights in a static gravitational field. If the frequency of a given type of atomic clock is the same when measured in a local, momentarily comoving freely falling frame (Lorentz frame), independent of the location or velocity of that frame, then the comparison of frequencies of two clocks at rest at different locations boils down to a comparison of the velocities of two local Lorentz frames, one at rest with respect to one clock at the moment of emission of its signal, the other at rest with respect to the other clock at the moment of reception of the signal. The frequency shift is then a consequence of the first-order Doppler shift between the frames. The result is a shift $Z = \frac{\Delta U}{c^2}$, where U is the difference in the Newtonian gravitational potential between the receiver and the emitter. If the frequency of the clocks had some dependence on their position, the shift could be written as $Z = (1+\alpha)\frac{\Delta U}{c^2}$. Comparison of a hydrogen-maser clock flown on a rocket to an altitude of about 10.000 km with a similar clock on the ground yielded a limit $\alpha < 2 \times 10^{-4}$.
Another important aspect of local position invariance is that if it is satisfied then the fundamental constants of non-gravitational physics should be constants in time. Though these tests are subject to many uncertainties and experimental limitations, there is no strong evidence for a possible spatial or temporal dependence of the fundamental constants.

2.4 Metric theories of gravity

The EEP is not just a verification that gravitation can be associated with a metric tensor which locally can be turned into the Minkowskian metric by a suitable choice of coordinates. If it is valid, then gravitation must be a curved space-time phenomenon, i.e., the effects of gravity must be equivalent to the effects of living in a curved space-time. For this reason, the only theories of gravity that have a hope of being viable are those that satisfy the following postulates (see [Will (1993)] and [Will (2005)]):

1. Spacetime is endowed with a symmetric metric.
2. The trajectories of freely-falling bodies are geodesics of that metric.
3. In local freely-falling reference frames, the non-gravitational laws of physics are those written in the language of special relativity.

Theories satisfying these postulates are known as *metric theories of gravity*, and their action can be written generically as

$$S_{MT} = S_G[g_{\mu\nu}, \phi, A_\mu, B_{\mu\nu}, \ldots] + S_m[g_{\mu\nu}, \psi_m] \,, \tag{18}$$

where $S_m[g_{\mu\nu}, \psi_m]$ represents the matter action, ψ_m denotes the matter and non-gravitational fields, and S_G is the gravitational action, which besides the metric $g_{\mu\nu}$ may depend on other gravitational fields (scalars, vectors, and tensors of different ranks). This form of the action guarantees that the non-gravitational fields of the standard model of elementary particles couple to gravitation only through the metric, which should allow to recover locally the non-gravitational physics of Minkowski space. The construction of $S_m[g_{\mu\nu}, \psi_m]$ can thus be carried out by just taking its Minkowski space form and going over to curved space-time using the methods of differential geometry. It should be noted that the EEP does neither point towards GR as the preferred theory of gravity nor provides any constraint or hint on the functional form of the gravitational part of the action. The functional S_G must provide

dynamical equations for the metric (and the other gravitational fields, if there are any) but its form must be obtained by theoretical reasoning and/or by experimental exploration.
It is worth noting that if S_G contains other long-range fields besides the metric, then **gravitational** experiments in a local, freely falling frame may depend on the location and velocity of the frame relative to the external environment. This is so because, unlike the metric, the boundary conditions induced by those fields cannot be trivialized by a suitable choice of coordinates. Of course, local **non-gravitational** experiments are unaffected since the gravitational fields they generate are assumed to be negligible, and since those experiments couple only to the metric, whose form can always be made locally Minkowskian at a given spacetime event.
Before concluding this section, it should be noted that string theories predict the existence of new kinds of fields with couplings to fermions and the interactions of the standard model in a way that breaks the simplicity of metric theories of gravity, i.e., they do not allow for a clean splitting of the action into a matter sector plus a gravitational sector. Such theories, therefore, must be regarded as non-metric. Improved tests of the EEP could be used to test the presence and/or intensity of such couplings, which are expected to represent short range interactions. These tests can be seen as a branch of high-energy physics not based on particle accelerators.

2.4.1 Two examples of metric theories: General relativity and Brans-Dicke theory.

The field equations of Einstein's theory of general relativity (GR) can be derived from the following action

$$S[g_{\mu\nu}, \psi_m] = \frac{1}{16\pi G}\int d^4x\sqrt{-g}R(g) + S_m[g_{\mu\nu}, \psi_m] \tag{19}$$

where R is the Ricci scalar defined below eq.(17). Variation of this action with respect to the metric leads to Einstein's field equations[10]

$$R_{\mu\nu} - \frac{1}{2}g_{\mu\nu}R = 8\pi G T_{\mu\nu} \tag{20}$$

In Einstein's theory, gravity is mediated by a rank-2 tensor field, the metric, and curvature is generated by the matter sources. Brans-Dicke theory introduces, besides the metric, a new gravitational field, which is a scalar. This scalar field is coupled to the curvature as follows

$$S[g_{\mu\nu}, \phi, \psi_m] = \frac{1}{16\pi}\int d^4x\sqrt{-g}\left[\phi R(g) - \frac{\omega}{\phi}(\partial_\mu\phi\partial^\mu\phi) - V(\phi)\right] + S_m[g_{\mu\nu}, \psi_m] \tag{21}$$

In the original Brans-Dicke theory, the potential was set to zero, $V(\phi) = 0$, so the theory had only one free parameter, the constant ω in front of the kinetic energy term, which had to be determined experimentally. Note that the Brans-Dicke scalar has the same dimensions as the inverse of Newton's constant and, therefore, can be seen as related to it. In Brans-Dicke theory, one can thus say that Newton's constant is no longer constant but is, in fact, a dynamical field. The field equations for the metric are

$$R_{\mu\nu}(g) - \frac{1}{2}g_{\mu\nu}R(g) = \frac{8\pi}{\phi}T_{\mu\nu} - \frac{1}{2\phi}g_{\mu\nu}V(\phi) + \frac{1}{\phi}\left[\nabla_\mu\nabla_\nu\phi - g_{\mu\nu}\Box\phi\right] + \frac{\omega}{\phi^2}\left[\partial_\mu\phi\partial_\nu\phi - \frac{1}{2}g_{\mu\nu}(\partial\phi)^2\right] \tag{22}$$

[10] Recall that $\delta\sqrt{-g} = -\frac{1}{2}\sqrt{-g}g_{\mu\nu}\delta g^{\mu\nu}$ and that $\delta R_{\mu\nu} = -\nabla_\mu\delta\Gamma^\lambda_{\lambda\nu} + \nabla_\lambda\delta\Gamma^\lambda_{\mu\nu}$.

The equation that governs the scalar field is

$$(3+2\omega)\Box\phi + 2V(\phi) - \phi\frac{dV}{d\phi} = \kappa^2 T \tag{23}$$

In this theory we observe that both the matter and the scalar field act as sources for the metric, which means that both the matter and the scalar field generate the spacetime curvature. In fact, even in vacuum the scalar field curves the spacetime. According to the way we wrote the metric field equations, it is tempting to identify the Brans-Dicke field with a new matter field. However, since the Brans-Dicke scalar is sourced by the energy-momentum tensor (via its trace, which is a scalar magnitude constructed out of the sources of energy and momentum), we say that it is a gravitational field. Note, in this sense, that standard matter fields, such as a Dirac field coupled to electromagnetism $(i\gamma^\mu\partial_\mu - m)\psi = e\gamma^\mu A_\mu\psi$, do not couple to energy and momentum.

3. Experimental determination of the gravity Lagrangian

Einstein's theory of general relativity (GR) represents one of the most impressive exercises of human intellect. As we have seen in previous sections, it implied a huge conceptual jump with respect to Newtonian gravity and, unlike the currently established standard model of elementary particles, no experiments were carried out to probe the structure of the theory. In spite of that, to date the theory has successfully passed all precision experimental tests. Its predictions are in agreement with experiments in scales that range from millimeters to astronomical units, scales in which weak and strong field phenomena can be observed [Will (2005)]. The theory is so successful in those regimes and scales that it is generally accepted that it should also work at larger and shorter scales, and at weaker and stronger regimes.

This extrapolation is, however, forcing us today to draw a picture of the universe that is not yet supported by other independent observations. For instance, to explain the rotation curves of spiral galaxies, we must accept the existence of vast amounts of unseen matter surrounding those galaxies. Additionally, to explain the luminosity-distance relation of distant type Ia supernovae and some properties of the distribution of matter and radiation at large scales, we must accept the existence of yet another source of energy with repulsive gravitational properties (see [Copeland et al. (2006)], [Padmanabhan (2003)], [Peebles and Ratra (2003)] for recent reviews on dark energy). Together those unseen (or dark) sources of matter and energy are found to make up to 96% of the total energy of the observable universe! This huge discrepancy between the gravitationally estimated amounts of matter and energy and the direct measurements via electromagnetic radiation motivates the search for alternative theories of gravity which can account for the large scale dynamics and structure without the need for dark matter and/or dark energy.

In this sense, there has been an enormous international effort in the last years to determine whether the gravity Lagrangian could depart from Einstein's one at cosmic scales in a way compatible with the cosmological observations that support the cosmic speedup. In particular, many authors have investigated the consequences of promoting the Hilbert-Einstein Lagrangian to an arbitrary function $f(R)$ of the scalar curvature (see [Olmo (2011)], [De Felice and Tsujikawa (2010)], [Sotiriou and Faraoni (2010)], [Capozziello and Francaviglia (2008)] for recent reviews). In this section we will show that the dynamics of the solar system can be used to set important constraints on the form of the function $f(R)$.

3.1 Field equations of $f(R)$ theories.

The action that defines $f(R)$ theories has the generic form

$$S = \frac{1}{2\kappa^2}\int d^4x\sqrt{-g}f(R) + S_m[g_{\mu\nu}, \psi_m] \,, \tag{24}$$

where $\kappa^2 = 8\pi G$, and we use the same notation introduced in previous sections. Variation of (24) leads to the following field equations for the metric

$$f_R R_{\mu\nu} - \frac{1}{2} f g_{\mu\nu} - \nabla_\mu\nabla_\nu f_R + g_{\mu\nu}\Box f_R = \kappa^2 T_{\mu\nu} \tag{25}$$

where $f_R \equiv df/dR$. According to (25), we see that, in general, the metric satisfies a system of fourth-order partial differential equations. The trace of (25) takes the form

$$3\Box f_R + R f_R - 2f = \kappa^2 T \tag{26}$$

If we take $f(R) = R - 2\Lambda$, (25) boils down to

$$R_{\mu\nu} - \frac{1}{2} g_{\mu\nu} R = \kappa^2 T_{\mu\nu} - \Lambda g_{\mu\nu} \,, \tag{27}$$

which represents GR with a cosmological constant. This is the only case in which an $f(R)$ Lagrangian yields second-order equations for the metric[11].

Let us now rewrite (25) in the form

$$R_{\mu\nu} - \frac{1}{2} g_{\mu\nu} R = \frac{\kappa^2}{f_R} T_{\mu\nu} - \frac{1}{2f_R} g_{\mu\nu}[R f_R - f] + \frac{1}{f_R}\left[\nabla_\mu\nabla_\nu f_R - g_{\mu\nu}\Box f_R\right] \tag{28}$$

The right hand side of this equation can now be seen as the source terms for the metric. This equation, therefore, tells us that the metric is generated by the matter and by terms related to the scalar curvature. If we now wonder about what generates the scalar curvature, the answer is in (26). That expression says that the scalar curvature satisfies a second-order differential equation with the trace T of the energy-momentum tensor of the matter and other curvature terms acting as sources. We have thus clarified the role of the higher-order derivative terms present in (25). The scalar curvature is now a dynamical entity which helps generate the space-time metric and whose dynamics is determined by (26).

At this point one should have noted the essential difference between a generic $f(R)$ theory and GR. In GR the only dynamical field is the metric and its form is fully characterized by the matter distribution through the equations $G_{\mu\nu} = \kappa^2 T_{\mu\nu}$, where $G_{\mu\nu} \equiv R_{\mu\nu} - \frac{1}{2} g_{\mu\nu} R$. The scalar curvature is also determined by the local matter distribution but through an algebraic equation, namely, $R = -\kappa^2 T$. In the $f(R)$ case both $g_{\mu\nu}$ and R are dynamical fields, i.e., they are governed by differential equations. Furthermore, the scalar curvature R, which can obviously be expressed in terms of the metric and its derivatives, now plays a non-trivial role in the determination of the metric itself.

[11] This is so only if the connection is assumed to be the Levi-Civita connection of the metric (metric formalism). If the connection is regarded as independent of the metric, Palatini formalism, then $f(R)$ theories lead to second-order equations. This point will be explained in detail later on.

The physical interpretation given above puts forward the central and active role played by the scalar curvature in the field equations of $f(R)$ theories. However, (26) suggests that the actual dynamical entity is f_R rather than R itself. This is so because, besides the metric, f_R is the only object acted on by differential operators in the field equations. Motivated by this, we can introduce the following alternative notation

$$\phi \equiv f_R \tag{29}$$

$$V(\phi) \equiv R(\phi) f_R - f(R(\phi)) \tag{30}$$

and rewrite eqs. (28) and (26) in the same form as (22) and (23) with the choice $w = 0$. This slight change of notation helps us identify the $f(R)$ theory in metric formalism with a scalar-tensor Brans-Dicke theory with parameter $\omega = 0$ and non-trivial potential $V(\phi)$, whose action was given in (21). In terms of this scalar-tensor representation our interpretation of the field equations of $f(R)$ theories is obvious, since both the matter and the scalar field help generate the metric. The scalar field is a dynamical object influenced by the matter and by self-interactions according to (23).

3.2 Spherically symmetric systems

A complete description of a physical system must take into account not only the system but also its interaction with the environment. In this sense, any physical system is surrounded by the rest of the universe. The relation of the local system with the rest of the universe manifests itself in a set of boundary conditions. In our case, according to (26) and (28), the metric and the function f_R (or, equivalently, R or ϕ) are subject to boundary conditions, since they are dynamical fields (they are governed by differential equations). The boundary conditions for the metric can be trivialized by a suitable choice of coordinates. In other words, we can make the metric Minkowskian in the asymptotic region and fix its first derivatives to zero (see chapter 4 of [Will (1993)] for details). The function f_R, on the other hand, should tend to the cosmic value f_{R_c} as we move away from the local system. The precise value of f_{R_c} is obtained by solving the equations of motion for the corresponding cosmology. According to this, the local system will interact with the asymptotic (or background) cosmology via the boundary value f_{R_c} and its cosmic-time derivative. Since the cosmic time-scale is much larger than the typical time-scale of local systems (billions of years versus years), we can assume an adiabatic interaction between the local system and the background cosmology. We can thus neglect terms such as $\dot{f}_{R_c}$, where dot denotes derivative with respect to the cosmic time.
The problem of finding solutions for the local system, therefore, reduces to solving (28) expanding about the Minkowski metric in the asymptotic region[12], and (26) tending to

$$3\Box_c f_{R_c} + R_c f_{R_c} - 2f(R_c) = \kappa^2 T_c \tag{31}$$

where the subscript c denotes cosmic value, far away from the system. In particular, if we consider a weakly gravitating local system, we can take $f_R = f_{R_c} + \varphi(x)$ and $g_{\mu\nu} = \eta_{\mu\nu} + h_{\mu\nu}$, with $|\varphi| \ll |f_{R_c}|$ and $|h_{\mu\nu}| \ll 1$ satisfying $\varphi \to 0$ and $h_{\mu\nu} \to 0$ in the asymptotic region. Note that should the local system represent a strongly gravitating system such as a neutron star or a black hole, the perturbative expansion would not be sufficient everywhere. In such cases, the

[12] Note that the expansion about the Minkowski metric does not imply the existence of global Minkowskian solutions. As we will see, the general solutions to our problem turn out to be asymptotically de Sitter spacetimes.

perturbative approach would only be valid in the far region. Nonetheless, the decomposition $f_R = f_{R_c} + \varphi(x)$ is still very useful because the equation for the local deviation $\varphi(x)$ can be written as

$$3\Box\varphi + W(f_{R_c} + \varphi) - W(f_{R_c}) = \kappa^2 T, \tag{32}$$

where T represents the trace of the local sources, we have defined $W(f_R) \equiv R(f_R)f_R - 2f(R[f_R])$, and $W(f_{R_c})$ is a slowly changing constant within the adiabatic approximation. In this case, φ needs not be small compared to f_{R_c} everywhere, only in the asymptotic regions.

3.2.1 Spherically symmetric solutions

Let us define the line element[13] [Olmo (2007)]

$$ds^2 = -A(r)e^{2\psi(r)}dt^2 + \frac{1}{A(r)}\left(dr^2 + r^2 d\Omega^2\right), \tag{33}$$

which, assuming a perfect fluid for the sources, leads to the following field equations

$$A_{rr} + A_r\left[\frac{2}{r} - \frac{5}{4}\frac{A_r}{A}\right] = \frac{\kappa^2\rho}{f_R} + \frac{Rf_R - f(R)}{2f_R} + \frac{A}{f_R}\left[f_{R_{rr}} + f_{R_r}\left(\frac{2}{r} - \frac{A_r}{2A}\right)\right] \tag{34}$$

$$A\psi_r\left[\frac{2}{r} + \frac{f_{R_r}}{f_R} - \frac{A_r}{A}\right] - \frac{A_r^2}{4A} = \frac{\kappa^2 P}{f_R} - \frac{Rf_R - f(R)}{2f_R} - A\frac{f_{R_r}}{f_R}\left[\frac{2}{r} - \frac{A_r}{2A}\right] \tag{35}$$

where $f_R = f_{R_c} + \varphi$, and the subscripts r in ψ_r, f_{R_r}, $f_{R_{rr}}$, M_r denote derivation with respect to the radial coordinate. Note also that $f_{R_r} = \varphi_r$ and $f_{R_{rr}} = \varphi_{rr}$. The equation for φ is, according to (32) and (33),

$$A\varphi_{rr} = -A\left(\frac{2}{r} + \psi_r\right)\varphi_r - \frac{W(f_{R_c} + \varphi) - W(f_{R_c})}{3} + \frac{\kappa^2}{3}(3P - \rho) \tag{36}$$

Equations (34), (35), and (36) can be used to work out the metric of any spherically symmetric system subject to the asymptotic boundary conditions discussed above. For weak sources, such as non-relativistic stars like the sun, it is convenient to expand them assuming $|\varphi| \ll f_{R_c}$ and $A = 1 - 2M(r)/r$, with $2M(r)/r \ll 1$. The result is

$$-\frac{2}{r}M_{rr}(r) = \frac{\kappa^2\rho}{f_{R_c}} + V_c + \frac{1}{f_{R_c}}\left[\varphi_{rr} + \frac{2}{r}\varphi_r\right] \tag{37}$$

$$\frac{2}{r}\left[\psi_r + \frac{\varphi_r}{f_{R_c}}\right] = \frac{\kappa^2}{f_{R_c}}P - V_c \tag{38}$$

$$\varphi_{rr} + \frac{2}{r}\varphi_r - m_c^2\varphi = \frac{\kappa^2}{3}(3P - \rho) \tag{39}$$

where we have defined

$$V_c \equiv \left.\frac{Rf_R - f}{2f_R}\right|_{R_c} \quad \text{and} \quad m_c^2 \equiv \left.\frac{f_R - Rf_{RR}}{3f_{RR}}\right|_{R_c}. \tag{40}$$

[13] As pointed out in [Olmo (2007)], solar system tests are conventionally described in isotropic coordinates rather than on Schwarzschild-like coordinates. This justifies our coordinate choice in (33).

This expression for m_c^2 was first found in [Olmo (2005a;b)] within the scalar-tensor approach. It was found there that $m_c^2 > 0$ is needed to have a well-behaved (non-oscillating) Newtonian limit. This expression and the conclusion $m_c^2 > 0$ were also reached in [Faraoni and Nadeau (2005)] by studying the stability of de Sitter space. The same expression has been rediscovered later several times.
Outside of the sources, the solutions of (37), (38) and (39) lead to

$$\varphi(r) = \frac{C_1}{r} e^{-m_c r} \tag{41}$$

$$A(r) = 1 - \frac{C_2}{r}\left(1 - \frac{C_1}{C_2 f_{R_c}} e^{-m_c r}\right) + \frac{V_c}{6} r^2 \tag{42}$$

$$A(r) e^{2\psi} = 1 - \frac{C_2}{r}\left(1 + \frac{C_1}{C_2 f_{R_c}} e^{-m_c r}\right) - \frac{V_c}{3} r^2 \tag{43}$$

where an integration constant ψ_0 has been absorbed in a redefinition of the time coordinate. The above solutions coincide, as expected, with those found in [Olmo (2005a;b)] for the Newtonian and post-Newtonian limits using the scalar-tensor representation and standard gauge choices in Cartesian coordinates. Comparing our solutions with those, we identify

$$C_2 \equiv \frac{\kappa^2}{4\pi f_{R_c}} M_\odot \text{ and } \frac{C_1}{f_{R_c} C_2} \equiv \frac{1}{3} \tag{44}$$

where $M_\odot = \int d^3x \rho(x)$. The line element (33) can thus be written as

$$ds^2 = -\left(1 - \frac{2GM_\odot}{r} - \frac{V_c}{3} r^2\right) dt^2 + \left(1 + \frac{2G\gamma M_\odot}{r} - \frac{V_c}{6} r^2\right)(dr^2 + r^2 d\Omega^2) \tag{45}$$

where we have defined the effective Newton's constant and post-Newtonian parameter γ as

$$G = \frac{\kappa^2}{8\pi f_{R_c}}\left(1 + \frac{e^{-m_c r}}{3}\right) \text{ and } \gamma = \frac{3 - e^{-m_c r}}{3 + e^{-m_c r}} \tag{46}$$

respectively. This completes the lowest-order solution in isotropic coordinates.

3.2.2 The gravity Lagrangian according to solar system experiments

From the definitions of Eq.(46) we see that the parameters G and γ that characterize the linearized metric depend on the effective mass m_c (or inverse length scale $\lambda_{m_c} \equiv m_c^{-1}$). Newton's constant, in addition, also depends on f_{R_c}. Since the value of the background cosmic curvature R_c changes with the cosmic expansion, it follows that f_{R_c} and m_c must also change. The variation in time of f_{R_c} induces a time variation in the effective Newton's constant which is just the well-known time dependence that exists in Brans-Dicke theories. The length scale λ_{m_c}, characteristic of $f(R)$ theories, does not appear in the original Brans-Dicke theories because in the latter the scalar potential was assumed to vanish, $V(\phi) \equiv 0$, in contrast with (30), which implies an infinite interaction range ($m_c = 0 \rightarrow \lambda_{m_c} = \infty$).
In order to have agreement with the observed properties of the solar system, the Lagrangian $f(R)$ must satisfy certain basic constraints. These constraints will be very useful to determine the viability of some families of models proposed to explain the cosmic speedup. A very representative family of such models, which do exhibit self-accelerating late-time cosmic

solutions, is given by $f(R) = R - R_0^{n+1}/R^n$, where R_0 is a very low curvature scale that sets the scale at which the model departs from GR, and n is assumed positive. At curvatures higher than R_0, the theory is expected to behave like GR while at late times, when the cosmic density decays due to the expansion and approaches the scale R_0, the modified dynamics becomes important and could explain the observed speedup.

In viable theories, the effective cosmological constant V_c must be negligible. Most importantly, the interaction range λ_{m_c} must be shorter than a few millimeters because such Yukawa-type corrections to the Newtonian potential have not been observed, and observations indicate that the parameter γ is very close to unity. This last constraint can be expressed as $(L_S^2/\lambda_{m_c}^2) \gg 1$, where L_S represents a (relatively short) length scale that can range from meters to planetary scales, depending on the particular test used to verify the theory. In terms of the Lagrangian, this constraint takes the form

$$\left.\frac{f_R - Rf_{RR}}{3f_{RR}}\right|_{R_c} L_S^2 \gg 1\,. \tag{47}$$

A qualitative analysis of this constraint can be used to argue that, in general, $f(R)$ theories with terms that become dominant at low cosmic curvatures, such as the models $f(R) = R - R_0^{n+1}/R^n$, are not viable theories in solar system scales and, therefore, cannot represent an acceptable mechanism for the cosmic expansion.

Roughly speaking, eq.(47) says that the smaller the term $f_{RR}(R_c)$, with $f_{RR}(R_c) > 0$ to guarantee $m_\varphi^2 > 0$, the heavier the scalar field. In other words, the smaller $f_{RR}(R_c)$, the shorter the interaction range of the field. In the limit $f_{RR}(R_c) \to 0$, corresponding to GR, the scalar interaction is completely suppressed. Thus, if the nonlinearity of the gravity Lagrangian had become dominant in the last few billions of years (at low cosmic curvatures), the scalar field interaction range λ_{m_c} would have increased accordingly. In consequence, gravitating systems such as the solar system, globular clusters, galaxies,...would have experienced (or will experience) observable changes in their gravitational dynamics. Since there is no experimental evidence supporting such a change[14] and all currently available solar system gravitational experiments are compatible with GR, it seems unlikely that the nonlinear corrections may be dominant at the current epoch.

Let us now analyze in detail the constraint given in eq.(47). That equation can be rewritten as follows

$$R_c\left[\left.\frac{f_R}{Rf_{RR}}\right|_{R_c} - 1\right] L_S^2 \gg 1 \tag{48}$$

We are interested in the form of the Lagrangian at intermediate and low cosmic curvatures, i.e., from the matter dominated to the vacuum dominated eras. We shall now demand that the interaction range of the scalar field remains as short as today or decreases with time so as to avoid dramatic modifications of the gravitational dynamics in post-Newtonian systems with the cosmic expansion. This can be implemented imposing

$$\left[\frac{f_R}{Rf_{RR}} - 1\right] \geq \frac{1}{l^2 R} \tag{49}$$

[14] As an example, note that the fifth-force effects of the Yukawa-type correction introduced by the scalar degree of freedom would have an effect on stellar structures and their evolution, which would lead to incompatibilities with current observations.

as $R \to 0$, where $l^2 \ll L_S^2$ represents a bound to the current interaction range of the scalar field. Thus, eq.(49) means that the interaction range of the field must decrease or remain short, $\sim l^2$, with the expansion of the universe. Manipulating this expression, we obtain

$$\frac{d\log[f_R]}{dR} \leq \frac{l^2}{1 + l^2 R} \tag{50}$$

which can be integrated twice to give the following inequality

$$f(R) \leq A + B\left(R + \frac{l^2 R^2}{2}\right) \tag{51}$$

where B is a positive constant, which can be set to unity without loss of generality. Since f_R and f_{RR} are positive, the Lagrangian is also bounded from below, i.e., $f(R) \geq A$. In addition, according to the cosmological data, $A \equiv -2\Lambda$ must be of order a cosmological constant $2\Lambda \sim 10^{-53}$ m^2. We thus conclude that the gravity Lagrangian at intermediate and low scalar curvatures is bounded by

$$-2\Lambda \leq f(R) \leq R - 2\Lambda + \frac{l^2 R^2}{2} \tag{52}$$

This result shows that a Lagrangian with nonlinear terms that grow with the cosmic expansion is not compatible with the current solar system gravitational tests, such as we argued above. Therefore, those theories cannot represent a valid mechanism to justify the observed cosmic speed-up. Additionally, our analysis has provided an empirical procedure to determine the form of the gravitational Lagrangian. The function $f(R)$ found here nicely recovers Einstein's gravity at low curvatures but allows for some quadratic corrections at higher curvatures, which is of interest in studies of the very early Universe.

4. Quantum gravity phenomenology and the early universe

The extrapolation of the dynamics of GR to the very strong field regime indicates that the Universe began at a singularity and that the death of a sufficiently massive star unavoidably leads to the formation of a black hole or a naked singularity. The existence of space-time singularities is one of the most impressive predictions of GR. This prediction, however, also represents the end of the theory, because the absence of a well-defined geometry implies the absence of physical laws and lack of predictability [Hawking (1975); Novello and Bergliaffa (2008)]. For this reason, it is generally accepted that the dynamics of GR must be changed at some point to avoid these problems. A widespread belief is that at sufficiently high energies the gravitational field must exhibit quantum properties that alter the dynamics and prevent the formation of singularities. However, a completely satisfactory quantum theory of gravity is not yet available. To make some progress in the qualitative understanding of how quantum gravity may affect the dynamics of the Universe near the big bang, in this section we show how certain modifications of GR may be able to capture some aspects of the expected phenomenology of quantum gravity.

We begin by noting that Newton's and Planck's constants may be combined with the speed of light to generate a length $l_P = \sqrt{\hbar G/c^3}$, which is known as the Planck length. The Planck length is usually interpreted as the scale at which quantum gravitational phenomena should play a non-negligible role. However, since lengths are not relativistic invariants, the existence of the Planck length raises doubts about the nature of the reference frame in which it should be measured and about the limits of validity of special relativity itself. This poses the following question: can we combine in the same framework the speed of light and the Planck length in such a way that both quantities appear as universal invariants to all observers? The solution to this problem will give us the key to consider quantum gravitational phenomena from a modified gravity perspective.

4.1 Palatini approach to modified gravity

To combine in the same framework the speed of light and the Planck length in a way that preserves the invariant and universal nature of both quantities, we first note that though c^2 has dimensions of squared velocity it represents a 4-dimensional Lorentz scalar rather than the squared of a privileged 3-velocity. Similarly, we may see l_P^2 as a 4-d invariant with dimensions of length squared that needs not be related with any privileged 3-length. Because of dimensional compatibility with a curvature, the invariant l_P^2 could be introduced in the theory via the gravitational sector by considering departures from GR at the Planck scale motivated by quantum effects. However, the situation is not as simple as it may seem at first. In fact, an action like the one we obtained in the last section[15],

$$S[g_{\mu\nu}, \psi] = \frac{\hbar}{16\pi l_P^2} \int d^4x\sqrt{-g}\left[R + l_P^2 R^2\right] + S_m[g_{\mu\nu}, \psi] , \tag{53}$$

where $S_m[g_{\mu\nu}, \psi]$ represents the matter sector, contains the scale l_P^2 but not in the invariant form that we wished. The reason is that the field equations that follow from (53) are equivalent to those of the following scalar-tensor theory

$$S[g_{\mu\nu}, \varphi, \psi] = \frac{\hbar}{16\pi l_P^2} \int d^4x\sqrt{-g}\left[(1+\varphi)R - \frac{1}{4l_P^2}\varphi^2\right] + S_m[g_{\mu\nu}, \psi] , \tag{54}$$

which given the identification $\phi = 1 + \varphi$ coincides with the case $w = 0$ of Brans-Dicke theory with a non-zero potential $V(\phi) = \frac{1}{4l_P^2}(\phi - 1)^2$. As is well-known and was explicitly shown in Section 3, in Brans-Dicke theory the observed Newton's constant is promoted to the status of field, $G_{eff} \sim G/\phi$. The scalar field allows the effective Newton's constant G_{eff} to dynamically change in time and in space. As a result the corresponding effective Planck length, $\tilde{l}_P^2 = l_P^2/\phi$, would also vary in space and time. This is quite different from the assumed constancy and universality of the speed of light in special relativity, which is implicit in our construction of the total action. In fact, our action has been constructed assuming the Einstein equivalence principle (EEP), whose validity guarantees that the observed speed of light is a true constant and universal invariant, not a field[16] like in varying speed of light theories [Magueijo (2003)] (recall also that Nordström's first scalar theory was motivated by the constancy of the speed of light). The situation does not improve if we introduce higher curvature invariants in (54). We thus see that the introduction of the Planck length in the gravitational sector in the form of a universal constant like the speed of light is not a trivial issue. The introduction of

[15] Restoring missing factors of c in (24), we find that $\frac{1}{16\pi G} = \frac{\hbar}{16\pi l_P^2}$ and, therefore, $\kappa^2 = 8\pi l_P^2/\hbar$.

[16] If the Einstein equivalence principle is true, then all the coupling constants of the standard model are constants, not fields [Will (2005)].

curvature invariants suppressed by powers of $R_P = 1/l_P^2$ unavoidably generates new degrees of freedom which turn Newton's constant into a dynamical field.
Is it then possible to modify the gravity Lagrangian adding Planck-scale corrected terms without turning Newton's constant into a dynamical field? The answer to this question is in the affirmative. One must first note that metricity and affinity are a priori logically independent concepts [Zanelli (2005)]. If we construct the theory *à la Palatini*, that is in terms of a connection not a priori constrained to be given by the Christoffel symbols, then the resulting equations do not necessarily contain new dynamical degrees of freedom (as compared to GR), and the Planck length may remain space-time independent in much the same way as the speed of light and the coupling constants of the standard model, as required by the EEP. A natural alternative, therefore, seems to be to consider (53) in the Palatini formulation. The field equations that follow from (53) when metric and connection are varied independently are [Olmo (2011)]

$$f_R R_{\mu\nu}(\Gamma) - \frac{1}{2} f g_{\mu\nu} = \kappa^2 T_{\mu\nu} \tag{55}$$

$$\nabla_\alpha \left(\sqrt{-g} f_R g^{\beta\gamma}\right) = 0 \,, \tag{56}$$

where $f = R + R^2/R_P$, $f_R \equiv \partial_R f = 1 + 2R/R_P$, $R_P = 1/l_P^2$, and $\kappa^2 = 8\pi l_P^2/\hbar$. The connection equation (56) can be easily solved after noticing that the trace of (55) with $g^{\mu\nu}$,

$$R f_R - 2f = \kappa^2 T \,, \tag{57}$$

represents an algebraic relation between $R \equiv g^{\mu\nu}R_{\mu\nu}(\Gamma)$ and T, which generically implies that $R = R(T)$ and hence $f_R = f_R(R(T))$ [from now on we denote $f_R(T) \equiv f_R(R(T))$]. For the particular Lagrangian (53), we find that $R = -\kappa^2 T$, like in GR. This relation implies that (56) is just a first order equation for the connection that involves the matter, via the trace T, and the metric. The connection turns out to be the Levi-Civita connection of an auxiliary metric,

$$\Gamma^\alpha_{\mu\nu} = \frac{h^{\alpha\beta}}{2}\left(\partial_\mu h_{\beta\nu} + \partial_\nu h_{\beta\mu} - \partial_\beta h_{\mu\nu}\right) \,, \tag{58}$$

which is conformally related with the physical metric, $h_{\mu\nu} = f_R(T) g_{\mu\nu}$. Now that the connection has been expressed in terms of $h_{\mu\nu}$, we can rewrite (55) as follows

$$G_{\mu\nu}(h) = \frac{\kappa^2}{f_R(T)} T_{\mu\nu} - \Lambda(T) h_{\mu\nu} \tag{59}$$

where $\Lambda(T) \equiv (R f_R - f)/(2 f_R^2) = (\kappa^2 T)^2/R_P$, and looks like Einstein's theory for the metric $h_{\mu\nu}$ with a slightly modified source. This set of equations can also be written in terms of the physical metric $g_{\mu\nu}$ as follows

$$R_{\mu\nu}(g) - \frac{1}{2} g_{\mu\nu} R(g) = \frac{\kappa^2}{f_R} T_{\mu\nu} - \frac{R f_R - f}{2 f_R} g_{\mu\nu} - \frac{3}{2(f_R)^2}\left[\partial_\mu f_R \partial_\nu f_R - \frac{1}{2} g_{\mu\nu} (\partial f_R)^2\right] + \frac{1}{f_R}\left[\nabla_\mu \nabla_\nu f_R - g_{\mu\nu} \Box f_R\right] \,. \tag{60}$$

In this last representation, one can use the notation introduced in (29) and (30) to show that these field equations coincide with those of a Brans-Dicke theory with parameter $w = -3/2$ (see eq.(22)). Note that all the functions $f(R)$, R, and f_R that appear on the right hand side of (60) are functions of the trace T. This means that the modified dynamics of (60) is due to the new matter terms induced by the trace T of the matter, not to the presence of new dynamical degrees of freedom. This also guarantees that, unlike for the $w \neq -3/2$ Brans-Dicke theories, for the $w = -3/2$ theory Newton's constant is indeed a constant.
From the structure of the field equations (59) and (60), and the relation $g_{\mu\nu} = (1/f_R)h_{\mu\nu}$, it follows that $g_{\mu\nu}$ is affected by the matter-energy in two different ways. The first contribution corresponds to the cumulative effects of matter, and the second contribution is due to the dependence on the local density distributions of energy and momentum. This can be seen by noticing that the structure of the equations (59) that determine $h_{\mu\nu}$ is similar to that of GR, which implies that $h_{\mu\nu}$ is determined by integrating over all the sources (gravity as a cumulative effect). Besides that, $g_{\mu\nu}$ is also affected by the local sources through the factor $f_R(T)$. To illustrate this point, consider a region of the spacetime containing a total mass M and filled with sources of low energy-density as compared to the Planck scale ($|\kappa^2 T/R_P| \ll 1$). For the quadratic model $f(R) = R + R^2/R_P$, in this region (59) boils down to $G_{\mu\nu}(h) = \kappa^2 T_{\mu\nu} + O(\kappa^2 T/R_P)$, and $h_{\mu\nu} \approx (1 + O(\kappa^2 T/R_P))g_{\mu\nu}$, which implies that the GR solution is a very good approximation. This confirms that $h_{\mu\nu}$ is determined by an integration over the sources, like in GR. Now, if this region is traversed by a particle of mass $m \ll M$ but with a non-negligible ratio $\kappa^2 T/R_P$, then the contribution of this particle to $h_{\mu\nu}$ can be neglected, but its effect on $g_{\mu\nu}$ via de factor $f_R = 1 - \kappa^2 T/R_P$ on the region that supports the particle (its classical trajectory) is important. This phenomenon is analogous to that described in the so-called Rainbow Gravity [Magueijo and Smolin (2004)], an approach to the phenomenology of quantum gravity based on a non-linear implementation of the Lorentz group to allow for the coexistence of a constant speed of light and a maximum energy scale (the flat space version of that theory is known as Doubly Special Relativity [Amelino-Camelia (2002); Amelino-Camelia and Smolin (2009); Magueijo and Smolin (2002; 2003)]). In Rainbow Gravity, particles of different energies (energy-densities in our case) perceive different metrics.

4.2 The early-time cosmology of Palatini $f(R)$ models.

The quadratic Palatini model introduced above turns out to be virtually indistinguishable from GR at energy densities well below the Planck scale. It is thus natural to ask if this theory presents any particularly interesting feature at Planck scale densities. A natural context where this question can be explored is found in the very early universe, when the matter energy-density tends to infinity as we approach the big bang.
In a spatially flat, homogeneous, and isotropic universe, with line element $ds^2 = -dt^2 + a^2(t)d\vec{x}^2$, filled with a perfect fluid with constant equation of state $P = w\rho$ and density ρ, the Hubble function that follows from (60) (or (59)) takes the form

$$H^2 = \frac{1}{6f_R}\frac{\left[f + (1+3w)\kappa^2\rho\right]}{\left[1 + \frac{3}{2}\Delta\right]^2}, \tag{61}$$

where $H = \dot{a}/a$, and $\Delta = -(1+w)\rho\partial_\rho f_R/f_R = (1+w)(1-3w)\kappa^2\rho f_{RR}/(f_R(Rf_{RR} - f_R))$. In GR, (61) boils down to $H^2 = \kappa^2\rho/3$. Since the matter conservation equation for constant w leads to $\rho = \rho_0/a^{3(1+w)}$, in GR we find that $a(t) = a_0 t^{\frac{2}{3(1+w)}}$, where ρ_0 and a_0 are constants.

This result indicates that if the universe is dominated by a matter source with $w > -1$, then at $t = 0$ the universe has zero physical volume, the density is infinite, and all curvature scalars blow up, which indicates the existence of a big bang singularity. The quadratic Palatini model introduced above, however, can avoid this situation. For that model, (61) becomes [Barragan et al. (2009a;b); Olmo (2010)]

$$H^2 = \frac{\kappa^3 \rho}{3} \frac{\left(1 + \frac{2R}{R_P}\right)\left(1 + \frac{1-3w}{2}\frac{R}{R_P}\right)}{\left[1 - (1+3w)\frac{R}{R_P}\right]^2} . \qquad (62)$$

This expression recovers the linear dependence on ρ of GR in the limit $|R/R_P| \ll 1$. However, if R reaches the value $R_b = -R_P/2$, then H^2 vanishes and the expansion factor $a(t)$ reaches a minimum. This occurs for $w > 1/3$ if $R_P > 0$ and for $w < 1/3$ if $R_P < 0$. The existence of a non-zero minimum for the expansion factor implies that the big bang singularity is avoided. The avoidance of the big bang singularity indicates that the time coordinate can be extended backwards in time beyond the instant $t = 0$. This means that in the past the universe was in a contracting phase which reached a minimum and bounced off to the expanding phase that we find in GR.

We mentioned at the beginning of this section that the avoidance of the big bang singularity is a basic requirement for any acceptable quantum theory of gravity. Our procedure to construct a quantum-corrected theory of gravity in which the Planck length were a universal invariant similar to the speed of light has led us to a cosmological model which replaces the big bang by a cosmic bounce. To obtain this result, it has been necessary to resort to the Palatini formulation of the theory. In this sense, it is important to note that the metric formulation of the quadratic curvature model discussed here, besides turning the Planck length into a dynamical field, is unable to avoid the big bang singularity. In fact, in metric formalism, all quadratic models of the form $R + (aR^2 + bR_{\mu\nu}R^{\mu\nu})/R_P$ that at late times tend to a standard Friedmann-Robertson-Walker cosmology begin with a big bang singularity. Palatini theories, therefore, appear as a potentially interesting framework to discuss quantum gravity phenomenology.

4.3 A Palatini action for loop quantum cosmoloy

Growing interest in the dynamics of the early-universe in Palatini theories has arisen, in part, from the observation that the effective equations of loop quantum cosmology (LQC) [Ashtekar et al. (2006a;b;c); Ashtekar (2007); Bojowald (2005); Szulc et al. (2007)], a Hamiltonian approach to quantum gravity based on the non-perturbative quantization techniques of loop quantum gravity [Rovelli (2004); Thiemann (2007)], could be exactly reproduced by a Palatini $f(R)$ Lagrangian [Olmo and Singh (2009)]. In LQC, non-perturbative quantum gravity effects lead to the resolution of the big bang singularity by a quantum bounce without introducing any new degrees of freedom. Though fundamentally discrete, the theory admits a continuum description in terms of an effective Hamiltonian that in the case of a homogeneous and isotropic universe filled with a massless scalar field leads to the following modified Friedmann equation

$$3H^2 = 8\pi G \rho \left(1 - \frac{\rho}{\rho_{crit}}\right) , \qquad (63)$$

where $\rho_{crit} \approx 0.41\rho_{Planck}$. At low densities, $\rho/\rho_{crit} \ll 1$, the background dynamics is the same as in GR, whereas at densities of order ρ_{crit} the non-linear new matter contribution forces the vanishing of H^2 and hence a cosmic bounce. This singularity avoidance seems to be a generic feature of loop-quantized universes [Singh (2009)].

Palatini $f(R)$ theories share with LQC an interesting property: the modified dynamics that they generate is not the result of the existence of new dynamical degrees of freedom but rather it manifests itself by means of non-linear contributions produced by the matter sources, which contrasts with other approaches to quantum gravity and to modified gravity. This similarity makes it tempting to put into correspondence Eq.(63) with the corresponding $f(R)$ equation (60). Taking into account the trace equation (57), which for a massless scalar becomes $Rf_R - 2f = 2\kappa^2\rho$ and implies that $\rho = \rho(R)$, one finds that a Palatini $f(R)$ theory able to reproduce the LQC dynamics (63) must satisfy the differential equation

$$f_{RR} = -f_R \left(\frac{Af_R - B}{2(Rf_R - 3f)A + RB} \right) \quad (64)$$

where $A = \sqrt{2(Rf_R - 2f)(2R_c - [Rf_R - 2f])}$, $B = 2\sqrt{R_c f_R (2Rf_R - 3f)}$, and $R_c \equiv \kappa^2\rho_c$. If one imposes the boundary condition $\lim_{R\to 0} f_R \to 1$ at low curvatures, and $\ddot{a}_{LQC} = \ddot{a}_{Pal}$ (where $\ddot{a}$ represents the acceleration of the expansion factor) at $\rho = \rho_c$, the solution to this equation is unique. The solution was found numerically [Olmo and Singh (2009)], though the following function can be regarded as a very accurate approximation to the LQC dynamics from the GR regime to the non-perturbative bouncing region (see Fig.1)

$$\frac{df}{dR} = -\tanh\left(\frac{5}{103} \ln\left[\left(\frac{R}{12R_c} \right)^2 \right] \right) \quad (65)$$

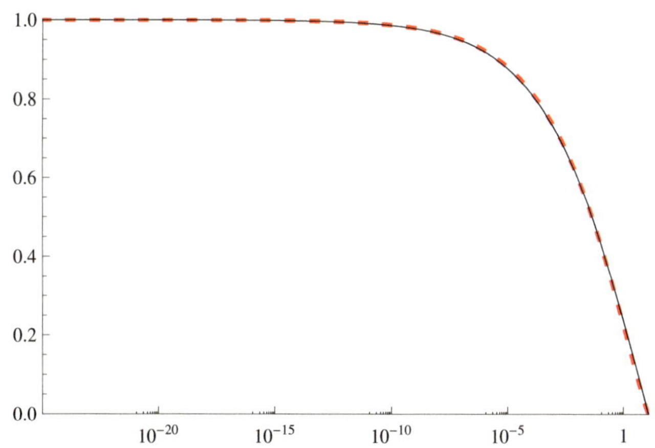

Fig. 1. Vertical axis: df/dR ; Horizontal axis: R/R_c. Comparison of the numerical solution with the interpolating function (65). The dashed line represents the numerical curve.

This result is particularly important because it establishes a direct link between the Palatini approach to modified gravity and a cosmological model derived from non-perturbative quantization techniques.

4.4 Beyond Palatini $f(R)$ theories.

Nordström's second theory was a very interesting theoretical exercise that successfully allowed to implement the Einstein equivalence principle in a relativistic scalar theory. However, among other limitations, that theory did not predict any new gravitational effect for the electromagnetic field. In a sense, Palatini $f(R)$ theories suffer from this same limitation. Since their modified dynamics is due to new matter contributions that depend on the trace of the stress-energy tensor, for traceless fields such as a radiation fluid or the electromagnetic field, the theory does not predict any new effect. This drawback can be avoided if one adds to the Palatini Lagrangian a new piece dependent on the squared Ricci tensor, $R_{\mu\nu}R^{\mu\nu}$, where we assume $R_{\mu\nu} = R_{\nu\mu}$ [Barragan and Olmo (2010); Olmo et al. (2009)]. In particular, the following action

$$S[g_{\mu\nu}, \Gamma^{\mu}_{\alpha\beta}, \psi] = \frac{1}{2\kappa^2}\int d^4x\sqrt{-g}\left[R + a\frac{R^2}{R_P} + \frac{R_{\mu\nu}R^{\mu\nu}}{R_P}\right] + S_m[g_{\mu\nu}, \psi]\,, \tag{66}$$

implies that $R = R(T)$ but $Q \equiv R_{\mu\nu}R^{\mu\nu} = Q(T_{\mu\nu})$, i.e., the scalar Q has a more complicated dependence on the stress-energy tensor of matter than the trace. For instance, for a perfect fluid, one finds

$$\frac{Q}{2R_P} = -\left(\kappa^2 P + \frac{\tilde{f}}{2} + \frac{R_P}{8}\tilde{f}_R^2\right) + \frac{R_P}{32}\left[3\left(\frac{R}{R_P} + \tilde{f}_R\right) - \sqrt{\left(\frac{R}{R_P} + \tilde{f}_R\right)^2 - \frac{4\kappa^2(\rho+P)}{R_P}}\right]^2, \tag{67}$$

where $\tilde{f} = R + aR^2/R_P$ and R is a solution of $R\tilde{f}_R - 2\tilde{f} = \kappa^2 T$. From this it follows that even if one deals with a radiation fluid ($P = \rho/3$) or with a traceless field, the Palatini action (66) generates modified gravity without introducing new degrees of freedom.

For this model, it has been shown that completely regular bouncing solutions exist for both isotropic and anisotropic homogeneous cosmologies filled with a perfect fluid. In particular, one finds that for $a < 0$ the interval $0 \leq w \leq 1/3$ is always included in the family of bouncing solutions, which contains the dust and radiation cases. For $a \geq 0$, the fluids yielding a non-singular evolution are restricted to $w > \frac{a}{2+3a}$, which implies that the radiation case $w = 1/3$ is always nonsingular. For a detailed discussion and classification of the non-singular solutions depending on the value of the parameter a and the equation of state w, see [Barragan and Olmo (2010)].

As an illustration, consider a universe filled with radiation, for which $R = 0$. In this case, the function Q boils down to [Barragan and Olmo (2010)]

$$Q = \frac{3R_P^2}{8}\left[1 - \frac{8\kappa^2\rho}{3R_P} - \sqrt{1 - \frac{16\kappa^2\rho}{3R_P}}\right]. \tag{68}$$

This expression recovers the GR value at low curvatures, $Q \approx 4(\kappa^2\rho)^2/3 + 32(\kappa^2\rho)^3/9R_P + \ldots$ but reaches a maximum $Q_{max} = 3R_P^2/16$ at $\kappa^2\rho_{max} = 3R_P/16$, where the squared root of (68) vanishes. It can be shown that at ρ_{max} the shear also takes its maximum, namely, $\sigma^2_{max} = \sqrt{3/16}R_P^{3/2}(C_{12}^2 + C_{23}^2 + C_{31}^2)$, which is always finite, and the expansion vanishes producing a cosmic bounce regardless of the amount of anisotropy (see Fig.2). The model (66), therefore, avoids the well-known problems of anisotropic universes in GR, where anisotropies grow faster than the energy density during the contraction phase leading to a singularity that can only be avoided by sources with $w > 1$.

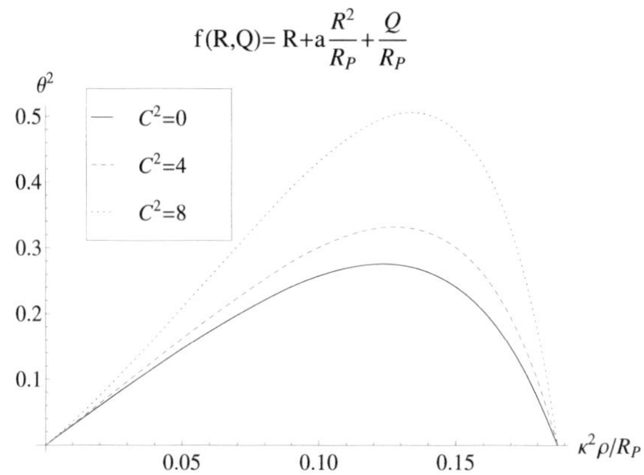

Fig. 2. Evolution of the expansion as a function of $\kappa^2\rho/R_P$ in radiation universes with low anisotropy, which is controlled by the combination $C^2 = C_{12}^2 + C_{23}^2 + C_{31}^2$. The case with $C^2 = 0$ corresponds to the isotropic flat case, $\theta^2 = 9H^2$.

5. References

Amelino-Camelia, G. (2002). *Int.J.Mod.Phys.* D 11, 35.

Amelino-Camelia, G. and Smolin, L. (2009). *Phys. Rev.* D 80, 084017.

Ashtekar, A. , Pawlowski,T., and Singh,P. (2006). *Phys. Rev. Lett.* 96, 141301.

Ashtekar, A. , Pawlowski,T., and Singh,P. (2006). *Phys. Rev.* D 73, 124038.

Ashtekar, A. , Pawlowski,T., and Singh,P. (2006). *Phys. Rev.* D 74, 084003.

Ashtekar, A. (2007) *Nuovo Cim.* 122 B, 135.

Barragan, C., Olmo, G. J., and Sanchis-Alepuz, H. (2009). *Phys. Rev.* D80, 024016, [arXiv:0907.0318 [gr-qc]].

Barragan, C., Olmo, G. J., and Sanchis-Alepuz, H. (2009). [arXiv:1002.3919 [gr-qc]].

Barragan, C., and Olmo, G.J. (2010). *Phys. Rev.* D 82, 084015.

Birkhoff, G.D. (1944). Proc. Natl. Acad. Sci. U. S. A. 30(10), 324Ű334.

Bojowald, M. (2005). *Living Rev. Rel.* 8, 11.

Capozziello, S. and Francaviglia, M. (2008). *Gen. Rel. Grav.* 40, 357.

Copeland, E. J. et al. (2006). *Int. J. Mod. Phys. D15*, 1753-1936, [hep-th/0603057].

De Felice, A. and Tsujikawa, S. (2010). *Living Rev. Rel.* 13, 3, [arXiv:1002.4928 [gr-qc]].

Dicke, R.H. (1964). *The Theoretical Interpretation of Experimental Relativity*, Gordon and Breach, New York, U.S.A.

Einstein, A. and Fokker, A.D. (1914). Ann. d. Phys. 44, 321.

Faraoni, V. and Nadeau, S. (2005). *Phys. Rev.* D 72, 124005.

Hawking, S.W. (1975). *Phys.Rev.* D 14, 2460.

Magueijo, J. (2003). *Rept. Prog. Phys.* 66, 2025.

Magueijo, J. and Smolin, L. (2002). *Phys.Rev.Lett.* 88, 190403.

Magueijo, J. and Smolin, L. (2003). *Phys.Rev.* D 67, 044017.

Magueijo, J. and Smolin, L. (2004). *Class. Quant. Grav.* 21, 1725.

Norton, J.D. (1992). Arch. Hist. Ex. Sci. 45, 17.

Nordström, G. (1912). Phys. Zeit. 13, 1126.

Nordström, G. (1913). Ann. d. Phys. 42, 533.

Novello, M. and Perez Bergliaffa, S.E. (2008). *Phys.Rep.* 463, 127-213.

Olmo, G.J. (2005), *Phys. Rev.* D72, 083505.
Olmo, G.J. (2005), *Phys. Rev. Lett.* 95, 261102.
Olmo, G.J. (2007). *Phys. Rev.* D75, 023511.
Olmo, G.J. and Singh, P. (2009). JCAP 0901, 030.
Olmo, G.J., Sanchis-Alepuz, H. , and Tripathi, S. (2009). *Phys. Rev.* D 80, 024013.
Olmo, G.J. (2010). AIP Conf. Proc. 1241, 1100-1107, [arXiv:0910.3734 [gr-qc]].
Olmo, G.J. (2011), *Int. J. Mod. Phys. D*, in press, [arXiv:1101.3864 [gr-qc]].
Padmanabhan, T. (2003). *Phys. Rep.* 380, 235.
Peebles, P. J. E. and Ratra, B. (2003). *Rev. Mod. Phys.* 75, 559.
Rovelli, C. (2004).*t Quantum Gravity*, Cambridge U. Press.
Singh, P. (2009). *Class. Quant. Grav.* 26, 125005, [arXiv:0901.2750 [gr-qc]].
Sotiriou, T.P. and Faraoni, V. (2010). *Rev. Mod. Phys.* 82, 451-497, arXiv:0805.1726 [gr-qc].
Szulc, L. , Kaminski, W. , and Lewandowski, J. (2007). *Class.Quant.Grav.* 24, 2621.
Thiemann, T. (2007). *Modern canonical quantum general relativity*, Cambridge U. Press.
Will, C.M. (1993). *Theory and Experiment in Gravitational Physics*, Cambridge University Press, Cambridge.
Will, C.M. (2005). *Living Rev.Rel.* 9,3,(2005), gr-qc/0510072 .
Zanelli, J. (2005). arXiv:hep-th/0502193v4 .

4

Duration, Systems and Cosmology

Robert Vallée
Université Paris-Nord
France

1. Introduction

In physics and other fields, the definition of duration is a fundamental problem. We are used to astronomical time based upon the idea that the rotation of Earth is perfectly regular, an assumption which nowadays we know to be slightly erroneous. A unit of time based upon the period of a chosen atomic vibration is preferred. In all cases we need a time which we accept as a standard, taken for granted. But there are cases where duration associated with such a universal time does not seem appropriate. At certain linstants it seems to an individual that time elapses more slowly or more quickly. This psychological time is subjective, it depends upon the person concerned and the circumstances. For a given individual it also depends upon age; at the end of life a day seems shorter than in youth. Of course this "relativity" of duration has nothing to do with relativity of time met in special relativity and is not at all in opposition to it. Before starting our presentation we must warn that we shall abundantly make use of mathematics as we believe they may be a help for thinking, despite the fact that in many cases only the qualitative aspects of the conclusions must be retained.

If we have chosen a standard or reference time t, such as for example the astronomical one, what is the most general time θ we can derive from it as a function $\theta(t)$? We assume that $\theta(t)$ must be a continuous function of t (though a discrete time could be proposed) and add that $\theta(t)$ must not decrease when t increases. More precisely we may write, if f(a,b) is the the duration of interval (a,b), and since f(a,b) must increase in the large with b and decrease in the large with a

$$f(a,b) + f(b,c) = f(a,c)$$

$$f(a,b) \geq 0, \; b \geq a$$

$$f(a,a) = 0.$$

If we add the hypothesis that f is differentiable, we have

$$f(a+da, b) + f(b, c+dc) = f(a+da, c+dc).$$

Replacing f(a+da,b) by $f(a,b)+\partial f(a,b)/\partial a \, da$ and the like for f(b, c+dc) and f(a+da, c+dc) we obtain

$$\partial f(a,b)/\partial a = \partial f(a,c)/\partial a.$$

So $\partial f(a,b)/\partial a$ is independent of b. Consequently, after integration with respect to a, we have

$$f(a,b) = F(a) + Cst,$$

the integration constant being a function of b or G(b). It gives

$$f(a,b) = F(a) + G(b).$$

But, since f(a,a) = 0, we have G = - F and so

$$f(a,b) = G(b) - G(a)$$

or

$$f(a,b) = \theta(b) - \theta(a),$$

function $\theta(t)$ being obviously continuous at not decreasing with t, as required.
We consider now a dynamical system. This involves in its evolution equation a reference time t which, in a way, is impartial. But since it has nothing to do with the considered system it does not take into account its intrinsic behaviour. Metaphorically all reference instants have not the same value. If the system were conscious, some instants, or short intervals of reference time, would have a greater importance than others. In the extreme case of very profound sleep or, better, of a coma, duration is not felt. This is close to the point of view of Aristotle in chapter IV of his "Physics" (Hussey, 1983): "When we feel no change in our thought, or we are unconscious of this change, or when we feel it without being aware of it, then it seems to us that no time have elapsed". Augustine in book XI of his "Confessions" expresses an opinion not far from that of Aritotle (Warner, 1963): "What is time? If nobody asks, I know; but if I want to explain, I do not know! Nevertheless – I tell it confidently – I know that if nothing happened, there would be no time passed…"Finally we are inclined to propose as a first approach that the more rapidly the state of the system changes, the more important are the corresponding reference instants.
We choose, as an index of importance of reference instant t, the scalar square of the speed of evolution of the state at this instant, that is to say $(dX(t)/dt)^2$. Of course many other indexes are possible, given for example by a strictly increasing function of the modulus of the speed. So we propose as an intrinsic or "internal duration" $d(t_1,t_2)$ of reference interval (t_1,t_2) the integral (Vallée, 1996, 2005)

$$\int_{t_1,t_2} (dX(t)/dt)^2\, dt,$$

the internal duration of infinitesimal interval (t,t+dt) being $(dXt)/dt)^2dt$. An "internal time", coherent with this duration and defined up to an additive constant, is given by

$$\theta(t) = d(t_0,t)$$

and we have

$$d(t_1,t_2) = \theta(t_2) - \theta(t_1),$$

a result which is in accordance with what we expected from the most general time we can derive from a given standard or reference time t.

2. Explosions and implosions

This internal time may be used for any dynamical system defined by a differential equation. We have particularly considered what we have called "elliptic explosion-implosion", "hyperbolic explosion" and, as an intermediary case,"parabolic explosion" (Vallée, 1996, 2005).

2.1 Elliptic explosion-implosion

In the case of an "elliptic explosion-implosion", the equation of evolution is given by

$$dX(t)/dt = q/p \ \ \mathrm{sgn}(p-t) \ \ (q^2 - X^2(t))^{1/2} / X(t) \tag{1}$$

where the state X(t) is a mere scalar with

$$X(0) = 0, \ \ p > 0, q > 0, \ \ 0 \leq t \leq 2p.$$

It is easy to see that

$$X(t) = q/p \ (p^2 - \ (p-t)^2)^{1/2} \tag{2}$$

since by derivation it gives

$$dX(t)/dt = q/p \ \ (p-t) \ / \ (p^2 - (p-t)^2)^{1/2} \tag{3}$$

which is the expression obtained from (1) if we replace X(t) by (2)
The graph of function X(t), which represents the evolution of state X(t)with reference time t, is the upper part of an ellipse of great axis 2p and small axis q .The absciss of the center is p and its ordinate is 0. X(t) starts from 0 at t=0, increases to its maximum value q at t= p, then decreases and attains 0 at t=2p. The speed at t= 0 is $+\infty$ and $-\infty$ at t= 2p. That is why we have an explosion at the beginning and an implosion at the end, and so what we can call an "elliptic explosion -implosion". The square of the speed of evolution is according to (3), after a very classical decomposition of

$$q^2/p^2 \ \ (t-p)^2/(p^2 - (p-t)^2)^{1/2} \ = \ q^2/p^2 \ \ (p-t)^2/t(2p-t),$$

given by

$$(d(X(t))/dt)^2 = q^2/p^2 \ \ (q^2 - X^2(t)) \ / \ X^2(t)$$

$$= q^2/2p \ (1/t - 2/p \ + 1/2p\text{-}t).$$

The "internal time" we can obtain by integration is defined up to an additive constant we can choose freely. The most simple choice gives

$$\theta(t) = q^2/2p \ \ (\mathrm{Log}t - 2t/p - \mathrm{Log}(2p\text{-}t) \tag{4}$$

We see that when the reference time t varies from 0 to 2p, the "internal time" varies from $-\infty$ to $+\infty$. Obviously this circumstance, the push back of t = 0 to $\theta = -\infty$ and the push forward of $t = 2p$ to $+\infty$, is linked to the behaviour of the square of the speed near $t = 0$ and near $t = 2p$ which generates logarithms.

2.2 Hyperbolic explosion

In the case of "hyperbolic explosion" the equation of evolution is

$$dX(t)/dt = q/p\ (q^2 + X^2(t)) / X(t)\,, \tag{5}$$

with

$$X(0) = 0\,,\ p>0,\ q>0,$$

$$0 \leq t$$

It is easy to verify that we have

$$X(t) = q/p\ ((p+t)^2 - p^2)^{1/2}\,, \tag{6}$$

since by derivation

$$dX(t)/dt = q/p\ (p+t) / ((p+t)^2 - p^2),$$

expression also obtained from (5) when we replace $X(t)$ by (6).
The graph of function $X(t)$ is the upper half right part of an hyperbola. The absciss of its center is $-p$ and its ordinate is 0. The asymptote associated with the graph of $X(t)$ has a slope equal to q/p. $X(t)$ starts from 0 at $t = 0$, then tends to $+\infty$ when t tends to $+\infty$. For great values of t, $X(t)$ behaves as $(q/p)\,t + q$. The square of the speed of evolution is

$$(dX(t)/dt)^2 = q^2/2p\ (1/t + 2/p - 1/2p{+}t).$$

This gives an "internal time" equal to

$$\theta(t) = q^2/2p\ (\mathrm{Log}t + 2t/p\ - \mathrm{Log}(2p+t)) \tag{7}$$

for which when reference time t varies from 0 to $+\infty$, "internal time" varies from $-\infty$ to $+\infty$.

2.3 Parabolic explosion

The "parabolic explosion" is an intermediary case, as parabola is "intermediary" between ellipse and hyperbola. Starting from equation (1), we shall make p tend to ∞ while keeping q^2/p equal to a constant h. We have

$$dX(t)/dt = (h/p)^{1/2}\ (hp - X^2(t))^{1/2} / X(t)$$

$$= (h^2 - h/p\ X^2(t))^{1/2} / X(t),$$

which gives, when p tends to ∞, the new equation of evolution

$$dX(t)/dt = h/X(t) \tag{8}$$

with

$$X(0) = 0,$$

and

$$h > 0.$$

So

$$X(t) = (2ht)^{1/2}. \qquad (9)$$

The graph of function X(t) is the upper part of a parabola of summit at t = 0 and having t axis as axis. It is the limit of the half ellipse seen in the elliptic case. We have an explosion at t = 0 with initial speed + ∞. This speed decreases with time and tends to 0 while X(t) tends to +∞. The square of the speed is

$$(dX(t)/dt)^2 = h/2t,$$

giving the "internal time"

$$\theta(t) = h/2 \ \mathrm{Log}t, \qquad (10)$$

which varies from -∞ to +∞ when t varies from 0 to +∞.

3. Infinite internal duration

In the three cases seen above we have observed the possibility of an infinite "internal duration" linked to the push back (or forward) of a particular reference instant. Obviously this is linked to the behaviour of $(dX(t)/dt)^2$ near this reference instant. We choose, to simplify the presentation, reference instant t = 0, and suppose that X(t) is an analytic function near this point. So X(t) behaves near t = 0 as t^n, dX(t)/dt as t^{n-1}, $(dX(t)/dt)^2$ as t^{2n-2} and so $\int (dX(t)/dt)^2\, dt$ as $t^{2n-1}/_{2n-1}$. If n is different from 1/2, there is no singularity and no push back of t = 0 to θ = -∞. But if n = ½ X(t) behaves as $t^{1/2}$, dX(t)/dt as $t^{-1/2}$, $(dX(t)/dt)^2$ as t^{-1} and $\int (d(X(t)/dt)^2\, dt$ as Log t. There is a push back of t = 0 to θ = - ∞ and possibility of infinite interval time . A push forward of reference instant t = a to θ = + ∞ happens if X(t) behaves as $(a-t)^{1/2}$ near t = a. In the case of "elliptic explosion-implosion", X(t) behaves as $t^{1/2}$ near t = 0 and as $(2p-t)^{1/2}$ near t = 2p. So, as we have seen, we have a push back and a push forward. For "hyperbolic explosion" as well as for "parabolic explosion" X(t) behaves as $t^{1/2}$ near t = 0 and there is a push back.

4. Equation of evolution in term of "internal time"

It may be of interest, for a given evolution of a system described by function X(t), to express X(t) in term of "internal time" θ instead of reference time t. Let us take as an example the case of "parabolic explosion". We have

$$X(t) = (2ht)^{1/2}$$

and

$$\theta(t) = h/2 \ \mathrm{Log}t,$$

$$t(\theta) = \exp(2\theta/h)$$

It gives

$$X(t(\theta)) = (2h)^{1/2} \exp(\theta/h).$$

So, in term of "internal time", the state varies exponentially from 0 to $+\infty$ while θ varies from $-\infty$ to $+\infty$, instead of growing as $t^{1/2}$ when t goes from 0 to $+\infty$.

5. Time and space

The system considered may, more generally, be defined by X(t,x), a scalar function of "reference time" t and space point x , satisfying a partial derivative equation. We consider, as the index of importance of reference instant t, the integral, supposed to be convergent, extended to whole space S, of the square of the speed of evolution $(\partial X(t,x)/\partial t)^2$, that is to say

$$\int_S (\partial X(t,x)/\partial t)^2\, dx. \qquad (11)$$

So the "internal duration" of interval (t_1,t_2) is given by

$$d(t_1,t_2) = \int_{t_1,t_2} \int_S (\partial X(t,x)/\partial t)^2\, dx\, dt,$$

and an « internal time » by

$$\theta(t) = d(t_0,t).$$

We shall apply this formalism to the dynamical system constituted by a space-time field of temperatures, in the case of heat diffusion, with $S = (-\infty,+\infty)$. Temperature at point x, at reference instant t, is u(t,x). The partial derivative equation of evolution is

$$\partial u(t,x)/\partial t - \partial^2 u(t,x)/\partial x^2 = 0.$$

If the repartition of temperatures at t = 0 is given by function (more generally distribution) $u_0(x)$, the solution of the above equation is

$$u(x,t) = \int_{-\infty}^{+\infty} 1/2(\pi t)^{1/2}\ \exp(-(x-s)^2/4t)\ u_0(s)\, ds.$$

At initial reference instant t = 0, we suppose that the field of temperatures is given by $\delta(x)$ or Dirac distribution centered at x= 0 (in a rather simplified language it is equal to 0 everywhere except at t = 0 where it is infinite, the integral being nevertheless equal to 1). The repartition of temperatures at reference instant t is, according to the precedent equation and the properties of $\delta(x)$, given classically by the Laplace-Gauss function

$$u(t,x) = (4\pi t)^{1/2} \exp(-x^2/4t).$$

When reference instant t tends to $+\infty$, this function "flattens" and tends to $\varepsilon(x)$ or "epsilon distribution" (Vallée, 1992), in short it is equal to zero everywhere, the integral being nevertheless equal to 1. We have

$$(\partial u(t,x)/\partial t)^2 = 1/16\pi\ (1+x^2/2t)^2/\ t^3\ \exp(-x^2/2t)$$

and

$$\int_R (\partial u(t,x)/\partial t)^2\, dx = 3\,(2\pi)^{1/2}/16\ \ t^{-5/2}.$$

So the « internal duration » of reference interval (t_1,t_2) is, by integration from t_1 to t_2,

$$d(t_1,t_2) = (2\pi)^{1/2}/8 \ (t_1^{-3/2} - t_2^{-3/2})$$

and an "internal time" is given by the following function (increasing with t)

$$\theta(t) = -(2\pi)^{1/2}/8 \ t^{-3/2}. \quad (12)$$

When reference time t varies from 0 to $+\infty$, "internal time" θ varies from $-\infty$ to 0. Initial reference instant $t = 0$ is pushed back to $-\infty$. "Internal duration" from $t > 0$ to $+\infty$, is finite and equal to $(2\pi)^{1/2}/8 \ t^{-3/2}$.

6. Time and cosmology

We shall now interpret the notion of "internal time" in the field of cosmology. We consider models for which the state of the universe, at reference instant t, is given by the so called scalar factor R(t). According to Lemaître, Friedman and Robertson (Berry, 1976) a possible equation of evolution is

$$(dR(t)/dt)^2 = 8\pi G/3 \ \rho(t) R^2(t) - kc^2 + \Lambda/3 \ R^2(t), \quad (13)$$

$$R(0) = 0,$$

G being the gravitational constant, c the speed of light, k the index of curvature (k= -1, space with negative curvature; k = 0, flat space; k =+1, space with positive curvature), Λ the cosmological constant, $\rho(t)$ the density of matter equal to $a/R^3(t))$ or its material equivalent $b/R^4(t)$ when there is only radiation, a and b being two constants. When k = +1, R(t) is interpreted as the radius of the universe.

6.1 Radiation with null cosmological constant

If we consider the case of positive curvature with null cosmological constant and density of matter negligible compared to the equivalent density of matter of pure radiation (k=+1, Λ=0, $\rho(t) = b/R^4(t)$), we have

$$(dR(t)/dt)^2 = 8\pi G/3 \ b/R^2(t) - c^2 \quad (14)$$

and

$$R(0) = 0.$$

This case corresponds to the « elliptic explosion-implosion" considered above where, after having taken the square of the two members of equation (1), we replace X(t) by R(t), choose

$$q = cp,$$

and

$$p = (b8\pi G/3)^{1/2} / c^2.$$

It gives, according to (2),

$$R(t) = q/p \ (2pt - t^2)^{1/2} = (2(8\pi Gb/3)^{1/2} \, t - c^2t^2)^{1/2}$$

and the graph of function R(t) is, as we know, elliptical.

The "internal time" of this cosmological system is, according to equation (3),

$$\theta(t) = c^2p/2\ \ (\text{Log } t - 2t/p\ - \text{Log}(2p - t)),$$

or

$$\theta(t)\ = (2\pi Gb/3)^{1/2}(\text{Log}t - tc^2/(2\pi Gb/3)^{1/2}\ -\ \text{Log}(2(2\pi Gb/3)^{1/2} - t). \qquad (15)$$

While reference time t goes from 0 (big bang) to t = 2p (big crunch) "internal time" θ goes from $-\infty$ to $+\infty$. In that case I propose to call θ "generalized cosmological time" (Vallée, 1996, 2005) in remembrance of "cosmological time" (Milne, 1948) given by

$$c^2p/2\ \ \text{Log}t$$

or

$$(2\pi Gb/3)^{1/2}\,\text{Log}t. \qquad (16)$$

which is approximately valid for t "small".
If we consider now the case of flat space, null cosmological constant and pure radiation

$$k = 0,\ \Lambda = 0,$$

$$\rho(t) = b/R^4(t)),$$

we have

$$(dR(t)/dt)^2 = 8\pi G/3\ \ b/R^2(t),$$

or

$$dR(t)/dt = 2\ (2\pi Gb/3)^{1/2}/R(t).$$

We recognize, according to (9), a "parabolic explosion" with $h=2(b2\pi G/3)^{1/2})$. We have

$$R(t) = 2\ (b2\pi G/3)^{1/4}\ \ t^{1/2}$$

and, according to (10), the "internal time" is given by

$$\theta(t) =\ (2\pi Gb/3)^{1/2}\ \text{Log}t, \qquad (17)$$

identical to the approximate formula (13). While t goes from 0 (big bang) to $+\infty$, θ varies from $-\infty$ to $+\infty$.

6.2 No matter nor radiation

There are other cases (Berry, 1989) for which we can introduce "internal time". Some of them may not be realistic, but due to the uncertainty concerning our conception of the universe and its evolution, they must not be discarded systematically.
For example we may have a universe with no matter and no radiation, at least as an approximation. As a first case we add that space has a negative curvature and a negative cosmological constant

$$k = -1,\ \Lambda < 0,$$

$$\rho(t) = 0.$$

We have

$$(dR(t)/dt)^2 = c^2 + \Lambda/3\ R^2(t),\ \ R(0) = 0,$$

$$dR(t)/dt = (c^2 + \Lambda/3\ R^2(t))^{1/2},$$

which gives

$$R(t) = c\ (3/|-\Lambda|)^{-1/2}\ \sin(t\ (|-\Lambda|/3)^{1/2}).$$

R(t) starts from 0 (big bang) reaches its maximum $c(3/|-\Lambda|)^{1/2}$,decreases and attains 0 at t = $2\pi\ (|-\Lambda|/3)^{-1/2}$ (big crunch). We have

$$(dR(t)/dt)^2 = c^2 \cos^2(t\ (|-\Lambda|/3)^{1/2})) = c^2/2\ (1+\cos 2t\ (|-\Lambda|/3)^{1/2})$$

which give after integration

$$\theta(t) = c^2/2\ \ (t + \sin 2t(|-\Lambda|/3)^{1/2})\ /2(|-\Lambda|/3)^{1/2}). \tag{18}$$

So θ varies from 0 to $c^2\pi/4\ (|-\Lambda|/3)^{-1/2}$ as t varies from 0 to $\pi/2\ (|-\Lambda|/3)^{-1/2}$. A finite reference duration gives here a finite "internal duration".
Another possibility is the case of a universe with no matter nor radiation as above but with positive curvature and positive cosmological constant

$$k = +1,\ \Lambda > 0,$$

$$\rho(t) = 0.$$

We have

$$(dR(t)/dt)^2 = -c^2 + \Lambda/3\ \ R^2(t),$$

$$dR(t)/dt = (-c^2 + \Lambda/3\ R^2(t))^{1/2},$$

and so

$$R(t) = c\ (3/\Lambda)^{1/2}\ \ \cosh(t(\Lambda/3)^{1/2}).$$

R(t) starts from $c(3/\Lambda)^{1/2}$ at t= 0 and tends, in a way closer and closer to an exponential to $+\infty$ as t tends to $+\infty$. We have

$$(dR(t)/dt)^2 = c^2 \sinh^2(t(\Lambda/3)^{1/2}),$$

and after integration

$$\theta(t) = c^2/2\ \ (-t + \cosh 2(t(\Lambda/3)^{1/2}/2(\Lambda/3)^{1/2}), \tag{19}$$

which varie from $c^2/2$ to $+\infty$ when t varies from 0 to $+\infty$ (not forgetting that θ is defined up to an arbitrary constant). An infinite reference duration gives an infinite "internal duration".

6.3 Matter but no radiation

We shall consider some other cases, also presented by Berry, where radiation is negligeable. We start with the hypothesis of a flat space and a negative cosmological constant

$$k=0,\ \Lambda<0,$$

$$\rho(t) = a/R^3(t).$$

We have

$$(dR(t)/dt)^2 = 8\pi Ga/3/R(t) + \Lambda/3\ R^2(t),$$

which gives, A being a constant,

$$R(t) = A \sin^{2/3} (t/2\ (3|\Lambda|)^{1/2}). \tag{20}$$

So R(t) starts from 0 at t = 0 with infinite speed (big bang) reaches its maximum, decreases and attains 0 again for $t = 2\pi/(3|\Lambda|)^{1/2}$ (big crunch). Near t = 0 R(t) behaves as $t^{2/3}$ which is different from $t^{1/2}$. So, as we have seen, there is no push back of reference instant t= 0 to $-\infty$ and, for analogous reasons, no push forward of instant $t = 2\pi/(3|\Lambda|)^{1/2}$ to $+\infty$.
We have now the intermediary case where the cosmological constant is equal to zero (k=0, Λ=0, $\rho(t) = a/R^3(t)$)., which gives

$$(dR(t)/dt)^2 = 8\pi Ga/3\ /R(t)$$

and

$$R^{1/2}(t)\ dR(t)/dt = (8\pi Ga/3)^{1/2}$$

or

$$R(t) = (8\pi Ga/3)^{1/3}\ t^{2/3}$$

When t varies from 0 to $+\infty$, R(t) increases from 0 with infinite speed (big bang) to $+\infty$.
We have

$$(dR(t)/dt)^2 = 4/3\ (8\pi Ga/3)^{2/3}\ t^{-2/3},$$

which gives after integration

$$\theta(t) = 3\ (8\pi Ga/3)^{2/3} (2/3)^{2/3}\ t^{1/3}. \tag{21}$$

There is no push back of reference instant t = 0.
Now we must see the case where the cosmological constant is positive (k=0, $\Lambda>0$, $\rho(t)$ = a/R). We have

$$(dR(t)/dt)^2 = 8\pi G/3\ a/R(t) + \Lambda/3\ R^2(t)$$

and B being a constant

$$R(t) = B \sinh^{2/3}(t/2\ (3\Lambda)^{1/2}). \tag{22}$$

R(t) starts from zero at t=0 with infinite speed (big bang) and tends to $+\infty$ exponentially. There is no push back of reference instant t=0 to $-\infty$.

6.4 Null cosmological constant and no radiation

There are other interesting cases with negligible radiation and null cosmological constant. We start with a space of negative cuvature (k=-1,Λ=0, $\rho(t) = a/R^3(t)$). We have

$$(dR(t)/dt)^2 = 8\pi Ga/3\ /R(t) + c^2. \tag{23}$$

It is easier to represent the graph of function R(t) parametrically than explicitely. This gives with $0 \leq u$

$$R(t) = 8\pi Ga/6\ (\cosh u - 1),$$

$$t = 8\pi Ga/6c\ (\sinh u - u).$$

R(t) starts from 0 at reference instant t= 0 with infinite speed (big bang), then tends to $+\infty$ asymptotically as ct. Near t =0, R(t) behaves as $t^{1/3}$. It proves as we have already seen that there is no push back of reference instant t =0 to $-\infty$.
If space is flat ($k=0$, $\Lambda=0$, $\rho(t) = a/R^3$) we find a case already studied above, we have

$$(dR(t)/dt)^2 = 8\pi Ga/3/R(t)$$

and

$$\theta(t)= 3\ (8\pi Ga/3)^{2/3}\ (2/3)^{2/3}\, t^{1/3}.$$

We also have the case of a space of positive curvature ($k=+1,\Lambda=0,\rho(t)=a/R^3$). We have

$$(dR(t)/dt)^2 = 8\pi Ga/3/R(t) - c^2. \qquad (24)$$

The graph of function R(t) is a cycloid represented parametrically by

$$R(t) = 8\pi Ga/6\ (1 - \cos v),$$

$$t = 8\pi Ga/6c\ (v - \sin v),$$

$$0 \leq v \leq \pi.$$

R(t) starts from 0, at reference instant t=0, with an infinite speed (big bang). It increases up to $8\pi Ga/6$ attained at $t = 8\pi Ga/6c\ (\pi/2 -1)$, then decreases to 0 attained at $t = 8\pi Ga/6\ (\pi - 2)$. Near t = 0, R(t) behaves as $t^{2/3}$ and so there is no push back of instant t = 0 to $-\infty$, and for similar reasons no push forward of instant $t = 8\pi Ga/6\ (\pi-2)$ to $+\infty$.

7. Another approach to "internal time"

This new approach will put, metaphorically speaking, emphasis on perception. The purpose being to propose a modelling of the perception duration. First we consider the linear differential equation

$$dx(t)/dt = - a(t)\, x(t) + v(t), \qquad (25)$$

where t is reference time, x(t) and v(t) two scalar functions . We have classically

$$x(t) = \varphi(t, t_0)\, x(t_0) + \int_{t_0,t} \varphi(t,\tau)\, v(\tau)\, d\tau \qquad (26)$$

where

$$\varphi(t, t_0) = \exp\left(- \int_{t_0, t} a(s)\, ds\right) \qquad (27)$$

with the hypothesis that $\varphi(t, t_0)$ tends to 0 if t tends to $+\infty$. When a(t) is a mere constant a, it means obviously that a is strictly positive. The sign – has been placed before the integral to make more evident that the positivity of a has this consequence. Let us remark that we have the following property of " transitivity"

$$\varphi(t',t)\,\varphi(t,\tau) = \varphi(t',\tau). \qquad (28)$$

We interpret v(t) as an external influence. In the most simple case we have

$$v(t) = b(t)\,u(t),$$

more generally we could have

$$v(t) = b_0(t)\,u(t) + b_1(t)\,du(t)/dt + \ldots$$

involving derivatives of u(t). This formulation is nor irrealistic and is well adapted to the modellisation of a tachymeter or an accelerometer if we consider mainly the first or the second derivatives the other terms being rather negligible. This formula which we may also write with the help of the Dirac distribution δ and its derivatives

$$v(t) = \int_{-\infty,+\infty} (\, b_0(\tau)\,\delta(t-\tau) + b_1(\tau)\,\delta'(t-\tau) + \ldots)\, u(\tau)\, d\tau$$

Each $b_{i(}(t)$ is a "factor of attention" concerning a particular derivative. The passage of function u to function v is made by what we call "observation operator" (Vallée, 1951, 2002). Here this operator acts in an instantaneous way, being purely local. The first factor $b_0(t)$, or more simply b(t), may be considered positive (when it is null there is no attention and so no perception at the considered instant t). This factor of attention has been pointed out (Condillac, 1754) : "…it remains an impression more or less strong according to the fact that the attention has been more or less intense". More generally the passage of u to v through v = O(x), O being an "observation operator", is not instantaneous but hereditary, that is to say involving the past and present of u. This has been observed (Bergson, 1939): "In fact 'pure' perception, that is to say instantaneous, is only an ideal, a limit. Every perception fills a certain length of duration, extends the past in the present…". For example we may have a convolution

$$v(t) = \int_{t_0,t} k(t-\tau)\, u(\tau)\, d\tau,$$

where v(t) depends upon the values of u on interval (t_0,t). More generally, if we do not leave the case of linear "observation operators" we have a Volterra composition giving

$$v(t) = \int_{t_0\, t} k(t,\tau)\, d\tau\, .$$

The formalism of "observation operators" permits to see in which cases such an operator does not alter the observed function u. We must have

$$O(u) = \lambda\, u,$$

so u must be an eigen function of operator O and λ is the associated eigen value which may be complex. If $\lambda = 1$, we have a fixed point. We give these details about "observation operators" because we shall meet them in several circumstances, the problem of appreciation of time having to do with observation and more generally with what we could call "mathematical epistemology" (Vallée, 2002).

Let us come back to the differential equation and its solution (26). We interpret $\varphi(t,\tau)\, v(\tau)$ as what remains at instant t of the perception $v(\tau)$ felt at anterior instant τ , it is the result of the transfer by memorization of $v(\tau)$ from τ to t. The property of transitivity (27) makes this transfer coherent. According to the hypothesis that $\varphi(t,\tau)$ tends to 0 if $t-\tau$ tends to $+\infty$, we may say that the transferred perception $\varphi(t,\tau)\, v(\tau)$ tends to 0 if τ tends to $-\infty$. In other words we may conclude that the more ancient is a perception the more feeble is its remembrance. If our differential system starts with the null state we have according to (26) a Volterra composition which reduces to convolution when a(s) is a constant a

$$x(t) = \int_{t_0,\, t} \varphi(t,\tau)\, v(\tau)\, d\tau.$$

We may interpret x(t) as the result of the superposition of all the successive perceptions, from t_0 to t, as they are transferred to t by memorization. The passage from v(t) to x(t) is given by a special type of linear "observation operator" which we call "memorization operator". It may be compared to the factor of forgetfulness (Vogel, 1965) or the memory coefficient (Allais, 1972).

We shall consider now differential equation (25), or more precisely its solution (26) where we replace x(t) by $\theta(t)$, as a model of perception and memorization of duration valid for a dynamical system considering t as "reference time" and $\theta(t)$ as a subjective or "internal time", even if these expressions are to be understood metaphorically. Taking into account (27) we have

$$\theta(t) = \int_{t_0,t} \exp\left(-\int_{\tau,t} a(s)ds\right)\, v(\tau)\, d\tau, \qquad (29)$$

to which we add

$$a(t) \geq 0, \qquad v(t) \geq 0\,.$$

While $t - t_0$ is the reference duration of interval (t_0, t), $\theta(t) - \theta(t_0) = \theta(t)$ is its subjective or internal duration. It depends upon the way the considered system perceives through v(t) and memorizes, not upon the way the state of the system evolves as it was the case in the first approach to "internal time".

Of course t is a good parametrisation of time, in the sense that two distinct instants are represented by two different values of t. If $\theta(t)$ is to be a good parametrisation of time, it must be a strictly increasing function of t, a condition realized if v(t) never vanishes on an interval not reduced to a mere instant. The hypothesis that $a(t) \geq 0$ has for consequence that the factor of memorization $\varphi(t,\tau)$ decreases in the large when τ diminishes.

We must now interpret function v(t) in the most simple case where

$$v(t) = b(t)\, u(t),$$

with

$$b(t) \geq 0\,, \quad u(t) \geq 0.$$

Of course more elaborated cases could be considered, making a full use of « observation operators ».

We consider that function u(t) gives the evolution of the weight of importance of reference instant t. The very simple "observation operator" represented by b(t) appears as a "factor of attention" given to u(t) which represents the intrinsic importance of reference instant t itself. Since t is by definition an objective time or "reference time", all instants t have an equal intrinsic importance which we may decide to be equal to 1. So we may write simply

$$vt) = b(t)$$

as the weight attributed to instant t is equal to the "factor of attention" b(t) at this instant.

7.1 Perception of duration with imperfect memorization

In our model we have an imperfect memorization when a(t) is not identical to 0. The factor of memorization

$$\varphi(t,\tau) = \exp\left(- \int_{\tau,t} a(s)ds\right)$$

is strictly inferior to 1 and tends to 0 when t – τ tends to + ∞, so when τ tends to - ∞. The remembrance of past perceptions vanishes with time. Since, as we have seen, v(t) is equal to b(t) we may write according to (28)

$$\theta(t) = \int_{t_0,\, t} \exp\left(- \int_{\tau,\, t} a(s)ds\right) b(\tau)d\tau,$$

or

$$d\theta(t) = (-a(t)\,\theta(t) + b(t))\, dt,$$

explaining the antagonistic roles of b(t) > 0 and -a(t) θ(t) < 0 , representing attention on one side and oblivion on the other. Or, if we consider the case where a(t) reduces to constant a>0,

$$\theta(t) = \exp(-at) \int_{\mathbf{to,t}} \exp a\tau \; b(\tau)\, d\tau.$$

It is interesting to see what happens when b(t) is a Dirac distribution δ(t-γ), centered on instant γ . It is an idealisation which gives

$$\theta(t) = \exp(-a\,(t-\gamma))\,,$$

$$t > \gamma.$$

We see that the reference duration of instant γ, obviously equal to 0, is perceived just after this instant, due to the infinite attention implied, as having a finite subjective or internal duration; It is perceived later has having a decreasing value tending to 0. If we replace the Dirac δ by a flash of attention, things are not so sharply defined but are of the same nature: a very short interval of reference time is perceived as middle sized interval of "internal time" and this impression diminishes with time and disappears. The reference time, 1/a, necessary to see the perceived duration divided by e is a measure of what we may call the "subjective duration of the present" or the "thickness of an instant", in accordance with Bergson's remark quoted above and close to the concept of constant of time familiar in dynamics.

7.2 Perception of duration with perfect memorization

Perfect memorization is obtained when the factor of memorization is always equal to 1, that is to say when a(t) is identical to 0. Then

$$d\theta(t)/dt = b(t),$$

with

$$b(t) \geq 0 \, , \theta(t_0) = 0.$$

We have

$$\theta(t) = \int_{t_0,t} b(\tau)\, d\tau, \tag{30}$$

it is the subjective or "internal duration" of interval (t_0,t). In other terms it is the subjective duration of instants of consciousness in this interval, instants where b(t) = 0 having no impact (metaphorically or not : deep sleep or coma). If b(t) takes only values 1 or 0, $\theta(t)$ is equal to the internal duration (as well as reference duration) of instants of consciousness. When b(t) takes value 1 at discrete instants it acts as a stroboscopic selecting device. When b(t) is constant $\theta(t)$ is proportional to t-t_0. If for example, the "factor of attention" decreases exponentially with time, that is to say if

$$b(t) = b \exp(-\lambda t),$$

with

$$b > 0, \lambda > 0.$$

We have, with $t_0 = 0$,

$$\theta t) = 1 - \exp(-\lambda t).$$

The "internal duration" or perceived duration since $t_0 = 0$, increases from 0 to 1 while "reference duration" tends to $+\infty$. This circumstance may be found during the observation of a disintegrating radioactive material if the factor of attention b(t) is proportional to disintegration activity which decreases exponentially.

8. Another approach to cosmology and other problems

First we consider the case of perfect memorization for a conscious being which may be human. It has often been remarked that the perceived or "internal duration" of the same interval of "reference time" diminishes with age. It may be interpreted by saying that the intrinsic importance of an instant in early age is much greater than later. Birth may be compared to a kind of biological big bang, particularly if we consider that life starts at the very moment of conception. Another argument is that an interval of reference time measured by comparison to the length life already elapsed, seems shorter and shorter. In other words we may say that the "factor of attention" given to a reference instant is proportional to the intrinsic importance of this instant. So if t = 0 is the reference instant of conception, an acceptable "factor of attention " b(t) adapted to this case may be given by a function decreasing with time and starting with a great value, we may even have $b(0) = +\infty$. The most simple example is given by

$$b(t) = b/t \, , \tag{31}$$

$$b > 0.$$

Then we have

$$\theta(t) - \theta(t_0) = b \operatorname{Log} (t/t_0) = b \operatorname{Log} t - b \operatorname{Log} t_0,$$

but since $\theta(t)$ is defined up to an arbitrary constant, we write

$$\theta(t) = b \,\mathrm{Log}\, t \tag{32}$$

We find here a logarithmic psychological time already proposed (Lecomte du Noüy, 1936) on the basis of the speed of cicatrisation of wounds which decreases with time. With this "internal time", the beginning of life is pushed back to $-\infty$, a feeling frequent among human beings. Considerations formally similar may be developed in the case of some thermodynamical systems where entropy decreases as $1/t$, generating an "entropic time" (Prigogine, 1947).

We had found the same result with "parabolic explosion" and with the case, seen at the end of section 6.1, which introduces Milne's cosmological time. Near reference instant $t = 0$ (big bang) the events affecting the universe are important to an extreme, it is quite acceptable to admit that $b(t)$, as a physical "factor of importance" (very metaphorically a "factor of attention" adapted to it), has a pole at $t = 0$ and may be represented by function $1/t$. Reference instant $t = 0$ is pushed back to "internal instant" $\theta = -\infty$. From the "internal time" point of view this universe has no beginning, the instant of big bang is not an "internal instant".

We may imagine now a more general case, of explosive-implosive type. The reference instants near the beginning of explosion ($t = 0$) or near the end of implosion ($t = \sigma$) have a tremendous importance and we may admit that function $b(t)$ has a pole at each of these points. The most simple example, if we affect the two poles of the same coefficient b, assuming that they are of the same importance, is given by

$$b(t) = b(1/t + 1/\sigma\text{-}t), \tag{33}$$

$$b > 0,\ \sigma > 0,$$

and then we have

$$\theta(t) = b \,\mathrm{Log}\,(t/t_0) - b \,\mathrm{Log}\,(\sigma\text{-}t/\sigma\text{-}t_0)$$

or, $\theta(t)$ being defined up to a constant,

$$\theta(t) = b\,(\mathrm{Log}\, t - \mathrm{Log}\,(\sigma\text{-}t)), \tag{34}$$

which varies from $-\infty$ (at $t = 0$) to $+\infty$ (at $t = \sigma$) and is equal to 0 for $t = \sigma/2$. It is the reciprocal of the logistic function

$$t(\theta) = \sigma \exp \theta/b \;/\; 1 + \exp \theta/b\,.$$

We already had a result close to (34) with the case of elliptic explosion-implosion which gave for "internal time"

$$\theta(t) = q^2/2p\ (\mathrm{Log} t - 2t/p - \mathrm{Log}(2p\text{-}t)) \tag{4}$$

met again in the cosmological case of radiation with null cosmological constant (15). According to (34) reference instants $t = 0$ and $t = \sigma$ are pushed back to $-\infty$ for the first one and to $+\infty$ for the second. There is no reference instants for big bang nor for big crunch So the "life of the universe" whose reference duration is finite and equal to σ has an infinite "internal duration". From the "internal time" point of view this universe has no beginning

nor end as well as in the elliptic explosion-implosion and cosmological case with no radiation and null cosmological constant.
The presence of term 2t/p may be surprising. In fact it does not exist in the new formulation (34). If we identify σ with 2p and $q^2/2p$ to b, to make the comparison more clear, (4) becomes

$$\theta(t) = b\,(\mathrm{Log}t - 4t/\sigma - \mathrm{Log}(\sigma - t)).$$

In the case of elliptic « explosion-implosion", according to the following expression given in section 2.1

$$(d(X(t)/dt)^2 = q^2/2p\,(1/t - 2/p + 1/2p\text{-}t)),$$

We see that the index of importance of reference instant t contains, apart from the terms involving 1/t and 1/2p-t , a constant negative term. The result is that for t = p, the index of importance is equal to 0. This is quite normal since at inst ant t = p the speed of evolution of the system is null. Since equation (33) gives

$$b(\sigma/2) = 4/\sigma,$$

we may change (33) into

$$b(t) = b\,(1/t - 4/\sigma + 1/\sigma\text{-}t), \tag{35}$$

the factor of importance or of attention b(t), always positive, being considered as a whole and not decomposable into parts.
We may also consider that the behaviour of the system near the end of its evolution is not necessarily symmetrical of its behaviour near the beginning, and propose

$$b(t) = b_1\, 1/t + b_2\, 1/\sigma\text{-}t,$$

$$b_1 > 0,\; b_2 > 0,$$

or even, if we want to have

$$b(\sigma/2) = 0,$$

$$b(t) = b_1\, 1/t - 2\,(b_1 + b_2)/\sigma + b_2 1/\sigma\text{-}t\,. \tag{36}$$

Since we made a comparison between an explosive system with an infinite speed of evolution at the beginning, which is the case of universe in certain modellisations, and the evolution of a human being (or another sort of conscious creature) as far as "internal time" is concerned. We could also make the same comparison between an explosive-implosive system with an infinite positive speed of evolution at the beginning and an infinite negative speed at the end, also a possible case for universe or a conscious creature.

9. Conclusion

The measure of time has always been a problem to philosophers and scientists. Early human beings of the paleolithic age founded it on the regular movements of moon and sun and the apparent rotation of heavens, opening the way to scientific thinking. It has been said that if the sky of earth had been extremely cloudy this sort of thinking would have been delayed or even would never have started. Space seems to have been less mysterious to human mind up to the point that the standard representation of time is a straight line, even if attempts to introduce n dimensional time have been made (Vallée, 1991).But aside from attempts to measure time in a way acceptable to all, time as it is felt by each individual is another problem. Each of us, by an "inverse transfer" (Vallée, 1974) is unconsciously ready to attribute his own intimate structures, also his feeling of time and duration to universe itself. A behaviour which, despite its obvious defect of giving a distorted image of the environment has the advantage to render time more familiar, introducing a kind of taming of the external world.

Influenced by the subjective apprehension of time we evoked, our aim is to propose, for a dynamical system, this expression being taken with its broadest meaning, a concept of "internal time". The system may be inanimate, in the sense that, for it, consciousness is meaningless, or it may be a conscious being. So in many cases certain concepts will be used metaphorically, particularly when applied to the inanimate. We used the concept of importance of an instant, giving two kinds of definitions. In the first case the degree of importance is directly linked to the intensity of change of the state of the system at this instant, a point of view which is more or less akin to that of Aristotle and Augustine. In the second case, relatively close to that of Condillac or even of Lecomte du Noüy, this importance is more or less "felt". It seems that this aspect of the problem of time has mostly interested philosophers and not scientists with the exception of economist Allais and cosmologist Milne.

One of the most interesting results obtained is that in certain cases the "reference duration" of the life (or a part of it) of the system is finite while its "internal duration" is infinite. Of course this is not the general case, but permits in some cases, mainly certain "explosion-implosions" to eliminate apparent paradoxes. Many people when a big bang or big crunch theory is presented to them do not see that the "instant" of the big bang or the "instant" of the big crunch are not instants. They do not belong to the set of instants of time as well as the degree 0 of Kelvin temperature scale is not a degree of temperature in fact unaccessible. Allusions made to psychological interpretation of some sort of "biological" explosion-implosion are obviously extremely controversial, particularly when a sort of big crunch is considered. These considerations must be seen from a rather metaphorical point of point of view, remembering that in many cases only the qualitative aspects of mathematical results must be retained., a viewpoint which is rather new, despite the progress of ideas such as fuzzy sets and fuzzy logiics.

Indeed in recent cosmological models with inflation, big bang is excluded. So what we have presented concerning big bang seems to loose a part of its interest. Nevertheless a universe starting with $R(0)$ strictly positive may have an "internal time" pushing back $t = 0$ to $\theta = -\infty$, since it is only the behaviour of $(dR(t)/dt)^2$ which is important. Moreover we do not know what will be the models of universe even in near future with the problem of "dark matter". So the idea of "internal time "remains, in cosmology. But another problem arises: before the so called Planck's time there are difficulties concerning laws of physics, and so the definition of time itself.

What about the future of "internal time" in cosmology and other fields? A first point is that, for the sake of simplicity, we considered only systems with scalar state. It would be good to generalize to vector state of dimension n. The definition of the index of importance of reference instant permits it, since it is the scalar square of a vector. But other indexes are possible, they just need to be a norm or any positive increasing function of a norm. The choice we made, the square of the Euclidian norm, has many advantages but others are eligible. Another direction of research would be to introduce some kind of "internal time" adapted to a quantum mechanical systems, despite obvious difficulties, or even to the conception of a discrete time having something to do with Planck's time as a unit of duration.

More generally a scientific introduction of subjectivities affecting systems, in certain cases metaphorically speaking, has a great interest. The consideration of concepts intrinsically linked to a system, instead of being imposed from the outside, is also very desirable. Equations of evolution or values of some traits expressed in term of such subjective or intrinsic features may be suggestive. An example is given by the length of life for certain species of animals measured by the number of their heartbeats, length which appears as of the same order of magnitude.

10. References

Allais, M (1972), Forgetfulness and interest, *Journal of Money, Credit and Banking,* (Feb.) pp.46-76.

Bergson, H (1939), Matière et mémoire, Presses Universitaires de France, Paris.

Berry, M (1989), Principles of Cosmology and Gravitation, Institute of Physics, ISBN 0-85274-037-9,Bristol.

Condillac, E Bonnot de (1754), Traité des sensations, Paris.

Hussey, E (1983), Physics, books III and IV, Oxford Press.

Lecomte du Noüy, P (1936), Le temps et la vie, Gallimard, Paris.

Milne, E A (1948), Kinematic Relativity, Clarendon Press, Oxford.

Prigogine, I (1947), Étude thermodynamique des processus irreversibles, Dunod, Paris.

Vallée, R (1951), Sur deux classes d'"opérateurs d'observation", *Comptes Rendus de l'Académie des Sciences* ,Vol. 233, pp. 1350-1351.

Vallée, R (1975), Observation, decision and structure transfer in systems theory, In: Progress in Cybernetics and Systems Research, R.Trappl, (Ed.), Vol.1, pp.15-20, Hemisphere Publishing Corporation, Washington.

Vallée, R (1991), Perception, memorisation and multidimensional time; *Kybernetes,* Vol. 20, No.6, ,pp. 15-28.

Vallée, R (1992), The 'epsilon distribution' or the antithesis of Dirac's delta, In: Cybernetics and Systems Research, R.Trappl, (Ed.), 97-102, World Scientific, Singapore.

Vallée, R (1996), Temps propre d'un système dynamique, cas d'un système explosif-implosif, Actes du 3ème Congrès International de Systémique, E. Pessa & M P Penna (Eds.), pp.967-970, Edizioni Kappa, Rome

Vallée, R (2002), Mathematical and fomalized epistemologies, In : Quantum Mechanics, Mathematics, Cognition and Action, Proposals for a Formalized Epistemology, M. Mugur-Schachter, A. van der Merwe (Eds.), pp. 309-324, Kulver Academic Publishers, Dordrecht.

Vogel, T (1965), Théorie des systèmes évolutifs, Gauthier-Villars, Paris.

Warner, R (1963), The Confessions of St. Augustine, Penguin Books, ISBN 0-451-62474-2, New York.

5

Revised Concepts for Cosmic Vacuum Energy and Binding Energy: Innovative Cosmology

Hans-Jörg Fahr and Michael Sokaliwska
Argelander Institut für Astronomie, Bonn
Germany

1. Introduction

Contemporary cosmology confronted with WMAP observations of the cosmic microwave background radiation and with distant supernova locations in the magnitude - redshift diagram obviously has to call for cosmic vacuum energy as a necessary prerequisite. Most often this vacuum energy is associated with the cosmological constant Λ, introduced by Einstein and presently experiencing a fantastic revival in form of "dark energy". Within the framework of General Relavity the term connected with Λ acts analogous to constant vacuum energy density. With a positive value, Λ describes an inflationary action on cosmic dynamics which in view of more recent cosmological data to most astronomers appears to be absolutely needed. In this article, however, we shall question this hypothesis of a constant vacuum energy density showing that it is not justifyable on physical grounds, because it claims for a physical reality that acts upon spacetime and matter dynamics without itself being acted upon by spacetime or matter.

In the past cosmic mass generation mechanisms have been formulated at different places in the literature and based on different physical concepts. A deeper study proves that these alternative theoretical forms of mass creation in the expanding universe all lead to terms in the GR field equations which can be shown to act analogously to terms arising from vacuum energy. In addition we also demonstrate that gravitational cosmic binding energy connected with structure formation acts identically to negative cosmic mass energy density, i.e. reducing the action of proper mass density. This again resembles an action of cosmic vacuum energy. Hence one is encouraged to believe that actions of cosmic vacuum energy, gravitational binding energy and mass creation are closely related to eachother, perhaps are even in some respect identical phenomena.

Based on results presented in this article we propose that the action of vacuum energy on cosmic spacetime dynamics inevitably leads to a decay of vacuum energy density. Connected with this decay is a decrease of cosmic binding energy and the appearance of new gravitating mass in the universe, identifyable with creation of newly appearing effective mass in the expanding universe. If this all is adequately taken into account by the energy-momentum tensor of the GR field equations, one is then led to non-standard cosmologies which for the first time can guarantee the conservation of the total energy both in static and expanding universes.

2. The concept of absolutely empty space

The question what means empty space , or synonymous for that - vacuum - , in fact is a very fundamental one and has already been put by mankind since the epoch of the greek natural philosophers till the present epoch of modern quantum field theoreticians. The changing opinions given in answers to this fundamental question over the changing epochs have been reviewed for example by Overduin & Fahr (2003) , but we do not want to repeat here all of these different answers that have been given in the past, but only to begin this article we want to emphasize a few fundamental aspects of our thinking of the physical constitution of empty space. Especially challenging in this respect is the possibility that empty space could be energy-charged. This we shall investigate further below in this article.

In our brief and first definition we want to denote empty space as a spacetime without any topified or localized energy representations, i.e.without energy singularities in form of point masses like baryons, leptons, darkions (i.e. dark matter particles) or photons, even without point-like quantum mechanical vacuum fluctuations. If then nevertheless it should be needed to discuss that such empty spaces could be still energy-loaded, then this energy of empty space has to be seen as a pure volume-energy, somehow connected with the magnitude of the volume or perhaps with a scalar quantity of spacetime metrics, like for instance the global curvature of this space. In a completely empty space of this virtue of course no spacepoints can be distinguished from others, and thus volume-energy or curvature, if existent, are numerically identical at all space coordinates.

Under these prerequisites it nevertheless would not be the most reasonable assumption, as many people believe, that vacuum energy density $\epsilon_{vac} = \rho_{vac}c^2$ needs to be considered as a constant quantity whatever spacetime does or is forced to do, i.e. whether it expands, collapses or stagnates. This is simply because the unit of volume is no cosmologically relevant quantity - and consequently vacuum energy density neither is. If at all, it would probably appear more reasonable to assume that the energy loading of a homologously comoving proper volume does not by its magnitude reflect the time that has passed in the cosmic evolution, i.e. perhaps that specific quantity has to be a constant. But this then, surprisingly enough, would mean that the enduring quantity, instead of the vacuum energy density ϵ_{vac}, is

$$e_{vac} = \epsilon_{vac}\sqrt{-g_3}d^3V \tag{1}$$

where g_3 is the determinant of the 3d-space metric which in case of a Robertson-Walker geometry is given by

$$g_3 = g_{11}g_{22}g_{33} = -\frac{1}{(1-Kr^2)}R^6r^4\sin^2\vartheta \tag{2}$$

with K denoting the curvature parameter, the function $R = R(t)$ determines the time-dependent scale of the universe and the differential 3-space volume element in normalized polar coordinates is given by

$$d^3V = drd\vartheta d\varphi \tag{3}$$

This then leads to the relation

$$e_{vac} = \epsilon_{vac}\sqrt{R^6r^4\sin^2\vartheta/(1-Kr^2)drd\vartheta d\varphi} = \epsilon_{vac}\frac{R^3}{\sqrt{1-Kr^2}}r^2\sin\vartheta drd\vartheta d\varphi$$

which shows that a postulated invariance of e_{vac} consequently and logically would lead to a variability of the vacuum energy density in the form

$$\epsilon_{vac} = \rho_{vac}c^2 \sim R(t)^{-3} \tag{4}$$

which for instance would already exclude that Einstein's cosmological constant could ever be treated as an equivalent to a vacuum energy density, since requiring the identity $\Lambda = 8\pi G\rho_{vac}/c^2$.

On the other hand the invariance of the vacuum energy per co-moving proper volume, e_{vac}, can of course only be expected with some physical sense, if this quantity does not do any work on the dynamics of the cosmic metrics, especially by physically or causally influencing the evolution of the scale factor $R(t)$ of the universe.

If on the other hand such a work is done and vacuum energy influences the dynamics of the cosmic spacetime, since it leads to a non-vanishing energy-momentum tensor, then thermodynamic requirements should be fulfilled, for example relating vacuum energy density and vacuum pressure by the standard thermodynamic relation (see Goenner (1997))

$$\frac{d}{dR}(\epsilon_{vac}R^3) = -p_{vac}\frac{d}{dR}R^3 \tag{5}$$

This equation is shown to be fulfilled by an expression of the form

$$p_{vac} = -\frac{3-n}{3}\epsilon_{vac} \tag{6}$$

if the vacuum energy density itself is represented by a scale-dependence $\epsilon_{vac} \sim R^n$. Then, however, it turns out that the above thermodynamic condition, besides for the trivial case $n = 3$ when the vacuum does not at all act as a pressure (since $p_{vac}(n = 3) = 0$) , is only non-trivially fulfilled for $n \lessgtr 3$ which would still allow for $n = 0$, i.e. a constant vacuum energy density $\epsilon_{vac} \sim R^0 = const.$

A much more rigorous, but highly interesting restriction for n is, however, obtained when one recognizes that the above thermodynamic expression (5) under cosmic conditions needs to be enlarged by the work that the expanding volume does against the inner gravitational binding in this volume. In mesoscale gas dynamics (aerodynamics, meteorology etc.)this term does generally not play a role, however, on cosmic scales there is a need to take into account this term. Under cosmic perspectives binding energy is an absolutely necessary quanity to be brought into the thermodynamical energy balance. As worked out in quantitative terms by Fahr & Heyl (2007a;b) this then leads to the following completed relation

$$\frac{d}{dR}(\epsilon_{vac}R^3) = -p_{vac}\frac{d}{dR}R^3 - \frac{8\pi^2 G}{15c^4}\frac{d}{dR}[(\epsilon_{vac} + 3p_{vac})^2R^5] \tag{7}$$

where the last term accounts for binding energy.

This completed equation, as one can easily show, is also solved by the relation of the form $p_{vac} = -\frac{3-n}{3}\epsilon_{vac}$,

but only if: $n = 2$!

meaning that the corresponding vacuum energy density must vary like

$$\epsilon_{vac} \sim R^{-2} \tag{8}$$

This thus means that, if it has to be taken into account that vacuum energy acts upon spacetime in a thermodynamical sense then the most reasonable assumption for the vacuum energy density would be to assume that it drops off with the expansion inversely proportional to the square of the cosmic scale - instead of it being a constant.

3. Philosophical perspectives of vacuum concepts and an effective vacuum-energy density

For fundamental conceptual reasons it may be necessary to explore why at all a vacuum should gravitate, since, when really being "nothing", then it should most probably not do anything. At least based on an understanding that the ancient greek atomists had, the vacuum is a complete emptiness simply offering empty places and thereby allowing atoms freely to move. One should then really not expect to have any gravitational action from such a vacuum. Aristotle, however, brought into this conceptual viewing his principle of nature's objection against emptiness ("horror vacui"). This is a new aspect realizing that empty space around matter particles is not as empty as without those particles, but is polarized by the existence or presence of real matter. This idea furtheron very much complicated the concept of vacuum making it a rather lengthy and even not yet finished story (see e.g. Barrow (2000); Fahr (2004); Wesson et al. (1996)). In the recent decades it became evident that vacuum must be energy-loaded (see e.g. Lamoreaux (2010); Streeruwitz (1975); Zeldovich (1981)) and by its energy it should hence also influence gravitational fields, even, if it is not clear in which concrete form.

Nowadays the GRT action of the vacuum is taken into account by an appropriately formulated, hydrodynamical energy-momentum tensor $T_{\mu\nu}^{vac}$, formulating the metrical source of the energy sitting in the vacuum as described by a fluid with vacuum pressure p_{vac} and equivalent vacuum mass energy density ρ_{vac}. Then with a constant vacuum energy density $\epsilon_{vac} = \rho_{vac}c^2$, as assumed in the present-day standard cosmology (Bennett et al., 2003), one obtains this tensor in the form (see e.g. Overduin and Fahr, 2001)

$$T_{\mu\nu}^{vac} = (\rho_{vac}c^2 + p_{vac})U_\mu U_\nu - p_{vac}g_{\mu\nu} = \rho_{vac}c^2 g_{\mu\nu} \tag{9}$$

where U_λ are the components of the vacuum fluid 4-velocity vector.

This term, taken together with Einstein's cosmological constant term Λ (Einstein, 1917), and placed on the right-hand side of the GRT field equations then leads to an effective cosmological constant given by

$$\Lambda_{eff} = \frac{8\pi G}{c^2}\rho_{vac} - \Lambda \tag{10}$$

The first problem always seen after Einstein (1917) is connected with the free choice one is left with concerning the numerical value of Λ. One way to obtain a first answer to that question, at least for the completely empty, i.e. matter-free space, is a rationally pragmatic and aprioristic definition, - namely an answer coming up from an apriori definition of how empty space should be constituted and should be manisfesting itself. If it is rationally postulated that empty space should be free of any spacetime-curving sources, and thus free of local or global curvature, if one requires that selfparallelity of 4-vectors at parallel transports along closed wordlines in this empty space should be guaranteed, and if one expects no action of empty space on freely propagating test photons in this empty space, then as shown by Overduin & Fahr (2003) or Fahr (2004) the only viable solution is $\Lambda_{eff,0} = 0!$, meaning that the cosmological constant should be fixed such that

$$\Lambda_0 = \Lambda - \frac{8\pi G}{c^2}\rho_{vac,0} \tag{11}$$

where $\rho_{vac,0}$ denotes the equivalent mass density of the vacuum of empty, i.e. matter-free space. Once fixed in this above form, the cosmological constant cannot be different from this value Λ_0 in a matter-filled universe, simply meaning that in a matter-filled universe the effective quantity representing the action of the vacuum energy density is given by:

$$\Lambda_{eff} = \frac{8\pi G}{c^2}(\rho_{vac} - \rho_{vac,0}) \tag{12}$$

expressing the interesting fact that in matter-filled universe only the difference between the values of the vacuum energy densities $\rho_{vac,0}$ of empty space and of matter-polarized space ρ_{vac} gravitates, i.e. influences the spacetime geometry. That could give an explanation why obviously the vacuum energy calculated by field theoreticians does not gravitate by its full magnitude.

This also points to the perhaps most astonishing fact that the geometrically relevant vacuum energy density depends on the matter distributed in space, and in a homogeneous universe this can only mean that: $\rho_{vac} = \rho_{vac}(\rho)$, an idea that deeply reminds to the views already developed by Aristotle at around 400 bC.

Though this idea of the vacuum state being influenced by the presence of matter in space appears to be reasonable in view of field sources polarizing space around them by acting on sporadic quantum fluctuations and partly screening off the strength of real field sources, it stays nevertheless hard to draw any quantitative conclusions from that context. For that reason we shall try another way below to find the unknown function $\rho_{vac} = \rho_{vac}(\rho)$.

4. The standard cosmology based on five cosmic scalar quantities

Standard cosmology is based on some basic scalar quantities that are treated as 3-spacecoordinate-independent, but time-dependent. Amongst these are matter density ρ, scalar pressure p, isotropic curvature characterized by a space-independent Riemann scalar R, and the cosmological constant Λ. These basic elements can be used, if the universe is treated as homogeneously filled with matter of a space-indendent scalar pressure and carries out a homologous expansion. Then Einstein's General relativistic field equations can be condensed to a set of only two cosmologically relevant linear differential equations of second order for the scale of the universe R and its first and second derivatives with respect to time, $\dot{R}$ and $\ddot{R}$, given in the form (see e.g. Goenner (1997))

$$\left(\frac{\dot{R}(t)}{R(t)}\right)^2 = H^2(t) = \frac{8\pi G}{3}\rho(t) - \frac{kc^2}{R^2(t)} + \frac{\Lambda c^2}{3} \tag{13}$$

and:

$$\frac{\ddot{R}(t)}{R(t)} = -\frac{4\pi G}{3c^2}(3p(t) + \rho(t)c^2) + \frac{\Lambda c^2}{3} \tag{14}$$

Here $H(t) = \dot{R}/R$ is the Hubble function that depends on the contributing densities ρ, the pressure and the curvature parameter k, attaining values of $k = 0$ (uncurved space); $k = +1$ (positively curved space) or $k = -1$ (negatively curved space).

The matter density ρ nowadays in cosmology is composed of baryonic and dark matter, i.e. $\rho = \rho_b + \rho_d$, where the two quantities vary identically with cosmic time or cosmic scale. At cosmic times greater than the recombination period $t \geq t_{rec}$ the associated pressures $p_{b,d}$

usually are neglected with respect to their corresponding rest mass densities $\rho_{b,d}c^2$. Then depending on selected values for the ratios $\Omega_b = \rho_b/\rho_c$, $\Omega_d = \rho_d/\rho_c$ and $\Omega_\Lambda = \rho_\Lambda/\rho_c$, with $\rho_\Lambda = \Lambda c^2/8\pi G$ and the critical density given by $\rho_c = 3H^2/8\pi G$, one obtains a manifold of different solutions $R = R(t)$ of the above system of differential equations, each belonging to a specific set of numerical values for the five cosmologically relevant parameters: $H_0 = H(t_0)$, k, $\Omega_{b,0} = \Omega_b(t_0)$, $\Omega_{d,0} = \Omega_d(t_0)$ and $\Omega_{\Lambda,0} = \Omega_\Lambda(t_0)$. To decide which of these parameter sets best fits cosmologically relevant observational data, like the WMAP data from the "Wilkinson Microwave Anisotropy Probe" survey (Bennett et al., 2003) or the distant supernova data (Perlmutter et al., 1999), multi-parameter fit procedures have recently been carried out. As the best-fitting consensus the following set of parameters thereby has been found: $H_0 = 71 km/s/Mpc$, $k = 0$, $\Omega_{b,0} = 0,046$, $\Omega_{d,0} = 0.23$ and $\Omega_{\Lambda,0} = 0.73$. These values are nowadays taken as result of modern precision cosmology, characterizing the facts of our actual universe. Perhaps, however, a reminder to weaknesses in the basic assumptions of such a form of precision cosmology may be in place here.

One most essential ingredience of standard cosmology is the assumption that the total, spacelike mass of the physical universe, conceivable for any spacepoint on the basis of a point-oriented spacetime metrics g^*_{ik} - irrespective of its dark or baryonic nature, is constant. This then is usually thought to imply that the corresponding matter densities $\rho_{b,d}$ in a homogeneous universe scale reversely proportional to the 3d- volume of the physical universe $V_3^* = \int^{x_\infty} d^3x \sqrt{\det^3 g^*}$, which in all cases of standard cosmology means inversely proportional to R^3.

Another essential point of standard cosmology is to assume a strict homogeneity of energy depositions in cosmic space connected with an isotropic homologous expansion of cosmic matter. Though these items seem to be cosmo-philosophically well supported by the so-called "cosmological principle" (see e.g. Stephani (1988)), one nevertheless has to recognize that the actual universe is very much different from expectations derived from this principle. In fact the actual universe is highly structured in forms of galaxies, galaxy clusters, superclusters, walls and voids (Ellis, 1983; Geller & Huchra, 1989) - perhaps one can call that a "structured homogeneity". Only on scales larger than several hundred million lightyears the universe seems to be nearly homogeneous. However if the structuring develops as function of cosmic time, then this actual universe does not expand like an equivalent one with homogeneously smeared out matter (Buchert, 2008; Wiltshire, 2007). Matter distribution had perhaps been very homogeneous, at least down to temperature fluctuations of the order of $\frac{\Delta T}{T} \approx 10^{-5}$ at the epoch of the last scattering of CMB photons when the cosmic microwave background was freezing out of cosmic matter distribution. In the cosmic eons after that phase in fact matter distribution, as evident in the appearance of the universe, must have become very inhomogeneous through gravitational growth of seed structures. Fitting a perfectly symmetrical spacetime geometry to a universe which , however, has a lumpy matter distribution up to largest scales (e.g. see Wu et al. (1999)) represents a highly questionable procedure as shown by Buchert (2001; 2005; 2008) or Wiltshire (2007) (see chapter 9.1).

Besides of the above, perhaps the even most problematic concept used in present-day standard cosmology is the application of a constant vacuum energy density $\epsilon_{vac} = \rho_{vac}c^2$. Historically and ideologically this originates from Einstein's introduction of a cosmological constant Λ (Einstein, 1917) emanating from application of the variational principle to the spacetime Lagrangian (Overduin & Fahr, 2003), appearing as such on the left, i.e. the "metrical" side of the GRT field equations, however, when transfered to the right side of these equations, is equivalent to a vacuum energy density $\rho_{vac} = c^2\Lambda/8\pi G$, also associated with a vacuum pressure $p_{vac} = -\rho_{vac}c^2$ (e.g. see Peebles & Ratra (2003)). In this form it has experienced a great importance in the present epoch of cosmology (Bennett et al., 2003; Perlmutter et al., 1999).

The problem with this concept of a constant vacuum energy density has already been adressed in the first section of this paper and here can be enlarged to the whole universe: At the expansion of the universe, connected with the increase of the cosmic 3-space volume V^3, consequently the total vacuum energy $E_{vac} = \int \rho_{vac}c^2 dV^3\sqrt{-g^3} \sim \int dV^3\sqrt{-g^3}$permanently increases. This could perhaps even be accepted, if vacuum energy is completely actionless as a cosmologically decoupled quantity with no backreaction to cosmic expansion. As we have shown before, constant vacuum energy density, however, **is** associated with a pressure $p_{vac} = -\rho_{vac}c^2$ that evidently acts on the cosmic expansion accelerating its rate. The purely geometrical increase of cosmic vacuum energy thus is untenable.

This is all the more true when matter density comoves with the cosmic scale expansion to configurations with permanently decreasing gravitational binding. Here it must appear as completely unphysical that an evolving cosmic system, at the same time, gains energy in form of increasing vacuum energy, while simultaneously it has to do work against the internal, intermaterial gravitational attractive forces. For instance for an uncurved universe (i.e. $k = 0$) and Λ put equal to zero, the first Friedmann equation (see Equ. (13)) simplifies into the form $\dot{R}^2 = (8\pi G/3)\rho R^2 = \Phi(R)$ and thus allows to identify a relevant cosmic gravitational potential $\Phi(R)$ in analogy to the one in Hamilton-Lagrangian dynamics (see Fahr & Heyl (2007a;b)). Therefore at the cosmic expansion permanently work has to be done by cosmic matter against an intermaterial force per mass which for $\rho \sim R^{-3}$ is given by

$$f(R) = -\frac{d\Phi}{dR} = \frac{8\pi G}{3}\rho_0 R_0 (\frac{R_0}{R})^2 \tag{15}$$

Instead of loosing energy by permanently doing work $dE/dt = -\dot{R}f(R)$ against this force per time unit, - and instead of decelerating its expansion due to that, the universe may even accelerate its expansion by $\ddot{R} = f(R) + \Lambda Rc^2/3$. With the action of a constant vacuum energy density ($\Lambda = const$) this universe even accumulates more and more energy in form of vacuum energy. This shows that the concept of constant vacuum energy density implies a physically highly implausible "perpetuum mobile" principle: The vacuum permanently acts upon matter and spacetime geometry, but is itself not acted upon by these latter quantities (see Fahr & Heyl (2007a;b), and Figure 1 for illustrative purposes).

This may raise the question whether at present with the form of the standard cosmology one may have a correct basis for a successful description of the given universe and its dynamics. Thus in the ongoing part of this article we shall investigate the following four fundamental, cosmologically relevant critical points:

1. **Is the mass of the universe constant?**
2. **What is metric-relevant cosmic mass density?**
3. **How is gravitational binding energy represented in the energy-momentum tensor?**
4. **How all of that is reflected in a variable vacuum energy density?**

With the arguments given below we demonstrate that an expanding universe with constant total energy, the so-called "economic universe" (also termed as a "coasting universe) is indicated as most probable in which both cosmic mass density and cosmic vacuum energy density are decreasing according to $(1/R^2)$, R being the characteristic scale of the universe. Under these conditions the origin of the present universe from an initially pure cosmic

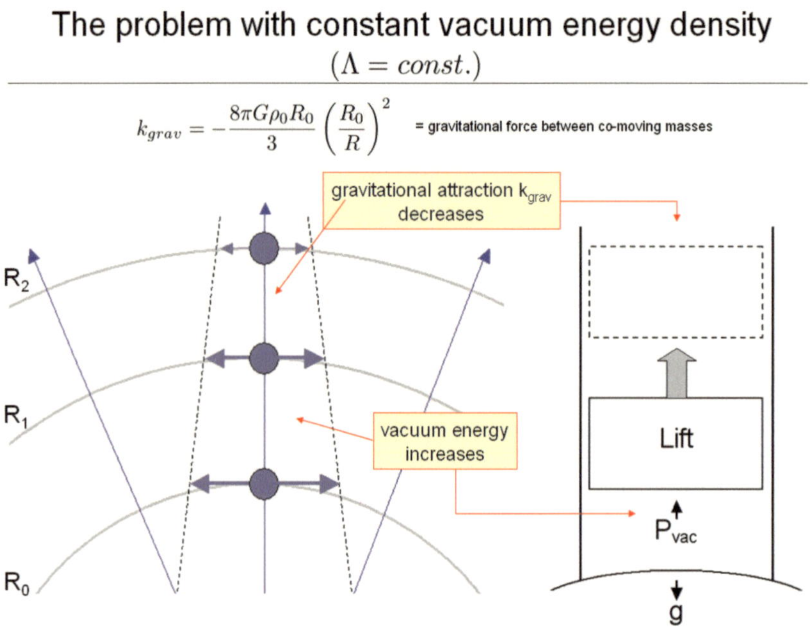

Fig. 1. Schematic illustration of the physical action of a constant vacuum energy density and of inter-material cosmic gravitational fields requiring work to be done, if co-moving matter is transfered to larger cosmic scales $S = R$.

vacuum state appears to be possible. This is because the incredibly huge vacuum energy density, derived by quantumfield theoreticians, in this economic universe decays during its expansion up to present-day scales to just the observationally permitted small value of the present universe, but its energy reappears in the energy density of created effective cosmic matter. It is interesting to see that very similar conclusions concerning the ratio of cosmic vacuum energy and cosmic matter density have been drawn from attempts to formulate the GRT equations in a scale-invariant, Weyl'ian form like recently tried in the Quasi-Steady-State-cosmology (QSSC) by Hoyle et al. (1993), or in conformal cosmological scalar-tensor theories by Mannheim (2000) or by Scholz (n.d.).

5. How to define the mass of the universe?

According to the famous Mach principle (Mach, 1883) inertial masses of cosmic particles are not particle-genuine quantities, but have a relational character being a functional of the spacetime constellation of other cosmic masses in the universe. Only with respect to other masses accelerations have physical relevance (see also Jammer & Bain (2000)). As a consequence, inertial particle masses, and, perhaps in the sense of the general relativistic equivalence principle, also heavy masses, should change their values when the spatial constellation of the surrounding cosmic masses changes - which is the case in an expanding universe with increase of its scale $R = R(t)$. This principle implies that inertia depends in some unclear way on the presence and distribution of other massive bodies in the universe, and has been seriously studied in its consequences (see reviews given in Barbour & Pfister (1995), or Barbour (1995),Wesson (2004),Jammer & Bain (2000)).

In the beginning even Einstein attempted to develop his GR field equations in full accordance to Mach's principle, however, in the later stages he recognized the non-Machian character of his GR theory and divorced from this principle (Holton, 1970). Experts of this field still today have controversial opinions whether or not Einstein's GR theory is "Machian" or "non-Machian". Nevertheless attempts have been made to develop an adequate form of a "relational", i.e. Machian mechanics (Goenner, 1995; Reissner, 1995). Especially the requested concrete scale-dependence of cosmic masses is unclear in its nature, though perhaps already suggested by conformal invariance requirements or general relativistic action principle arguments given by early arguments developed in Hoyle (1990; 1992); Hoyle et al. (1994a;b) along the line of the general relativistic action principle.
We study this relation a little deeper here starting from the question what at all should and could be called in a physically relevant, conceptually meaningful sense "the mass of the universe M_u" and how then it could be understood, if this quantity increases with the universal scale R**?** According to the most logical concept, this mass M_u should represent the spacelike sum over all masses distributed in the universe at some event of time, judged from some arbitrary cosmic vantage point, i.e. the space-like sum of all masses within the mass horizon associated to this point. One way to define such a quantity has been mathematically carried out by Fahr & Heyl (2007b) and leads to the following mathematical expression of cosmic mass

$$M_u c^2 = 4\pi\rho_0 c^2 \int_0^{R_u} \frac{\exp(\lambda(r)/2) r^2 dr}{\sqrt{1-(\frac{H_0 r}{c})^2}} \tag{16}$$

where the function in the numerator of the integrand is given by the following metrical expression

$$\exp(\lambda(r)) = \frac{1}{1 - \frac{8\pi G}{rc^2}\rho_0 \int_0^r \frac{x^2 dx}{\sqrt{1-(\frac{H_0 x}{c})^2}}} \tag{17}$$

The reason behind this above expression is that the environment around an arbitrary vantage point is described analogous to a point in the center of a star surrounded by stellar matter distribution, the difference in this case being only that the metric in this cosmic case also is of the inner Schwarzschild form, however, with the matter density given by the cosmic density ρ_o taking into account the additional fact that matter in the surroundings of a homologously expanding universe is equipped with the Hubble dynamics of the expanding universe.
As evident from the above expression no real matter can be summed-up anymore from beyond the "local Schwarzschild infinity" (i.e. "point-associated Schwarzschild mass horizon", see Fahr & Heyl (2006)) which is at a distance

$$R_u = \frac{1}{\pi}\sqrt{\frac{c^2}{2G\rho_0}} \tag{18}$$

which, however, also means that the mass horizon distance is related to the cosmic mass density by

$$\rho_0(R_u) = \frac{c^2}{2\pi^2 G R_u^2} \tag{19}$$

and naturally leads to a point-associated mass of the universe given by Fahr & Heyl (2006)

$$M_u = \frac{3\pi c^2}{8G} R_u \tag{20}$$

This scale dependence of cosmic mass, does not only point to the fact that Mach's relation is fulfilled for the mass of the universe in the above definition of M_u. It in addition also proves that Thirring's relation derived from a completely different context (see Mashhoon et al. (1984), and also Fahr & Zoennchen (2006)) in the form

$$M_u = \frac{3c^2 R_u}{4G} \tag{21}$$

is also fulfilled up to the factor $(\pi/2)$.

6. Gravitational binding energy reflected in an effective mass density

In a completely different approach Fischer (1993) may be giving from a new aspect of physics an explanation for this change of cosmic mass M_u with scale R coming to conclusions very similar to the above ones. He makes an attempt to include the gravitational binding energy into the energy-momentum tensor $T_{\mu\nu}$ of the GRT field equations. Interestingly enough his derivations lead to the result, that in a positively curved universe the corresponding term for the binding, or potential energy density $T^p_{\mu\nu}$ has to be introduced into the GRT equations by

$$T^p_{\mu\nu} = -C\frac{\rho}{\Gamma} g_{\mu\nu} \tag{22}$$

where $g_{\mu\nu}$ denotes the metric tensor, C is an appropriately defined constant which amongst other factors contains the gravitational constant G, and Γ is the actual curvature radius of the positively curved universe.

In this formulation two things are perhaps eye-catching: At first this term again contains a proportionality to the density ρ , and at second this term has a negative sign and has $g_{\mu\nu}$ as a factor, thus in the GRT field equations formally it has the same action as that term connected with the action of vacuum energy density formulated with the quantity Λ_{eff}. This points to an interesting physical connection between vacuum energy and gravitational binding energy. Obtaining its space-like components as vanishing and adding up the time-like tensor components T_{00} and T^p_{00} of cosmic matter und cosmic binding energy then shows a very surprising connection between creation of matter and binding energy given in the form

$$\hat{T}_{00} = T_{00} + T^p_{00} = (\rho - C\frac{\rho}{\Gamma}) g_{00} \tag{23}$$

This can thus be interpreted as saying that the intermaterial, gravitational binding energy reduces the cosmologically, i.e. geometrically acting, relevant, effective cosmic matter density to $\rho^* \leq \rho$, where ρ should be called the "proper density" given in uncurved spacetimes, by the following amount

$$\rho^* = \rho(1 - C\frac{1}{\Gamma}) \tag{24}$$

If in the course of the cosmic expansion the cosmic curvature radius Γ increases, it thus means that gravitational binding energy, and, equivalent to that, the cosmic vacuum energy should decrease, while at the same time the effective density changes in time in a Machian form with a rate

$$\dot{\rho}^* = \frac{d}{dt}[\rho(1 - C\frac{1}{\Gamma})] \tag{25}$$

It is perhaps interesting to recognize that for instance for a universe with Hoyle's "steady state requirement", i.e. with $d\rho/dt = 0!$, this then evidently would require

$$\dot{\rho}^* = \rho C \frac{1}{\Gamma^2}\dot{\Gamma} \tag{26}$$

This means a mass creation rate proportional to the matter density ρ itself which is positive for increasing cosmic curvature radius Γ. In other words: At decreasing cosmic binding energy the effective density increases by the rate $\dot{\rho}^*$which , as will be shown further down in this paper, is identical to that one obtained by Hoyle (1948).
It is interesting to notice that an introduction of the gravitational binding energy according to the suggestion by Fischer (1993) leads to two differential equations that can be combined to

$$\ddot{S} = \frac{C\rho_c}{6\Gamma}(S_0 - S) \tag{27}$$

which leads to cosmological solutions for positively curved universes representing an oscillatory behaviour of the cosmic scale parameter R around an equilibrium value R_0 with positively valued ($R \leq R_0$) und negatively valued ($R \geq R_0$) vacuum energy densities in the successive half-phases of the oscillation. It is perhaps challenging to conjecture that the action of vacuum energy, binding energy and creation of effective matter density could be closely related to eachother and perhaps even be identical.

A similar connection between vacuum energy and mass density was also pointed out by **?** who showed that the cosmological term connected with the quantity Λ should be coupled to matter density ρ and, concretely spoken, should in fact be proportional to it.
The problem of what should be called cosmic matter density thereby is by far not a trivial one, because the "matter density" is intrinsically connected with the prevailing spacetime geometry. The latter, however, only aposteriori is obtained from solutions of the GRT field equations after putting the right mass density into the energy-momentum tensor. The usual definition of matter density as "mass per unit volume" is in fact problematic in curved spaces. Usually the density is identified with what one should call the "proper density", i.e. mass within a free-falling unit volume, i.e. within a reference system without internal tidal gravitational accelerations. Of course in the universe one finds co-moving inertial restframes, nevertheless even in such systems tidal accelerations are acting over finite dimensions of a Finite 3d-space volume, causing for a metrical distortion of unit volumes. The effect of this metrical distortion reduces the proper density ρ as has been discussed by Fahr & Heyl (2007a;b) and for the low-density limit $\rho_0 \ll \rho_c$ (with ρ_c denoting the Schwarzschild density on a scale $R_{ES}(M) = \sqrt[3]{3M/4\pi\rho_0}$ (see Einstein and Straus, 1945) given by $\rho_c = (3/4\pi)(c^2/2G)^3 M^{-2}$) also leads to a reduction of the proper density given by an expression

$$\rho^* = \rho_0(1 - (\rho_0/\rho_c)^{1/3}) \tag{28}$$

7. Effective mass change as equivalence to cosmic mass generation

Early attempts to describe universes with mass creation like those presented by Hoyle (1948) show very interesting relations between this form of matter creation and the change of effective cosmic matter density. To describe a steady-state universe Hoyle (1948) introduced

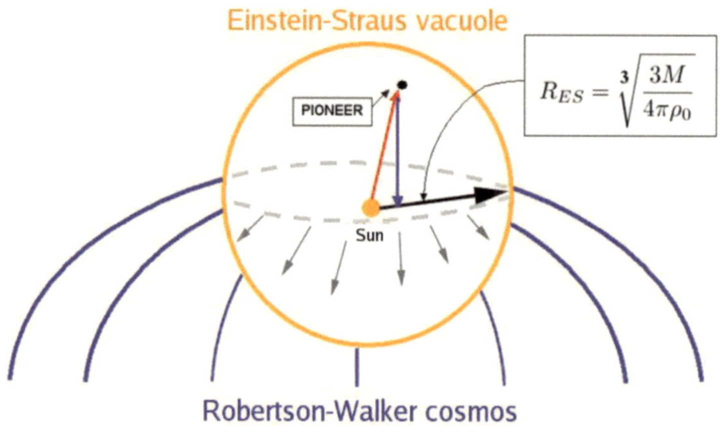

Fig. 2. Visualization of the Einstein-Straus globule surrounding a mass M within the expanding Robertson-Walker universe.

a divergence-free mass-creation tensor $C_{\mu\nu} = -3R\dot{R}\delta_{\mu\nu}/cA$ into the GR field equations, with A being a constant curvature scale. With the introduction of this term he can describe a universe with constant mass density $\rho = \rho_0 = const$, an inflationary expansion $R = R_0 \exp[c(t - t_0)/A]$, and a mass creation rate given by $\dot{\rho} = \frac{c}{A}\rho_0$. As we have recently shown (Fahr & Heyl, 2007a;b) an identical inflationary expansion is also described by an Einstein-de Sitter cosmological model of an empty universe, however, under the action of a cosmological constant Λ. This is true, if this constant Λ is related to Hoyle's creation rate by

$$\Lambda^{3/2} = \frac{8\pi G\sqrt{3}}{c^5}\dot{\rho} \tag{29}$$

This points to the fact that cosmologically analogous phenomena can be described by the action either of mass creation $\dot{\rho}$ or of a cosmological constant $\Lambda = 8\pi G\rho_{vac}/c^2$, i.e. by a vacuum energy density. It may furthermore be of interest to recognize that Hoyle's creation rate automatically leads to the fulfillment of a quasi-Machian relation between mass and radius of the universe, which has already been mentioned before, and here reappears from this context in the form

$$M_u = M_{u0}\exp[\frac{c(t-t_0)}{A}]^3 = M_{u0}\left(\frac{R(t)}{R_0}\right)^3 \tag{30}$$

The above analysis came along the early mass-creation theory published by Hoyle (1948). This early theoretical approach has, however, been consequently extended by Hoyle and his co-workers and has meanwhile been put into a larger astrophysical framework (see Hoyle et al. (1993; 1994a;b; 1997) where individual strong gravity centers in an expanding universe are considered that act as centers of mass creation called "Quasi-steady state cosmologies" (QSSC-models). Later in this paper we discuss these QSSC-models in a broader context, since these models are connected with more general scale-invariance requirements in the GRT field equations. We want, however, to emphasize already here that the above-revealed evidence (29), here derived from Hoyle's early creation theory and revealing a close relation between mass creation rate, vacuum energy density and actual cosmic mass density, is again equally retained in these later QSSC-models as we shall show later in this paper.

8. Mass increase on local scales

According to Einstein & Straus (1945) a locally realized mass M is surrounded by a spherical shell with a radius $R_{ES}(M) = \sqrt[3]{3M/4\pi\rho_0}$. At this shell surface a steady and differentiable transition from the inner Schwarzschild metric into the outer Robertson-Walker metric of a homologously expanding universe is possible. This also implies that spacepoints on the Einstein-Straus shell are expanding with respect to the center of the shell as Robertson-Walker spacepoints do, i.e. like

$$\dot{R}_{ES}/R_{ES} = \dot{R}_0/R_0 = H_0 \tag{31}$$

with H_0 denoting the Hubble constant.
Adopting vacuum energy as being ubiquitously active in the universe one can ask, what amount of work the pressure connected with this vacuum energy does at the expansion of the local Einstein-Straus globule. For the inside of this globule this work is positively valued, and due to energy conservation reasons, it should thus lead to an increase of the energy constituted by this globule. Ascribing this energy gain to the internal mass of the globule then delivers the interesting result (Fahr & Heyl, 2007a;b)) that

$$\frac{\dot{M}}{M} = \frac{\rho_{0,vac}}{\rho_{0,mat}} H_0 \tag{32}$$

where $\rho_{0,vac}$ and $\rho_{0,mat}$ denote the densities of the present mass equivalent of the vacuum energy and of the cosmic matter. For a constant ratio of these energy densities the above relation simply expresses, - since $\dot{M}/M \sim \dot{R}/R$ - (i.e. the economical universe, see further down), a proportionality of the globular mass M, - and, if generalized to the scale R_u, of the mass M_u of the universe - , with the radius in the form

$$M/R_{ES}(M) \sim M_u/R_u = const \tag{33}$$

again as already envisioned by Mach (1883), but here proven as being valid also on local scales.

9. Why structure formation accelerates the cosmic expansion rate

Here we want to start with an easyminded exercise showing that gravitational structure formation in the universe may have the quite unexpected tendency to accelerate, like a force would do, the Hubble flow velocity, a virtue that is nowadays all over in the astrophysical literature ascribed to the action of the vacuum pressure p_{vac}. Let us assume that structure formation has developed at some epoch of cosmic evolution to some organized state such that not anymore a homogeneous matter density distribution prevails, but instead a homogeneous distribution of hierarchically organized matter distribution. From galactic number count statistics one knows that this expresses itself in observed local two-point correlation functions $\xi(l)$ expressing the probability to find another galaxy at a distance l from the local space point. For completely homogeneous matter distribution the function ξ would be constant. In cosmic reality, however, this two-point correlation probability over wide ranges of scales is shown to fall off by

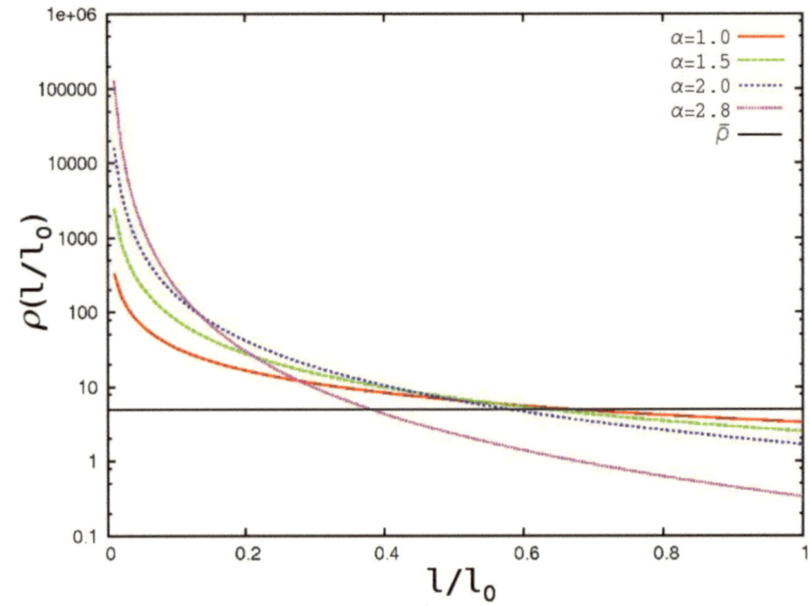

Fig. 3. Dependence of ρ for different values of α. The black solid line represents the case of a homogeneous density $\bar{\rho}$

$$\xi(l) = \xi(l_0) \cdot (\frac{l_0}{l})^{\alpha} \tag{34}$$

with the power index $\alpha \simeq 1.8$ and some inner scale l_0 typical for galaxies (see Bahcall (1988); Bahcall & Chokshi (1992)). In terms of matter density this expresses the fact that cosmic matter distribution has been organized, so that the mean density has not changed, but a density clustering has appeared at each local environment. This clustering is associated with a more pronounced gravitational binding of this organized matter, i..e. more negative potential energy has developed during the process of structuring.

To calculate the latter we start from a local density distribution corresponding to the probability function given by Eqn.(34) and write the clustered density in the form $\rho(l) = \rho_0$ $(l/l_0)^{-\alpha}$. In order to conserve the initial mass at the structuring process the central density ρ_0 has to be defined as

$$\rho_0 = \frac{3-\alpha}{3}\bar{\rho} \cdot (l_m/l_0)^{\alpha} \tag{35}$$

with l_m as an outer integration scale. Figure 3 shows the dependence of $\rho(l)$ on the power-law index α.

Now the potential energy of this organized, clustered matter can be calculated according to Fahr and Heyl (2007b)

$$\epsilon_{pot} = G\rho_0^2 l_0^5 \int_1^{x_m} 4\pi x^2 dx x^{-\alpha} \frac{1}{x} \int_1^{x} 4\pi x'^2 dx' x'^{-\alpha} \tag{36}$$

where the normalized distance scale has been defined by $x = l/l_0$. Thus one obtains

$$\epsilon_{pot} = (4\pi)^2 G\rho_0^2 l_0^5 \int_1^{x_m} x dx x^{-\alpha} [\frac{1}{3-\alpha}(x^{3-\alpha} - 1)] \tag{37}$$

which leads to

$$\epsilon_{pot} = \frac{(4\pi)^2}{3-\alpha} G\rho_0^2 l_0^5 \int_1^{x_m} dx[(x^{4-2\alpha} - x^{1-\alpha})] \tag{38}$$

and

$$\epsilon_{pot} = \frac{(4\pi)^2}{3-\alpha} G\rho_0^2 l_0^5 \left| \frac{x^{5-2\alpha}}{5-2\alpha} - \frac{x^{2-\alpha}}{2-\alpha} \right|_1^{x_{ma}} \tag{39}$$

which when taking $x_m \gg 1$ leads to

$$\epsilon_{pot} = \frac{(4\pi)^2}{3-\alpha} G\rho_0^2 l_0^5 x_m^{5-2\alpha} \left(\frac{1}{5-2\alpha} - \frac{x_m^{-3}}{2-\alpha}\right) \simeq \frac{(4\pi)^2}{(3-\alpha)(5-2\alpha)} G\rho_0^2 l_0^5 x_m^{5-2\alpha} \tag{40}$$

and reminding the requirement $\rho_0 = \frac{3-\alpha}{3}\bar{\rho} x_m^{\alpha}$ finally leads to

$$\epsilon_{pot} = \frac{(4\pi)^2(3-\alpha)}{9(5-2\alpha)} G\bar{\rho}^2 l_0^5 x_m^5 \tag{41}$$

Now it is interesting to recognize that for $\alpha = 0$ (i.e. homogeneous matter distribution) in fact again the potential energy of a homogeneously filled sphere with radius l_m is found, namely $\epsilon_{pot}(\alpha = 0) = \frac{(4\pi)^2}{15} G\bar{\rho}^2 l_m^5$ (see Fahr and Heyl, 2007). $\epsilon_{pot}(\alpha = 0)$ serves as reference value for the potential energy in the associated re-homogenized universe.

9.1 A one-dimensional analogue

Now imagine a one-dimensional, unidirectional cosmological matter flow as an easy-minded representation of the cosmic Hubble-flow, then one should trust the validity of the following set of equations due to mass-, momentum-, and energy-flow conservation

$$\rho U = \Phi_1$$

$$\rho(\dot{U} + U\frac{d}{dz}U) = F$$

$$\rho U \left(\frac{U^2}{2} + \bar{\epsilon}_{pot}\right) = \Phi_2$$

Here Φ_1 and Φ_2 denote constant mass and energy flows, U is the flow velocity and $\bar{\epsilon}_{pot} = \epsilon_{pot}/(4\pi\rho l_m^3/3)$ denotes the potential energy per mass. F is a force per volume that we want to find, but do not know yet. Now, neglecting explicit local time-dependence (i.e. $\dot{U} = 0$) one finds from the third equation

$$\left(\frac{U^2}{2} + \bar{\epsilon}_{pot}\right) = \Phi_2/\Phi_1 = const \tag{42}$$

which leads to

$$\frac{d}{dz}\left(\frac{U^2}{2} + \bar{\epsilon}_{pot}\right) = \frac{F}{\rho} - \frac{d}{dz}\left[\frac{(4\pi)(3-\alpha)}{3(5-2\alpha)} G\bar{\rho} l_m^2\right] = 0 \tag{43}$$

Describing the ongoing of cosmic structuring purely by a change in time of the power index α, this then delivers the interesting result

$$\frac{F}{\rho} = \frac{d}{dz}\left[\frac{(4\pi)(3-\alpha)}{3(5-2\alpha)}G\bar{\rho}l_m^2\right] = -\frac{4\pi}{3}G\rho l_m^2\frac{3-2\alpha}{(5-2\alpha)^2}\frac{d\alpha}{dz} \quad (44)$$

expressing the fact that for values $\alpha \geq 1.5$ further increase of the structuring index α manifests a positive force F that accelerates the cosmic mass flow. For us this seems the first time it has been shown that gravitational structuring in a moving cosmic flow implies an acceleration of the flow velocity, inditcating that analogously in an expanding universe this might aswell induce an acceleration of the cosmic expansion as usually ascribed to the action of vacuum-energy.

9.2 Structured universes

An independent consideration perhaps points into the same direction as derived above allowing to conclude that cosmic binding energy acts as if it would reduce the effectively gravitating matter density, hence like a form of positive vacuum energy density. It namely turns out that a structured universe expands differently from a homogenized universe with identical total mass (see Buchert (2001; 2005; 2008); Räsänen (2006); Wiltshire (2007); Zalaletdinov (1992)). Quantitatively this was especially shown by Wiltshire (2007) for a 2-phase toy-model of the universe representing the distribution of cosmic matter in form of non-homologously expanding low-density voids and high-density walls. Describing for this purpose this cosmic matter structure by so-called volume-filling factors f_v and f_w and defining the phasestructure densities by $\rho_{v,w} = \int_{V_{v,w}} d^3x\sqrt{\det^3 g}\rho(t,\vec{x})/V_{v,w}$ with $V_{v,w}$ denoting the void- and wall-volume respectively, one obtains the following relation

$$\bar{\rho}_2 = \rho_v f_v + \rho_w f_w = \rho_v f_v + \rho_w(1-f_v) \quad (45)$$

Introducing typical phase scales $R_{v,w}$ and describing their temporal variations with phase-averaged GRT field equations, one obtains the phase densities for the voids and the walls, respectively, as given by

$$\rho_v = \bar{\rho}_2(R/R_v)^3 \quad (46)$$

and:

$$\rho_w = \bar{\rho}_2(R/R_w)^3 \quad (47)$$

Reminding that the acceleration parameter, generally defined by $q = -\ddot{R}R/\dot{R}^2$, for the homogenized, above mentioned 2-phase universe turns out to be obtainable in the following form (Wiltshire, 2007)

$$\bar{q}_2(f_v) = \frac{-(1-f_v)(8f_v^3+39f_v^2-12f_v-8)}{(4+f_v+4f_v^2)^2} \quad (48)$$

then proves that in a globally uncurved universe the structure function f_v causes a term in the GRT field equations which is analogous to that describing the action of a vacuum energy density ρ_{vac} of the value

$$\rho_{vac} = \frac{\bar{\rho}_2(1-2\bar{q}_2)}{2(\bar{q}_2+1)} \quad (49)$$

This shows that in a nearly void-dominated universe, i.e. with $f_v \simeq 1$ and $\bar{q}_2(f_v \simeq 1) = 0$, one would find a well-tuned constant expansion dynamics (i.e. a "coasting universe"; Fahr

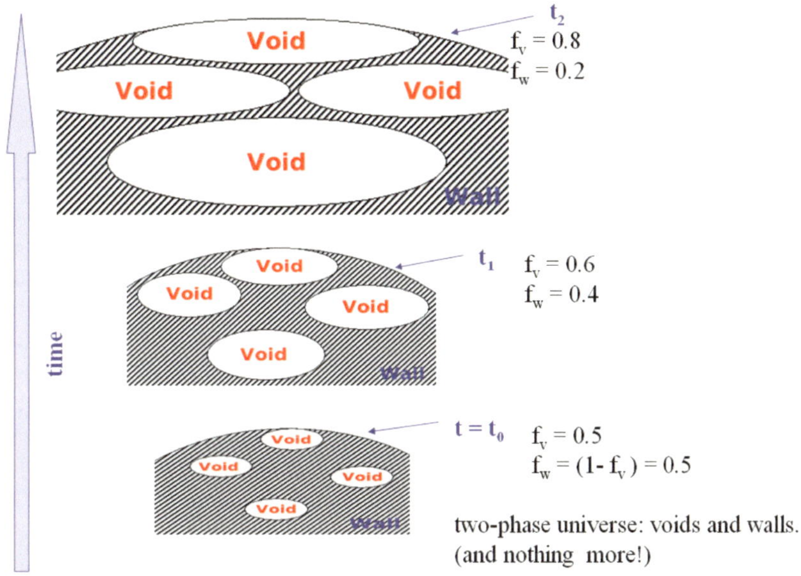

Fig. 4. Illustration of the non-homologous expansion of a two phase universe with void and wall regions having different matter densities.

& Heyl (2007a); Fahr (2006); Fahr & Heyl (2007b); Kolb (1989)) analogous to the action of a vacuum energy density given by $\rho_{vac}(f_v \simeq 1) \simeq (1/2)\bar{\rho}_2$. For phase-structures as they may come up during the non-homologous expansion of the two-phase universe (i.e. with $\dot{R}_w \leq \dot{R}_v$) characterized by a structure function $f_v \geq f_{vc} = 0.57$, where f_{vc} denotes the critical void-volume fill factor $\bar{q}_2$ changes its sign and one obtains $\bar{q}_2 \leq 0$, i.e. an accelerated expansion of the universe which is conventionally ascribed to the action of a vacuum energy $\rho_{vac}(f_v \geq f_{vc}) \geq \bar{\rho}_2/2$. In these phases, one could as well state it like that, the average density $\bar{\rho}_2$ in such a universe appears to be reduced to an effective density given by

$$\bar{\rho}_2(f_v \geq f_{vc}) = \bar{\rho}_2 - \bar{\rho}_{vac}(f_v \geq f_{vc}) = \bar{\rho}_2(1 - \frac{1-2\bar{q}_2}{2(\bar{q}_2+1)}) \tag{50}$$

This shows that in that phase of non-homologous structure evolution characterized by $f_v \geq f_{vc} = 0.57$ the average cosmic density appears to be reduced by more than 50 percent due to gravitational binding energies sitting in the wall-structured, dense matter formations.

Some caution, however, in advertizing this result too much, is perhaps in place. This is due to the fact that Wiltshire in his analysis starts out from the scalar differential equations given by Eqns. (13) and (14) and in these only treats cosmic averages of the remaining scalar quantities $R = g^{ij}R_{ij}$, denoting the Riemann scalar as contraction and the Ricci tensor R_{ij} by the metric tensor g^{ij}, and ρ. Thereby it turns out that when going back from his 2-phase universe to an averaged homogeneous replace-universe some back-reaction terms $Q = Q(\langle\langle\rho\rangle\rangle,\langle\langle R\rangle\rangle)$ are obtained, entering the two scalar differential equations of the Einstein field equations, which are left from the homogenization. A correct treatment of spacetime inhomogeneities would, however, require the calculation of 'back-reaction' terms starting from the level of nonlinear, second-order partial differential equations coming from the tensor formulation of the GRT field equations. This calculation has up to now not been carried out, and thus Wiltshire's results should at present not be over-emphasized, but taken with some scepticism (Buchert, 2008).

10. The universe as energy-less system

Is it imaginable that the universe, enormously large and extended as it is, nevertheless does not represent huge amounts of energy, to the contrary perhaps is a system of vanishing energy. If not representing any real, countable energy, it then might be understandable that such a universe, despite its evolution, can actually even originate from nothing, since permanently constituting nothing. But how can all what we see in the universe, when added up, represent a vanishing amount of energy?
This could in fact be possible, because in physics one knows that there exist positively and negatively valued energies, so that their sum can cancel. If all the positively valued energies in the universe accumulate to E and the negatively valued energies , i.e. the gravitational binding energies in the universe, accumulate to U , then it might turn out that the sum of both, i.e. $L = E + U$, vanishes. In the following we shall show that the "$L = 0$" - universe is actually possible, if matter density and vacuum energy density vary in specific forms with the scale of the universe.
As we have shown in Fahr and Heyl (2007a/b) the total energy $E = E(R)$ of an uncurved universe can be calculated as the spacelike sum over all energies given by the following expression

$$E(R) = \int^{V^3} (\hat{\rho}c^2 + 3\hat{p})\sqrt{-g_3}d^3V = \frac{4\pi}{3}R^3(\hat{\rho}c^2 + 3\hat{p}) \tag{51}$$

For a complete sum all mass densities have been subsummed by the quantity $\hat{\rho}$ which comprehends baryonic matter, dark matter and vacuum equivalent mass density, i.e. is given in the form $\hat{\rho} = \rho_b + \rho_d + \rho_{vac}$, as well all pressures constituting energy densities are subsummed by the quantity $\hat{p} = p_b + p_d + p_{vac}$. As one can see from the above expression, the total energy $E(R)$ is proportional to R^3.
In that phase of the universe which we try to energetically balance here pressures of baryonic and dark matter may be assumed to be negligible with respect to their corresponding rest mass energy densities. In addition, a polytropic relation between ρ_{vac} and p_{vac} can be used in the form

$$p_{vac} = -\frac{(3-n)}{3}\rho_{vac}c^2 \tag{52}$$

since for the most general case a scale-dependent vacuum energy density in the form $\rho_{vac} \sim R^{-n}$ must be admitted (see Fahr and Heyl, 2007b).
In a similar way one can also calculate the total gravitational binding energy $U(R)$ in this universe as the spacelike sum over the total potential energy and obtains the following expression

$$U(R) = \int_0^R 4\pi r^2(\rho_b + \rho_d + (n-2)\rho_{vac})\Phi(r)dr \tag{53}$$

where $\Phi(r) = -(2/3)\pi G(\rho_b + \rho_d + (n-2)\rho_{vac})r^2$ is the internal cosmic gravitational potential. This then leads to

$$U(R) = -\frac{8\pi^2 G}{15}(\rho_b + \rho_d + (n-2)\rho_{vac})^2 R^5 \tag{54}$$

Now the No-energy-requirement $L = E + U = 0$ simply leads to the following relation

$$\frac{3c^2}{2\pi G R^2} = (\rho_b + \rho_d + (n-2)\rho_{vac}) \tag{55}$$

with n being the unknown polytropic constant in the relation between vacuum pressure and vacuum mass density $p_{vac} = -\frac{(3-n)}{3}\rho_{vac}c^2$. As evident from the above relation, the requirement $L = 0$ is only fulfilled, if all mass densities in the universe scale as R^{-2}, identical to the scale-dependence already derived at different places and within different contexts presented further above in this article. The pressing question, how this mass creation could be explained, can now easily be answered on the basis of the above deduced context, namely because now vacuum energy density, different from the assumptions in the standard cosmology, is not anymore taken as constant, but turns out to be variable and decaying at the expansion of the universe with $\rho_{vac} \sim R^{-2}$ with the selfsuggesting solution $\dot{\rho}_{vac} \propto \dot{\rho}$. The most encouraging point in this view now is that the universe can start from a Planck volume V_{pl} with a Planck scale $R = r_{pl} = \sqrt{Gh/2\pi c}$ with the initial vacuum energy density of $\rho_{vac}(r_{pl}) = m_{pl}/(4\pi r_{pl}^3/3)$ (just the value calculated by field theoreticians) and then only later at our present epoch has dropped down to the accepted astrophysical values of the present universe corresponding to $\rho_{vac,0} = 0.73\rho_{c,0} \simeq 10^{-29} g/cm^3$ (see Fahr and Heyl, 2007b).

11. Discussion and outlook

We would like to finish this article reminding the readers to a series of more recent papers in which the conclusion of a scale-variability of cosmic masses, reached in this paper here, also is drawn, however, from quite independent theoretical views connected with general symmetry or invariance principles valid in a generalized form of Einstein's general relativistic field theory. The latter theory is not conformally scale-invariant as was emphasized by Hoyle (1990; 1992). Einstein's field equations can be derived from a variational principle applied to the following universal action function

$$S_{0,1} = -\sum_a \int_1^2 m_a da + \frac{1}{12} M_p^2 \int_1^2 \int R\sqrt{-g} d^4x \tag{56}$$

where the Planck mass has been defined by:

$$M_p = \frac{3ch}{4\pi G} = 1.06 \cdot 10^{-6} g \simeq 10^{19} \frac{GeV}{c^2} \tag{57}$$

Here m_a and da are the masses and worldline increments of the particles in the universe, and R and g are the Riemann scalar and the determinant of the metric tensor g_{ik}. The quantity d^4x is the differential 4D spacetime volume element. As Hoyle pointed out, if one measures the action in units of the Planck constant h, and all velocities in units of the velocity of light c, then masses attain the dimension $[1/L]$ where L is a cosmic length scale. Hoyle furtheron emphasizes in his articles - Maxwell's theory, quantum theory and Dirac's theory - they are all conformally invariant, but Einstein's theory is not.

Conformal invariance (invariance with respect to local scale-recalibrations) according to H. Weyl should also be fulfilled by the theory of general relativity. Following this conceptual view of Weyl (1961) also the field theory like GRT should fulfill conformal scale-invariance. This requirement when connected with the general request of the minimum action principle then as can be seen from Equ. (3) automatically requires that mass is created at geodetic motions of comoving cosmic masses. To respect these theoretical prerequisites would mean that the field equations should be invariant with respect to local recalibrations of the worldline element according to:

$$da^{*2} = \Psi^2(\overrightarrow{A})g_{ik}(\overrightarrow{A})da^i da^k = L(\overrightarrow{A})^{-2}da^2 \tag{58}$$

This is now only fulfilled in connection with the cosmic action minimum, if at the same time where the above relation holds the masses in the universe do also scale by:

$$m_a^* = m_a \frac{1}{\Psi(\overrightarrow{A})} = L(\overrightarrow{A})m_a \tag{59}$$

Taking creation of matter as concequence of a scale-invariant GRT action principle Hoyle et al. (1993) have developed their Quasi-Steady-State cosmology (QSSC) deriving a scalar mass creation field $C(X)$ which is obtained as solution of a wave equation given by

$$\Box_X C(X) + \frac{1}{6}R(X)C(X) = f^{-1} \sum_{A_0} \frac{\delta_4(X - A_0)}{\sqrt{-g(A_0)}} \tag{60}$$

where $\Box_X$ is the 4-d Laplace operator, X denotes a 4-d spacetime point, $R(X)$ is the Riemannian scalar at X, and A_0 are 4-d spacetime positions of real particles in the universe. The function f is needed as a positive coupling constant. At the place of a particle A_0 one obtains the gradient components of the creation field by

$$C_i(A_0) = \left[\frac{\partial C(X)}{\partial x^i}\right]_{A_0} \tag{61}$$

and is lead to a scalar mass creation bound by the relation

$$\frac{\partial}{\partial t}\left(C_i C^i\right)_{A_0} = \dot{m}_a^2(A_0) \tag{62}$$

where m_a is the mass of the particle at A_0. As the authors analyse further down in their article (Hoyle et al., 1993) creation of field bosons can only occur in connection with massive particles at places A_0 , and becomes effective only where strong gradients of the $C(X)-$ fields due to strong Riemannian scalar curvatures $R(X)$ are established in the universe, i.e. near already existing strong mass concentrations. A steady-state form of creation, like that required by Hoyle (1948), under these restricting auspices is unlikely. Mass generation in this QSSC does only happen when particles come close to cosmic mass concentrations or cosmic black holes. But from localized creation rates an average cosmic creation rate $\langle \dot{C}^2 \rangle_4$ can be derived which then instead of Eqns. (1) and (2) can be brought into the field equations of QSSC yielding the following form

$$\left(\frac{\dot{R}(t)}{R(t)}\right)^2 = \frac{8\pi G}{3}\rho(t) - \frac{kc^2}{R^2(t)} + \frac{\Lambda c^2}{3} - \frac{4\pi G}{3} f \left\langle \dot{C}^2 \right\rangle_3 \tag{63}$$

and:

$$\frac{\ddot{R}(t)}{R(t)} = -\frac{4\pi G}{3c^2}(3p(t) + \rho(t)c^2) + \frac{\Lambda c^2}{3} + \frac{8\pi}{3} G f \left\langle \dot{C}^2 \right\rangle_3 \tag{64}$$

This system of equations has been solved by Sachs et al. (1996) in the following form

$$R_{QSSC}(t) = \exp[(\frac{t}{P})\{1 + \eta \cos\theta(t)\}] \quad (65)$$

where P is a constant and $\theta(t)$ is a known periodic function with a period $Q \ll P$ and $\eta \leq 1$ as a constant parameter. It turns out that the envellope of the above solution behaves like a solution of the standard cosmology, however, with a vacuum energy density given by

$$\Lambda_{QSSC} = -6\pi G f \left\langle \dot{C}^2 \right\rangle_3 \quad (66)$$

The above demonstrates that QSSC cosmological theories, taking general-relativistic scale invariance as a serious request, will automatically lead to cosmic mass creation and to a fake form of negative vacuum energy density.

There are also recent studies by Mannheim (2001; 2003; 2006) in the literature which point into a similar direction. Mannheim (2006) investigates the logical independence of the general covariance principle, the equivalence principle and the Einstein GRT field equations and manifests several restrictions in the present-day formulation of the energy-momentum tensor which can shed light to why at present the standard cosmology is in troubles. As we do in this article here, he also argues that to solve the outstanding present-day cosmological constant problem with the enormous discrepancy of field-theoretical and astrophysically admittable vacuum energy density, it is not necessary to quench the vacuum energy term itself, but only to find out, by what amount the vacuum energy actually gravitates. His answer is going into the same direction than the one given in this article here culminating in the claim that most of the field-theoretical vacuum energy does not gravitate since it is just compensated by the action of the cosmological constant Λ leading to the fact that for empty space $\Lambda_{eff,0} = 0!$. The gravitationally relevant part of vacuum energy only is due to the matter-polarized vacuum.

To reach this conclusion he carefully checks all the ingredients of all terms on the RHS and LHS of the Einstein GRT equations. He identifies, as one of problems, the conventional formulation of the energy-momentum tensor T_{ik} based on the assumption of geodetic motions of massive, singular particles with invariant masses m which first leads to the expression

$$T^{ik} = \frac{mc}{\sqrt{-g}} \int d\tau \cdot \delta^4(x - y(\tau)) \frac{dy^i}{d\tau} \frac{dy^k}{d\tau} \quad (67)$$

which is covariantly conserved and systematically leads to the corresponding hydrodynamical expression for T^{ik} that is generally used in present-day cosmology. This formulation is used despite the modern understanding that particles are far from being kinematic objects with invariant masses, but are thought to realize their masses dynamically by means of spontaneous symmetry breaking, and despite the fact that the standard SU(3)xSU(2)xU(1) - field unification theory ascribes the basic level of material energy representation to scalar wave fields rather than to particles. The variational principle, if applied to the scalar wave action, then leads to the following equation of motion for the scalar wave field S given by

$$S^{;\mu}_{;\mu} + \frac{\xi}{6} S R^{\mu}_{\mu} - m^2 S = 0 \quad (68)$$

This equation is very similar to the one derived by Hoyle et al. (1993), except that in the latter the mass creation is connected with the existing particle motions.

Mannheim discusses several possibilities to change Einstein's GRT equations in order to absorb the concept of dynamical masses from field theoretical considerations as discussed above. Seeking, however, for alternatives to Einstein's GRT equations by looking for generalizations, one should always take care that in these generalizations the Einstein equations are contained as a special case. Amongst the general covariant pure metric theories of gravity the most convincing generalization, as it appears to Mannheim, is to complement the Einstein Hilbert action by additional coordinate-invariant pure metric terms which, in the Newton limit, do not perturb the validity of Newtons gravity on the scale of the solar system. Also he discusses additional macroscopic gravitational fields as a company of the metric tensor g_{ik}. Here the most suggestive step would be to introduce scalar fields. As also taken up by Scholz (n.d.), the idea from H.Weyl to start from conformal gravity theories is discussed by Mannheim (2006). Weyl developing his metrical gravity theory recognized an enlarged Riemann tensor, the conformal, so-called Weyl tensor $C_{\lambda\mu\nu\kappa}$, with remarkable symmetry properties. It namely invariantly transforms under the conformal transformation $g_{\mu\nu}(x) \rightarrow \exp[2\alpha(x)]g_{\mu\nu}(x)$ as $C^{\lambda}_{\mu\nu\kappa}(x) \rightarrow C^{\lambda}_{\mu\nu\kappa}(x)$, since all derivatives of the function $\alpha(x)$ drop out identically. Due to this property the Weyl tensor manifests the same relation to conformal transformations as does the Maxwell tensor to gauge transformations. This can be used to introduce the Weyl action function

$$I_W = -\alpha_g \int d^4x \sqrt{-g} C_{\lambda\mu\nu\kappa}(x) C^{\lambda\mu\nu\kappa}(x) \tag{69}$$

which is invariant under conformal transformations. Here α_g is a dimensionless constant controling conformal cosmology by a theory-immanent effective coupling quantity, obviously replacing Newton´s gravitational constant G in Einstein´s GRT equations. This Weyl action I_W forbids interestingly enough the appearance of any fundamental integration constant like the cosmological constant Λ, as it is admitted at the application of the action-minimizing variational principle to the Einstein-Hilbert action function. The GRT field equations derived on the basis of the Weyl action I_W lead to a new energy momentum tensor of conformal cosmology given by

$$T^{\mu\nu} = T^{\mu\nu}_{kin} - \frac{1}{6}S_0^2(R^{\mu\nu} - \frac{1}{2}g^{\mu\nu}R^{\alpha}_{\alpha}) - g^{\mu\nu}\lambda S_0^4 = 0 \tag{70}$$

where the first term on the RHS is the conventional energy momentum tensor of the moving matter particles which is fully compensated by a second part connected with the spacetime geometry and the scalar function S_0. In this conformal theory there is energy not just in the matter fields, but in the spacetime geometry as well. As Mannheim (2006) can show the associated generalized conformal field equation can be brought into the form

$$R^{\mu\nu} - \frac{1}{2}g^{\mu\nu}R^{\alpha}_{\alpha} = \frac{6}{S_0^2}(T^{\mu\nu}_{kin} - g^{\mu\nu}\lambda S_0^4) \tag{71}$$

revealing that this conformal cosmology equation is analogous to the Einstein GRT equations with the difference of an effective dynamically induced gravitational coupling function given by $G_{eff} = -\frac{3c^2}{4\pi S_0^2}$ (see also Mannheim (1992) and the conformal analogue of Einstein's Λ given by $\bar{\Lambda} = \lambda S_0^4$. When solving the above equation for a Robertson-Walker symmetrical geometry, and introducing as conformal analogues to Einstein's GRT the quantities

$$\bar{\Omega}_m = \frac{8\pi G_{eff}\rho_m}{3c^2H^2} \tag{72}$$

$$\bar{\Omega}_\Lambda = \frac{\bar{\Lambda}}{3cH^2} \tag{73}$$

then Mannheim (2000) obtains the following result for the acceleration parameter

$$q = \frac{1}{2}(1 + \frac{3p_m}{\rho_m})\,\bar{\Omega}_m - \bar{\Omega}_\Lambda \tag{74}$$

again demonstrating from the basis of this conformal cosmology that something analogous to vacuum energy is operating and causing an accelerated expansion but physically connected with nothing like an energy-loaded vacuum but with a scalar field S_0.
At the end of this article we would like to conclude from all what has been analysed in original studies presented in this article here and from companying literature discussed in this article, that vacuum energy density as it is treated in standard cosmology, i.e. treated as a constant quantity, does not appear to be physically justified, but a generalized representation of this term should be further discussed in cosmology which, however, is of a completely different nature and is variable in magnitude depending on geometrical properties or scalar field properties in the universe.
Although the standard model of cosmology, the ΛCDM-model celebrated big successes in the past and most of the astronomers believe in it, it seems that reality behaves a bit different. Recent investigations by Kroupa et al. (2010) have shown that ΛCDM fails, since on scales of the Local group no dark matter action can be admitted, and so the standard model is faced with a big problem. Therefore it is convenient to consider also alternative models, like the ones presented in this article in order to develop a model of the universe that reflects cosmic reality better than ΛCDM. Nevertheless these kinds of models will have to prove themselves when they are applied to modern cosmological observations like the Supernova Ia data or the anisotropies of the Cosmic Microwave Backround (CMB). However the question remains if the CMB actually represents the matter distribution for a time of about 300000 years after the big bang, or if they should be interpreted in a different way under the conditions of mass-creating models (Fahr & Zoennchen, 2009)?

12. References

Bahcall, N. A. (1988). Large-scale structure in the universe indicated by galaxy clusters, *ARAA* 26: 631–686.

Bahcall, N. A. & Chokshi, A. (1992). The clustering of radio galaxies, *APJL* 385: L33–L36.

Barbour, J. B. (1995). General Relativity as a Perfectly Machian Theory, *in* J. B. Barbour & H. Pfister (ed.), *Mach's Principle: From Newton's Bucket to Quantum Gravity*, pp. 214–+.

Barbour, J. B. & Pfister, H. (eds) (1995). *Mach's Principle: From Newton's Bucket to Quantum Gravity*.

Barrow, J. D. (2000). *The book of Nothing: Vacuum, voids and the latest ideas about the origin of the universe*.

Bennett, C. L., Hill, R. S., Hinshaw, G., Nolta, M. R., Odegard, N., Page, L., Spergel, D. N., Weiland, J. L., Wright, E. L., Halpern, M., Jarosik, N., Kogut, A., Limon, M., Meyer, S. S., Tucker, G. S. & Wollack, E. (2003). First-Year Wilkinson Microwave Anisotropy Probe (WMAP) Observations: Foreground Emission, *APJS* 148: 97–117.

Buchert, T. (2001). On Average Properties of Inhomogeneous Fluids in General Relativity: Perfect Fluid Cosmologies, *General Relativity and Gravitation* 33: 1381–1405.

Buchert, T. (2005). LETTER TO THE EDITOR: A cosmic equation of state for the inhomogeneous universe: can a global far-from-equilibrium state explain dark energy?, *Classical and Quantum Gravity* 22: L113–L119.

Buchert, T. (2008). Dark Energy from structure: a status report, *General Relativity and Gravitation* 40: 467–527.

Einstein, A. (1917). Kosmologische Betrachtungen zur allgemeinen Relativitätstheorie, *Sitzungsberichte der Königlich Preußischen Akademie der Wissenschaften (Berlin), Seite 142-152.* pp. 142–152.

Einstein, A. & Straus, E. G. (1945). The Influence of the Expansion of Space on the Gravitation Fields Surrounding the Individual Stars, *Reviews of Modern Physics* 17: 120–124.

Ellis, G. F. R. (1983). Relativistic Cosmology: its Nature, Aims and Problems, *in* B. Bertotti, F. de Felice, & A. Pascolini (ed.), *General Relativity and Gravitation, Volume 1*, pp. 668–+.

Fahr, H. & Heyl, M. (2007a). Cosmic vacuum energy decay and creation of cosmic matter, *Naturwissenschaften* 94: 709–724.

Fahr, H. J. (2004). The cosmology of empty space: How heavy is the vacuum?, *Philosophy of Natural Sciences* 33: 339–353.

Fahr, H. J. (2006). Cosmological consequences of scale-related comoving masses for cosmic pressure, matter and vacuum energy densities, *Found.Phys.Lett.* 19.

Fahr, H. J. & Heyl, M. (2006). Concerning the instantaneous mass and the extent of an expanding universe, *Astronomische Nachrichten* 327: 733–+.

Fahr, H. J. & Heyl, M. (2007b). About universes with scale-related total masses and their abolition of presently outstanding cosmological problems, *Astronomische Nachrichten* 328: 192–+.

Fahr, H. J. & Zoennchen, J. H. (2006). Cosmological implications of the Machian principle, *Naturwissenschaften* 93: 577–587.

Fahr, H. J. & Zoennchen, J. H. (2009). The writing on the cosmic wall: Is there a straightforward explanation of the cosmic microwave backround?, *Annalen der Physik* 18(10). 699–721

Fischer, E. (1993). A cosmological model without singularity, *APSS* 207: 203–219.

Geller, M. J. & Huchra, J. P. (1989). Mapping the universe, *Science* 246: 897–903.

Goenner, H. F. M. (1995). Mach's Principle and Theories of Gravitation, *in* J. B. Barbour and H. Pfister (ed.), *Mach's Principle: From Newton's Bucket to Quantum Gravity*, pp. 442–+.

Goenner, H. F. M. (1997).

Holtonl, G. J. (1970). *Einstein and the search for reality*, Vol. 6.

Hoyle, F. (1948). A New Model for the Expanding Universe, *MNRAS* 108: 372–+.

Hoyle, F. (1990). On the relation of the large numbers problem to the nature of mass, *APSS* 168: 59–88.

Hoyle, F. (1992). Mathematical theory of the origin of matter, *APSS* 198: 195–230.

Hoyle, F., Burbidge, G. & Narlikar, J. V. (1993). A quasi-steady-state cosmological model with creation of matter, *Astrophys. J.* 410: 437–457.

Hoyle, F., Burbidge, G. & Narlikar, J. V. (1994a). Astrophysical Deductions from the Quasi Steadystate Cosmology, *MNRAS* 267: 1007–+.

Hoyle, F., Burbidge, G. & Narlikar, J. V. (1994b). Further astrophysical quantities expected in a quasi-steady state Universe, *AAP* 289: 729–739.

Hoyle, F., Burbidge, G. & Narlikar, J. V. (1997). On the Hubble constant and the cosmological constant, *MNRAS* 286: 173–182.

Jammer, M. & Bain, J. (2000). Concepts of Mass in Contemporary Physics and Philosophy, *Physics Today* 53(12): 120000–68.

Kolb, E. W. (1989). A coasting cosmology, *APJ* 344: 543–550.

Kroupa, P., Famaey, B., de Boer, K. S., Dabringhausen, J., Pawlowski, M. S., Boily, C. M., Jerjen, H., Forbes, D., Hensler, G. & Metz, M. (2010). Local-Group tests of dark-matter concordance cosmology . Towards a new paradigm for structure formation, *AAP* 523: A32+.

Lamoreaux, S. K. (2010). Systematic Correction for "Demonstration of the Casimir Force in the 0.6 to 6 micrometer Range", *ArXiv e-prints* .

Mach, E. (1883). *Die Mechanik in ihrer Entiwcklung: Eine historitsch kritische Darstellung*.

Mannheim, P. D. (1992). Conformal gravity and the flatness problem, *APJ* 391: 429–432.

Mannheim, P. D. (2000). Attractive and repulsive gravity., *Foundations of Physics* 30: 709–746.

Mannheim, P. D. (2001). Cosmic Acceleration as the Solution to the Cosmological Constant Problem, *APJ* 561: 1–12.

Mannheim, P. D. (2003). How Recent is Cosmic Acceleration?, *International Journal of Modern Physics D* 12: 893–904.

Mannheim, P. D. (2006). Alternatives to dark matter and dark energy, *Progress in Particle and Nuclear Physics* 56: 340–445.

Mashhoon, B., Hehl, F. W. & Theiss, D. S. (1984). On the gravitational effects of rotating masses - The Thirring-Lense Papers, *General Relativity and Gravitation* 16: 711–750.

Overduin, J. M. & Fahr, H. (2003). Vacuum energy and the economical universe, *Found. Physics Letters* 16.

Peebles, P. J. & Ratra, B. (2003). The cosmological constant and dark energy, *Reviews of Modern Physics* 75: 559–606.

Perlmutter, S., Aldering, G., Goldhaber, G., Knop, R. A., Nugent, P., Castro, P. G., Deustua, S., Fabbro, S., Goobar, A., Groom, D. E., Hook, I. M., Kim, A. G., Kim, M. Y., Lee, J. C., Nunes, N. J., Pain, R., Pennypacker, C. R., Quimby, R., Lidman, C., Ellis, R. S., Irwin, M., McMahon, R. G., Ruiz-Lapuente, P., Walton, N., Schaefer, B., Boyle, B. J., Filippenko, A. V., Matheson, T., Fruchter, A. S., Panagia, N., Newberg, H. J. M., Couch, W. J. & The Supernova Cosmology Project (1999). Measurements of Omega and Lambda from 42 High-Redshift Supernovae, *APJ* 517: 565–586.

Räsänen, S. (2006). Accelerated expansion from structure formation, *JCAP* 11: 3–+.

Reissner, H. (1995). On the Relativity of Accelerations in Mechanics, *in* J. B. Barbour & H. Pfister (ed.), *Mach's Principle: From Newton's Bucket to Quantum Gravity*, pp. 134–+.

Scholz, E. (n.d.). Cosmological spacetime balanced by a scale-covariant scalar field, *Found. of Physics Lett.* .

Stephani, H. (1988). *Allgemeine Relativitätstheorie. Eine Einführung in die Theorie des Gravitationsfeldes*.

Streeruwitz, E. (1975). Vacuum fluctuations of a quantized scalar field in a Robertson-Walker universe, *PRD* 11: 3378–3383.

Wesson, P. S. (2004). Vacuum Instability, *ArXiv General Relativity and Quantum Cosmology e-prints* .

Wesson, P. S., Leon, J. P. D., Liu, H., Mashhoon, B., Kalligas, D., Everitt, C. W. F., Billyard, A., Lim, P. & Overduin, J. (1996). a Theory of Space, Time and Matter, *International Journal of Modern Physics A* 11: 3247–3255.

Weyl, H. (1961). *Raum, Zeit, Materie*.

Wiltshire, D. L. (2007). Cosmic clocks, cosmic variance and cosmic averages, *New Journal of Physics* 9: 377–+.

Wu, K. K. S., Lahav, O. & Rees, M. J. (1999). The large-scale smoothness of the Universe, *NAT* 397: 225–230.

Zalaletdinov, R. M. (1992). Averaging out the Einstein equations, *General Relativity and Gravitation* 24: 1015–1031.

Zeldovich, I. B. (1981). Vacuum theory - A possible solution to the singularity problem of cosmology, *Soviet Physics Uspekhi* 133: 479–503.

Part 3

Dark Matter, Dark Energy

6

Doubts About Big Bang Cosmology

Disney, M. J.
Physics and Astronomy, Cardiff University, Cardiff, Wales, UK

1. Introduction

Only beasts could remain indifferent to questions about the origin, structure and fate of the cosmos in which they live. Only saints could resign themselves to never knowing the answers. The upshot has been that every civilization known to anthropology has put together such meagre observations as it possesses, has interpreted them in the light of currently fashionable ideas, and then manufactured as plausible a cosmological story as it can to tell its students and its children. The trouble is that none of those cosmologies have stood the test of time. Have we any reason to be more confident in the Big Bang Cosmology (BBC) which is fashionable today?

There are many good reasons to be sceptical of cosmology *as a subject*. For instance:

(A) There is only one universe! At a stroke this removes from our armoury as scientists all the statistical tools that have proved indispensable for understanding most of astronomy.

(B) The universe has been opaque to electromagnetic radiation for all but 4 of the 60 decades of time which stretch from the Planck era (dex -43 sec) to today (dex +17 sec.) Since as much interesting physics could have occurred in each logarithmic decade, it seems foolish to hope the we will ever know much about the origin of the cosmos, which is lost too far back in the logarithmic mists of Time. Even the Large Hadron Collider will probe the microscopic physics back only as far as dex (-10) secs [1].

(C) Cosmology requires us to extrapolate what physics we know over huge ranges in space and time, where such extrapolations have rarely, if ever, worked in physics before. Take gravitation for instance. When we extrapolate the Inverse Square Law (dress it up how you will as G.R.) from the Solar System where it was established, out to galaxies and clusters of galaxies, it simply never works. We cover up this scandal by professing to believe in "Dark Matter" - for which independent evidence is lacking.

(D) The human and historical time frame is so short compared to the cosmic one that we have in effect only a few still shots of a dynamical universe, with no proper (oblique) motions. It's as if we had to deduce not only the final score, but the rules of a football match from a few still photos.

(E) By cosmical (i.e. intergalactic) standards our local background is very bright. For instance the extra-galactic universe contributes less than one percent of the optical photons even at a dark mountain site on a moonless night. Much of the universe must therefore, and at many wavelengths, still lie hidden below the sky, even from space, because of the problem of *contrast*. And according to Tolman [2] distant extended objects like galaxies will be dimmed by (1+z) to the fourth power, an enormous factor at the kind of redshifts (z~10) where galaxies are supposed to form. Many galaxies, even nearby, will be sunk below the sky.

Even so cosmology is such a fascinating subject that I for one would like to believe that progress can and is being made. But how could one tell? Just because large teams are dedicated to working out the details of BBC doesn't mean that the underlying paradigm is secure. Although Hubble is widely and incorrectly credited with the discovery of expansion back near 1930, in fact he died in 1956 still sceptical, as were many of his contemporaries, of the dramatic notion that redshift implies expansion. Today the opposite attitude prevails where expansion, and all that it implies, goes virtually unquestioned. To be sure there is more evidence, but not all that evidence points in the same direction. Scientific history is littered with theories which once fitted many facts - Newtonian gravitation for instance. In the end though it is the discrepancies which signify more, even where they are relatively minor (e.g. the perihelion of Mercury). And as a galaxy astronomer I can see many worrying discrepancies between BBC as it stands now and the galaxies we can observe so minutely in our neighbourhood. We do BBC no favours by accepting it without question. We only blind ourselves to other truths or modifications that might be staring us in the face.

Here I discuss BBC mainly from an epistemic point of view and in particular try to answer two questions:

(1) Do we have enough evidence to be confident that BBC is broadly right?

(2) Where the evidence is contradictory, as it certainly is in the case of BBC, can one nevertheless come to a rational verdict on its soundness, taking into account the whole surrounding network of interlocking clues?

As to the first question I will suggest that the answer is ' Probably No' because BBC appears to have more Free Parameters than relevant observations. As to the second there is a Bayesian way to summarize contradictory evidence, but one's final verdict necessarily depends on the rather arbitrary weights (Likelihoods) one must attach to some of the contradictory clues. There is a great deal of room for debate here but I contend that it is a debate that needs to be held, and discussed openly.

2. BBC's lack of evidential significance

We question the significance of BBC by looking at the difference between the number of measurements with cosmological relevance that have been made, and the number of Free Parameters (FPs) introduced by BB theory to fit those same measurements. Where that difference is comfortably positive, one might regard cosmological theory as "significant" in the sense that the fit may be better, perhaps much better than one could have expected by chance. But where it is zero or negative there is no such balance of probabilities to recommend it.

Precisely which, and how many FPs are regarded as 'Cosmological', as distinct from more widely 'Astrophysical', is to some extent a question of taste, but it does not matter much so long as we treat them consistently, i.e. if included for fitting they also be included for measurement.

We proceed by means of an historical table (Table 1) where each line introduces either new FPs (column 3) widely touted then as being of cosmological significance or the first (seldom the best) claimed measurement of them (column 4), with the concurrent difference in number between the two i.e. the concurrent "Significance", in column 5. This is purely a counting exercise with no real need to understand what the parameters are, or how they have been measured. Readers interested in such details can however follow them up in Ratra and Vogeley's excellent recent review 'The Beginning and Evolution of the Universe' [7]. I deliberately halted the survey after the first account of WMAP's findings in 2003 in order to let the dust settle but have used the same ensemble of parameters as they did. No doubt more recent, and probably more controversial additions (or subtractions) could be made, according to taste.

	(1)	(2)	(3)	(4)	(5)
	DATE	NEW STEP	NEW FREE PARAMS	NEW MEASUREMENTS	CURRENT SIGNIFICANCE.
1	1917	Einstein's model	H_0, k_0, Ω_0	One equation between them.	-2
2	1921	Cosmological constant	Ω_Λ		-3
3	1929	Galaxy Redshifts		H_0	-2
4	1965	Cosmic Background Radiation(CBR)	η	η	-2
5	1970's	Big Bang Nucleosynthesis	Ω_b	Ω_b	-2
6	1974	Cosmochronology		$(\sim 1/H_0)$	-2
7	1978	Dark Matter	Ω_M		-3
8	1970,s	Initial Seeds	A,n_s		-5
9	1978	Gravitational Waves	r		-6
10	1981	Inflation	N		-7
11	1980's	Large Scale Structure	b, σ_8, ξ		-10
12	1990	COBE		A	-9
13	1998	Supernovae	w	Ω_Λ	-9
14	1998	Clustering		σ_8	-8
15	2000	Galaxy Infall		ξ	-7
16	2000	BOOMERANG		n_s, Ω_M, Ω_0 (k_0 inferred from equation in row 1)	-4
17	2003	WMAP	dn_s/dlogk, τ, τ_0	τ,dn_s/dlogk, b	-4

Table 1. Cosmological parameters

The main conclusion to note is the large number of Free Parameters that have, over the years, been widely and variously allowed into the discussion of cosmology. Many have been measured (column 4) with varying degrees of reliability. But at no stage, so far as I can see, has there been an excess of independent measurements over FPs. Nor is the trend a healthy one (col. 5). The Significance there, which is what matters in the end, is no better now than it was back in 1917. Of course we've got more measurements, far more, but so have we got a far more elaborate theory, one covered all over with sticking plasters such as Inflation, Dark Matter, and Dark Energy designed to stick poor Humpty Dumpty together again. Even the three successful predictions (of apparently flat space, by Inflation; of the Light Element abundances, by nuclear theory (retrodicted); of the maximum ages of the oldest star-clusters, by Expansion) are overbalanced by at least half a dozen unpredicted surprises (redshifts, CBR, Dark Matter, Inflation, Dark Energy and no CBR quadrupole).

Of course there are many caveats, some pro-cosmology, some anti. On the pro-side, the counting of *independent* measurements is not trivial. Modern instruments make measurements not in a single channel but in a spectrum of channels within a given dimension (e.g. wavelength). This could increase the information returned by as much as the logarithm of the number of such channels i.e. by several. On the anti- side note that we have been counting only the FPs *explicitly admitted* within the theory. But BBC is not a single theory any more but 5 separate sub-theories constructed on top of one another. The ground floor is a theory, historically but not fundamentally grounded in General Relativity, to explain the redshifts - this is Expansion, which happily also accounts for the Cosmic Background Radiation. The second floor is Inflation - needed to solve the horizon and 'flatness' problems of the Big Bang. The third floor is the Dark Matter hypothesis required to explain the existence of contemporary visible structures, such as galaxies and clusters, which otherwise would never condense within the expanding fireball. The fourth floor is some kind of description for the 'seeds' from which such structure is to grow. And the fifth and topmost floor is the mysterious Dark Energy idea needed to allow for the recent acceleration of the Expansion, apparently detected in supernova observations. Each new super incumbent theory was selected out of an essentially infinite set of alternatives, to fit the observations as they were known at the time. By rejecting the alternatives one is, in effect, fitting several extra *implicit* FPs in each case. These extra "conceptual" FPs should arguably be added to the totals in Table 1, perhaps 2 or 3 for each sub-theory, reducing the total Significance by 10 or more. This is why such a counting exercise can never be precise.

These caveats are however arguments at the margin. A healthy theory, with a large positive Significance, could afford to ignore them. BBC, with its formally negative Significance, must remain for now a bloody tilting ground for its protagonists and sceptics.

3. Contradictory evidence

Some aspects of the BBC scenario are better supported than others. The existence of Cosmic Background Radiation (CBR) with a Black Body spectrum speaks strongly in favour of an early dense hot phase, the essential feature of a Big Bang cosmology, and that state offers a plausible womb for gestating the light elements that cannot be manufactured in stars. However if redshift is truly evidence of Expansion it should dim distant galaxies out of sight in a most dramatic way (The Tolman Effect]. But we can see redshift 7 galaxies all too easily - an inconvenience which can only be explained by assuming an equally dramatic rate of galaxy evolution which fortuitously cancels Tolman.

On the other hand the universe seems to be highly isotropic - not what one expects of a monotonically expanding cosmos in which new, causally disconnected material, continuously appears over the horizon. This stumbling block of isotropy was solved by 'Inflation', a vague concept in which it is assumed that once-upon-a-time the universe was small enough and static enough for causal contacts to propagate, after which it 'inflated' exponentially to its present configuration. Apart from being ad-hoc it is extremely ugly in that it precludes us from ever deciding whether the cosmos is spatially finite or infinite. Thus it throws out most of the cosmological baby with the bath water.

In a hot high-pressure cosmos, structure will only form late - after radiation and matter have decoupled, and then only slowly, so it is difficult to explain the rich world of clustered galaxies we observe today. The structure problem was neatly solved by hypothesizing the existence of overwhelming amounts of Cold Dark Matter (CDM), that is to say dark matter with a low velocity dispersion which doesn't interact with radiation.[e.g.4] Thus it would condense through much of the radiation era and then act as a focus for the lower amounts of ordinary (baryonic) matter to coagulate around. And it wasn't *ad hoc* because there already existed strong observational evidence that galaxies were dominated dynamically by unseen matter 10 to 100 times more massive than the ordinary baryons which make up their stars and gas.

The CDM provides a natural scenario, called Hierarchical Galaxy Formation [HGF], for forming galaxies by the merger of smaller objects into larger. Unfortunately the observations reveal that galaxies don't form in this manner. They appear to evolve in the reverse order, big ones first ('downsizing') and to be far too regular to have formed by random mergers in this hierarchical manner (later).

So the evidence is contradictory, as it often is in a developing and perhaps primitive science. In a recent open minded review of BBC Peebles and Nusser [5], while pointing to some serious cracks in the edifice, particularly with regard to structure- formation, nevertheless concluded; "We do not anticipate that this debate will lead to a substantial departure from the standard picture of cosmic evolution from a hot Big Bang, because the picture passes a tight network of tests…."

Fair enough, but surely a convincing discussion demands a quantitative measure of the combined strength of such a network, of the jigsaw of interlocking bonds between the hypothesis and its surrounding evidential support - both bonds which fit and bonds which don't. That we try to supply next.

4. Evaluating a network of evidence

We here assemble a tool for evaluating a jigsaw of contradictory evidence then apply it to the BBC, less in the hope of immediately convincing the reader than in demonstrating how simply and powerfully the tool can work. The conclusions it will lead to will necessarily rely on the Likelihoods (weights) that any user must attach to the various pieces of evidence, either for or against, that go to make his jigsaw. In the case of BBC it is hard to see how many of those weights can be other than rough and ready. Thus so must be one's final conclusion.

The only permit we know of for Induction is Bayes' Theorem [e.g.6]:

$$P(H \mid E_1) = P(E_1 \mid H) \times P(H) / P(E_1)$$

which gives the Probability of Hypothesis H, given Evidence E_1. Rewriting in terms of $\bar{H}$ ('not-H')

$$P(\bar{H} \mid E_1) = P(E_1 \mid \bar{H}) \times P(\bar{H}) / P(E_1)$$

and dividing through

$$O(H \mid E_1) \equiv \frac{P(H \mid E_1)}{P(\bar{H} \mid E_1)} = \frac{P(E_1 \mid H)}{P(E_1 \mid \bar{H})} \times \frac{P(H)}{P(\bar{H})} \equiv L(E_1 \mid H) \times O(H) \tag{1}$$

which yields the Odds-*on* H, given E_1, in terms of O(H) - the Odds-on H *prior* to considering E_1, and $L(E_1 | H)$ the 'Likelihood-Odds', [Sometimes called the 'Bayes' Factor'] the Probability of E_1 *if* H is true, divided by the Probability if it is *not*. Written thus (1) is no more than self-evident common sense.
Next consider a second clue E_2 ; by an identical argument:

$$O(H | E_2) = L(E_2 | H) \times O(H | E_1) = L(E_2 | H) \times L(E_1 | H) \times O(H)$$

and so on, so that finally, considering all n clues:

$$O(H | E_1, E_2, ...E_n) = L(E_1 | H) \times L(E_2 | H) .... \times L(E_n | H) \times O(H) \tag{2}$$

which we henceforth label 'The Detective's Equation' because it formalizes the procedure a rational detective would use to combine all the clues, and the Likelihoods she attaches to them, to reach some final measure of her conviction in Hypothesis H. [The Equation is presumably well known but we could find no reference to it in the literature]
The Detective's Equation does exactly what we want. Each Likelihood-Odds $L(E_i | H)$ is a measure of the strength we assign to the fit between one piece of evidence E_i in the jigsaw and the hypothesis H we are trying to fit. The combined strength is multiplicative so that several weaker fits may nevertheless combine to equal the strength of a single strong bond. This suggests that all the evidence must be included, even where it is rather weak, or hard to weigh. A bad fit is characterised by its odds *against* H, so that its Likelihood - Odds L(E|H) is fractional, thus detracting from the strength of the final result. Equivocal evidence obviously has a Likelihood of 1 and could be ignored. [Lacking any precise theory of the errors involved in quantitative data then if a Normal distribution is adopted, as least contentious, an error of 0.1 sigma corresponds to odds of 12 to 1 *on*; of 2 sigma 44 to 1 *against*, and so on.]
Now let us apply it to BBC, piece of evidence by piece, assembling the running results as we go along in Table 2. If the BBC is true:
(A) Nothing should be older than the expansion age τ_E essentially distance divided by recession velocity. This appears to be obeyed because, where ages can be determined, for instance for star-clusters, for white-dwarf stars and for certain radio-active elements, they all appear younger than the expansion-age of about 14 billion years [7]. By definition the Likelihood - Odds of this evidence is

$$L(E_A | H) = \frac{P(E_A | H)}{P(E_A | \bar{H})}$$

$P(E_A | H)$ is obviously 1 but $P(E_A | \bar{H})$ is certainly not zero. Indeed galaxies, the building blocks of the cosmos, are notoriously difficult to age and some could well be much older than τ_E. Thus $L(E_A | H)$ presumably increases the odds on H, but by how much? There is no obvious or objective answer. It would be unwise to either ignore the evidence altogether, or to give it too much weight. A reasonable compromise *might* be to assign the Likelihood a value of 5 say, i.e. assume that the observed ages increase the odds-on H by 5.
(B) Firmer support comes from evidence that the expanding Universe should have emerged from an earlier dense and hot state. Thus the discovery of the Cosmic Background Radiation

(CBR) and its Black Body Spectrum provide, in the absence of any alternative explanation (i.e. $P(E_B \mid \bar{H})$) strong evidence in favour of H with a Likelihood of $L(E_B \mid H) = 50$ say. (I am reluctant to use Likelihoods more than 50, or less than 1/50, in a subject which has so often proved wrong.)
Combining (A) and (B) using the Detectives Equation (2)

$$O(H \mid E_A, E_B) = 5 \times 50 \times O(H)$$

i.e. between them they have increased the odds on BBC by no less than 250.
(C) The cosmos on its largest scale ought to look highly anisotropic - whereas the very reverse is observed; the CBR temperatures at the antipodes being identical to a few parts per million. This is serious evidence against BBC and might reduce the Likelihood in its favour by as much as the CBR argued for it; i.e. one might justifiably assume that $L(E_C \mid H) = 1/50$ (but see Inflation later).
Thus one might proceed through the list of clues (Table 2) assigning Likelihoods in each case as follows: (D) Because of gravity between its parts the cosmic expansion ought to be decelerating - but it is not. (E) If redshift is truly evidence of Expansion it should dim distant galaxies in a dramatic way [8]. But we can see high redshift galaxies all too easily - an inconvenience which can only be explained by assuming a rate of galaxy evolution which fortuitously cancels [8]. (F) As mentioned, light elements like Helium and Deuterium whose abundances cannot be otherwise explained could have been synthesized in approximately the right amounts in the Big Bang. (G) There are minute but measurable irregularities in the CBR with a scale naturally explained in terms of expansion. (H) Expansion naturally suppresses condensation into structures such as the galaxies which surround us on all sides. The problem is that radiation pressure in the early universe would have smoothed out any irregularities in baryonic matter so that by the time the two decoupled there would have been no 'seeds' from which such irregularities could naturally grow by self-gravitation.
Taken together, and with the crude Likelihoods I have assigned them in Table 2:

$$O(H \mid E....) = 5 \times 50 \times 1/50 \times 1/2 \times 1/10 \times 10 \times 20 \times 1/50 = 1 \times O(H)$$

In other words, by chance, all the 8 clues used so far have cancelled out so that they neither favour nor disfavour BBC.
Next add some modern refinements and observations:
(I) The structure problem was neatly solved by hypothesizing the existence of overwhelming amounts of Cold Dark Matter (CDM), that is to say dark matter with a low velocity dispersion which doesn't interact with radiation. Thus it would condense through much of the radiation era and then act as a focus for the lower amounts of ordinary (baryonic) matter to coagulate around. And it wasn't *ad hoc* because there already existed strong observational evidence that galaxies were dominated dynamically by unseen matter 10 to 100 times more massive than the ordinary baryons which make up their stars and gas [9].
(J) CDM provides a natural scenario, called Hierarchical Galaxy Formation, for forming galaxies by the merger of smaller objects into larger . Unfortunately it doesn't seem any longer to be the mode by which observed galaxies formed. Big galaxies evolved first, small ones later [10].
(K) The stumbling block of isotropy was solved by 'Inflation', a vague concept in which it is assumed that once-upon-a-time the universe was small enough and static enough for causal contacts to propagate, after which it 'inflated' exponentially to its present configuration [11, 12

(L) Most surprisingly, recent attempts to measure deceleration using exploding stars lead to the unpredicted discovery that the universal expansion appears to have accelerated recently [13, 14]. Sometimes called 'Dark Energy' this phenomenon has not been plausibly explained.
Table 2 shows the above clues, their associated Likelihoods, and in the last column the Running Odds as one multiplies those Likelihoods together row after row, not counting any prior O(H).

CLUE		Likelihood $L(C_i \mid H)$	Running Odds
A) Nothing older than expansion age	Yes	5	5
B) Earlier dense state	Yes	50:1=50	250
C) Universe should be anisotropic	No	1:50=1/50	5
D) Universe should decelerate	No	1:2=1/2	5/2
E) Galaxies should dim with redshift	No	1/10	1/4
F) Could produce light element abundances	Yes	20	5
G) Predicts CBR structure (First Peak)	Yes	10	50
H) Can't produce observed matter structure		1/50	1
I) But CDM can produce such structure		25	1/2
J) But real gals very unlike CDM models		1/10	5/2
K) Inflation may explain isotropy		2	5
L) Recent acceleration unexplained		1/20	1/4

Table 2. Big bang cosmology likelihoods

The end result, seen at the bottom of the last column, appears, to say the least of it, thoroughly unconvincing. The combined odds of all the above evidence yields odds of 4 to 1 *against* BBC. However that result relies on a number of Likelihoods whose evaluation is bound to be contentious, but which no honest thinker can evade if they are to come to a defensible conclusion.
My conclusion is as tentative as the Likelihoods I have declared. At 4 to 1 against at least it agrees with my uneasy feeling that BBC, once rather beautiful and economic, has grown uglier and more ad hoc in recent years.
My point is not to persuade readers of my own particular viewpoint but to persuade them to subject their own convictions on this matter to the same Bayesian analysis. If nothing else it should encourage tolerance of dissent, badly needed in this field, or so it seems to me.

5. Science or folk tale?

If cosmology is to be a science then the arguments of the last two sections, in so far as they are right, suggest that BBC may be in a sickly state. There is much anecdotal evidence to support this suspicion. For instance, after publishing a previous sceptical article on this topic [15], I received hundreds of e-mails from professional astronomers saying 'Thank God somebody is saying these things at last – but don't quote me'. Then again many younger astronomers will privately admit that they don't believe a word of Lambda-CDM, 'But if I don't acknowledge it in my grant and observing proposals then I don't succeed.'
Anecdotes aside let us look at some symptoms of BBC's malaise.
(A) When the supernova results came out, cosmology should have stopped in its stride. BBC had utterly failed to predict such a thing. But what happened instead? BBC jumped on its

horse and galloped off in chase of yet another free parameter, Lambda, based on zero physics but with the catchy title 'Dark Energy'.

(B) 'Multiverses', much discussed by certain cosmologists, are not science. What can never be detected is not physics, but metaphysics [26].

(C) Computer simulations have much to answer for. So they produce 'filaments'? So what. Look at star-formation simulations on a sub-parsec scale: they produce beautiful filaments too [27]. It has nothing to do with cosmology. As Zeldovich explained long ago filaments are the natural outcome when gravity overwhelms internal pressure. In any case computer simulations, and the Scientific Method, have yet to take the full measure of one another. Until they do, arguments based on simulations should be given a low weight. Computer simulations have a very mixed record in Astrophysics.

(D) So much is made to hang from the WMAP data. But it is just another still, and a very messy picture, of a single moment in time. Maybe it's an earlier moment, but not so much earlier in the relevant logarithmic sense. And most of the non-Galactic structure lies in the First Peak - which has only an oblique bearing on cosmology - and whose position can be derived from dimensional considerations alone.

(E) Most unhealthy is the present comedy surrounding CDM. Galaxies, near or far, simply do not conform to the dictates of this once attractive theory. Bigger galaxies seem to evolve before small ones - 'down-sizing' [10]. There aren't hundreds of dwarfs for every giant [17]. Galaxies don't have cuspy cores [18]. Mergers are rare and cannot lead to the thin discs we see on every side [19]. There is little or no correlation between the properties of exponential galaxies and their environments [20, 21]. Finally galaxies exhibit a drastic and puzzling degree of self-organisation (forming a 1-parameter set) that is totally at odds with Hierarchical Galaxy Formation, the child of CDM [22, 23]. And of course we haven't found the DM, cold or otherwise. So CDM is in tatters - but somehow, like the Emperors New Clothes, lives on. Why? Presumably because without it BBC has lost a vital prop - a means for forming structure. What is so bizarre is the asymmetry between galaxy astronomy, which is rich with Information, and cosmology, which is not. And yet the cart is pushing the horse here. [24, 25].

Some opine that one shouldn't criticise an hypothesis without offering an alternative. I do not agree. Publishing its weaknesses ought to encourage alternatives, even where the critic cannot find one himself. However it is interesting to note that all the dynamical discrepancies that call for Dark Matter could as well point elsewhere. Wherever large lumps of matter are accelerated by gravitation (e.g. in clusters) the acceleration is always too large; it's as if each accelerated lump is dragged along by its neighbours. That is Mach's Principle. And if there is something to it extragalactic astronomers would be the first to know.

Given its rickety state one wonders at the hushed respect in which BBC is still widely held (*'Cosmology Deference'*). Had the subject matter been less momentous one feels that parts of it at least would have been discarded some time ago. But there lies its singularity, its difference from the rest of science. Mankind seems to need a cosmology, and just now BBC is the only one he's got. But for this observer at least, something even more mysterious and interesting appears to be going on out there. The last thing we should be doing is trying to force it into an old-fashioned corset that doesn't seem to fit. Scepticism is the portal to progress. Science risks discredit if it isn't willing to apply to cosmology the same sceptical attitude that it does to all other supplicants for its approval. Is BBC really a science, or is it a just-so folk tale heavily disguised as a science? One cosmologist [28] said: *"Cosmology is the dot com of the sciences. Boom or bust. It is about nothing less than the origin and evolution of the Universe, the all of everything. It is the boldest of enterprises and not for the fainthearted. Cosmologists are the flyboys of astrophysics, and they often live up to all that image conjures up".* That sounds to me like special pleading.. If so then science should certainly turn it down.

The highest compliment we can pay BBC is to treat it as a scientific hypothesis, like any other, and weigh up its pros and cons.

6. References

[1] Rees,M.,1995, Perspectives in Modern Cosmology, CUP,109
[2] Tolman,R.C.,1930, Proc.Nat.Acad.Sciences, 511 Nos. 5 and 7
[3] Spergel D.N., et al., 2003, Astrophys.J.Suppl., 148,175
[4] Peebles,P.J.E.,1993, Principles of Physical Cosmology, Princeton Univ.Press.
[5] Peebles,P.J.E. and Nusser,A., 2010, Nature, 465,565-9
[6] Jaynes,E.T, 2003, Probability Theory, CUP
[7] Ratra,B., Vogely,M.S., 2008, Publ.Astr.Soc.Pacific, 235-265
[8] Bryn-Jones & Disney M.J., 1997, The Hubble Space Telescope and the High Redshift Universe, World Scientific Pr., 151
[9] Faber S.M. and Gallagher J.S., 1979, ARA&A, 17,135
[10] Noeske,H.G. et al. 2007, Ap.J., 660, L47
[11] Brout.R.,Englert and Gunzig,E., 1978,Ann.Phys.,115, 78
[12] Guth,A.H.,1981, Phys.Rev.D, 23, 347
[13] Riess,A.J., et al., 1998, A.J., 116,1009
[14] Perlmutter,S., et al., 1999, Ap.J., 517,565
[15] Disney,M.J., 2000, Genl. Relativity and Grav.,32,1125
[16] Gauch,H.G. Jnr., 2005, 'Scientific Method in Practice',Chapt.8, CUP
[17] Moore,B. et al., 1999, MNRAS, 310,1147
[18] de Blok, W.J.G., 2010, Adv.Astr.Astrophys., Art. 789293
[19] van der Kruit, P., Freeman K.C., 2011, ARA&A, 'Disk Galaxies', (in press)
[20] Gavazzi.G.,et al., 2003, A&A, 400, 451
[21] Nair P.B., et al, 2010, Ap.J., 715, 606-22
[22] Disney, M.J. et al, 2008, Nature, 455, 1082-4
[23] Garcia-Appadoo,D., et al., 2009, MNRAS, 394, 340-356
[24] Kroupa P.,et al, 2010. A&A, 532..32K
[24] McGaugh, Stacy.S., 2005, Phys. Rev. Lett., 19,17,171302
[26] Ellis, G.F.R., 2011, Nature, 469, 294-5
[27] Whitworth,A.P., et al., 1994, A&A, 290,421
[28] Turner, M. S., 2001, arXiv:astro-ph/0102057

7

Applications of Nash's Theorem to Cosmology

Abraão J S Capistrano[1] and Marcos D Maia[2]
[1]*Universidade Federal do Tocantins*
[2]*Universidade de Brasília*
Brazil

1. Introduction

The understanding of gravitational phenomena has been considered a fundamental problem in modern Cosmology. Recent observations of the CMBR power spectrum in the 7-year data from WMAP (Komatsu et.al., 2011; Jarosik et.al., 2011) tell that the gravitational field perturbations amplify the higher acoustic modes due to the gravitational field of baryons and mainly on the influence of Dark matter. Dark matter has been regarded as to be responsible for inducing a strong gravitational effect on cosmological scale that would lead the young universe to form large scale structures. Such perturbations are also verified at the local scales of galaxies and clusters of galaxies. Moreover, the gravitational perturbations also play an important role in the acceleration of the universe. Due to the cosmological constant paradigm, modifications of gravity have been studied as a alternative route to obtain the require correction for Friedman's equations.

In this sense, Nash's theorem on gravitational perturbations along extra dimensions has been revealed to be an appropriated tool in a manner of dealing with such perturbations. In our present discussion, we seek such explanation within the foundations of geometry, notably using the notion of geometric or gravitational flow, determined by the extrinsic curvature. In order to understand the concept of geometric flow, we give a brief review of the problem of embedding space-times and of its compatibility with the observational aspects of physics.

We discuss the structure and concepts related to the embedding theory as the basis for a more general theory of gravitation. In this framework, for instance, the cosmological constant problem is seen as a symptom of the ambiguity of the Riemann curvature in general relativity. The solution of that ambiguity provided by Nash's theorem eliminates the direct comparison between the vacuum energy density and Einstein's cosmological constant, besides being compatible with the formation of structures and the accelerated expansion of the universe. Moreover, it is shown how space-times solutions of Einstein's equations can be smoothly deformed along the extra dimensions of an embedding space and how the deformation, described by the extrinsic curvature, produces an observable effect of topological character in the universe.

In the following section, we begin reviewing the brane-world program motivated by the problem of unification of the fundamental interactions. The third section is devoted to Nash's embedding theorem and its relation to the gravitational perturbations. The correct embedding structure of space-time is present here without using junction conditions. In the fourth section, we show some of the cosmological applications when considering a correct embedding structure of the space-time. Hence, final remarks are commented in the Conclusion section.

2. On the gravitational constant and Brane-world program

As well known, the gravitational constant in the Newton's Law given by

$$\vec{F} = m\vec{a} = G\frac{mm'}{r^2}\frac{\vec{r}}{r} \ , \tag{1}$$

was introduced to convert the physical dimensions $[M^2]/[L^2]$ to the dimensions of force $[M][L]/[T^2]$. It has the value $G = 6,67 \times 10^{-8}$ $\text{cm}^3/\text{g.sec}^2$, with the same value in a wide range of applications of (1). In 1914, Max Planck suggested a natural units system in which $G = c = \hbar = 1$ and everything else would be measured in centimeters. For that purpose it was assumed that Newton's equation (1) also holds at quantum level. Under this condition, comparing the gravitational energy for $m = m'$ with the quantum energy for a wavelength $\lambda \sim r$, it follows that

$$E =< \vec{F}.\vec{r} >= G\frac{m^2}{\lambda} = \frac{\hbar c}{\lambda} \ .$$

Together with Maxwell equations and the laws of thermodynamics, this leads to three quantities which characterize the so-called Planck regime:

$$m_{pl} = \sqrt{\frac{\hbar c}{G}} \sim 10^{19}\text{Gev}, \ \lambda_{pl} = \sqrt{\frac{\hbar G}{c^3}} \sim 10^{-33}\text{cm}, \ t_{pl} = \sqrt{\frac{\hbar G}{c^5}} \sim 10^{-44}\text{sec}. \tag{2}$$

Planck's conclusion established a landmark in the development of modern physics:

> "These quantities retain their natural significance as long as the law of gravitation and that of the propagation of light in a vacuum and the two principles of thermodynamics remain valid; they therefore must be found always the same, when measured by the most widely different intelligences according to the most different methods" (Planck, 1914)

Today, we can safely say that electrodynamics, actually all known gauge theories, and the laws of thermodynamics remain solid. However, the validity of Newton's law at 10^{-33}cm has not been experimentally confirmed. It has been recently shown to hold at 10^{-3}cm, but with strong hints that it breaks down at 10^{-4}cm (Decca et al., 2007). It should be noted also that the constant G is valid for the Newtonian space-time which has the product topology $\Sigma_3 \times \mathbb{R}$, where Σ_3 denotes the 3-dimensional simultaneity sections, implying that the gravitational constant has the physical dimensions $[G] = [L]^3/[M][T]^2$, appropriate for 3-dimensional manifolds only.

In 1916, Newton's gravitational law changed dramatically to General Relativity, including the principles of equivalence, the general covariance and Einstein's equations in a 4-dimensional space-time

$$R_{\mu\nu} - \frac{1}{2}Rg_{\mu\nu} = 8\pi G T_{\mu\nu} \ . \tag{3}$$

The Newtonian gravitational constant G, was retained in (3), to guarantee that the theory would reproduce the Newtonian theory in its weak field limit, without the need to change constants. However, the consequences of this are quite embarrassing: indeed, the maintenance of G in (3) originates the hierarchy problem of the fundamental interactions. While all relativistic gauge interactions are quantized at the Tev scales of energies, gravitation would be quantized only at $\sim 10^{19}$Gev, which, as we have seen, coincide with the level predicted by Planck for Newtonian quantum gravity which is the weak field limit of General

Relativity. Furthermore, the relativistic quantum gravitational theory compatible with the physical dimensions of G would be defined only in a 3-dimensional foliation of the space-time, as originally conceived by Dirac (Dirac, 1959), Arnowitt, Deser and Misner (Arnowitt et al., 1962). However, such foliation is not consistent with the diffeomorphism invariance of General Relativity (Kuchar, 1992).

The criticism on the validity of Planck's regime for quantum gravity is the basis of the brane-world program by Arkani-Hamed, G. Dvali and S. Dimopolous (ADD for short) (Arkani-Hamed et al., 1998) proposing a solution of the hierarchy problem of the two fundamental energy scales in nature, namely, the electroweak and Planck scales $[M_{Pl}/m_{EW} \sim 10^{16}]$ (Carter, 2001). It contains essentially three fundamental postulates:

1. the space-time or brane-world is an embedded differentiable sub manifold of another space (the bulk) whose geometry is defined by the Einstein-Hilbert action (therefore this should not be confused with the "brane" of string/M-theory);
2. all gauge interactions are confined to the four-dimensional brane-world (this is a consequence of the poincaré symmetry of the electromagnetic field and in general of the dualities of yang-mills fields, which are consistent in four-dimensional space-time only);
3. gravitation is defined by Einstein's equations for the bulk, propagating along the extra dimensions at Tev energy scale.

It follows from (2) that all ordinary matter fields interacting with gauge fields must also be confined to the same space-time; the original ADD paper refers to graviton probes to the extra dimensions, but classically it means that the bulk is locally foliated by a family brane-world sub-manifolds, whose metric depend on the extra-dimensional coordinates in the bulk.

The impact of such program in theoretical and observational cosmology has been discussed at length as, e.g., in Refs. (Randall, 1999, a;b; Dvali, 2000; Sahni, 2002; 2003; Shiromizu, 2000; Dick, 2001; Hogan, 2001; Deffayet, 2002; Alcaniz, 2002; Jain et al., 2002; Lue, 2006). For instance, concerning the dark matter problem, just like the gravitational field of ordinary matter, dark matter gravity could also propagate in the bulk and in principle should be derived from the same bulk gravitational equations. When considering the acceleration expansion problem, modifications of gravity at very large scales also have been regarded as an alternative route to deal with the accelerated expansion of the universe, often described by something called dark energy. That route in turn has been predominantly associated with the existence of extra-dimensions which a modified friedman's equation can be obtained and provide the correct acceleration expansion.

Some popular brane-world models use Strings/M-theory motivations and use additional postulates such as a z_2 symmetry across the brane-world (or d-brane-world) as in the Randall-Sundrum models (Randall, 1999, b). This symmetry was not considered here essentially because the z_2 symmetry breaks the regularity of the embedding, thus preventing the use of the perturbation mechanism which is the essential feature in our arguments.

To be free from these limitations we require a model independent formulation based on the perturbational theory of embedded submanifolds as stated in (Maia et al., 2005; 2007), rather than particular junction conditions that we discuss more details in the next section.

3. The embedding problem

The embedding of a manifold into another is a non-trivial problem and has its roots in the classic problem in differential geometry, originated in the early days of the Riemannian

geometry. The curvature tensor defined by Riemann can describe the local shape of a Riemannian manifold only up to the condition that it does not "stretch".
Reviewing the concept, given a basis $\{e_\mu\}$ the Riemann tensor describes the curvature of a manifold by displacing a vector field e_ρ along a closed parallelogram defined by e_μ and e_ν and comparing the result with the original vector obtaining:

$$R(e_\mu, e_\nu)e_\rho = R_{\mu\nu\rho\sigma}e^\sigma = [\nabla_\mu, \nabla_\rho]e^\sigma \ .$$

When the difference is zero, the manifold is said to be flat. Such Riemannian flat space is not necessarily equal to a flat space in Euclidean geometry. For instance, it could likewise be a cylinder or a helicoid. After Riemann conceptualized a manifold intrinsically, the question if the geometry of a Riemannian manifold has the same geometry of a manifold embedded in an Euclidean soon arose. Today we know that every Riemannian manifold defined intrinsically can be embedded isometrically, locally or globally, in a Euclidean space with *appropriate* dimensions (Odon, 2010).
Nonetheless, the existence of a background geometry is necessary to fix the ambiguity of the Riemann curvature of a given manifold, without a reference structure. General Relativity solves this ambiguity problem by specifying that the tangent Minkowski space is a flat plane, as decided by the Poincaré symmetry, and not by the Riemann geometry itself. The same space-time is chosen as the ground state for the gravitational field, where particles and quantum field are defined. This choice would be fine, were not for the experimental evidences of a small but non-zero cosmological constant. Since the presence of this constant is not compatible with the Minkowski space-time, we face a conflicting situation: Either we define particles, quantum fields and their vacua states in the Minkowski space-time using the Poincaré group, or else these properties should be defined in a De Sitter space-time using the De Sitter group (Maia et al., 2009). The cosmological constant and the vacuum energy density based on the Poincaré symmetry cannot be present simultaneously in Einstein's equations, without bringing up the current cosmological constant issue.
The ambiguity of the curvature tensor was known by Riemann himself, when he acknowledged that his curvature tensor defines a class of objects and not just one (Riemann, 1854). This is explicit in Riemann's words when he states *"by considering arbitrary bendings -without stretching"* of such surfaces which are equivalent to a plane due to the lines on the surfaces remain unaltered even when bending. It imposes a serious constraint on the dynamics of the geometry itself. This means that the Riemann curvature has a degree of ambiguity, characterizing classes of equivalence of manifolds which would otherwise have different shapes or topologies where it cannot evolve nor stretch. In particular, there are infinite many flat Riemannian manifolds, all with zero Riemann curvature, but with different shapes.
A solution of such ambiguity was conjectured by L. Schlaefli in 1871, proposing that *all Riemannian manifolds must be embedded in a larger space*, so that the components of the extrinsic curvature may decide the difference between two Riemann-flat geometries (Schlaefli, 1873). However, the embedding depend on the solution of the Gauss-Codazzi-Ricci equations, involving the metric, the extrinsic curvature and the third fundamental form as independent variables. They provide the necessary and sufficient conditions for the existence of the embedded manifold (Eisenhart, 1966). Until recently those equations could be solved only with the help of positive power series expansions of the embedding functions (that is, they must be analytic functions), and so each embedding had to be examined separately.

The proof that all differentiable Riemannian manifolds can be embedded in a space with sufficient number of dimensions using exclusively smooth functions was given by Nash (Nash, 1956) in 1956, when he introduced the notion of smoothing operators in Riemannian geometry, leading to the geometric flow condition

$$k_{\mu\nu} = -\frac{1}{2}\frac{\partial g_{\mu\nu}}{\partial y} \tag{4}$$

where $k_{\mu\nu}$ denotes the extrinsic curvature and y represents a coordinate on a direction orthogonal to the embedded geometry.

In the following we derive the condition (4) in the simple case of just one extra dimension. Higher dimensional cases were also implicit in Nash's paper and this was applied as a possible extension of the ADM quantization of the gravitational field (Maia et al., 2007).

4. Geometric flow

Consider a Riemannian manifold $\bar{V}_n$ with metric $\bar{g}_{\mu\nu}$, and its local isometric embedding in a D-dimensional Riemannian manifold V_D, $D = n+1$, given by a differentiable and regular map $\mathcal{X} : \bar{V}_n \rightarrow V_D$ satisfying the embedding

$$g_{\mu\nu} = \mathcal{G}_{AB}\mathcal{X}^A_{,\mu}\mathcal{X}^B_{,\nu};\ \mathcal{G}_{AB}\mathcal{X}^A_{,\mu}\eta^B_b = 0;\ \mathcal{G}_{AB}\eta^A_a\eta^B_b = g_{ab} = \pm\delta_{ab}\,. \tag{5}$$

where we have denoted by $\mathcal{G}_{AB}$ the metric components of V_D in arbitrary coordinates, and where $\bar{\eta}$ denotes the unit vector field orthogonal to $\bar{V}_n$. The extrinsic curvature of $\bar{V}_n$ is by definition the projection of the variation of η on the tangent plane (Eisenhart, 1966)

$$\bar{k}_{\mu\nu} = -\mathcal{X}^A_{,\mu}\bar{\eta}^B_{,\nu}\mathcal{G}_{AB} = \mathcal{X}^A_{,\mu\nu}\bar{\eta}^B\mathcal{G}_{AB}\,. \tag{6}$$

The integration of the system of equations gives the required embedding map $\mathcal{X}$.

In order to understand the meaning of the extrinsic curvature, construct the one-parameter group of diffeomorphisms defined by the map $h_y(p) : V_D \rightarrow V_D$, describing a continuous curve $\alpha(y) = h_y(p)$, passing through the point $p \in \bar{V}_n$, with unit normal vector $\alpha'(p) = \eta(p)$ (Crampin, 1986). The group is characterized by the composition $h_y \circ h_{\pm y'}(p) \overset{def}{=} h_{y\pm y'}(p)$, $h_0(p) \overset{def}{=} p$. Applying this diffeomorphisms to all points of a small neighborhood of p, we obtain a congruence of curves (or orbits) orthogonal to $\bar{V}_n$. It does not matter if the parameter y is time-like or not, nor if it is positive or negative.

Given a geometric object $\bar{\omega}$ in $\bar{V}_n$, its Lie transport along the flow for a small distance δy is given by $\Omega = \bar{\Omega} + \delta y \pounds_\eta \bar{\Omega}$, where $\pounds_\eta$ denotes the Lie derivative with respect to η Crampin (1986). In particular, the Lie transport of the Gaussian frame $\{\mathcal{X}^A_\mu, \bar{\eta}^A_a\}$, defined on $\bar{V}_n$ gives

$$\mathcal{Z}^A_{,\mu} = \mathcal{X}^A_{,\mu} + \delta y\, \pounds_\eta \mathcal{X}^A_{,\mu} = \mathcal{X}^A_{,\mu} + \delta y\; \eta^A_{,\mu} \tag{7}$$

$$\eta^A = \bar{\eta}^A + \delta y\, [\bar{\eta}, \bar{\eta}]^A \quad = \bar{\eta}^A \tag{8}$$

However, from (6) we note that in general $\eta_{,\mu} \neq \bar{\eta}_{,\mu}$.

It is important to note that the set of coordinates $\mathcal{Z}^A$ obtained by integrating these equations *does not necessarily describe another manifold.* In order to be so, they need to satisfy embedding equations similar to (5):

$$\mathcal{Z}^A_{,\mu}\mathcal{Z}^B_{,\nu}\mathcal{G}_{AB} = g_{\mu\nu},\ \mathcal{Z}^A_{,\mu}\eta^B\mathcal{G}_{AB} = 0,\ \eta^A\eta^B\mathcal{G}_{AB} = 1\,. \tag{9}$$

Replacing (7) and (8) in (9) and using the definition (6) we obtain the metric and the extrinsic curvature of the new manifold

$$g_{\mu\nu} = \bar{g}_{\mu\nu} - 2y\bar{k}_{\mu\nu} + y^2\bar{g}^{\rho\sigma}\bar{k}_{\mu\rho}\bar{k}_{\nu\sigma} \tag{10}$$

$$k_{\mu\nu} = \bar{k}_{\mu\nu} - 2y\bar{g}^{\rho\sigma}\bar{k}_{\mu\rho}\bar{k}_{\nu\sigma}\ . \tag{11}$$

Taking the derivative of (10) with respect to y we obtain Nash's deformation condition (4). The analogy of geometry with fluid flows is similar but different from the Ricci flow proposed by R. Hamilton using the caloric fluid and Fourier's heat flux to obtain the expression

$$R_{\mu\nu} = -\frac{1}{2}\frac{\partial g_{\mu\nu}}{\partial y}$$

that resembles (4) (Hamilton, 1982). This result was subsequently applied with enormous success by G. Perelman to solve the Poincaré conjecture (Perelman, 2002). Unfortunately the Ricci-flow is not relativistic and it is not compatible with Einstein's equations or with relativistic cosmology.

The equations (9) need to be integrated so define a new manifold. The integrability conditions for these equations are given by the non-trivial components of the Riemann tensor of the embedding space[1], expressed in the frame $\{\mathcal{Z}^A_\mu, \eta^A\}$ as

$${}^5\mathcal{R}_{ABCD}Z^A{}_{,\alpha}Z^B{}_{,\beta}Z^C{}_{,\gamma}Z^D{}_{,\delta} = R_{\alpha\beta\gamma\delta} + (k_{\alpha\gamma}k_{\beta\delta} - k_{\alpha\delta}k_{\beta\gamma}) \tag{12}$$

$${}^5\mathcal{R}_{ABCD}Z^A{}_{,\alpha}Z^B{}_{,\beta}Z^C{}_{,\gamma}\eta^D = k_{\alpha[\beta;\gamma]} \tag{13}$$

These are the mentioned Gauss-Codazzi equations (the third equation -the Ricci equation- does not appear in the case of just one extra dimension) (Eisenhart, 1966). The first of these equation (Gauss) shows that the Riemann curvature of the embedding space acts as a reference for the Riemann curvature of the embedded space-time. Both Riemann curvatures are ambiguous in the sense described by Riemann, but Gauss' equation (12) shows that their difference is given by the extrinsic curvature, completing the proof of the Schlaefli embedding conjecture by use of Nash's deformation condition (4). The second equation (Codazzi) complements this interpretation, stating that the projection of the Riemann tensor of the embedding space along the normal direction is given by the tangent variation of the extrinsic curvature.

Equations (10) and (11) describe the metric and extrinsic curvature of the deformed geometry V_4. By varying y they describe a continuous sequence of deformations in the the embedding space. The existence of these deformations are given by the integrability conditions (12) and (13) which are therefore not dynamical equations.

As in Kaluza-Klein and in the brane-world theories, the embedding space V_5 has a metric geometry defined by the higher-dimensional Einstein's equations

$${}^5\mathcal{R}_{AB} - \frac{1}{2}{}^5\mathcal{R}\mathcal{G}_{AB} = G_* T^*_{AB}\ . \tag{14}$$

where G_* is the new gravitational constant and where T^*_{AB} denotes the components of the energy-momentum tensor of the known gauge fields and material sources. From these

[1] To avoid confusion with the four dimensional Riemann tensor $R_{\alpha\beta\gamma\delta}$, the five-dimensional Riemann tensor is denoted by ${}^5\mathcal{R}_{ABCD}$. The extrinsic curvature terms in these equations follows from the five-dimensional Christoffel symbols together with the use of (4).

dynamical equations we may derive the gravitational field in the embedded space-times. Taking the tangent, vector and scalar components[2] of (14) and using the previous confinement conditions (19) one can obtain

$$R_{\mu\nu} - \frac{1}{2}Rg_{\mu\nu} - Q_{\mu\nu} = 8\pi G T_{\mu\nu} \tag{15}$$

$$k^{\rho}_{\mu;\rho} - h_{,\mu} = 0\,, \tag{16}$$

where the term $Q_{\mu\nu}$ in the first equation results from the expression of R_{AB} in (14), involving the orthogonal and mixed components of the Christoffel symbols for the metric $\mathcal{G}_{AB}$. Explicitly this new term is

$$Q_{\mu\nu} = g^{\rho\sigma}k_{\mu\rho}k_{\nu\sigma} - k_{\mu\nu}h - \frac{1}{2}\left(K^2 - h^2\right)g_{\mu\nu}\,, \tag{17}$$

where $h^2 = g^{\mu\nu}k_{\mu\nu}$ is the squared mean curvature and $K^2 = k^{\mu\nu}k_{\mu\nu}$ is the squared Gauss curvature. This quantity is therefore entirely geometrical and it is conserved in the sense of

$$Q^{\mu\nu}{}_{;\nu} = 0\,. \tag{18}$$

Therefore we may derive observable effects associated with the extrinsic curvature capable to be seen by four-dimensional observers in space-times.
With all these tools at hand, modern Cosmology has been investigated and represents an important source of data that can provide a deeper comprehension of the gravitational structure and evolution of the universe. Not only this, but it calls for new gravitational theories far beyond Einstein's approach. Even though we are long way from a concrete fully-developed theory, dark matter and dark energy play a major role on this quest, representing fundamental constraints to these new gravitational models. It is also important to make the following observations:
1) A cosmological constant was not included in the equation for the higher dimensional space V_5 in (14), so that the cosmological constant problem does not appear. With this choice we also ensure the existence of an embedded 4-dimensional Minkowski space-time (a cosmological constant was included in (Maia et al., 2005), but here we see no reason for it).

2) In contrast with the extra dimensional perturbative behaviour of the gravitational field, all gauge fields of the standard model remain confined to the four-dimensional space-time. This is a direct consequence of the gauge field structure. Just as a reminder, the Yang- Mills equations can be written as $D \wedge F = 0$, $D \wedge F^* = 4\pi J^*$, where $F = F_{\mu\nu}dx^\mu \wedge dx^\nu$, $F_{\rho\sigma} = [D_\rho, D_\sigma]$, $D_\mu = I\partial_\mu + A_\mu$, $F^* = F^*_{\mu\nu}dx^\mu \wedge dx^\nu$ and $F^*_{\mu\nu} = \epsilon_{\mu\nu\rho\sigma}F^{*\;\rho\sigma}$. The duality operation $F \to F^*$ requires the existence of an isomorphism between 3-forms and 1-forms, which can only be realized in a four dimensional space-time manifold. Therefore, the confinement of gauge fields, matter and vacuum states is a property that is independent of the perturbation of the brane-world geometry.
There are two relevant consequences of the confinement. In the first place, it implies that all ordinary matter which interacts with the gauge fields, and also the vacuum states and its energy-momentum tensor associated with the confined fields also remain confined to the four-dimensional brane-world. Secondly, the diffeomorphism invariance of General

[2] The third gravitational equation was omitted here due to the fact that it vanishes in 5-D, but when the higher dimensional space-time is considered, one can obtain the equation $R - (K^2 - H^2) + \mathcal{R}(D-5) = 0$, sometimes called gravitational scalar equation.

Relativity cannot apply to the bulk manifold V_D, for it would imply in breaking the confinement. Of course, such limitation could be fixed by applying a coordinate gauge, but then we will be imposing a modification to Nash's theorem. Nash's theorem demands the embedding to be differentiable and regular, so that there is a 4×4 non-singular sub-matrix of the Jacobian determinant of the embedding map, thus guaranteeing the diffeomorphism invariance in the four-dimensional embedded submanifold only. Admitting that the original (on-embedded) space-time is a solution of Einstein's equations, the gauge fields, matter and its vacuum states keep a 1 : 1 correspondence with the source fields in the embedded space-time structure. Consequently, the confinement can be generally set as a condition on the embedding map such that

$$8\pi G T_{\mu\nu} = G_* Z^A_{,\mu} Z^B_{,\nu} T^*_{AB}, \quad Z^A_{,\mu} \eta^B T^*_{AB} = 0, \quad \text{and} \quad \eta^A \eta^B T^*_{AB} = 0 \tag{19}$$

3) Einstein's equations can be written as

$${}^5\mathcal{R}_{AB} = G_* \left(T^*_{AB} - \frac{1}{3} T^* \mathcal{G}_{AB} \right)$$

The tensor ${}^5\mathcal{R}_{AB}$ may be evaluated in the embedded space-times by contracting it with $Z^A_{,\mu}$, $Z^B_{,\nu}$, $Z^A_{\mu}\eta^B$ and $\eta^A\eta^B$. Using (4), (9) and the confinement conditions (19), Einstein's equations become

$${}^5\mathcal{R}_{\mu\nu} = R_{\mu\nu} + \frac{\partial k_{\mu\nu}}{\partial y} - 2k_{\mu\rho}k^{\rho}_{\nu} + h h_{\mu\nu} \tag{20}$$

$${}^5\mathcal{R}_{\mu 5} = k^{\rho}_{\mu;\rho} + \frac{\partial \Gamma^{\rho}_{\mu 5}}{\partial y} \tag{21}$$

It follows that the Israel-Lanczos condition does not follow from Einstein's equations (3) by themselves. It becomes necessary that the embedded geometry does satisfy particular conditions such that the Ricci curvature of the embedding space coincide with the extrinsic curvature of the embedded space-time, that is ${}^5\mathcal{R}_{\mu\nu} = k_{\mu\nu}$, which is not generally true. One of these conditions is that the embedded space-time acts as a mirror boundary between two regions of the embedding space (see e.g. (Israel, 1966)). In this case we may evaluate the difference of ${}^5\mathcal{R}_{\mu\nu}$ from both sides of the space-times and the above mentioned boundary condition holds. However, in doing so the deformation given by (4) ceases to be. Therefore, to find the deformations caused by the extrinsic curvature, such special conditions are not applied and they are not needed. To make it clear how it works, one can first take (14) and contracting with the metric $\mathcal{G}^{AB}$ and using the confinement conditions in (19) and (14), one can find

$$\mathcal{R} = -\frac{2}{3}\alpha_* T^* , \tag{22}$$

and also

$$\mathcal{R}_{AB} = \alpha_* \left(T^*_{AB} - \frac{1}{3} T^* \mathcal{G}_{AB} \right) , \tag{23}$$

where the components can be obtained in the Gaussian frame $\{Z^A_{,\mu}, \eta^A\}$. Hence, we have

$$\mathcal{R}_{AB} Z^A_{,\mu} Z^B_{,\nu} = \alpha_* \left(T^*_{AB} - \frac{1}{3} T^* \mathcal{G}_{AB} \right) Z^A_{,\mu} Z^B_{,\nu} = \alpha_* \left(T^*_{\mu\nu} - \frac{1}{3} T^* g_{\mu\nu} \right) .$$

As we can see, the right side of the previous equation is the same expression as that verified in the IDL condition which must coincide with the extrinsic curvature in the brane-world. However, this is not true inasmuch as the left side of the equation is the contracted form of Gauss equations. We may check it writing the components in the Gaussian frame of (14) and obtain (15). As a consequence of Gauss-Codazzi-Ricci equations, in the higher dimensional space-time structure, the direct contraction of the Ricci equation gives

$$R = \mathcal{R} - (K^2 + H^2) + 2\frac{\partial h}{\partial y}\,, \tag{24}$$

where $\mathcal{R}_{AB}\eta^A\eta^B = \frac{\partial h}{\partial y} + K^2$.
Taking (22) and (24), and applying in (15), one can find

$$R_{\mu\nu} - \frac{\partial k_{\mu\nu}}{\partial y} - 2k^{\rho}_{\mu}k_{\rho\nu} + hk_{\mu\nu} = \alpha_* \left(T^*_{\mu\nu} - \frac{1}{3}T^* g_{\mu\nu}\right)\,. \tag{25}$$

In fact, it shows that the IDL condition only can be obtained by imposing some serious constraints on the embedding process. Still, if we want to insist on obtaining the IDL condition, we must assume some simplifying conditions. Let the brane-world has a boundary such that it separated into two sides labeled (+) and (-) regions. The difference calculated in each side of the brane-world is zero when $y \to 0$. In other words, we have the same equation obtained in (25) the more we approach $y = 0$ from each side inasmuch as there is not a effective distinction in the riemannian geometry when evaluated from each side to the other. This situation turns to be quite different when the Z_2 is considered. In this case, the extrinsic curvature (or any object that could access extra-dimensions) has its image mirrored in the brane-world (which acts as a mirror). For instance, if we have $k^+_{\mu\nu} = -k^-_{\mu\nu}$, the derivatives $\left[-\left(\frac{\partial k_{\mu\nu}}{\partial y}\right)\right] = \alpha_* \left(T^*_{\mu\nu} - \frac{1}{3}T^* g_{\mu\nu}\right)$ constantly change when they approach $y \to 0$. By using the mean value theorem in the interval $[-y, y]$, we can evaluate the difference between both sides and obtain

$$\left[-\left(\frac{\partial k_{\mu\nu}}{\partial y}\right)\right] = \frac{-k^+_{\mu\nu} + k^-_{\mu\nu}}{y}.$$

Denoting $[X] = X^+ - X^-$ and $X = \bar{X}(x)\delta(y)$, we have

$$y[X] = \int_{-y}^{y} \frac{d}{d\xi}(|\xi|X)d\xi = \int_{-y}^{y} \frac{\partial|\xi|}{\partial\xi}Xd\xi + \int_{-y}^{y} |\xi|\frac{dX}{d\xi}d\xi$$

$$= \int_{-y}^{y} \frac{\partial|\xi|}{\partial\xi}\bar{X}\delta(\xi)d\xi + \int_{-y}^{y} |\xi|\frac{\partial\delta(\xi)}{\partial\xi}\bar{X}d\xi = 2\bar{X}\,.$$

In the case that $[X] = \alpha_* \left(T^*_{\mu\nu} - \frac{1}{3}T^* g_{\mu\nu}\right)$, we obtain Lanczos equation

$$k^+_{\mu\nu} - k^-_{\mu\nu} = -2\alpha_* \left(T^*_{\mu\nu} - \frac{1}{3}T^* g_{\mu\nu}\right)\,, \tag{26}$$

that describes the jump of the extrinsic curvature in the background separation point $y = 0$. Hence, the IDL condition is obtained when the Z_2 symmetry is applied to (26) obtaining

$$k_{\mu\nu} = \alpha_* \left(T^*_{\mu\nu} - \frac{1}{3}T^* g_{\mu\nu}\right)\,. \tag{27}$$

The use of Z_2 symmetry induces a serious constraint on the embedding differentiable structure. Once a perturbation occurs in a point of the background it is mirrored in the brane-world background and two tangent vectors on each side can be defined. The projections of these vectors point in opposite directions which means that the embedding differentiable functions cannot be properly defined (Maia, 2004).
In summary, the theoretical scheme presented here are consequence of a fundamental perturbational process stated by Nash's embedding theorem. Nash's perturbation method innovates in two basic aspects: first, there is no need to apply the restrictive convergent series power of analytical function hypothesis to make an embedding between two manifolds. Secondly, the perturbational nature of the process we can obtain dynamical equations as well as integrating them such as in Cauchy's problem in Mechanics and it also gives a prescription on how to construct geometrical structures by deforming simpler ones. It seems that this geometric perturbation process has to do with the formation of structures in the early universe. When Nash's theorem is applied to physics, it provides a general mathematical tool appropriated to the brane-world program. In the model independent covariant formulation the extrinsic curvature appears as an independent symmetric tensor field which evolves together with the brane-world dynamics. Interestingly, the presence of the independent symmetric rank-two tensor field has been considered long before the observation of the accelerated expansion of the universe under different motivations and circumstances as a possible repulsive gravitational field (Isham et al., 1971).

5. Cosmological applications

After all these geometrical considerations, in the following we summarize important ideas of works on the applications of Nash's theorem to Cosmology as seen in (Maia et al., 2009; 2005; Odon, 2010; Capistrano, 2010). The first step to do is to defined the background geometry. The standard Friedman-Lemaître-Robertson-Walker(FLRW) model is sufficiently simple to make it locally embedded in a 5-dimensional flat space, satisfying Nash's differentiable conditions. Therefore, it can be taken as a background cosmology, which can be deformed along the fifth-dimension. However, here the effects of the extrinsic geometry are shown in the FLWR background only (that is without perturbations).

5.1 The Cosmological Constant problem

The so-called *Cosmological Constant problem* had its first seeds planted in 1916, with the ideas of Nernst (Nernst, 1916). He studied the non-vanishing vacuum energy density that was fulfilled with radiation-only content, which was confirmed by the Casimir effect in 1948 (Casimir, 1948; Mostepanenko, 1997; Jaffe, 2005). In late 1920's, Pauli (Pauli, 1933; Straumann, 2002; Rugh, 2002) made studies about the gravitational influence of the vacuum energy density of the radiation field, suggesting a conflict between the vacuum energy density and gravitation. If vacuum energy density is considered, then gravity should be dispensed. Intriguingly, the conflicting Pauli's results passed unnoticed by scientific community. Only on subsequent decades, the observations of quasars in the mid-late of the 1960's suggested the reconsideration of Λ (Petrosian, 1974).
Here we refer to the cosmological constant problem described in (Weinberg, 1989). Using the semiclassical Einstein's equations in General Relativity the quantum vacuum can be described as a perfect fluid with state equation $p_v = -<\rho>_v=$ constant (Zel'dovich, 1967):

$$R_{\mu\nu} - \frac{1}{2}Rg_{\mu\nu} + \Lambda g_{\mu\nu} = 8\pi G T^{m}_{\mu\nu} + 8\pi G <\rho>_v g_{\mu\nu} \,, \tag{28}$$

where $T^{m}_{\mu\nu}$ stands for the classical sources. Comparing the constant terms in both sides of this equation we obtain $\Lambda/8\pi G =< \rho >$, or as it is commonly stated, the cosmological constant is the vacuum energy density. However, current observations tell that $\Lambda/8\pi G \sim 10{-}47(\text{Gev})^4$ (here, $c = 1$). On the other hand, admitting that quantum field theory holds up to the Planck scale, the vacuum energy density would be $< \rho >_v \sim (10^{19}\text{Gev})^4 = 10^{76}(\text{Gev})^4$. This difference cannot be resolved by any known theoretical procedure in quantum field theory. Even supposing that quantum field theory holds to the Tev scale or less, the difference would be still too large to compensate. This difficulty has become to known as the cosmological constant problem.

In one proposal to solve this problem, a scalar field is added to the right hand side of Einstein's equations, such that it adjusts the difference between the two constants (Chen & Wu, 1990; Waga, 1993; Caldwell & Linder, 2005; Lima, 2004; Padmanabham, 2007). Of course, this scalar field must also agree the other cosmological conditions, such as the structure formation, the past and present inflationary periods, and the smooth transition to and from the standard cosmology period. The adjustments of this field to such conditions have proven to be not so simple. A more geometrical approach to the problem, the Einstein-Hilbert action principle has been tentatively modified, using for example higher derivative Lagrangians, or more generally a Lagrangean defined by an arbitrary function of the Ricci curvature, in the so called f(R) theories (Capozziello et.al., 1998). However, it becomes a necessity to give a meaning to the resulting action principle, which is after all a fundamental principle. In comparison, the Einstein-Hilbert principle has a specific meaning, stating that the geometry of the space-time must be as smooth as possible. Furthermore, it comes after Newton's gravitational law, when it is expressed geometrically, so that at the end, it is founded in experimental facts. In this respect, given the arbitrariness of f(R), it is not at all clear that the present astrophysical observations are sufficient to decide on such function (Sokolowski, 2007). Another fine-tuning approach suggests new two fundamental scales (Alfonso-Faus, 2009), the cosmological quantum black hole (CQBH) and the quantum black hole (QBH) in order to solve the ambiguity of Λ in the cosmological problem by using an appropriate choice of parameters, e.g $\hbar \sim 10^{-122}$ that lead from the Planck scale to the Cosmological scale without conflicting with $\Lambda\hbar \sim 1$, instead of using $G = c = \hbar = 1$.

As also suggest in (Alfonso-Faus, 2009), we must emphasize that the previous difference in the cosmological problem is not only numerical, but it is mainly conceptual, resulting from the superposition of two incompatible ground states for the gravitational field in General Relativity: The flat Minkowski ground state was chosen to be the reference of curvature, but the experimental evidences of $\Lambda/8\pi G \neq 0$ however small, point to a De Sitter ground state, which is conceptually incompatible with the Minkowski's choice. The implications being that particles and fields, their masses and spins defined by the Casimir operators of the De Sitter group are different from those defined by the Poincaré group, and they coincide only when Λ vanishes. The above numerical and conceptual conflicts can be resolved with the Schlaefli embedding conjecture as implemented by Nash, where the De Sitter and Minkowski space-times may coexist. Indeed, in (15), $\Lambda/8\pi G$ is a gravitational component resulting from the gravitational equations in the embedding space. However, the vacuum energy density $< \rho >_v$ is a confined quantity in the space-time, regardless of the perturbations of its metric. Finally, the presence of the extrinsic curvature $k_{\mu\nu}$ in the conserved quantity $Q_{\mu\nu}$ of (15), imply that those constants cannot be canceled without imposing a constraint on the extrinsic curvature, which is now part of the gravitational dynamics in the embedding space (Maia et al., 2009; Capistrano & Odon, 2010).

5.2 The accelerated expansion

A interesting situation occurs when Nash's theorem is applied to the Dark energy problem as proposed in (Maia et al., 2005). One of the most known brane-world models is the Randall-Sundrum type II (RSII) (Randall, 1999, b). When applied to Cosmology, the vacuum energy density in a 3-brane is still smaller than the one predicted by quantum field theory, which means that the cosmological constant problem persists, even though the fundamental Tev scale energy is preserved. A similar situation occurs when dealing with the Dark energy problem in which the RS model II provides the following modified Friedmann equation

$$\left(\frac{\dot{a}}{a}\right)^2 = \frac{8\pi}{3m_{pl}^2}\rho + \frac{16\pi^2}{9m_5^6}\rho^2 , \tag{29}$$

where m_5 is the 5-dimensional planck scale, m_{pl} is the 4-dimensional planck scale. The correction term corresponds to the square of the energy density ρ^2 of the confined matter (Tujikawa, 2004; Tujikawa et.al., 2004; Maia, 2004). As it is well known, this result is not compatible with recent observational data (Komatsu et.al., 2011; Jarosik et.al., 2011) since the additional term on Friedmann's equation, i.e, the energy density ρ^2, provides a deceleration scenario of the universe, besides affecting the nucleosynthesis of large structures. To remedy this situation, other attempts have been studied, such as particular classes of bulk and brane scalar potentials (Langlois, 2001), notwithstanding they lead to a fine-tuning mechanisms.

In (Maia et al., 2005), the Friedmann-Lemaître-Robertson-Walker (FLRW) line element was embedded in a 5-dimensional space with constant curvature bulk space whose geometry satisfy Einstein's equations with a cosmological constant given by (14). When the equations are written in the Gaussian frame defined by the embedded space-time, we obtain a larger set of gravitational field equations. The general solution of (16) for the FLRW geometry was found to be

$$k_{ij} = \frac{b}{a^2}g_{ij}, \quad i,j = 1,2,3, \quad k_{44} = \frac{-1}{\dot{a}}\frac{d}{dt}\frac{b}{a} , \tag{30}$$

where we notice that the function $b(t) = k_{11}$ remains an arbitrary function of time. As a direct consequence of the confinement of the gauge fields, equation (16) is homogeneous, meaning that one component $k_{11} = b(t)$ remains arbitrary. Denoting the Hubble and the extrinsic parameters by $H = \dot{a}/a$ and $B = \dot{b}/b$, respectively, we may write all components of the extrinsic geometry in terms of B/H as follows

$$k_{44} = -\frac{b}{a^2}(\frac{B}{H} - 1)g_{44}, \tag{31}$$

$$K^2 = \frac{b^2}{a^4}\left(\frac{B^2}{H^2} - 2\frac{B}{H} + 4\right), \quad h = \frac{b}{a^2}(\frac{B}{H} + 2) \tag{32}$$

$$Q_{ij} = \frac{b^2}{a^4}\left(2\frac{B}{H} - 1\right)g_{ij}, \qquad Q_{44} = -\frac{3b^2}{a^4}, \tag{33}$$

$$Q = -(K^2 - h^2) = \frac{6b^2}{a^4}\frac{B}{H} , \tag{34}$$

Next, by replacing the above results in (15) and applying the conservation laws, we obtain the Friedmann equation modified by the presence of the extrinsic curvature, i.e.,

$$\left(\frac{\dot{a}}{a}\right)^2 + \frac{\kappa}{a^2} = \frac{4}{3}\pi G\rho + \frac{\Lambda_*}{3} + \frac{b^2}{a^4} . \tag{35}$$

When compared with the phenomenological quintessence phenomenology with constant EoS we have found a very close match with the golden set of cosmological data on the accelerated expansion of the universe.

Notice that we have not used the Israel-Lanczos condition (27) as used in (Randall, 1999, b). If we do so, in the case of the usual perfect fluid matter, then we obtain in (35) a term proportional to ρ^2. It is possible to argue that the above energy-momentum tensor $T_{\mu\nu}$ also include a dark energy component in the energy density ρ. However, in this case we gain nothing because we will be still in darkness concerning the nature of this energy. Finally, as it was shown in the previous section, the Israel-Lanczos condition requires that the four-dimensional space-time behaves like a boundary brane-world, with a mirror symmetry on it, which is not compatible with the regularity condition for local and differentiable embedding.

Therefore, the conclusion from (Maia et al., 2005) is that the extrinsic curvature is a good candidate for the universe accelerator. In the next section we start anew, with a mathematical explanation on why only gravitation access the extra dimensions using the mentioned theorem of Nash on local embeddings, and the geometric properties of spin-2 fields defined on space-times.

5.3 The dynamics of extrinsic curvature

Hitherto, we did not have at the time any previous information on the dynamics of the extrinsic curvature. The only widely accepted relation of that curvature with matter sources is the Israel-Lanczos boundary condition, as applied to the Randall-Sundrum brane-world cosmology. However, this condition fixes once for all the extrinsic curvature, so that it does not follow the dynamics of the brane-world. Thus, a more fundamental explanation for the dynamics of the extrinsic curvature is required. In the purpose of complementing the study shown in (Maia et al., 2005) is to show that the extrinsic curvature behaves as an independent spin-2 field whose effect on the gravitational field is precisely the observed accelerated expansion.

From the theoretical point of view, it would be a satisfactory solution for the dark energy problem if the $b(t)$ (35) function was a unique solution, but, in fact, it depends on a choice of a family of solutions for the extrinsic curvature induced by the homogeneity of the Codazzi equation (16) which is well-known equation in differential geometry. Thus, to be free from these pathologies a proper mechanism or an additional dynamical equation for extrinsic curvature should be implemented. In spite of Brane-world models get some attention on recent years due to several options for dark energy, their mechanisms are still not completely understood or justified. These are mostly based on specific models using special conditions. For such large scale phenomenology as the expansion of the universe, a general theory based on fundamental principles and on solid mathematical foundations is still lacking.

Another aspect of Nash's theorem is that the extrinsic curvature are the generator of the perturbations of the gravitational field along the extra dimensions. The symmetric rank-2 tensor structure of the extrinsic curvature lends the physical interpretation of an independent spin-2 field on the embedded space-time. The study of linear massless spin-2 fields in Minkowski space-time dates back to late 1930s (Pauli, 1939). Some years later, Gupta (Gupta, 1954) noted that the Fierz-Pauli equation has a remarkable resemblance with the linear approximation of Einstein's equations for the gravitational field, suggesting that such equation could be just the linear approximation of a more general, non-linear equation for massless spin-2 fields. In reality, he also found that any spin-2 field in Minkowski space-time must satisfy an equation that has the same formal structure as Einstein's equations. This amounts to saying that, in the same way as Einstein's equations can be obtained by an infinite sequence of infinitesimal perturbations of the linear gravitational equation, it is possible to obtain a non-linear equation for any spin-2 field by applying an infinite sequence of infinitesimal perturbations to the Fierz-Pauli equations. The result is an Einstein-like system of equations, the Gupta equations (Gupta, 1954; Fronsdal, 1978).

In order to write the Gupta equations for the extrinsic curvature $k_{\mu\nu}$ of an embedded Riemannian geometry with metric $g_{\mu\nu}$, we may use an analogy with the derivation of the Riemann tensor, defining the "connection" associated with $k_{\mu\nu}$ and then the corresponding Riemann tensor, but keeping in mind that the geometry of the embedded space-time is already defined by the metric tensor $g_{\mu\nu}$. Let us define the tensor

$$f_{\mu\nu} = \frac{2}{K} k_{\mu\nu}, \text{ and } f^{\mu\nu} = \frac{2}{K} k^{\mu\nu}, \tag{36}$$

so that $f^{\mu\rho} f_{\rho\nu} = \delta^{\mu}_{\nu}$. Subsequently, we construct the "Levi-Civita connection" associated with $f_{\mu\nu}$, based on the analogy with the "metricity condition". Let us denote by $||$ the covariant derivative with respect to $f_{\mu\nu}$ (while keeping the usual $(;)$ notation for the covariant derivative with respect to $g_{\mu\nu}$), so that $f_{\mu\nu||\rho} = 0$. With this condition we obtain the "f-connection"

$$\Upsilon_{\mu\nu\sigma} = \frac{1}{2}\left(\partial_\mu f_{\sigma\nu} + \partial_\nu f_{\sigma\mu} - \partial_\sigma f_{\mu\nu}\right)$$

and

$$\Upsilon_{\mu\nu}{}^{\lambda} = f^{\lambda\sigma}\,\Upsilon_{\mu\nu\sigma}$$

The "f-Riemann tensor" associated with this f-connection is

$$\mathcal{F}_{\nu\alpha\lambda\mu} = \partial_\alpha \Upsilon_{\mu\lambda\nu} - \partial_\lambda \Upsilon_{\mu\alpha\nu} + \Upsilon_{\alpha\sigma\mu}\Upsilon^{\sigma}_{\lambda\nu} - \Upsilon_{\lambda\sigma\mu}\Upsilon^{\sigma}_{\alpha\nu}$$

and the "f-Ricci tensor" and the "f-Ricci scalar", defined with $f_{\mu\nu}$ are, respectively,

$$\mathcal{F}_{\mu\nu} = f^{\alpha\lambda}\mathcal{F}_{\nu\alpha\lambda\mu} \text{ and } \mathcal{F} = f^{\mu\nu}\mathcal{F}_{\mu\nu}$$

Finally, write the Gupta equations for the $f_{\mu\nu}$ field

$$\mathcal{F}_{\mu\nu} - \frac{1}{2}\mathcal{F} f_{\mu\nu} = \alpha_f \tau_{\mu\nu} \tag{37}$$

where $\tau_{\mu\nu}$ stands for the source of the f-field, with coupling constant α_f. Note that the above equation can be derived from the action

$$\delta \int \mathcal{F}\sqrt{|f|}dv$$

Note also that, unlike the case of Einstein's equations, here we have not the equivalent to the Newtonian weak field limit, so that we cannot tell about the nature of the source term $\tau_{\mu\nu}$. For this reason, we start with the simplest Ricci-flat-like equation for $f_{\mu\nu}$, i.e.,

$$\mathcal{F}_{\mu\nu} = 0\,. \tag{38}$$

For simplicity, the equations were written in 5-d but it remains valid for a higher dimensional bulk. With this new set of equations, in principle the homogeneity of Codazzi equations can be lift. The work on Gupta's theorem is currently on progress and applications to the Dark energy problem have been recently investigated. A more detailed discussion can be found in (Maia et.al., 2011; Capistrano, 2010)

5.4 Local gravity and structure formation

Current local dark matter observations based on gravitational micro-lensing, optical and x-ray astronomical observations tell that the local dark matter phenomenology is different from that in cosmology. In fact, there is no evidence that the same structure formation caused by geometric perturbations similar to the cosmological situation is still present around the already formed structures, at least at the same rate. Gravitational lensing evidences a gravitational field with a certain metric symmetry. In some cases the dark matter gravitational field is anchored to an observed structure (spiral galaxies, gravitational halos in clusters etc.) and its metric symmetry is the same as that of the observed structure. Until very recently these observations indicated that the source of the local dark matter gravitation (that is, the dark matter itself) was usually attached to galaxies and clusters. In other cases, as in the example of the Abell 520 cluster (MS0451+02), the dark matter gravitational field seems to be away from any baryon substructures. Another recent evidence of the local dark matter gravity is observed through x-ray astronomy in near colliding clusters (exemplified by the bullet cluster 1E0657-558). The observed effect is the formation of a sonic bullet-like substructure moving through the intercluster plasma, long before the cluster themselves collide. This is attributed to the collision of the real dark matter halos assumed to be around the colliding clusters. Admitting Newtonian gravity, the center of mass of the moving object coincide with the Newtonian halos. Such wide range of experimental evidences from cosmology to local gravity suggests the necessity of a comprehensive analysis of the dark matter gravitational field *per se*, regardless of any other attributes that dark matter may eventually possess. Therefore, it is possible that the theoretical power spectrum obtained from (35) coincide with the observed one. In a preliminary analysis, we obtained a power spectrum which is similar to the power spectrum from the cosmic microwave background radiation obtained from the WMAP experiment. On the other hand, Nash's geometric perturbations may be present as a local

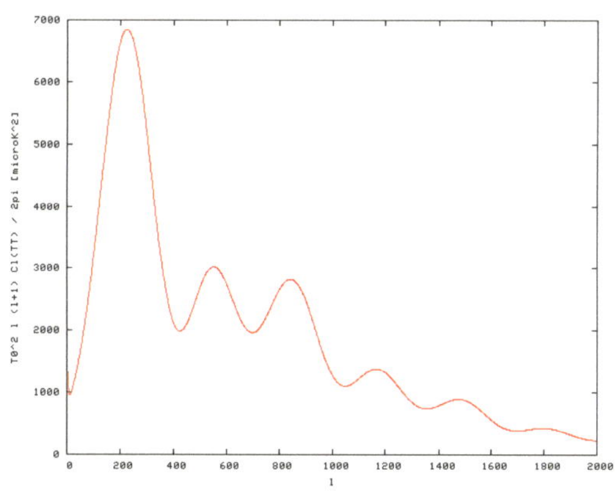

Fig. 1. The theoretical power spectrum calculated with the CAMB for $-1 \leq \omega_0 \leq -1/3$, Massive Neutrinos=1, massless neutrinos =3.04.

process, as for example in young galaxies and in cluster collisions. However, in most other cases there are not sufficient experimental evidences that it is still going on. The formation of large structures in the early universe has been mostly attributed to gravitational perturbations produced by other than baryons sources, generally referred to as the dark matter component of the universe. In the present case, the extrinsic curvature solution of (37) should have an observable effect in space-time, independently of the perturbations.

6. Conclusions

The fundamental problems of Modern cosmology are three-fold: the Λ paradigm, dark energy and dark matter. With the high developing of the observational methods and devices, these problems have demanded a series of theoretical needs also stimulating the development of theories beyond Einstein's. Our approach here was to stress the study of the embedding process between manifolds and its necessity for the contemporary physics. By its own nature, the embedding between manifolds is a perturbational process of geometry and the recent fundamental problems on Cosmology seem to point to the same question: what is gravity and how it can be perturbed? The studies on the extrinsic curvature have been made at length in the literature but with no the required accuracy by using junction conditions that induce the extrinsic geometry to be minimized to gauge fields and matter. Since we understand the embedding conditions, the using of any junction condition can be dispensed and the geometrical limitation for the embedding can be lifted.

In the early days of Riemannian geometry, the embedding between two Riemannian geometries was such a problem due to the fact the need of a relative geometric reference was missing. The existence of a background geometry is necessary to fix the ambiguity of the Riemann curvature of a given manifold, without a reference structure. General relativity solves this ambiguity problem by specifying that the tangent Minkowski space is a flat plane, as decided by the Poincaré symmetry, and not by the Riemann geometry itself. Such difficulty was known by Riemann himself, when he acknowledged that his curvature tensor defines a class of objects and not just one (Riemann, 1854). Unlike the case of string theory the bulk geometry is a solution of Einstein's equations, acting as a dynamic reference of shape for all embedded Riemann geometries. This generality follows from the remarkable accomplishment of Nash's theorem on embedded geometries. Nash showed that any Riemannian geometry can be generated by continuous sequence of infinitesimal perturbations defined by the extrinsic curvature. It seems natural that this result provides the required geometrical structure to describe a dynamically changing universe. This plays an essential feature for a new gravitational theory.

The four-dimensionality of the embedded space-times is determined by the dualities of the gauge fields, which corresponds to the equivalent concept of confinement gauge fields and ordinary matter in the brane-world program. However, this confinement implies that the extrinsic curvature cannot be completely determined, simply because Codazzi's equations becomes homogeneous. Incidently, the Randall-Sundrum model avoids this problem by imposing the Israel-Lanczos condition on a fixed boundary-like brane-world. Since the extrinsic curvature assumes a fundamental role in Nash's theorem, an additional equation is required. Recently, works on the subject noted that the extrinsic curvature is an independent rank-2 symmetric tensor, which corresponds to a spin-2 field defined on the embedded space-time. However, as it was demonstrated by Gupta, any spin-2 field satisfy an Einstein-like equation. After the due adaption to an embedded space-time, the analysis of Gupta's equations for the extrinsic curvature of the FLWR geometry and the study of the behavior of the extrinsic curvature at the various stages of the evolution of the universe is still an open question and the works on the subject are currently on progress.

The embedding of a space-time manifold into another defined by the Einstein-Hilbert principle may lead to an interesting gravitational theory, not only because its mathematical consistency provided by the Schlaefli conjecture as resolved by Nash's theorem, but mainly because it can meet the demands of modern cosmology, with the minimum of additional assumptions which can be fundamental for the development of a soft-after gravitational quantum field theory.

7. References

Komatsu, E.; Smith, K.M; Dunkley, J.; Bennett, C.L.; Gold, B; Hinshaw, G.; Jarosik, N.; Larson, D.; Nolta, M.R.; Page, L.; Spergel, D.N.; Halpern, M.; Hill, R.S.; Kogut, A.; Limon, M.; Meyer, S.S.; Odegard, N.; Tucker, G.S; Weiland, J.L; Wollack, E. & Wright, E.L. (2011). Seven-Year Wilkinson Microwave Anisotropy Probe (WMAP) Observations: Cosmological Interpretation. *The Astrophysical Journal Suppement Series*, Vol. 192, No. 18, (Feb 2011) 47pp., ISSN: 1538-4365 (online).

Jarosik, N.; Bennett, C.L; Dunkley, J.; Gold, B.; Greason, M.R; Halpern, M.; Hill, R.S.; Hinshaw, G.; Kogut, A.; Komatsu, E.; Larson, D.; Limon, M.; Meyer, S.S.; Nolta, M.R; Odegard, N.; Page, L.; Smith, K.M.; Spergel, D.N.; Tucker, G.S.; Weiland, J.L; Wollack, E. & Wright, E.L. (2011). Seven-Year Wilkinson Microwave Anisotropy Probe (WMAP) Observations: Sky Maps, Systematic Errors, and Basic Results. *The Astrophysical Journal Suppement Series*, Vol. 192, No. 14, (Feb 2011), 15pp., ISSN: 1538-4365 (online).

Planck, M. (1914). *The theory of Heat Radiation*, P. Blakiston Sons & Co., Philadelphia.

Decca, R.S; López, D.; Fischbach, E.; Klimchitskaya, G.L; Krause, D.E & Mostepanenko, V.M . (2007). Novel constraints on light elementary particles and extra-dimensional physics from the Casimir effect. *The European Physical Journal C*, Vol. 51, No. 4, (Jul 2007) 963-975, ISSN: 1434-6052.

Dirac, P.A.M. (1959). Fixation of Coordinates in the Hamiltonian Theory of Gravitation. *Physical Review*, Vol. 114, No. 3, (May 1959) 924-930.

Arnowitt, R; Deser, S. & Misner, C. (1962). *The dynamics of General Relativity*, In: *Gravitation: An introduction to current research*, L. Witten, (2nd Ed.), John Wiley & Sons, p.(227), ISBN: 0036-8075 (print), New York.

Kuchar, K.V. (1991). Time and Interpretations of Quantum Gravity, *4th Canadian Conference on general Relativity and relativistic Astrophysics*, ISBN: , Canada, May 1991, World Scientific, Winnipeg.

Arkani Hamed, N.; Dimopoulos, S. & Dvali, G. The hierarchy problem and new dimensions at a millimeter. *Physical Letters B*, Vol. 429, No. 3/4, (June 1998) p. 263-272, ISSN: 0370-2693.

Carter, B & Uzan, J.P. (2001). Reflection symmetry breaking scenarios with minimal gauge form coupling in brane world cosmology. *Nuclear Physics*, vol. 606, No. 1/2 , (Jul 2001) p. 45-58, ISSN: 0550-3213.

Randall, L. & Sundrum, R. (1999). Large Mass Hierarchy from a Small Extra Dimension. *Physical Review Letters*, vol. 83, No. 17, (Oct 1999) p. 3370-3373, ISSN: 1079-7114 (online).

Randall, L. & Sundrum, R. (1999). An alternative to compactification. *Physical Review Letters*, vol. 83, No. 23, (Dec 1999) p. 4690-4693, ISSN: 1079-7114 (online).

Dvali, G; Gabadadze, G. & Porrati, P. (2000). 4D gravity on a brane in 5D Minkowski space. *Physics Letters B*, vol. 485, No. 1-3, (Jul 2000) p. 208-214, ISSN: 0370-2693.

Sahni, V. & Shtanov, Y. (2002). New vistas in Braneworld Cosmology. *International Journal of Modern Physics D*, vol. 11, No. 10, (May 2002) p. 1515-1521, ISSN: 1793-6594 (online)

Sahni, V. & Shtanov, Y. (2003). Braneworld models of Dark Energy. *Journal of Cosmology and Astroparticle Physics*, vol. 2003, No. 14, (Nov 2003), ISSN: 1475-7516.

Shiromizu, T.; Maeda, K. & Sasaki, M. (2000). The Einstein equations on the 3-brane world. *Physical Review D*, vol. 62, No. 2, (June 2000) 024012 (6 pages), ISSN: 1550-2368 (online).

Dick, R. (2001). Brane worlds. *Classical and Quantum Gravity*, vol. 18, No. 17, (Sept 2001) R1, ISSN: 1361-6382 (online).

Hogan, C.J. (2001). Classical gravitational-wave backgrounds from formation of the brane world.*Classical and Quantum Gravity*, vol. 18, No. 19, (Oct 2001) 4039, ISSN: 1361-6382 (online).

Deffayet, C.; Dvali, G. & Gabadadze, G. (2002). Accelerated universe from gravity leaking to extra dimensions. *Physical Review D*, vol. 65, No. 4, (Jan 2002) 044023 (9 pages), ISSN: 1550-2368 (online).

Alcaniz, J.S. (2002). Some observational consequences of brane world cosmologies *Physical Review D*, vol. 65, No. 12, (Jun 2002) 123514 (6 pages), ISSN: 1550-2368 (online).

Jain, D.; Dev, A. & Alcaniz, J.S. (2002). Brane world cosmologies and statistical properties of gravitational lenses. *Physical Review D*, vol. 66, No. 8, (Oct 2002) 083511 (6 pages), ISSN: 1550-2368 (online).

Lue, A. (2006). The phenomenology of Dvali-Gabadadze-Porrati cosmologies. *Physics Reports*, vol. 423, No. 1, (Jan 2006) p. 1-48, ISSN: 0370-1573.

Maia, M.D.; Capistrano, A.J.S & Monte, E.M. (2009). The Nature of the Cosmological Constant problem. *International Journal of Modern Physics A*, Vol. 24, No. 08-09, (Sep 2009) p. 1545-1548, ISSN: 1793-656X (online).

Maia, M.D.; Monte, E.M.; Maia, J.M.F. & Alcaniz, J.S. (2005). On the geometry of Dark Energy. *Classical and Quantum Gravity*, vol. 22, No. 9, (April 2005) p. 1623-1636, ISSN: 1361-6382 (online)

Maia, M.D.; Silva, N. & Fernandes, M.C.B. (2007). Brane-world Quantum Gravity. *Journal of High Energy Physics*, vol. 2007, No. 0407:047, (April 2007) (13 pages), ISSN: 1029-8479.

Maia, M. D. ; Capistrano, A. J. S. ; Muller, D. (2009). Perturbations of Dark matter Gravity. *International Journal of Modern Physics D*, vol. 18, No. 8, (Aug 2009), p. 1273-1289, ISSN: 1793-6594.

Odon, P. I & Capistrano, A. J. S. (2010). Remarks on the foundations of geometry and immersion theory. Physica Scripta, vol. 81, No. 4, (March 2010) p. 045101, ISSN: 1402-4896 (Online).

Riemann, B. (1854). On the Hypotheses that Lie at the Bases of Geometry (1868), English translation by W. K. Clifford, *Nature*, vol. 8, No. 183, (May 1873) p. 14-17, ISSN: 0028-0836 (online)

Schlaefli, L. (1873). Sull'uso delle linee lungo le quali il valore assoluto di una funzione è costante. *Annali di Matematica pura ed applicata*, Vol. 6, No. 1, (1873) p.1-20, ISSN: 1618-1891 (online)

Eisenhart, L.P. (1997). *Riemannian Geometry*, Princeton U.P., 8th ed. (1997), ISBN: 0691-08026-7, New Jersey.

Nash, J. (1956). The imbedding problem for Riemannian manifolds. *Annals of mathematics*, Vol. 63, No. 01, (Jan 1956) p. 20-63, ISSN: 0003-486X (online)

Isham, C.J; Salam, A. & Strathdee, J. (1971). F-Dominance of Gravity. *Physical Review D*, vol. 3, No. 4, (Feb 1971) p. 867-873, ISSN: 1550-2368 (online)

Maia, M.D.; A. J. S. Capistrano, A.J.S.; Alcaniz, J. S. & Monte, E. M. (2011). The Deformable Universe. ArXiv:1101.3951 (to appear in *General Relativity and Gravitation*)

Capistrano, A. J. S. (2010). On the relativity of shapes. *Apeiron (montreal)*, vol. 17, No. 2, p. 42-58, (April 2010), ISSN: 0843-6061 (Online).
Crampin, M. & Pirani, F.A.E. (1986). *Applicable Differential Geometry*, Cambridge U.P., ISBN: 0521-23190-6 , New York.
Hamilton, R. (1982). Three Manifolds with positive Ricci curvature. *Journal of Differential Geometry*, Vol. 17, No. 2, (1982) p. 255-306, ISSN: 0022-040X.
Perelman, G. (2002). The entropy formula for the Ricci flow and its geometric applications. ArXiv:math/0211159 [math.DG].
Israel, W. (1966). Singular hypersurfaces and thin shells in general relativity. *Il Nuovo Cimento*, Vol. 44, No. 2, (July 1966) p. 1-14, ISSN: 1826-9877 (online)
Nernst, W. (1916). Über einen Versuch von quantentheoretischen Betrachtungen zur Annahme stetiger Energieänderungen zurückzukehren, *Verhandlungen der deutche physikalische Gesellschaft*, Vol. 18, No. 4, (1916) p. 83-116, ISBN-10: 1148398287, ISBN-13: 9781148398280
Casimir, H.B.G. (1948). On the attraction of two perfectly conducting plates. *Proceedings of the Koninklijke Nederlandse Akademie van Wetenschappen*, Vol. 51, No. 7, (1948) p.793-795, ISSN: 0920-2250.
Mostepanenko, V.M. & Trunov, N.N. (1997). *The Casimir effect and its application*, Oxford Claredon Press, ISBN-10: 0198539983, ISBN-13: 978-0198539988, New York.
Jaffe, R.L. (2005). Casimir effect and quantum vaccum. *Physical Review D*, vol. 72, No. 2, (July 2005) (5 pages) 021301(R), ISSN: 1550-2368 (online)
Pauli, W. (1933). Die allgemeinen Prinzipien der Wellmechanik, *Handbuch der Physik*, Springer-Verlag, Verlagsbuchhandlung, Vol. 24, No. 1, (1933) p. 221, ISSN: 0085-140X.
Straumann, N. (2002). The history of the cosmological constant problem, Arxiv: 0208027[gr-qc].
Rugh, S.E & Zinkernagel, H. (2002). The quantum vacuum and the cosmological constant problem. *Studies in History and Philosophy of Modern Physics*, Vol. 33, No. 4, (Dec 2002) p. 663-705, ISSN: 1355-2198.
Petrosian, V. (1974). Confrontation of Lemaître Models and the Cosmological Constant with Observations, In: Confrontation of Cosmological Theories with Observational Data, M.S Longair (edt.), p. 31-46, ISBN: 90-227-0457-0, D. Reidel Publishing, Boston MA.
Weinberg, S. (1989). The Cosmological constant problem. *Reviews of Modern Physics*, Vol. 61, No.1, (jan 1989) p. 1-23, ISSN: 1539-0756 (online)
Zel'dovich, Y.B. (1967). Cosmological Constant and Elementary Particles. *Journal of Experimental and Theoretical Physics Letters*, Vol. 6, No.9, (Nov 1967), p. 316-317, ISSN: 0021-3640.
Chen, W. & Wu, Yong-Shi. (1990). Implications of a cosmological constant varying as R^{-2}. *Physical Review D*, vol. 41, No. 2, (Jan 1990) p.695-698, ISSN: 1550-2368 (online)
Waga, I. (1993). Decaying vacuum flat cosmological models - Expressions for some observable quantities and their properties *The Astrophysical Journal*. Vol. 414, No. 2, (Sept 1993) p. 436-448, ISSN: 0004-637X.
Caldwell, R.R. & Linder, E. V. (2005). Limits of Quintessence. *Physical Review Letters*, Vol. 95, No. 14, (Sept 2005) (4 pages) 141301, ISSN: 1079-7114 (online).
Lima, J.A.S. (2004). Alternative dark energy models: an overview. *Brazilian Journal of Physics*, Vol. 34, No. 1a, (Mar 2004) p. 194-200, ISSN: 0103-9733.
Padmanabham, T. (2007). Dark energy and gravity. ArXiv:0705.2533 [gr-qc]

Capozziello, S.; De Ritis, R. & Marino, A. A. (1998). The effective cosmological constant in higher order gravity theories, Arxiv:9806043 [gr-qc]

Sokolowski, L.M. (2007). Metric gravity theories and cosmology: I. Physical interpretation and viability. *Classical and Quantum Gravity*, vol. 24, No. 13, (Jun 2007) p. 3391, ISSN: 1361-6382 (online)

Alfonso-Faus, A. (2009). Artificial contradiction between cosmology and particle physics: the lambda problem. *Journal of Astrophysics and Space Science*, Vol. 321, No. 1, (Fev 2009) p. 69-72, ISSN: 1572-946X

Tsujikawa, S. & Liddle, A. R. (2004). Constraints on braneworld inflation from CMB anisotropies. *Journal of Cosmology and Astroparticle Physics*, vol. 2004, No. 3, (March 2004), ISSN: 1475-7516.

Tsujikawa, S.; Sami, M. & Maartens, R. (2004). Observational constraints on braneworld inflation: The effect of a Gauss-Bonnet term. *Physical Review D*, vol. 70, No. 6, (Sept 2004) 063525 (10 pages), ISSN: 1550-2368 (online).

Maia, M. D. (2004). Covariant analysis of Experimental Constraints on the Brane-World. Arxiv: astro-ph/0404370v1.

Langlois, D. & Rodriguez-Martinez, M. (2001). Brane cosmology with a bulk scalar field. *Physics Review D*, Vol. 64, No. 12, (Nov 2001) 123507 (9 pages), ISSN: 1550-2368 (online).

Capistrano, A. J. S. & Odon, P. I. (2010). On the Cosmological Problem and the Brane-world geometry. *Central European Journal of Physics*, vol. 8, No. 1, (Feb 2010) p. 189-197, ISSN: 1644-3608 (Online).

Pauli, W. & Fierz, M. (1939). On Relativistic Wave Equations for Particles of Arbitrary Spin in an Electromagnetic Field. *Proccedings of Royal Society of London A*, Vol. 173, No. 953, (Nov 1939) p. 211-232, ISSN: 1471-2946 (online).

Gupta, S.N. (1954). Gravitation and Electromagnetism. *Physical Review*, vol. 96, No. 6, (Dec 1954) p.1683-1685, ISSN: 1550-2368 (online).

Fronsdal, C. (1978). Massless fiels with integer spin. *Physical Review D*, vol. 18, No. 10, (Nov 1978) p. 3624-3629, ISSN: 1550-2368 (online).

8

Modeling Light Cold Dark Matter

Abdessamad Abada[1] and Salah Nasri[2]
[1] *Laboratoire de Physique des Particules et Physique Statistique, Ecole Normale Supérieure*
[2]*Physics Department, UAE University, Al Ain*
[1]*Algeria*
[2]*United Arab Emirates*

1. Introduction

Dark matter is known to contribute about 22% to the total mass density in the Universe. Its existence started to be noticed in 1933, when Fritz Zwicky made an estimate of the total mass of the Coma cluster of galaxies outside our local group (Zwicky, 1933). Assuming that the galaxies in that cluster form a gravitationally bound system, he measured the cluster's geometrical size and the velocity dispersion of galaxies in it via Doppler redshift. He found that the mass of the Coma cluster had to be about 400 times larger than the estimate based on the number of galaxies and the total brightness of the cluster. He concluded that there must be some 'non-visible' form of matter which would provide enough gravity to hold the cluster gravitationally bound. This non-visible mass is called 'dark matter'.

There is by now extensive astronomical evidence supporting the existence of dark matter. The strongest such evidence comes from the measurements of the circular velocity of stars and gas in spiral galaxies versus their radial distance. If one assumes that the bulb in the center of a typical spiral galaxy is spherically symmetric, then one would expect the orbital velocity $v(r)$ outside the disk to behave like $1/\sqrt{r}$. Instead, the study of thousands of rotation curves of spiral galaxies shows that the orbital velocity rises from the center until it reaches a limiting value $v_C \sim (100 - 200)$ km/s, and then stays flat outside the galaxy core (Persic & Salucci & Stel, 1996). For example, the observed velocity of the rotation curve of the spiral galaxy M33, one of the brightest spiral galaxies in our local group, at $r \simeq 10$kpc is $v_C \simeq 120$km/s, whereas the expected velocity is $v \simeq 40$km/s. One infers from this that the total mass in the galaxy is about nine times the luminous matter ($\Omega_{\text{lum}} \sim 10\%$). This implies that there is about ten times more mass in the halo of spiral galaxies than in the disk.

There is also evidence of dark matter in elliptic galaxies and cluster of galaxies. This comes from the observation of X-rays emitted via the bremsstrahlung process $e + p \to e + p + \gamma$ from the intergalactic gas in the cluster. Assuming hydrostatic equilibrium, we can deduce from the measurement of the X-ray luminosity and the shape of its spectrum, assumed isothermal, the mass distribution in the galaxy that is necessary to bind the hot gas. The observations indicate that the total mass associated with these systems is considerably larger than the luminous component (Fabricant & Gorenstein, 1983; Stewart et al., 1984). Note that cluster masses can also be determined from their lensing effect on light from distant sources (Mellier, 1999).

Furthermore, during the past few years, data from the WMAP satellite has provided us with the most precise measurements yet of the cosmological parameters (Spergel et al., 2007; Pope et al., 2004). By analyzing the location and the height of the acoustic peaks of the temperature fluctuations, one can extract the contribution of the different species to the critical energy density of the Universe. For instance, the height of the first peak relative to the second one gives a baryon density of about 4%, which is consistent with the predictions of the primordial theory of Big Bang nucleosynthesis (Steigman, 2010). The third peak is sensitive to the amount of total matter density in the Universe and can be used to extract the energy density $\Omega_{\rm DM}$ of dark matter in the Universe. The best fit is (Komatsu, 2010):

$$\Omega_{\rm DM}\bar{h}^2 = 0.1123 \pm 0.0035, \qquad (1)$$

where $\bar{h}$ is the Hubble constant in units of $100{\rm km} \times {\rm s}^{-1} \times {\rm Mpc}^{-1}$.

Yet, even though dark matter dominates the matter mass 'budget' of the Universe, its very nature remains elusive. Indeed, what are its quantum numbers, its mass? How does it interact with the Standard Model particles? One should also say that up until now, it has not been *directly* detected. But there is a number of basic properties that any candidate for dark matter should have[1]. First of all, it must be massive, and this is because of the non-relativistic velocities involved. Second, it must be stable so that it would survive until today, which means it must have a lifetime larger than that of the Universe. Third, it must be electrically neutral, otherwise it would have been very likely seen via its electromagnetic interaction with visible matter. Also, the abundance of such stable charged massive particles would be severely constrained, in particular from searches in the deep sea water (Amsler et al., 2008). Fourth, a dark-matter candidate should not interact strongly. Indeed, if such a massive stable particle could do so, it would be able to bind and form anomalously heavy nuclei. But the resulting number of such anomalously heavy nuclei that would be present today is shown to be excluded by existing searches (Javorsek, 2001; 2002). Fifth, for a dark matter candidate to act as a seed for structure formation, it must decouple at a temperature of the order of its mass. Such a candidate is known as "cold dark matter". Sixth, it must give the right relic dark-matter density, which, by the latest astrophysical observations, is about 22% of the total energy density in the Universe (Komatsu, 2010).

While the Standard Model of elementary particle Physics is very successful at describing the interactions between 'visible' particles, it cannot accommodate for a weakly interacting massive particle (WIMP) as a suitable candidate for dark matter. Hence, extensions of the Standard Model are inevitable and, given the elusiveness of dark matter, modeling becomes a necessity. In this framework, the most popular candidate for dark matter is the neutralino, a neutral R-odd supersymmetric particle. Indeed, neutralinos are produced or destroyed in pairs only, thus rendering the lightest SUSY particle (LSP) stable (Ellis et al., 1984). In the minimal version of the supersymmetric extension of the Standard Model, the neutralino χ_1^0 is a linear combination of the fermionic partners of the neutral electroweak gauge bosons (gauginos) and the neutral Higgs bosons (higgsinos). It can annihilate through a t-channel sfermion exchange into Standard-Model fermions, or via a t-channel chargino-mediated process into W^+W^-, or through an s-channel pseudoscalar Higgs exchange into fermion pairs. Also, it can undergo elastic scattering with nuclei through mainly a scalar Higgs exchange (Jungman, 1996).

However, having a neutralino as a candidate for *light* dark matter can be a real challenge. For example, in mSUGRA, the constraint from WMAP observations and the bound on the pseudo-scalar Higgs mass from LEP give neutralino mass $m_{\chi_1^0} \geq 50{\rm GeV}$ (Belanger et al., 2009; Akrami et al., 2010). If one allows the gaugino masses M_1 and M_2 to be free parameters

[1] We implicitly mean a candidate from the realm of elementary particles.

and the gluino mass to satisfy the universal condition at some grand unification scale, that is, $M_3 = 3M_2$, then the LSP should be heavier than about 28GeV (Vasquez et al., 2010), see also (Feldman, 2010; Kuflik, 2010). A similar analysis is done in (Fornengo et al., 2011) with the gluino mass taken as a free parameter, and it is concluded that the lower limit on the neutralino mass varies between about 7GeV and 12GeV, depending on the gluino mass and the degeneracy of the squarks. In the extension of the MSSM with an extra singlet chiral superfield (NMSSM), a model with 11 input parameters, it is found that a neutralino with a mass of the order of a few GeVs is possible, with a higher likelihood peaked at around 15GeV (Vasquez et al., 2010).

Therefore, with the aim of modeling dark matter that could be as light as a few GeVs and maybe lighter, and with no clear clue yet as to what the internal structure of the WIMP is, if any, a 'pedestrian' approach can be attractive. In this logic, the simplest of models is to extend the Standard Model with a real scalar field, the dark matter, a Standard-Model gauge singlet that interacts with visible particles via the Higgs field only. To ensure stability, it is endowed with a discrete $\mathbb{Z}_2$ symmetry that does not break spontaneously. Such a model can be seen as a low-energy remnant of some higher-energy physics waiting to be understood. In this cosmological setting, such an extension has first been proposed in (Silveira, 1985) and further studied in (McDonald, 1994) where the unbroken $\mathbb{Z}_2$ symmetry is extended to a global U(1) symmetry. A more extensive exploration of the model and its implications was done in (Burgess et al., 2011), specific implications on Higgs detection and LHC physics discussed in (Barger et al., 2008) and one-loop vacuum stability looked into and perturbativity bounds obtained in (Gonderinger et al., 2010). However, the work (He et al., 2009; Asano & Kitano, 2010) considers this minimal extension too and uses constraints from the direct-detection experiments XENON10 (Angle et al., 2008) and CDMSII (Ahmed et al., 2009) to exclude dark matter masses smaller than 50, 70 and 75GeV for Higgs masses equal to 120, 200 and 350GeV respectively. Furthermore, it was recently shown that the Fermi-LAT data on the isotropic diffuse gamma-ray emission can potentially exclude this one-singlet dark-matter model for masses as low as 6GeV, assuming a NFW profile for the dark-matter distribution (Arina & Tytgat, 2011).

So, in order to allow for light dark matter in this 'bottom-up' approach, the natural step forward is to add another real scalar field, endowed with a $\mathbb{Z}_2$ symmetry too, but one which is spontaneously broken so that new channels for dark matter annihilation are opened, increasing this way the annihilation cross-section, hence allowing smaller masses for the WIMP. This auxiliary field must also be a Standard-Model gauge singlet. The present chapter introduces this extension and presents some of its aspects. The aim is to use this example as a generic prototype in order to show how modeling of cold dark matter can be done and what are the main steps to follow. Most of the technical material used here is drawn from (Abada, 2011).

This chapter is organized as follows. After this introduction, we present the model in the next section. The spontaneous breaking of the electroweak and the additional $\mathbb{Z}_2$ symmetries is performed in the usual way and the physical modes as well as the physical parameters are explained. There is mixing between the physical new scalar field and the Higgs, and this is one of the quantities parametrizing the subsequent physics. We discuss in section three the imposition of the constraint from the dark matter relic density on the dark-matter annihilation cross-section and study its effects. Of course, as we will see, the space of parameters is quite large and cannot be covered in its entirety in any study of reasonable size. Representative values have to be selected and the behavior of the model, as well as its capabilities, are described accordingly. Though our main interest in this work is light dark matter, yet we allow the dark-matter mass to vary from 0.1GeV to 100GeV, sometimes higher. We find that the model is rich enough to bear dark matter for most of these masses, including those in the very light sector. In section four, we determine the total cross section $\sigma_{\rm det}$ for non-relativistic elastic scattering of dark matter off a nucleon target and compare it to the current direct-detection experimental bounds and projected sensitivity. For this, we choose the results of CDMSII (Ahmed et al., 2009) and XENON100 (April et al., 2010), as well as the projections of SuperCDMS (Schnee et al., 2005) and XENON1T (April et al., 2010). Here too we cannot cover all of the parameters' space nor are we going to give a detailed account of the behavior of $\sigma_{\rm det}$ as a function of the dark matter mass, but general trends are mentioned. In section five, we show how low-energy particle phenomenology can constrain the various parameters of the model. We have space for only one typical example, namely, the decay of the B_s meson into a pair of $\mu^-\mu^+$. Here, we take from the start light dark matter, with a mass in the range 0.1GeV $-$ 10GeV. Finally, in the last section, we finish the chapter with a number of concluding remarks.

2. A two-singlet extension to the Standard Model

The Standard Model is extended by two real, scalar, and $\mathbb{Z}_2$ -symmetric fields. One is the dark matter field S_0 for which the $\mathbb{Z}_2$ symmetry is unbroken while the other field χ_1 undergoes spontaneous symmetry breaking. Both fields are Standard-Model gauge singlets and hence, can interact with the other sectors of the Standard Model only via the Higgs doublet H. This latter is taken in the unitary gauge such that $H^\dagger = 1/\sqrt{2}\,(0 \;\; h')$, where h' is a real scalar. The potential function involving S_0, h' and χ_1 is given by the following expression:

$$U = \frac{\tilde{m}_0^2}{2}S_0^2 - \frac{\mu^2}{2}h'^2 - \frac{\mu_1^2}{2}\chi_1^2 + \frac{\eta_0}{24}S_0^4 + \frac{\lambda}{24}h'^4 + \frac{\eta_1}{24}\chi_1^4 + \frac{\lambda_0}{4}S_0^2h'^2 + \frac{\eta_{01}}{4}S_0^2\chi_1^2 + \frac{\lambda_1}{4}h'^2\chi_1^2, \quad (2)$$

where the mass-squared parameters $\tilde{m}_0^2$, $\mu^2$and μ_1^2 and all the coupling constants are real positive numbers. The Higgs field undergoes spontaneous electroweak symmetry breaking and oscillates around the vacuum expectation value $v = 246$GeV (Nakamura et al., 2010). The field χ_1 will oscillate around the vacuum expectation value $v_1 > 0$. Both v and v_1 are related to the parameters of the theory by the two relations:

$$v^2 = 6\frac{\mu^2\eta_1 - 6\mu_1^2\lambda_1}{\lambda\eta_1 - 36\lambda_1^2}; \qquad v_1^2 = 6\frac{\mu_1^2\lambda - 6\mu^2\lambda_1}{\lambda\eta_1 - 36\lambda_1^2}. \quad (3)$$

The self-coupling constants are assumed sufficiently larger than the mutual ones and perturbation theory is assumed applicable throughout.

Writing $h' = v + \tilde{h}$ and $\chi_1 = v_1 + \tilde{S}_1$, the potential function becomes, up to an irrelevant zero-field energy:

$$U = U_{\rm quad} + U_{\rm cub} + U_{\rm quar}, \quad (4)$$

where the mass-squared (quadratic) terms are gathered in $U_{\rm quad}$, the cubic interactions in $U_{\rm cub}$ and the quartic ones in $U_{\rm quar}$. The quadratic terms are given by:

$$U_{\rm quad} = \frac{1}{2}m_0^2S_0^2 + \frac{1}{2}M_{\tilde{h}}^2\tilde{h}^2 + \frac{1}{2}M_1^2\tilde{S}_1^2 + M_{1h}^2\tilde{h}\tilde{S}_1, \quad (5)$$

where the mass-squared coefficients are related to the original parameters of the theory by the following relations:

$$m_0^2 = \tilde{m}_0^2 + \frac{\lambda_0}{2}v^2 + \frac{\eta_{01}}{2}v_1^2; \quad M_h^2 = -\mu^2 + \frac{\lambda}{2}v^2 + \frac{\lambda_1}{2}v_1^2;$$
$$M_1^2 = -\mu_1^2 + \frac{\lambda_1}{2}v^2 + \frac{\eta_1}{2}v_1^2; \quad M_{1h}^2 = \lambda_1 v v_1. \tag{6}$$

As we see, in this basis, the mass-squared matrix is not diagonal: there is mixing between the fields $\tilde{h}$ and $\tilde{S}_1$. Denoting by h and $S_1$the physical field eigenmodes of the mass-squared matrix, we rewrite:

$$U_{\text{quad}} = \frac{1}{2}m_0^2 S_0^2 + \frac{1}{2}m_h^2 h^2 + \frac{1}{2}m_1^2 S_1^2, \tag{7}$$

where the physical fields are related to the mixed ones by a 2×2 rotation:

$$\begin{pmatrix} h \\ S_1 \end{pmatrix} = \begin{pmatrix} \cos\theta & \sin\theta \\ -\sin\theta & \cos\theta \end{pmatrix} \begin{pmatrix} \tilde{h} \\ \tilde{S}_1 \end{pmatrix}. \tag{8}$$

Here θ is the mixing angle, related to the original mass-squared parameters by the relation:

$$\tan 2\theta = \frac{2M_{1h}^2}{M_1^2 - M_h^2}, \tag{9}$$

and the physical masses in (7) by the two relations:

$$m_h^2 = \frac{1}{2}\left[M_h^2 + M_1^2 + \varepsilon\left(M_h^2 - M_1^2\right)\sqrt{\left(M_h^2 - M_1^2\right)^2 + 4M_{1h}^4}\right];$$
$$m_1^2 = \frac{1}{2}\left[M_h^2 + M_1^2 - \varepsilon\left(M_h^2 - M_1^2\right)\sqrt{\left(M_h^2 - M_1^2\right)^2 + 4M_{1h}^4}\right], \tag{10}$$

where ε is the sign function.
Written now directly in terms of the physical fields, the cubic interactions are expressed as follows:

$$U_{\text{cub}} = \frac{\lambda_0^{(3)}}{2}S_0^2 h + \frac{\eta_{01}^{(3)}}{2}S_0^2 S_1 + \frac{\lambda^{(3)}}{6}h^3 + \frac{\eta_1^{(3)}}{6}S_1^3 + \frac{\lambda_1^{(3)}}{2}h^2 S_1 + \frac{\lambda_2^{(3)}}{2}hS_1^2, \tag{11}$$

where the cubic physical coupling constants are related to the original parameters via the following relations:

$$\begin{aligned}
\lambda_0^{(3)} &= \lambda_0 v \cos\theta + \eta_{01} v_1 \sin\theta, \quad \eta_{01}^{(3)} = \eta_{01} v_1 \cos\theta - \lambda_0 v \sin\theta; \\
\lambda^{(3)} &= \lambda v \cos^3\theta + \frac{3}{2}\lambda_1 \sin 2\theta \left(v_1 \cos\theta + v \sin\theta\right) + \eta_1 v_1 \sin^3\theta; \\
\eta_1^{(3)} &= \eta_1 v_1 \cos^3\theta - \frac{3}{2}\lambda_1 \sin 2\theta \left(v \cos\theta - v_1 \sin\theta\right) - \lambda v \sin^3\theta; \\
\lambda_1^{(3)} &= \lambda_1 v_1 \cos^3\theta + \frac{1}{2}\sin 2\theta \left[(2\lambda_1 - \lambda)\, v \cos\theta - (2\lambda_1 - \eta_1)\, v_1 \sin\theta\right] - \lambda_1 v \sin^3\theta; \\
\lambda_2^{(3)} &= \lambda_1 v \cos^3\theta - \frac{1}{2}\sin 2\theta \left[(2\lambda_1 - \eta_1)\, v_1 \cos\theta + (2\lambda_1 - \lambda)\, v \sin\theta\right] + \lambda_1 v_1 \sin^3\theta.
\end{aligned} \tag{12}$$

Written too directly in terms of the physical fields, the quartic interactions are given by:

$$U_{\rm quar} = \frac{\eta_0}{24}S_0^4 + \frac{\lambda^{(4)}}{24}h^4 + \frac{\eta_1^{(4)}}{24}S_1^4 + \frac{\lambda_0^{(4)}}{4}S_0^2h^2 + \frac{\eta_{01}^{(4)}}{4}S_0^2S_1^2 + \frac{\lambda_{01}^{(4)}}{2}S_0^2hS_1 + \frac{\lambda_1^{(4)}}{6}h^3S_1 + \frac{\lambda_2^{(4)}}{4}h^2S_1^2 + \frac{\lambda_3^{(4)}}{6}hS_1^3, \qquad (13)$$

where the physical quartic coupling constants are written in terms of the original parameters of the theory as follows:

$$\begin{aligned}
\lambda^{(4)} &= \lambda\cos^4\theta + \frac{3}{2}\lambda_1\sin^2 2\theta + \eta_1\sin^4\theta, \quad \eta_1^{(4)} = \eta_1\cos^4\theta + \frac{3}{2}\lambda_1\sin^2 2\theta + \lambda\sin^4\theta;\\
\lambda_0^{(4)} &= \lambda_0\cos^2\theta + \eta_{01}\sin^2\theta, \quad \eta_{01}^{(4)} = \eta_{01}\cos^2\theta + \lambda_0\sin^2\theta, \quad \lambda_{01}^{(4)} = \frac{1}{2}(\eta_{01}-\lambda_0)\sin 2\theta,\\
\lambda_1^{(4)} &= \frac{1}{2}\left[(3\lambda_1-\lambda)\cos^2\theta - (3\lambda_1-\eta_1)\sin^2\theta\right]\sin 2\theta;\\
\lambda_2^{(4)} &= \lambda_1\cos^2 2\theta - \frac{1}{4}(2\lambda_1-\eta_1-\lambda)\sin^2 2\theta;\\
\lambda_3^{(4)} &= \frac{1}{2}\left[(\eta_1-3\lambda_1)\cos^2\theta - (\lambda-3\lambda_1)\sin^2\theta\right]\sin 2\theta. \qquad (14)
\end{aligned}$$

In addition to the above sector and after spontaneous breaking of the electroweak and $\mathbb{Z}_2$ symmetries, we need to rewrite the part of the Standard Model lagrangian affected by the mixing angle θ. We thus have:

$$\begin{aligned}
U_{\rm SM} &= \sum_f\left(\lambda_{hf}h\bar{f}f + \lambda_{1f}S_1\bar{f}f\right) + \lambda_{hw}^{(3)}hW_\mu^-W^{+\mu} + \lambda_{1w}^{(3)}S_1W_\mu^-W^{+\mu}\\
&+\lambda_{hz}^{(3)}h\left(Z_\mu\right)^2 + \lambda_{1z}^{(3)}S_1\left(Z_\mu\right)^2 + \lambda_{hw}^{(4)}h^2W_\mu^-W^{+\mu} + \lambda_{1w}^{(4)}S_1^2W_\mu^-W^{+\mu}\\
&+\lambda_{h1w}hS_1W_\mu^-W^{+\mu} + \lambda_{hz}^{(4)}h^2\left(Z_\mu\right)^2 + \lambda_{1z}^{(4)}S_1^2\left(Z_\mu\right)^2 + \lambda_{h1z}hS_1\left(Z_\mu\right)^2. \qquad (15)
\end{aligned}$$

The quantities m_f, m_w and m_z are the masses of the fermion f, the W and the Z gauge bosons respectively, and the above coupling constants are given by the following relations:

$$\begin{aligned}
\lambda_{hf} &= -\frac{m_f}{v}\cos\theta; \qquad \lambda_{1f} = \frac{m_f}{v}\sin\theta;\\
\lambda_{hw}^{(3)} &= 2\frac{m_w^2}{v}\cos\theta; \qquad \lambda_{1w}^{(3)} = -2\frac{m_w^2}{v}\sin\theta;\\
\lambda_{hz}^{(3)} &= \frac{m_z^2}{v}\cos\theta; \qquad \lambda_{1z}^{(3)} = -\frac{m_z^2}{v}\sin\theta;\\
\lambda_{hw}^{(4)} &= \frac{m_w^2}{v^2}\cos^2\theta; \quad \lambda_{1w}^{(4)} = \frac{m_w^2}{v^2}\sin^2\theta; \quad \lambda_{h1w} = -\frac{m_w^2}{v^2}\sin 2\theta;\\
\lambda_{hz}^{(4)} &= \frac{m_z^2}{2v^2}\cos^2\theta; \quad \lambda_{1z}^{(4)} = \frac{m_z^2}{2v^2}\sin^2\theta; \quad \lambda_{h1z} = -\frac{m_z^2}{2v^2}\sin 2\theta. \qquad (16)
\end{aligned}$$

3. Effects of the relic density constraint

The original theory (2) has nine parameters: three mass parameters $(\tilde{m}_0, \mu, \mu_1)$, three self-coupling constants $(\eta_0, \lambda, \eta_1)$ and three mutual coupling constants $(\lambda_0, \eta_{01}, \lambda_1)$. The dark-matter self-coupling constant η_0 does not enter the calculations of the lowest-order processes to come **?**, so effectively, one is left with eight parameters. The spontaneous breaking of the electroweak and $\mathbb{Z}_2$ symmetries for the Higgs and χ_1 fields respectively introduces the two vacuum expectation values v and v_1 given to lowest order in (3). The value of v is fixed experimentally to be 246GeV and we fix the value of v_1 at the order of the electroweak scale, say 100GeV. So now six parameters left. It is natural to choose four of these the three physical masses m_0 (dark matter), m_1 (S_1 field) and m_h (Higgs), plus the mixing angle θ between S_1 and h. We give the Higgs mass the value $m_h = 138$GeV, compatible with current experimental bounds. The two last parameters one chooses to work with are the two physical mutual coupling constants $\lambda_0^{(4)}$ (dark matter – Higgs) and $\eta_{01}^{(4)}$ (dark matter – S_1 particle), see (13).
The thermal dynamics of the Universe within the standard cosmological model Kolb & Turner (1998) relates the WIMP relic density $\Omega_{\rm DM}$ to its annihilation rate by two relations, which are essentially model independent:

$$\Omega_{\rm DM}\bar{h}^2 \simeq \frac{1.07\times 10^9 x_f}{\sqrt{g_*}m_{\rm Pl}\langle v_{12}\sigma_{\rm ann}\rangle\,{\rm GeV}}; \quad x_f \simeq \ln\frac{0.038 m_{\rm Pl} m_0 \langle v_{12}\sigma_{\rm ann}\rangle}{\sqrt{g_* x_f}}. \tag{17}$$

The notation is as follows: the quantity $\bar{h}$ is the Hubble constant in units of $100{\rm km}\times{\rm s}^{-1}\times{\rm Mpc}^{-1}$, the quantity $m_{\rm Pl} = 1.22\times 10^{19}$GeV the Planck mass, m_0 the WIMP (dark matter) mass, $x_f = m_0/T_f$ the ratio of the WIMP mass to the freeze-out temperature T_f and g_* the number of relativistic degrees of freedom with mass less than T_f. The quantity $\langle v_{12}\sigma_{\rm ann}\rangle$ is the thermally averaged annihilation cross-section of a pair of two dark-matter particles multiplied by their relative speed in the center-of-mass reference frame. Solving (17) with the current accepted value (1) for $\Omega_{\rm DM}$ yields a constraint on the annihilation cross-section, i.e.:

$$\langle v_{12}\sigma_{\rm ann}\rangle \simeq (1.9\pm 0.2)\times 10^{-9}{\rm GeV}^{-2}. \tag{18}$$

In a given model like the one presented here, the above constraint translates into a relation between the parameters of the theory entering the calculated expression of $\langle v_{12}\sigma_{\rm ann}\rangle$, hence limiting the intervals of possible dark matter masses. This constraint can also be exploited in order to examine aspects of the theory like perturbativity, while at the same time reducing the number of parameters by one. For example, in this model, we can use (18) to obtain the mutual coupling constant $\eta_{01}^{(4)}$ as a function of the remaining four parameters $\left(m_0, m_1, \theta, \lambda_0^{(4)}\right)$ and study aspects of the model through its behavior. For example, we can ask which dark-matter mass regions are consistent with perturbativity. Note that through the relations (12) and (14), once the two mutual coupling constants $\lambda_0^{(4)}$ and $\eta_{01}^{(4)}$ are perturbative, all the other physical coupling constants will be.
The dark-matter annihilation cross sections (times the relative speed) through all possible channels within the model can be calculated in the usual manner to lowest order in perturbation theory Abada (2011). The quantity $\langle v_{12}\sigma_{\rm ann}\rangle$ is the sum of all these contributions. Imposing $\langle v_{12}\sigma_{\rm ann}\rangle = 1.9\times 10^{-9}{\rm GeV}^{-2}$ dictates the behavior of $\eta_{01}^{(4)}$, which is displayed as a function of the dark matter mass m_0. Of course, as there are four free parameters, the behavior is bound to be rich and diverse and we cannot describe every bit of it in such a small space.

Also, importantly enough, one has to note from the outset that for a given set of values of the parameters, the solution to the relic-density constraint is not unique: besides positive real solutions (when they exist), we may find negative real or even complex solutions. Indeed, from the physical coefficients in (12) and (14), one can show that $\langle v_{12}\sigma_{\text{ann}}\rangle$ is a sum of quotients of up-to-quartic polynomials in $\eta_{01}^{(4)}$. This means that, ultimately, the relic-density constraint is going to be an algebraic equation in $\eta_{01}^{(4)}$, which has always solutions in the complex plane, but not necessarily on the positive real axis. In our context, we are only interested in finding the smallest of the positive real solutions in $\eta_{01}^{(4)}$ when they exist, looking at its behavior and finding out in which mass regions it is small enough to be perturbative.
We start the description with a small mixing angle, say $\theta = 10\,^{\circ}$, and a very weak mutual S_0 – Higgs coupling constant, say $\lambda_0^{(4)} = 0.01$. The behavior of $\eta_{01}^{(4)}$ versus m_0 for the S_1 mass $m_1 = 10\text{GeV}$ is displayed in Fig. 1. The range of m_0 shown is wide, from 0.1GeV to

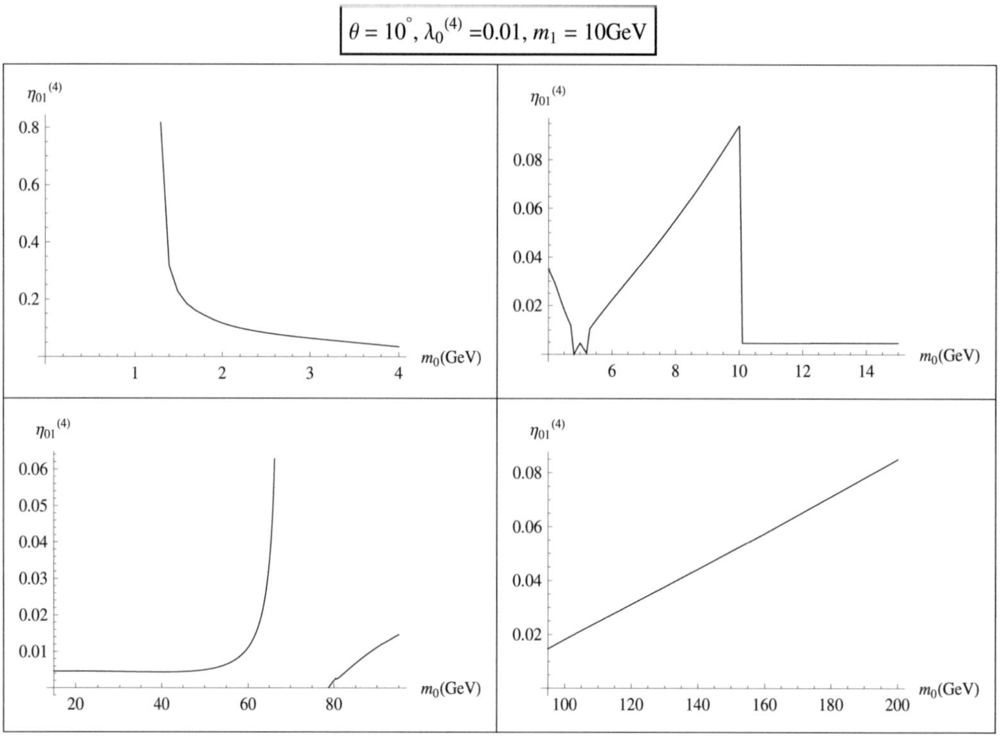

Fig. 1. $\eta_{01}^{(4)}$ vs m_0 for small m_1, small mixing and very small WIMP-Higgs coupling.

200GeV, cut in four intervals to allow for 'local' features to be displayed. We see that the relic-density constraint on S_0 annihilation has no positive real solution for $m_0 \lesssim 1.3\text{GeV}$, and so, with these very small masses, S_0 cannot be a dark matter candidate. In other words, for $m_1 = 10\text{GeV}$, the particle S_0 cannot annihilate into the lightest fermions only in a way compatible with the relic-density constraint; inclusion of the c-quark is necessary. Note that right about $m_0 \simeq 1.3\text{GeV}$, the c threshold, the mutual coupling constant $\eta_{01}^{(4)}$ starts at about 0.8, a value, while perturbative, that is roughly eighty-fold larger than the mutual S_0 – Higgs coupling constant $\lambda_0^{(4)}$. Then $\eta_{01}^{(4)}$ decreases, steeply first, more slowly as we cross the τ mass towards the b mass. Just before $m_1/2$, the coupling $\eta_{01}^{(4)}$ hops onto another solution branch that is just emerging from negative territory, gets back to the first one at precisely $m_1/2$ as this latter carries now smaller values, and then jumps up again onto the second branch as the first crosses the m_0-axis down. It goes up this branch with a moderate slope until m_0 becomes equal to m_1, a value at which the S_1 annihilation channel opens. Right beyond m_1, there is a sudden fall to a value $\eta_{01}^{(4)} \simeq 0.0046$ that is about half the value of $\lambda_0^{(4)}$, and $\eta_{01}^{(4)}$ stays flat till $m_0 \simeq 45$GeV where it starts increasing, sharply after 60GeV. In the mass interval $m_0 \simeq 66$GeV $-$ 79GeV, there is a 'desert' with no positive real solutions to the relic-density constraint, hence no viable dark matter candidate. Beyond $m_0 \simeq 79$GeV, the mutual coupling constant $\eta_{01}^{(4)}$ keeps increasing monotonously, with a small notch at the W mass and a less noticeable one at the Z mass.

For this value of m_1 (10GeV), all values reached by $\eta_{01}^{(4)}$ in the mass range considered are perturbativily acceptable. This may not be the case for larger values of m_1. For example, for $m_1 = 30$GeV while keeping $\theta = 10^o$ and $\lambda_0^{(4)} = 0.01$, the mutual coupling constant $\eta_{01}^{(4)}$ starts at $m_0 \simeq 1.5$GeV with the very large value 89.8 and decreases very sharply right after, to 2.04 at about 1.6GeV. The other overall features are similar to the case $m_1 = 10$GeV.

One important question to ask is whether the model ever allows for very light dark matter. To look into this matter, one fixes m_0 at a small value, say $m_0 = 0.2$GeV, and let m_1 vary. The behavior of $\eta_{01}^{(4)}$ is displayed in Fig. 2. The allowed S_0 annihilation channels are the very light fermions e, u, d, μ and s, plus S_1 when $m_1 < m_0$. Qualitatively, we notice that in fact, there are no solutions for $m_1 < m_0$, a mass at which $\eta_{01}^{(4)}$ takes the very small value $\simeq 0.003$. It goes up a solution branch and leaves it at $m_1 \simeq 0.4$GeV to descend on a second branch that enters negative territory at $m_1 \simeq 0.7$GeV, forcing $\eta_{01}^{(4)}$ to return onto the first branch. There is an accelerated increase till $m_1 \simeq 5$GeV, a value at which $\eta_{01}^{(4)} \simeq 0.5$. And then a desert, no positive real solutions, no viable dark matter.

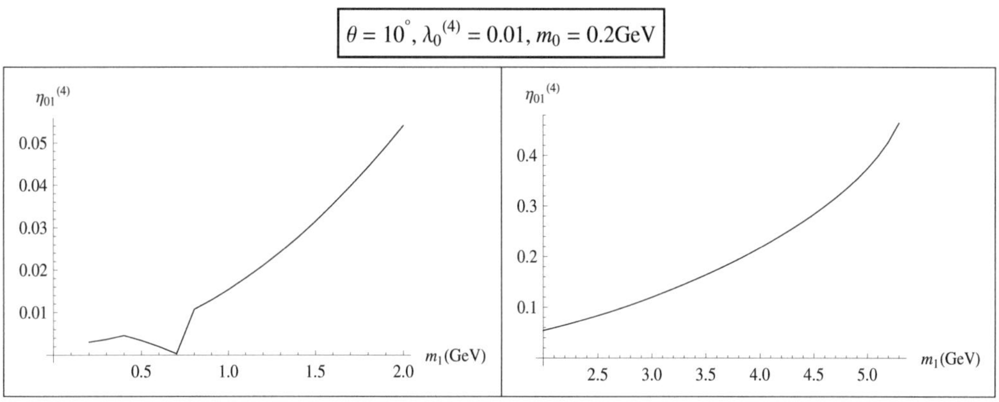

Fig. 2. $\eta_{01}^{(4)}$ vs m_1 for very light S_1, small mixing and very small WIMP-Higgs coupling.

Increasing m_0 until about 1.3GeV does not change these overall features: some 'movement' for very small values of m_1 and then an accelerated increase till reaching a desert with a lower bound that changes with m_0. Note that in all these cases where $m_0 \lesssim 1.3$GeV, all values of $\eta_{01}^{(4)}$ are perturbative. Therefore, the model can very well accommodate very light dark matter with a restricted range of S_1 masses. However, the situation changes after the inclusion of the τ annihilation channel. Indeed, though the overall shape of the behavior of $\eta_{01}^{(4)}$ as a function of m_1 is qualitatively the same, the desert threshold is pushed significantly higher, and more significantly, $\eta_{01}^{(4)}$ starts to be larger than one already at moderately small values of m_1, therefore loosing perturbativity. In fact, for $m_0 = 1.5$GeV already, the desert is effectively erased as we have a sudden jump to highly non-perturbative values of $\eta_{01}^{(4)}$ right after $m_1 \simeq 28$GeV **?**. However, for m_1 moderately small, for example $\lesssim 20$GeV in the case $m_0 = 1.5$GeV, the values of $\eta_{01}^{(4)}$ are smaller than one and physical use of the model is possible if needed.

Some new features come when increasing the value of the mutual coupling constant $\lambda_0^{(4)}$. Figure 3 shows the behavior of $\eta_{01}^{(4)}$ as a function of the dark matter mass m_0 when $\lambda_0^{(4)} = 0.2$, $\theta = 10^{\mathrm{o}}$ and $m_1 = 20$GeV. We see that $\eta_{01}^{(4)}$ starts at $m_0 \simeq 1.4$GeV with a value of about 1.95. It decreases with a sharp change of slope at the b threshold, then makes a sudden dive at about 5GeV, a change of branch at $m_1/2$ down till about 12GeV where it jumps up back onto the previous branch just before going to cross into negative territory. It drops sharply at $m_0 = m_1$ and then increases slowly until $m_0 \simeq 43.3$GeV. Beyond, there is nothing, a desert. This is of course different from the situation of very small $\lambda_0^{(4)}$ like in Fig. 1 above: here we see some kind of natural dark-matter mass 'confinement' to small-moderate viable[2] values.

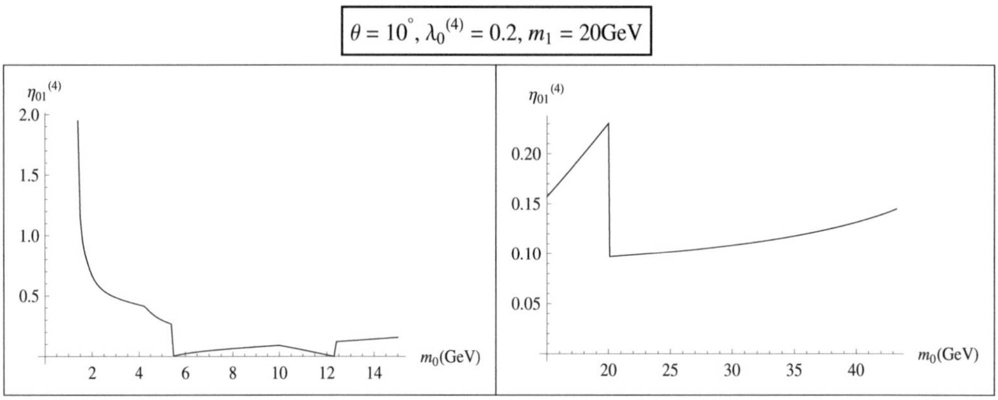

Fig. 3. $\eta_{01}^{(4)}$ vs m_0 for small mixing, moderate m_1 and WIMP-Higgs coupling.

For larger values of m_1 with moderate $\lambda_0^{(4)} = 0.2$, one obtains roughly the same behavior but here too not all values of $\eta_{01}^{(4)}$ are perturbative. For example, for $m_1 = 60$GeV, the mutual coupling $\eta_{01}^{(4)}$ starts very high ($\simeq 85$) at $m_0 \simeq 1.5$GeV, and then decreases rapidly. There is a usual change of branches and a desert starting at about 49GeV. However, what is interesting here is that, in contrast with previous situations, the desert starts at a mass $m_0 < m_1$, i.e., before the opening of the S_1 annihilation channel. In other words, the dark matter is annihilating into the light fermions only and the model is perturbatively viable in the range 20GeV – 49GeV.

Larger values of $\lambda_0^{(4)}$ can also be studied. For $\lambda_0^{(4)} = 1$ and as long as $m_1 \lesssim 79.2$GeV, one finds the usual small m_0-deserts as well as the familar action at the different mass thresholds, with nothing suprisingly new. However, for $m_1 \gtrsim 79.3$GeV, there is a highly non-perturbative

[2] Note that the values of $\eta_{01}^{(4)}$ for $1.6\text{GeV} \lesssim m_0 \lesssim 43.3$GeV are all perturbative.

branch $\eta_{01}^{(4)}$ jumps onto at small and moderate values of m_0 **?**. This highly non-perturbative region stretches in size as m_1 increases.
Increasing the S_1 – Higgs mixing angle θ can bring new features too. Figure 4 shows the behavior of $\eta_{01}^{(4)}$ as a function of m_0 for $\theta = 40^o$, $\lambda_0^{(4)} = 0.01$ and $m_1 = 20$GeV. One recognizes features similar to those of the case $\theta = 10^o$, though coming in different relative sizes. The very-small-m_0 desert ends at about 0.3GeV. There are by-now familiar features at the c and b masses, $m_1/2$ and m_1. Two relatively small forbidden intervals (deserts) appear for relatively large values of the dark matter mass: 67.3GeV – 70.9GeV and 79.4GeV – 90.8 GeV. The W mass is in the forbidden region but there is action as we cross the Z mass. Other values of m_1 behave similarly with an additional effect, namely, a sudden drop in slope at $m_0 = (m_h + m_1)/2$ coming from the ignition of S_0 annihilation into S_1 and Higgs.

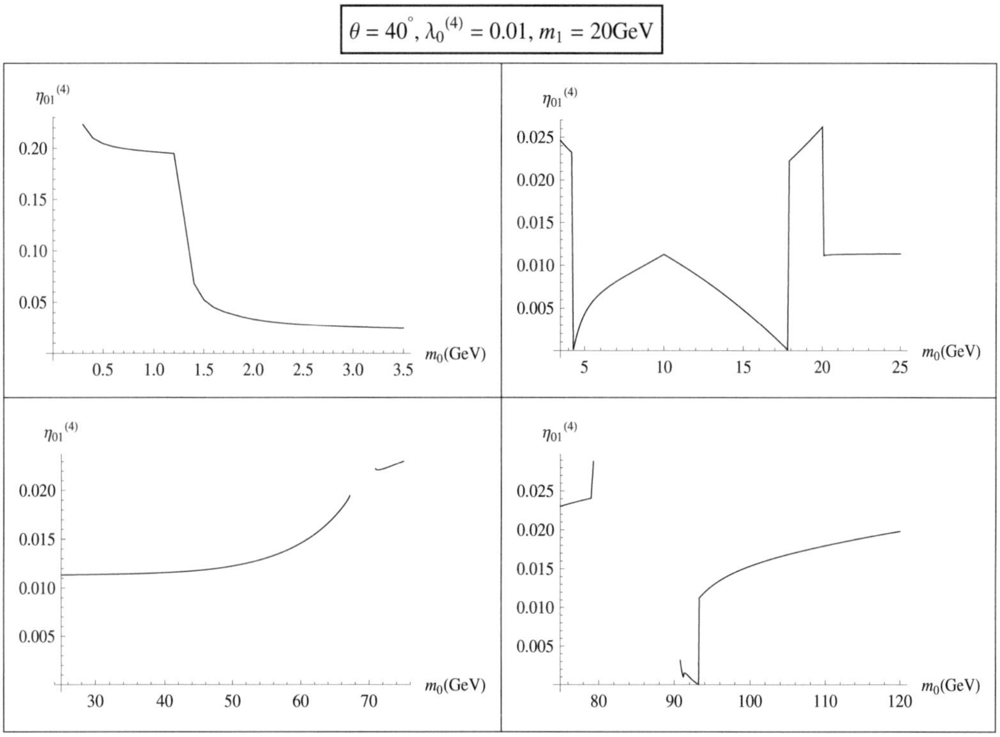

Fig. 4. $\eta_{01}^{(4)}$ versus m_0 for moderate m_1, moderate mixing and small WIMP-Higgs coupling.

Increasing the value of $\lambda_0^{(4)}$ for larger values of θ has the effect of making the behavior of $\eta_{01}^{(4)}$ smoother while keeping the same overall features like the confining of the mass of a viable dark matter to small-moderate values, a dark matter particle annihilating into light fermions only. It has also the effect of eliminating those highly non-perturbative regions discussed above.

4. Dark-matter direct detection

Experiments like CDMS II Ahmed et al. (2009), XENON 10/100 Angle et al. (2008); **?**, DAMA/LIBRA Bernabei et al. (2010) and CoGeNT Aalseth et al. (2010) carry a direct search

for a dark matter signal. Such a signal would typically come from the elastic scattering of a dark matter WIMP off a non-relativistic nucleon target. However, such experiments have not yet detected an unambiguous signal, but rather yielded increasingly stringent exclusion bounds on the dark matter – nucleon elastic-scattering total cross-section σ_{det} in terms of the dark matter mass m_0.
Therefore, in order to see if the present two-singlet extension of the Standard Model is a viable dark matter model, we have to calculate σ_{det} as a function of m_0 for different values of the parameters $(\theta, \lambda_0^{(4)}, m_1)$ and project its behavior against the experimental bounds. We will limit ourselves to the region 0.1GeV – 100GeV as we are interested in light dark matter only. As experimental bounds, we will use the results from CDMSII and XENON100, as well as the future projections of SuperCDMS Schnee et al. (2005) and XENON1T April et al. (2010). As the figures below show (Gaitskell et al., 2011), in the region of our interest, XENON100 is only slightly tighter than CDMSII, SuperCDMS significantly lower and XENON1T the most stringent by far. But it is important to note that all these results loose reasonable predictability in the very light sector, say below 5GeV.
The scattering of S_0 off a SM fermion f occurs via the t-channel exchange of the SM Higgs and S_1. In the non-relativistic limit, the effective Lagrangian describing this scattering reads:

$$\mathcal{L}^{(\text{eff})}_{S_0-f} = a_f S_0^2 \bar{f} f, \tag{19}$$

where the coupling constant a_f is related to the physical parameters of the theory by the following relation:

$$a_f = -\frac{m_f}{2v}\left[\frac{\lambda_0^{(3)}\cos\theta}{m_h^2} - \frac{\eta_{01}^{(3)}\sin\theta}{m_1^2}\right]. \tag{20}$$

From this interaction, we calculate the total cross-section for this scattering process and find:

$$\sigma_{S_0 f\to S_0 f} = \frac{m_f^4}{4\pi\left(m_f + m_0\right)^2 v^2}\left[\frac{\lambda_0^{(3)}\cos\theta}{m_h^2} - \frac{\eta_{01}^{(3)}\sin\theta}{m_1^2}\right]^2. \tag{21}$$

At the nucleon level, the effective interaction Lagrangian between S_0 and a nucleon $N = p$ or n has the form:

$$\mathcal{L}^{(\text{eff})}_{S_0-N} = a_N S_0^2 \bar{N} N, \tag{22}$$

where the effective S_0- nucleon coupling constant a_N is given by the relation:

$$a_N = \frac{\left(m_N - \frac{7}{9}m_B\right)}{v}\left[\frac{\lambda_0^{(3)}\cos\theta}{m_h^2} - \frac{\eta_{01}^{(3)}\sin\theta}{m_1^2}\right], \tag{23}$$

where m_N is the nucleon mass and m_B the baryon mass in the chiral limit **?**. The total cross section for non-relativistic $S_0 - N$ elastic scattering is therefore:

$$\sigma_{\text{det}} \equiv \sigma_{S_0 N\to S_0 N} = \frac{m_N^2\left(m_N - \frac{7}{9}m_B\right)^2}{4\pi\left(m_N + m_0\right)^2 v^2}\left[\frac{\lambda_0^{(3)}\cos\theta}{m_h^2} - \frac{\eta_{01}^{(3)}\sin\theta}{m_1^2}\right]^2. \tag{24}$$

Let us briefly discuss the behavior of σ_{det} as a function of m_0 for an indicative set of values of the parameters $(\theta, \lambda_0^{(4)}, m_1)$. Of course, we have to impose systematically the relic-density constraint on the dark matter annihilation cross-section (18). But in addition, we will require here that the coupling constants are perturbative, and so impose the additional requirement $0 \leq \eta_{01}^{(4)} \leq 1$. Also, before getting into some details, let us quickly mention some global trends in the behavior of the detection cross-section. Generally, as m_0 increases, the detection cross-section σ_{det} starts from high values, slopes down to minima that depend on the parameters and then picks up moderately. There are features and action at the usual mass thresholds, with varying sizes and shapes. Excluded regions are there, those coming from the relic-density constraint and new ones originating from the additional perturbativity requirement. Close to the upper boundary of the mass interval considered in this study, there is no universal behavior to mention as in some cases σ_{det} will increase monotonously and, in some others, it will decrease or 'not be there' at all.

For a small Higgs – S_1 mixing angle, say $\theta = 10^{\text{o}}$, and a very weak mutual S_0 – Higgs coupling, $\lambda_0^{(4)} = 0.01$, the behavior of σ_{det} is displayed in figure 5 where $m_1 = 20\text{GeV}$. We see that for the two mass intervals 20GeV – 65GeV and 75GeV – 100GeV, plus an almost singled-out dip at $m_0 = m_1/2$, the elastic scattering cross section is below the projected sensitivity of SuperCDMS. However, XENON1T will probe all these masses except for $m_0 \simeq 58\text{GeV}$ and 85GeV.

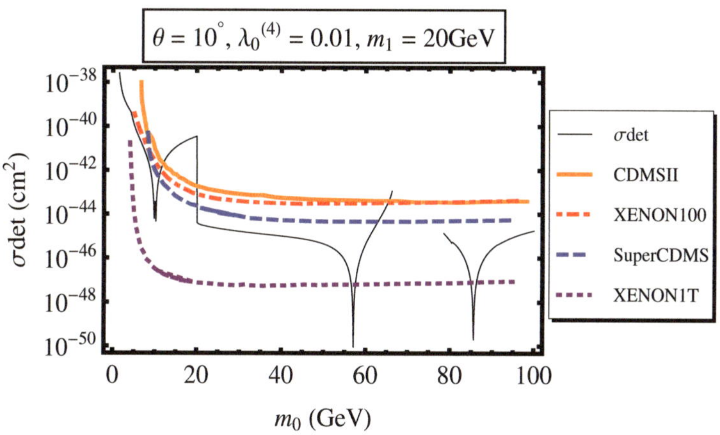

Fig. 5. Elastic $N - S_0$ scattering cross-section as a function of m_0 for moderate m_1, small mixing and small WIMP-Higgs coupling.

Also, as we see in Fig. 5, most of the mass range for very light dark matter is excluded for these values of the parameters. Is this systematic? In general, smaller values of m_1 drive the predictability ranges to the lighter sector of the dark matter masses. Figure 6 illustrates this pattern. We have taken $m_1 = 5\text{GeV}$, just above the lighter-quarks threshold. In the small-mass region, we see that SuperCDMS is passed in the range 5GeV – 30GeV. Here too, all this mass range will be probed by the XENON1T experiment, except a sharp dip at $m_0 = m_1/2 = 2.5\text{GeV}$, but for such a very light mass, the experimental results are not without ambiguity. Reversely, increasing m_1 shuts down possibilities for very light dark matter and thins the intervals as it drives the predicted masses to larger values Abada (2011).

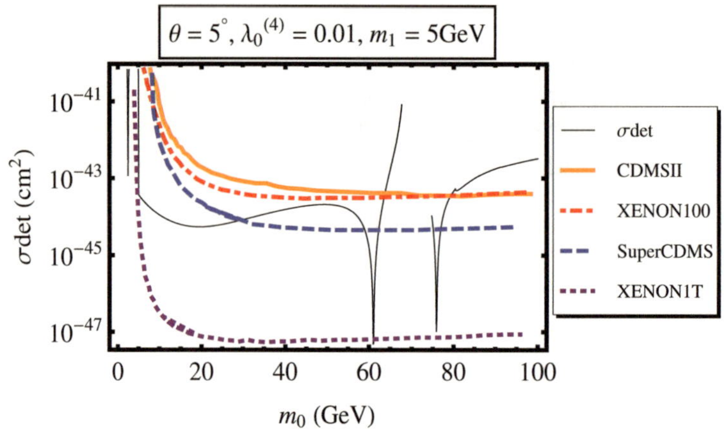

Fig. 6. Elastic $N - S_0$ scattering cross-section as a function of m_0 for light S_1, small mixing and small WIMP-Higgs coupling.

A larger mutual coupling constant $\lambda_0^{(4)}$ has the general effect of squeezing the acceptable intervals of m_0 by pushing the values of σ_{det} up. Also, increasing the mixing angle θ has the general effect of increasing the value of σ_{det}. Figure 7 shows this trend for $\theta = 40^{\text{o}}$; compare with Fig. 5. The only allowed masses by the current bounds of CDMSII and XENON100 are between 20GeV and 50GeV, the narrow interval around $m_1/2$, and another very sharp one, at about 94GeV. The projected sensitivity of XENON1T will probe all these mases except those at $m_0 \simeq$ 30GeV and 94GeV. Finally, there are regions of the parameters for which the model has no predictability. This can happen when we combine the effects of increasing the values of the two parameters $\lambda_0^{(4)}$ and m_1.

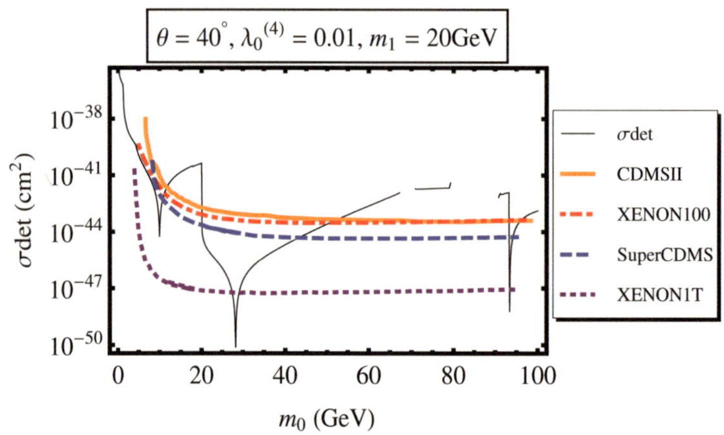

Fig. 7. Elastic $N - S_0$ scattering cross-section as a function of m_0 for moderate m_1, large mixing and small WIMP-Higgs coupling.

5. Constraints from phenomenology

Besides its direct scattering off a nucleon, a light dark-matter WIMP can manifest itself in various low-energy processes. Possible delectability puts restrictions on the various parameters of a model like the one presented here. In this section, we illustrate this mechanism with an example **?** and limit ourselves to low dark-matter masses, say from 0.1GeV – 10GeV. To ensure applicability of perturbation theory, the requirement $\eta_{01}^{(4)} < 1$ is here too imposed throughout, together with a choice of weak values for $\lambda_0^{(4)}$. Finally, all particle data used in the sequel is taken from (Nakamura et al., 2010).

The process we consider is the decay of the $\bar{b}s$ bound state B_s into, predominately, a pair of $\mu^+\mu^-$. The two corresponding SM diagrams sum up to yield a branching ratio $\mathrm{B}^{(\mathrm{SM})}\left(B_s \to \mu^+\mu^-\right) = (3.4 \pm 0.5) \times 10^{-9}$, whereas the experimental value is $\mathrm{B}^{(\exp)}\left(B_s \to \mu^+\mu^-\right) \simeq 4.7 \times 10^{-8}$. It means there is room for non-SM (invisible) processes to consider. In this two-singlet extension of the Standard Model, two additional decay diagrams occur, both via S_1 exchange, yielding together the branching ratio:

$$\mathrm{B}^{(S_1)}(B_s \to \mu^+\mu^-) = \frac{9\tau_{B_s} G_F^4 f_{B_s}^2 m_{B_s}^5}{2048\pi^5} m_\mu^2 m_t^4 \left|V_{tb}V_{ts}^*\right|^2 \frac{\left(1 - 4m_\mu^2/m_{B_s}^2\right)^{3/2}}{\left(m_{B_s}^2 - m_1^2\right)^2 + m_1^2\Gamma_1^2} \sin^4\theta. \qquad (25)$$

The particle data appearing in this expression are the B_s life-time $\tau_{B_s} = 1.43$ps, its mass $m_{B_s} =$ 5.366GeV, the Fermi coupling constant G_F, the B_s form factor f_{B_s} that we take to be 210MeV, the muon (t-quark) mass $m_{\mu(t)}$, and the CKM elements V_{tb} and V_{ts}. The quantity Γ_1 is the decay rate of the particle S_1.

This process depends directly on m_1 and the mixing angle θ, whereas m_0 and the mutual coupling $\lambda_0^{(4)}$ enter (25) via the decay rate Γ_1. A generic behavior of $\mathrm{B}^{(S_1)}(B_s \to \mu^+\mu^-)$ is shown in Fig. 8. Figure (L) shows the region (in gray) in the (m_1, θ) plane for which $\mathrm{B}^{(S_1)}$ is below the experimental value. The white narrow band about m_{B_s} is what is excluded by $\mathrm{B}^{(\exp)}$, whereas the white zone on the left is lost to the relic-density constraint and perturbativity requirement. Varying m_0 in the range 0.1GeV – 10GeV and $\lambda_0^{(4)}$ in the interval 0.01 – 0.9 has little direct effect on the behavior of $\mathrm{B}^{(S_1)}$ as a function of m_1 and θ, but does affect the relic-density constraint and perturbativity exclusion zones in their shapes, sizes and positions. Aside from these exclusion zones, most of the rest of the area is generically within the experimental bound, which means, in this sense, this process is not very restrictive by itself.

Figure (R) on the right in Fig. 8 shows the regions (in gray) for which $\mathrm{B}^{(S_1)}$ is squeezed between $\mathrm{B}^{(\exp)}$ from above and the Standard Model prediction $\mathrm{B}^{(\mathrm{SM})} + 3\sigma$ from below, thus targetting an unambiguous signal if any. The behavior we see in this figure is generic across the ranges of m_0 and $\lambda_0^{(4)}$: the V-shape structure in gray developing from $m_1 = m_{B_s}$ is the allowed region. The white region in the middle is due the $\mathrm{B}^{(\exp)}$ and the white region outside is due to $\mathrm{B}^{(\mathrm{SM})} + 3\sigma$. It can happen that some of the gray V is eaten up by the relic-density constraint and perturbativity requirement for larger values of $\lambda_0^{(4)}$.

Once a region is gray on figure (R), one has to check whether the dark-matter direct detection is allowed for the corresponding parameters. Remember that the constraint from relic density is applied systematically. Bearing in mind that the existing and predicted experimental bounds have no predictability for masses $0.1\text{GeV} \leq m_0 \leq 5\text{GeV}$, we have checked that the direct

detection cross-section is between SuperCDMS and Xenon1T for all gray points in figure (R), and this stays true for most values of m_0 and small $\lambda_0^{(4)}$. Therefore, there is no significant additional exclusion from direct detection.
From this process, there is probably one element to retain if we want the model to contribute a distinct signal to $B_s \to \mu^+\mu^-$ for the range of m_0 chosen, and that it to restrict 4.8GeV $\lesssim m_1 \lesssim$ 6.2GeV and $\theta \gtrsim 8^o$. No additional constraint on m_0 is necessary while keeping $\lambda_0^{(4)} \lesssim 0.1$ to avoid systematic exclusion from direct detection is safe.

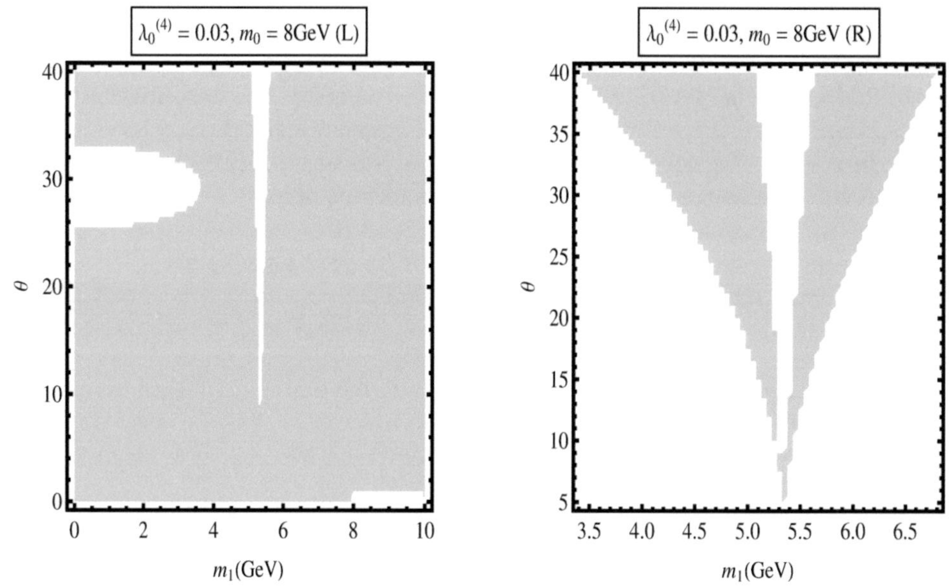

Fig. 8. The branching ratio (in gray) $B^{(S_1)}(B_s \to \mu^+\mu^-) \leq B^{(\exp)}$ in the left figure (L), and $B^{(SM)} + 3\sigma \leq B^{(S_1)}(B_s \to \mu^+\mu^-) \leq B^{(\exp)}$ in (R). The angle θ is in degrees.

6. Concluding remarks

In this chapter, we have tried to show how a plausible scenario can model light cold dark matter. The model consists in enlarging the Standard Model with two gauge-singlet $\mathbb{Z}_2$-symmetric scalar fields. One is the dark matter field S_0, stable, while the other undergoes spontaneous symmetry breaking, resulting in the physical field S_1. The goal is to open additional channels through which S_0 can annihilate, hence reducing its number density. The model is parametrized by three quantities: the physical mutual coupling constant $\lambda_0^{(4)}$ between S_0 and the Higgs, the mixing angle θ between S_1 and the Higgs and the mass m_1 of the particle S_1.
We have carried our analysis in three steps. First we have imposed on the annihilation cross-section of S_0 the constraint from the observed dark-matter relic density and looked at its effects through the behavior of the physical mutual coupling constant $\eta_{01}^{(4)}$ between S_0 and S_1 as a function of the dark matter mass m_0. Apart from forbidden regions (deserts) and others where perturbativity is lost, we find that for most values of the three parameters, there

are viable solutions in the small-moderate mass ranges of the dark matter sector. Deserts are found for most of the ranges of the parameters whereas perturbativity is lost mainly for larger values of m_1. Through the behavior of $\eta_{01}^{(4)}$, we could see the mass thresholds which mostly affect the annihilation of dark matter, and these are at the c, τ and b masses, as well as $m_1/2$ and m_1. Also, we have seen that for small values of m_1, very light dark matter is viable, with a mass as small as 1GeV. This is of course useful for understanding the results of the experiments DAMA/LIBRA, CoGeNT, CRESST Seidel (2010) as well as the recent data of the Fermi Gamma Ray Space Telescope.

The next step was to analyze dark-matter direct detection in the context of this model. We have imposed systematically the relic-density constraint and, in addition, restricted the dark-matter mass regions to be consistent with perturbativity ($\eta_{01}^{(4)} \leq 1$). We have found that the model survives current experimental bounds for a wide range of the parameter space, while at the same time recongnizing that most of the allowed mass regions will be probed by the XENON1T experiment.

The last step was to use an example to see how low-energy phenomenology can restrain the paramaters' space. We have analysed the decay of the meson B_s into a pair of $\mu^+\mu^-$ and saw how this could constrain significantly the S_1 mass and the S_1- Higgs mixing angle θ. Other processes can be envisaged, and further constraints should be expected (abada & Nasri, 2011). Implications on the Higgs detection through the measurable channels should also be considered as current experimental bounds from LEP II data can be used to constrain the mixing angle θ and possibly other parameters.

This model can be investigated in other directions. For example, the S_1 vacuum expectation value v_1 was taken equal to 100GeV, but a priori, nothing prevents us from considering other scales. However, taking v_1 much larger than the electro-weak scale requires $\eta_{01}^{(4)}$ to be very small, which will result in the suppression of the crucial annihilation channel $S_0S_0 \rightarrow S_1S_1$. Also, we have fixed the Higgs mass to $m_h = 138$GeV, which is consistent with the current acceptable experimental bounds (Nakamura et al., 2010). Yet, it can be useful to ask here too what the effect of changing this mass would be. Finally, in this study, besides the dark matter field S_0, only one extra field has been considered. Naturally, one can generalize the investigation to include N such fields and discuss the cosmology and particle phenomenology in terms of N. It just happens that the model is rich enough to open new possibilities in the quest for dark matter worth pursuing. At the same time, it tells us that modeling cold dark matter is as challenging as it is exciting.

7. References

Zwicky, T.; Helv. Phys. Acta 6, 124 (1933).

Persic, M., Salucci, P., Stel, F.; Mon. Not. Roy. Astron. Soc. 281, 27 (1996) (arXiv:astro-ph/9506004).

Fabricant, D., Gorenstein, P.; Ap. J. 267 (1983) 535.

Stewart, G.C., Canizares, C.R., Fabian, A.C., and Nilsen, P.E.J.; Ap. J. 278 (1984) 53.

Mellier, Y.; Ann. Rev. Ast. Astr. 37 (1999) 127.

Spergel, D.N. *et al.* [WMAP Collaboration]; Astrophys. J. Suppl. 170 (2007) 377 [arXiv:astro-ph/0603449].

Pope, A. *et al.* [The SDSS Collaboration]; Astrophys. J. 607 (2004) 655 (arXiv:astro-ph/0401249).

Steigman, G.; arXiv:1008.4765 [astro-ph.CO].

Komatsu. E. *et al.*; arXiv:1001.4538 [astro-ph.CO].
Amsler, C., *et al.* [Particle Data Group]; Phys. Lett. B 667, 1 (2008).
Javorsek, D. *et al.*; Phys. Rev. Lett. 87, 231804 (2001).
Javorsek, D. *et al.*; Phys. Rev. D 65, 072003 (2002).
Ellis, J., Hagelin, J., Nanopoulos, D., Olive, K. and Srednicki, M.; Nucl. Phys. B238 (1984) 453.
Jungman, G., Kamionkowski, M. and Griest, K.; Phys. Rept. 267 (1996) 195 (arXiv:hep-ph/9506380).
Belanger, G., Boudjema, F., Pukhov, A. and Singh, R.; JHEP 0911, 026 (2009).
Akrami, Y., Scott, P., Edsjo, J., Conrad, J. and Bergstrom, L.; JHEP 1004, 057 (2010).
Vasquez, D., Belanger, G., Boehm, C., Pukhov, A.and Silk, J.; Phys. Rev. D 82, 115027 (2010).
Feldman, D., Liu, Z. and Nath, P.; Phys. Rev. D81 (2010) 117701 (arXiv:1003.0437 [hep-ph]).
Kuflik, E., Pierce, A. and Zurek, K.; Phys. Rev. D81 (2010) 111701.
Fornengo, N., Scopel, S. and Bottino, A.; Phys. Rev. D 83, 015001 (2011).
Silveira, V. and Zee, A.; Phys. Lett. B161 (1985) 136.
McDonald, J.; Phys. Rev. D50 (1994) 3637.
Burgess, C., Pospelov, M. and ter Veldhuis, T.; Nucl. Phys. B619 (2001) 709.
Barger, V., Langacker, P., McCaskey, M., Ramsey-Musolf, M. and Shaughnessy, G.; Phys. Rev. D77 (2008) 035005 (arXiv:0706.4311 [hep-ph]).
Gonderinger, M., Li, Y., Patel, H. and Ramsey-Musolf, M.; JHEP 053 (2010) 1001, 2010 (arXiv:0910.3167 [hep-ph]).
He, X., Li, T., Li, X., Tandean, J. and Tsai, H.; Phys. Rev. D79 (2009) 023521 (arXiv:0811.0658 [hep-ph]).
Asano, M. and Kitano, R.; Phys. Rev. D 81, 054506 (2010).
Angle, J. *et al.* [XENON Collaboration]; Phys. Rev. Lett. 100 (2008) 021303 (arXiv:0706.0039 [astro-ph]).
Ahmed, Z. *et al.* [CDMS Collaboration]; Phys. Rev. Lett. 102 (2009) 011301 (arXiv:0802.3530 [astro-ph]).
Arina, C. and Tytgat, M.; JCAP 1101, 011 (2011).
Abada, A., Nasri. S. and Ghaffor, D.; Phys. Rev. D83, 095021 (2011).
Aprile, E. *et al.* [XENON100 Collaboration]; Phys. Rev. Lett. 105 (2010) 131302 (arXiv:1005.0380 [astro-ph.CO]).
Schnee, R. *et al.* [The SuperCDMS Collaboration]; arXiv:astro-ph/0502435.
Aprile, E. *et al.* [Xenon Collaboration]; J. Phys. Conf. Ser. 203 (2010) 012005.
Nakamura, K. *et al.* [Particle Data Group]; J. Phys. G37 (2010) 075021.
The effect of η_0 in the one-real-scalar extension of the Standard Model is discussed in Spergel, D. and Steinhardt, P.; Phys. Rev. Lett. 84 (2000) 3760 (astro-ph/9909386).
Kolb, E. and Turner, M.; *'The Early Universe'*, Addison-Wesley, (1998).
Bernabei, R., Belli, P., Cappella, F. *et al.* [DAMA/LIBRA Collaboration]; Eur. Phys. J. C67 (2010) 39 (arXiv:1002.1028 [astro-ph.GA]).
Aalseth, C.E. *et al.* [CoGeNT Collaboration]; arXiv:1002.4703 [astro-ph.CO].
Gaitskell, R., Mandic, V. and Filippini, J.; SUSY Dark Matter/Interactive Direct Detection Limit Plotter; http://dmtools.berkeley.edu/limitplots.
Abada, A. and Nasri, S.; work in progress.
Seidel, W.; WONDER 2010 Workshop, Laboratory Nazionali del Gran Sasso, Italy, March 22-23, 2010; IDM 2010 Workshop, Montpellier, France, July 26-30, 2010.

Part 4

New Cosmological Models

9

Higher Dimensional Cosmological Model of the Universe with Variable Equation of State Parameter in the Presence of G and Λ

G S Khadekar[1], Vaishali Kamdi[1] and V G Miskin[2]
[1]*Department of Mathematics, Rashtrasant Tukadoji Maharaj Nagpur University, Mahatma Jyotiba Phule Educational Campus, Amravati Road, Nagpur-440033*
[2]*Department of Mathematics, Yeshwantrao Chavan College of Engineering (YCCE), Hingna Road, Wanadongri, Nagpur- 441110*
India

1. Introduction

The Kaluza-Klein theory has a long and venerable history. However, the original Kaluza version of this theory suffered from the assumption that the 5-dimensional metric does not depend on the extra coordinate (the cylinder condition). Hence the proliferation in recent years of various versions of Kaluza-Klein theory, supergravity and superstrings. The number of authors (Wesson (1992), Chatterjee et al. (1994a), Chatterjee (1994b), Chakraborty and Roy (1999)) have considered multi dimensional cosmological model. Kaluza-Klein achievements is shown that five dimensional general relativity contains both Einstein's four-dimensional theory of gravity and Maxwell's theory of electromagnetism.
Chatterjee and Banerjee (1993) and Banerjee et al. (1995) have studied Kaluza-Klein inhomogeneous cosmological model with and without cosmological constants respectively. So far there has been many cosmological solution dealing with higher dimensional model containing a variety of matter field. However, there is a few work in a literature where variable G and Λ have been consider in higher dimension.
Beesham (1986a, 1986b) and Abdel-Rahman (1990) used a theory of gravitation using G and Λ as no constant coupling scalars. Its motivation was to include a G-coupling 'constant' of gravity as pioneered by Dirac (1937). Since the similar papers by Dirac (1938), a possible variation of G has been investigated with no success by several teams, through geophysical and astronomical observations, at the scale of solar system and with binary systems (Uzan (2003)). However, it should be stressed that we are talking here about time variations at a cosmological scale and cosmological observations still can not put strong limits on such a variation, specially at the late times of the evolution. In any case the strongest constraints are the presently observed G_0 value and observational limits of Λ_0. Sistero (1991) found exact solution for zero pressure models satisfying $G = G_0(\frac{R}{R_0})^m$. Barrow (1996) formulated and studied the problem of varying G in Newtonian Gravitation and Cosmology. Exact solutions and all asymptotic cosmological behaviour are found for universe with $G \propto t^m$.
A key object in dark energy investigation is the equation of state parameter ω, which relates pressure and density through an equation of state of the form $p = \omega\rho$. Due to lack of

observational evidence in making a distinction between constant and variable ω, usually the equation of state parameter is considered as a constant (Kujat et al. (2002), Bartelmann et al. (2005)) with values $0, \frac{1}{3}, -1$ and $+1$ for dust, radiation, vacuum fluid and stiff fluid dominated Universe respectively. But in general, ω is a function of time or redshift (Chevron and Zhuravlev (2000), Zhuravlev (2001), Peebles and Ratra (2003), Das et al. (2005)). For instance, quintessence models involving scalar fields give rise to time-dependent ω (Ratra and Peebles (1988), Turner and White (1997), Caldwell et al. (1998), Liddle and Scherrer (1999), Steinhardt et al. (1999)). So, there is enough ground for considering ω as time-dependent for a better understanding of the cosmic evolution.
A number of authors have argued in favor of the dependence $\Lambda \sim t^{-2}$ first expressed by Bertolami (1986) and later by several authors (Berman (1990), Beesham (1986b), Singh et al. (1998), Gasperini (1987), Khadekar et al. (2006)) in different context. Motivation with the work of Ibotombi (2007) and Mukhopadhyay et al. (arXiv:0711.4800v1, (2010)), in this work we have studied 5D Kaluza-Klein type metric with perfect fluid and variable G and Λ.
Recently the cosmological implication of a variable speed of light (VSL) during the early evolution of the universe have been considered by [Belincho and Chakrabarty (2003), Belincho (2004)]. Varying speed of light (VSL) model proposed by Moffat (1993) and Albrecht and Maguejio (1999) in which light was traveling faster in the early periods of the existence of the universe, might solve the same problems as inflation. Einstein's field equations for Friedmann-Roberton-Walker (FRW) space time in the VSL theory have been solved by Barrow (1999), who also obtained the rate of variation of speed of light required to solve the flatness and cosmological constant problem for a review of these theories.
We have obtained exact solutions for Zeldovich fluid models satisfying $G = G_0(\frac{R}{R_0})^m$ with global equation of state of the form $p = \frac{1}{3}\phi\rho$, where ϕ is a function of scale factor R. In section 2 and 3 of the chapter we have studied two variable Λ model of the form $\Lambda \sim (\frac{\dot{R}}{R})^2$ and $\Lambda \sim \rho$ under the assumption that the equation of state parameter ω is a function of time. It is shown that possibility of signature flip of the deceleration parameter q. In section 4 of the chapter we have examined the perfect fluid cosmological model by considering the equation of state parameter ω is constant with varying G, c and Λ by using Lie method given by Ibrabimov (1999) and find the possible forms of the constants G, Λ and c that integrable the field equations in the framework of Kaluza-Klein theory of gravitation.

2. Field equations

We consider the 5D Robertson-Walker metric

$$ds^2 = c^2(t)dt^2 - R^2(t)\left[\frac{dr^2}{(1-kr^2)} + r^2(d\theta^2 + sin^2\theta d\phi^2)\right] + A^2(t)d\psi^2, \tag{1}$$

where $R(t)$ is the scale factor, $A(t) = R^n$ and $k = 0,\ -1\ \ or + 1$ is the curvature parameter for flat, open and closed universe, respectively. The universe is assumed to be filled with distribution of matter represented by energy-momentum tensor of a perfect fluid

$$T_{ij} = (p+\rho)u_i u_j - p g_{ij}, \tag{2}$$

where, ρ is the energy density of the cosmic matter and p is its pressure and u_i is the five velocity vector such that $u_i u^j = 1$.

The Einstein field equations are given by

$$R_{ij} - \frac{1}{2} g_{ij} R = -8\pi G(t) \left[\frac{T_{ij}}{c^4} - \frac{\Lambda(t)}{8\pi G} g_{ij} \right], \tag{3}$$

where the cosmological term Λ is time-dependent and c, the velocity of light in vacuum. In the follwing two section we have assumed the velocity of light is unity i.e. $c = 1$. The conservation equation for variable G and Λ is given by

$$\dot{\rho} + (3+n)\frac{\dot{R}}{R}(p+\rho) = -\left(\frac{\dot{G}}{G}\rho + \frac{\dot{\Lambda}}{8\pi G} \right). \tag{4}$$

Using co-moving co-ordinates $u^i = (1,0,0,0,0)$ in (2) and with metric (1), the Einstein field equations become

$$8\pi G\rho = 3\left[(n+1)\frac{\dot{R}^2}{R^2} + \frac{k}{R^2} \right] - \Lambda(t), \tag{5}$$

$$8\pi G p = -(n+2)\frac{\ddot{R}}{R} - (n^2+n+1)\frac{\dot{R}^2}{R^2} - \frac{k}{R^2} + \Lambda(t), \tag{6}$$

$$8\pi G p = -3\left(\frac{\ddot{R}}{R} + \frac{\dot{R}^2}{R^2} + \frac{k}{R^2} \right) + \Lambda(t). \tag{7}$$

where dot $(\cdot)$ denotes derivative with respective to t.
The usual conservation law yields (i.e. $T^{ij}_{;j} = 0$)

$$\dot{\rho} + (3+n)(\rho+p)\frac{\dot{R}}{R} = 0. \tag{8}$$

Using Eq.(8) in Eq.(4)we have,

$$8\pi\dot{G}\rho + \dot{\Lambda} = 0. \tag{9}$$

Equations (5), (6) and (9) are the fundamental equations and they reduce to standard Friedmann cosmology when G and Λ are constants. Equations (5) and (6) may be written as

$$3(n+2)\ddot{R} = -8\pi G R(3p+\rho) - 3n^2\frac{\dot{R}^2}{R} + 2\Lambda R, \tag{10}$$

$$3(n+1)\dot{R}^2 = 8\pi G R^2\left[\rho + \frac{\Lambda}{8\pi G} \right] - 3k. \tag{11}$$

Eq.(8) can also be expressed as

$$\frac{d}{dt}(\rho R^{n+3}) + p\frac{d}{dt}(R^{n+3}) = 0. \tag{12}$$

Equations (5), (9) and (12) are independent and they will be used as fundamental. Once the problem is determined, the integration constants are characterized by the observable parameters

$$H_0 = \frac{\dot{R}_0}{R_0}, \tag{13}$$

$$\sigma_0 = \frac{4\pi}{3}\frac{G_0\rho_0}{H_0^2}, \tag{14}$$

$$q_0 = -\frac{\ddot{R}_0}{R_0 H_0^2}, \tag{15}$$

$$\epsilon_0 = \frac{p_0}{\rho_0}, \tag{16}$$

which must satisfy Einstein's equations at present cosmic time t_0:

$$\Lambda_0 = 3H_0^2\left[\sigma_0(3\epsilon_0+1) - \frac{(n+2)}{2}q_0 + \frac{n^2}{2}\right], \tag{17}$$

$$\frac{k}{R_0^2} = H_0^2\left[3(1+\epsilon_0)\sigma_0 - \frac{(n+2)}{2}q_0 + \frac{(n^2-2n-2)}{2}\right], \tag{18}$$

and the conservation Eq. (9) can be written as

$$\dot{\Lambda}_0 G_0 + 6\dot{G}_0 H_0^2\sigma_0 = 0. \tag{19}$$

3. Solutions of field equations

We find out the solutions of the field equations for two different equation of state: (i) $p = \frac{1}{3}\rho\phi$ and (ii) $p = \omega(t)\rho$

3.1 Case (I):

We assume the global equation of state

$$p = \frac{1}{3}\rho\phi, \tag{20}$$

where ϕ is a function of the scale factor R.
From Eq.(12) and Eq.(20) we obtain

$$\frac{1}{\psi}\frac{d\psi}{dR} + \frac{(n+3)}{3}\frac{\phi}{R} = 0, \tag{21}$$

where

$$\psi = \rho R^{n+3}. \tag{22}$$

Equation (21) be the first condition to determine the problem; either ϕ or ψ may be in term of arbitrary function. If ϕ is a given explicit function of R, then Eq.(20) is determined and ψ follows from Eq.(21)

$$\psi = \psi_0 \ exp\left[-\int \frac{(n+3)}{3}\frac{\phi}{R}dR\right]. \tag{23}$$

If ψ is given function, from Eq.(20) we get ϕ as

$$\phi = -\frac{3}{(n+3)}\frac{R}{\psi}\frac{d\psi}{dR}. \tag{24}$$

Substitute the value of ψ from Eq.(22) in the Friedmann's Eq.(5) we get

$$3(n+1)\dot{R}^2 = 8\pi G\psi R^{-(n+1)} + \Lambda R^2 - 3k. \tag{25}$$

Eqs. (9) and (22) with $\frac{d}{dt} = (\dot{R}\frac{d}{dR})$ give

$$8\pi\frac{dG}{dR} + \psi^{-1}R^{n+3}\frac{d\Lambda}{dR} = 0. \tag{26}$$

If $G(R)$ is given then after integrating from Eq.(26) we get $\Lambda(R)$ and from Eq.(25) we get $R = R(t)$ and the problem is solved. Similarly if $\Lambda(R)$ may be given instead of $G(R)$ derives from Eq.(26) we get $G(R)$ and then from Eq. (25) we get $R = R(t)$.

3.1.1 Zeldovich fluid satisfying $G = G_0(\frac{R}{R_0})^m$

To solve Eq.(26) for Zeldovich fluid with $\phi = 3$. In this case (23) gives,

$$\psi = \rho_0\left(\frac{R_0}{R}\right)^{n+3}. \tag{27}$$

Substituting ψ from (26) into (25), we have

$$\Lambda = \Lambda_0 + B_m\left[1 - \left(\frac{R}{R_0}\right)^{[m-2(n+3)]}\right]R_0^{-(n+3)}, \tag{28}$$

where,

$$B_m = \frac{6m}{[m-2(n+3)]}\sigma_0 H_0^2, \tag{29}$$

for $m \neq 2(n+3)$, B_m is a parameter related to the integration constant of Eq.(25). From Eq.(17),

$$\Lambda_0 = 3H_0^2\left[4\sigma_0 - \frac{(n+2)}{2}q_0 + \frac{n^2}{2}\right]. \tag{30}$$

Taking into account Eqs.(26 & 28), Friedmann's Eq. (24) takes the form

$$\dot{R}^2 = \alpha_n R^{m-2(n+2)} + \beta_n R^2 - \frac{1}{(n+1)}k, \tag{31}$$

where

$$\alpha_n = \frac{-4(n+3)}{(n+1)(m-2(n+3))}H_0^2\sigma_0 R_0^{(n+3)-m}, \tag{32}$$

$$\beta_n = \frac{H_0^2}{(n+1)}\left[\left(4 + \frac{2m}{(m-2(n+3))}R_0^{-(n+3)}\right)\sigma_0 - \frac{(n+2)}{2}q_0\right]. \tag{33}$$

Finally the equation for the parameter (18) reduces to

$$\frac{k}{R_0^2} = H_0^2\left[6\sigma_0 - \frac{(n+2)}{2}q_0 + \frac{(n^2-2n-2)}{2}\right]. \tag{34}$$

and (19) is also satisfied.

$$\Lambda_0 G_0 + 6\dot{G}_0 H_0^2 \sigma_0, \tag{35}$$

The model is characterized by the set of parameters $(H_0, G_0, \sigma_0, q_0, m)$ with $m \neq 2(n+3)$. The case $m < 2(n+3)$ implies $B_m < 0$ in Eq.(29) and $\alpha_n > 0$ in Eq.(32) and vice-versa; $\beta_n < (\geq)0$ according to $m\ n, \sigma_0$ and q_0 combine in Eq.(33); $\Lambda_0 < (\geq)0$ as $\sigma_0 < (\geq)(\frac{(n+2)}{2}q_0 - \frac{n^2}{2})$ as given by Eq. (30). From Eq.(34) it is observed that for the curvature parameter $k = +1,\ 0, -1$ we get $[6\sigma_0 - \frac{(n+2)}{2}q_0 + \frac{(n^2-2n-2)}{2}] < (\geq)0$. The models are completely characterized by the set of parameters $(H_0, G_0, \sigma_0, q_0, m)$ with $m \neq 2(n+3)$.

3.2 Case (II):

Let us choose the barotropic equation of state

$$p = \omega\rho. \tag{36}$$

Here, we assume that the equation of state parameter ω is time-dependent i.e. $\omega = \omega(t)$ such that $\omega = (\frac{t}{\tau})^a - 1$ where τ is a constant having dimension of time.
Field equations (5-7) can also be expressed as

$$3(n+1)H^2 + \frac{3k}{R^2} = 8\pi G\rho + \Lambda(t), \tag{37}$$

$$3(n+1)H^2 + 3(n+1)\dot{H} = -8\pi G[(n+1)p + \rho] - \frac{3nk}{R^2} + n\Lambda(t). \tag{38}$$

From Eq. (37), for flat universe ($k = 0$), we get

$$\rho = \frac{3(n+1)H^2 - \Lambda(t)}{8\pi G}. \tag{39}$$

Using Eq. (37) and Eq.(38) with Eq. (36) we get the differential equation of the form

$$\frac{dH}{dt} = \frac{(1+\omega)\Lambda}{3} + [(n+1)\omega - 2]H^2. \tag{40}$$

To solve Eq. (40) we assume two variable Λ model: $\Lambda = 3\alpha H^2$ and $\Lambda = 8\pi G\gamma\rho$.

3.2.1 Case (i): $\Lambda = 3\alpha H^2$

For this case Eq. (40) reduces to

$$\frac{dH}{H^2} = \left[\frac{(n+\alpha+1)t^a}{\tau^a} - (n+3)\right] dt. \tag{41}$$

After solving equation (41) we get,

$$H = \frac{(a+1)\tau^a}{[(n+3)(a+1)t\tau^a - (n+\alpha+1)t^{(a+1)}]}, \tag{42}$$

writing $H = \frac{\dot{R}}{R}$ in Eq. (42) and integrating it further we get the solution set as

$$R(t) = C_2\left[(n+3)(a+1)\tau^a t^{-a} - (n+\alpha+1)\right]^{-\frac{1}{a(n+3)}}, \tag{43}$$

$$\rho(t) = \frac{3(n-\alpha+1)(a+1)^2\tau^{2a}}{8\pi G}\left[(n+3)(a+1)\tau^a t-(n+\alpha+1)t^{(a+1)}\right]^{-2}, \tag{44}$$

$$\Lambda(t) = \frac{3\alpha(a+1)^2\tau^{2a}}{\left[(n+3)(a+1)\tau^a t-(n+\alpha+1)t^{(a+1)}\right]^2}, \tag{45}$$

where C_2 is an integration constant.
If $a = 0$ then $\omega = 0$ and $\tau = 1$ but Eq. (42) indicates that a can not be equal to zero for physical validity.
Again, using Eqs. (39) and (42) we get

$$\frac{\alpha}{(n+1)} = 1-\Omega_m = \Omega_\Lambda, \tag{46}$$

where, in absence of any curvature, matter density Ω_m and dark energy density Ω_Λ are related by the equation

$$\Omega_m + \Omega_\Lambda = 1. \tag{47}$$

3.2.2 Case (ii): $\Lambda = 8\pi G\gamma\rho$
For this case Eq. (40) can be written as

$$\frac{dH}{dt} = \left[\left(\frac{1+2\gamma}{1+\gamma}\right)\left(\frac{t}{\tau}\right)^a (n+1)-(n+3)\right]H^2. \tag{48}$$

After solving Eq. (48) we get,

$$H = \frac{(1+\gamma)(a+1)\tau^a}{\left[(n+3)(1+\gamma)(a+1)\tau^a t-(1+2\gamma)(n+1)t^{a+1}\right]}. \tag{49}$$

Using $H = \frac{\dot{R}}{R}$ in Eq. (49) and integrating we get

$$R(t) = C_3\left[(n+3)(1+\gamma)(a+1)\tau^a t^{-a}-(1+2\gamma)(n+1)\right]^{-\frac{1}{a(n+3)}}, \tag{50}$$

$$\rho(t) = \frac{3(n+1)}{8\pi G}\frac{(1+\gamma)(a+1)^2\tau^{2a}}{\left[(n+3)(1+\gamma)(a+1)\tau^a t-(1+2\gamma)(n+1)t^{(a+1)}\right]^2}, \tag{51}$$

$$\Lambda(t) = \frac{3(n+1)\gamma(1+\gamma)(a+1)^2\tau^{2a}}{\left[(n+3)(1+\gamma)(a+1)\tau^a t-(1+2\gamma)(n+1)t^{(a+1)}\right]^2}, \tag{52}$$

where C_3 is an integration constant.
Eq. (50) shows that for physical validity $a \neq 0$. Again from the field equations we can easily find that γ is related to Ω_m and Ω_Λ through the relation

$$\gamma = \frac{\Omega_\Lambda}{\Omega_m}. \tag{53}$$

4. Solution of the field equations by using Lie method

The Einstein's filed equations (3) with varying G, c and Λ for the flat model (1) when for $R = A$ i.e. $n = 1$ and $k = 0$ can be written as

$$8\pi G\rho + \Lambda c^2 = 6H^2, \tag{54}$$

$$-8\pi G p + \Lambda c^2 = 3\left(\frac{\ddot{R}}{R} + H^2\right), \tag{55}$$

$$\dot{\rho} + 4(\rho + p)H = -\frac{\dot{\Lambda}c^4}{8\pi G\rho} - \frac{\dot{G}}{G} + 4\frac{\dot{c}}{c}. \tag{56}$$

We assume that div $(T^i_j) = 0$, then with $p = \omega\rho$, where $\omega = constant$ then Eq. (56) reduces to

$$\dot{\rho} + 4(1+\omega)\rho H = 0, \tag{57}$$

$$-\frac{\dot{\Lambda}c^4}{8\pi G\rho} - \frac{\dot{G}}{G} + 4\frac{\dot{c}}{c} = 0. \tag{58}$$

In this section we shall study the Kaluza-Klein type cosmological model through the method of Lie group symmetries, showing that under the assumed hypothesis there are other solutions of the field equations. We shall show how the Lie method allow us to obtain different solutions for the field equations.

In order to use the Lie method, we can write the field equations: from Eqs. (54)-(55) we obtain

$$\frac{\ddot{R}}{R} - \frac{\dot{R}^2}{R^2} = -\frac{8\pi G}{3c^2}(\omega+1)\rho, \tag{59}$$

and therefore,

$$\dot{H} = -\frac{8\pi G}{3c^2}(\omega+1)\rho. \tag{60}$$

From equation (57), we can obtain

$$H = -\frac{1}{4(\omega+1)}\frac{\dot{\rho}}{\rho}. \tag{61}$$

Hence

$$\dot{H} = -\frac{1}{4(\omega+1)}(\frac{\dot{\rho}}{\rho}\dot{)}. \tag{62}$$

Hence from Eq. (60)

$$(\frac{\dot{\rho}}{\rho}\dot{)} = \frac{16\pi G}{3c^2}(\omega+1)^2\rho. \tag{63}$$

By taking $A_0 = -\frac{16\pi}{3}(\omega+1)^2$, we get

$$(\frac{\dot{\rho}}{\rho}\dot{)} = \frac{A_0 G}{3c^2}\rho. \tag{64}$$

After expanding Eq. (64) we get

$$\ddot{\rho} = \frac{\dot{\rho}^2}{\rho} + \frac{AG}{3c^2}\rho. \tag{65}$$

We are going now to apply the standard procedure of Lie group analysis to this equation [see Ibragimov (1999) for details and notation].
A vector field X

$$X = \zeta(t,\rho)\partial_t + \eta(t,\rho)\partial_\rho, \tag{66}$$

is a symmetry of equation (65) iff

$$-\zeta f_t - \eta f_\rho + \eta_{tt} + (2\eta_{t\rho} - \zeta_{tt}) + (\eta_{\rho\rho} - 2\zeta_{t\rho})\dot{\rho}^{-2} - \zeta_{\rho\rho}\dot{\rho}^3$$
$$+ (\eta_\rho - 2\zeta_t - 3\dot{\rho}\zeta_\rho)f - \left[\eta t + (\eta_\rho - \zeta_t)\dot{\rho} - \dot{\rho}^2\zeta_\rho\right] f_{\dot{\rho}} = 0, \tag{67}$$

where $f(t,\rho,\dot{\rho}) = \frac{\dot{\rho}^2}{\rho} + \frac{A_0 G}{3c^2}\rho$.
By expanding and separating (67) with respect to power of $\dot{\rho}$ we obtain the overdetermined system:

$$\zeta_{\rho\rho} + \rho^{-1}\zeta_\rho = 0, \tag{68}$$

$$\eta_{\rho\rho} - 2\zeta_{t\rho} + \rho^{-2}\eta - \rho^{-1}\eta_\rho = 0, \tag{69}$$

$$2\eta_{t\rho} - \zeta_{tt} - 3A\frac{G}{c^2}\rho^2\zeta_\rho - 2\rho^{-1}\eta_t = 0, \tag{70}$$

$$\eta_{tt} - A(\frac{\dot{G}}{c^2} - 2G\frac{\dot{c}}{c^3})\rho^2\zeta - 2A\eta\frac{G}{c^2} + (\eta_\rho - 2\zeta_t)A\frac{G}{c^2}\rho^2 = 0. \tag{71}$$

Solving (68) - (71), we find that

$$\zeta(t,\rho) = 2et + a, \qquad \eta(t,\rho) = (bt + d_0)\rho, \tag{72}$$

subject to the constrain

$$\frac{\dot{G}}{G} = 2\frac{\dot{c}}{c} + \frac{bt + d_0 - 4e}{2et - a}, \tag{73}$$

where a, b, e, d_0 are all constants.
In order to solve Eq. (73) we consider the case $b = 0$ and $d_0 - 4e = 0$. In this case the solution (73) reduces to

$$\frac{\dot{G}}{G} = 2\frac{\dot{c}}{c} \Rightarrow \frac{G}{c^2} = B = Constant, \tag{74}$$

which means that constant G and c vary but in such a way that the relation $\frac{G}{c^2}$ is constant.
The solution of the type

$$\frac{dt}{\zeta(t,\rho)} = \frac{d\rho}{\eta(t,\rho)}, \tag{75}$$

is called invariant solution, therefore, from (72) with $b = 0$ and $d_0 - 4e = 0$, the energy density is obtained as:

$$\frac{dt}{-2et + a} = \frac{d\rho}{4e\rho}, \tag{76}$$

$$\Rightarrow \rho = \frac{\rho_0}{(2et - a)^2}, \tag{77}$$

for simplicity we adopt

$$\Rightarrow \rho = \rho_0 t^{-2}, \tag{78}$$

where ρ_0 is constant of integration.
From the value of ρ we can easily obtained the scale factor R as: from (61) after integration we get

$$\rho = A_\omega R^{-4(\omega+1)}, \tag{79}$$

$$\Rightarrow R = (A^*_\omega t)^{1/2(\omega+1)}, \tag{80}$$

where A_ω and A^*_ω are constants.
From this value of R we can easily find H and from equation (54) we obtain the behaviour of cosmological constant Λ.

$$c^2\Lambda = 6H^2 - \frac{8\pi G\rho}{c^2}\rho, \tag{81}$$

we get

$$\Lambda = (3\beta^2 - 8\pi B\rho_0)\frac{1}{t^2c^2} = \frac{L_0}{t^2c^2}, \tag{82}$$

where $L_0 = (3\beta^2 - 8\pi B\rho_0)$.
Put all the above results in (58) we get the exact behaviour for c:

$$\frac{\dot{c}}{c} + \lambda(\frac{\dot{c}}{c} + \frac{1}{t}) = 0, \tag{83}$$

where $\lambda = \frac{L_0}{8\pi B\rho_0}$ with $\lambda \in R^+$ i.e. a positive real number and thus we get from (83) after integration

$$c = c_0 t^{-\alpha}, \tag{84}$$

where $\alpha = \frac{\lambda}{1+\lambda}$.

Also from

$$\frac{\dot{G}}{G} = 2\frac{\dot{c}}{c} \Rightarrow G = G_0 t^{-2\alpha}. \tag{85}$$

Hence we get the following solutions in the framework of Kaluza-Klien theory of gravitation.

$$G = G_0 t^{-2\alpha} \quad c = c_0 t^{-\alpha}, \quad \Lambda = \Lambda_0 t^{-2(1-\alpha)},$$

$$R = (A^*_\omega t)^{1/2(\omega+1)}, \quad \rho = \rho_0 t^{-2}. \tag{86}$$

This type of solutions are obtained by Belinchon (2004) in the context of general theory of relativity.

5. Conclusion

In this chapter by considering the gravity with G and Λ a coupling constant of Einstein field equations with usual conservation laws ($T^{ij}_{;j} = 0$), we obtained the exact solution of the field equations. It is shown that the field equations for perfect fluid cosmology are identical to Eisenstein equations for G and Λ including Eq. (12). It is also observed that, the additional conservation Eq. (9) gives the coupling of scalar field with matter.
In the case (I) by introducing the general method of solving the cosmological field equations using a global equation of state of the form $p = \frac{1}{3}\rho\phi$, without loss of generality, we find the exact solutions for Zeldovich matter distribution. It is observed that from Eq. (29) $B_m < 0$ for

the case $m < 2(n+3)$ and $\Lambda_0 < (\geq)0$ as $\sigma_0 < (\geq)(\frac{(n+2)}{2}q_0 - \frac{n^2}{2})$. Similarly $[6\sigma_0 - \frac{(n+2)}{2}q_0 + \frac{(n^2-2n-2)}{2}] < (\geq)0$ depends on the value of curvature parameter k.
In the case (II), by using equation of state of the form $p = \omega(t)\rho$, we again find out the exact solutions of the field equations for two different cases: $\Lambda = 3\alpha H^2$ and $\Lambda = 8\pi G\gamma\rho$. By selecting a simple power law expression of t for the equation of state parameter ω, equivalence of model $\Lambda \sim (\frac{\dot{R}}{R})^2$ and $\Lambda \sim \rho$ have been established in the frame work of Kaluza-Klein theory of gravitation. With the help of Eqs. (45) and (52) it is easy to show that Eqs. (42) and (49) are differ by constant while Eqs. (43) and (44) become identical with the Eqs. (50) and (51) respectively. This implies that $\Lambda \sim (\frac{\dot{R}}{R})^2$ and $\Lambda \sim \rho$ are equivalent for five dimensional space time.
Using Eq. (42) and Eq. (46), we obtain

$$q = -\left[1 - \left[(n+3) - (n+1)(2-\Omega_m)\left(\frac{t}{\tau}\right)^a\right]\right]. \tag{87}$$

Eq. (87) shows that q is time dependent and hence may be change its sign during cosmic evolution. It has also been possible to show that the sought for signature flipping of deceleration parameter q can be obtained by a suitable choice of a.
In the last section of the chapter we have studied the behaviour of time varying constants G, c and Λ in a perfect fluid model. To obtain the solution we imposed the assumption, $div(T^i_j) = 0$, from which we obtained the dimensional constant A_ω that relates $\rho \propto R^{-4(\omega+1)}$and the relationship $\frac{G}{c^2} = B = constant$ for all value of t, i.e. G and c vary but in such a way that $\frac{G}{c^2}$ remain constant. It is also observed that G, c and Λ are decreasing function of t. The Lie method maybe the most powerful but has drawbacks, it is very complicate.

6. References

Wesson, P. S. (1992), *Astrophys. J.*, Vol. 394, 19.
Chatterjee, S., Panigrahi, D. and Banerjee, A. (1994a), *Class. Quantum Grav.* , Vol. 11, 371
Chatterjee, S., Bhui, B., Basu, M. B. and Banerjee, A. (1994b), *Phys. Rev.* , Vol. D 50, 2924
Chakraborty, S. and Roy, A. (1999), *Int. J. Mod. Phys.*, Vol. D 8, 645.
Chatterjee, S. and Banerjee, A. (1993), *Class. Quantum Grav.* , Vol. 10, L1
Banerjee, A., Panigrahi, D. and Chatterjee, S. (1995), *J. Math. Phys.* , Vol. 36, 3619.
Beesham, A. (1986a), *Nuovo Cimento*, Vol. B96, 19.
Beesham, A. (1986b), *Int. J. Theor. Phys.*, Vol. 25, 1295.
Abdel-Rahman, A.-M.M. (1990). *Gen. Relativ. Gravit.* , Vol. 22, 655.
Dirac, P. A. M. (1937), *Nature* , Vol. 139, 323.
Dirac, P. A. M. (1938), *Proc. Roy. Soc. Lond.* , Vol. 165, 199.
Uzan, J. P. (2003), *Rev. Mod. Phys.* , Vol. 75, 403.
Sistero, R. F. (1991), *Gen. Relativ. Gravit.* , Vol. 23, 11.
Barrow, J. D. (1996), *Roy. Astron. Soc.* , Vol. 282, 1397.
Kujat, J. et al. (2002), *Astrophys. J.* , Vol. 572, 1.
Bartelmann, M. et al. (2005), *New Astron. Rev.* , Vol. 49, 199.
Chevron, S. V. and Zhuravlev, V. M. (2000), *Zh. Eksp. Teor. Fiz.* , Vol. 118, 259.
Peebles, P. J. E. and Ratra, B. (2003), *Rev. Mod. Phys.*, Vol. 75, 559.
Das, A. et al. (2005), *Phys. Rev. D* , Vol. 72, 043528.

Ratra, B. and Peebles, P. J. E. (1988), *Phys. Rev. D* , Vol. 37, 3406.
Turner, T. S. and White, M. (1997), *Phys. Rev. D*, Vol. 56, R4439.
Caldwell et al. (1998), *Phys. Rev. Lett.* , Vol. 80, 1582.
Liddle, A. R. and Scherrer, R. J. (1999), *Phys. Rev. D* , Vol. 59, 023509.
Steinhardt, P. J. et al. (1999), *Phys. Rev. D* , Vol. 59, 123504.
Bertolami, O. (1986), *Nuovo Cimento* , Vol. 93, 36.
Berman, M. S. (1990), *Int. J. Theor. Phys.*, Vol. 29, 567.
Singh, T., Beesham, A. and Mbokazi, W. S. (1998), *Gen. Relativ. Gravit.* , Vol. 30, 537.
Gasperini, M. (1987), *Phys. Lett. B*, Vol. 194, 347.
Khadekar, G. S. et al. (2000), *Int. Jou. Mod. Phys. D*, Vol. 15, 1.
Singh, N. I. and Sorokhaibam, A. (2007), *Astrophys. Space Sci.*, Vol. 310, 131.
Mukhopadhyay, U., Ray, S. and Dutta Choudhury, S. B. (2010), *arXiv:0711.4800v3* .
Belinchon, J. A., and Chakrabarty, I. (2003), *Int.Jour. Moder. Phys.*, Vol.D12, 1113, *gr-qc/044046*.
Belinchon, J.A., (2004), *An excuse for revising a theroy of time-varying constant, gr-qc/044026*.
Moffat, J.W., (1993), *Int.Jour. Moder. Phys.D*, Vol.2, 351.
Albrechet, A., and Magueijo, J., (1999), *Phys. Rev. D*, vol.59, 043516.
Barrow, J.D., (1999), *Phys. Rev. D*,Vol.59, 043515.
Ibravimov, N.H., (1999), *Elementary Lie group Analysis and Ordinary Differential Equations, Jhon Wiley and Sons*.

10

Cosmological Bianchi Class A Models in Sáez-Ballester Theory

J. Socorro[1], Paulo A. Rodríguez[1], Abraham Espinoza-García[1], Luis O. Pimentel[2] and Priscila Romero[3]
[1]*Departamento de Física de la DCeI de la Universidad de Guanajuato-Campus León, Guanajuato*
[2]*Departamento de Física de la Universidad Autónoma Metropolitana*
[3]*Facultad de Ciencias de la Universidad Autónoma del Estado de México, Instituto Literario No. 100, Toluca, Edo de Mex México*

1. Introduction

Several observations suggest that in galaxies and galaxy clusters there is an important quantity of matter that is not interacting electromagnetically, but only through gravitation. This is the well known dark matter problem. Several solutions have been consider for this problem, modifying the gravitational theory or introducing new forms of matter and interaccions. To address the dark matter problem Saez and Ballester (SB) (Saez & Ballester, 1986) formulated a scalar-tensor theory of gravitation in which the metric is coupled with a dimensionless scalar field. In a recent analysis using the standard scalar field cosmological models (Socorro et al., 2010; 2011), contrary to claims in the specialized literature, it is shown that the SB theory cannot provide a realistic solution to the dark matter problem of Cosmology for the dust epoch, because the contribution of the scalar field is equivalent to stiff matter. We can reinterpret this result in a sense that the galaxy halo was formed during this primigenius epoch and its evolution until the dust era using the standard scalar field cosmological theory. In this theory the strength of the coupling between gravity and the scalar field is determined by an arbitrary coupling constant ω. This constant ω can be used to have a lorenzian (-1,1,1,1) or seudo-lorenzian (-1,-1,1,1) signature when we build the Wheeler-DeWitt equation. The values for this constant, in the classical regime, are dictated by the condition to have real functions. Other problem inherent to this theory is that not exist how build the invariants with this field as in the case to scalar curvature. So, was necessary to reinterpret the formalism where this field is considered as matter content in the theory in the Einstein frame.
On the other hand, this approach is classified with another name, by instant, Armendariz-Picon et al, called this formalism as K-essence (Armendariz et al., 2000), as a dynamical solution for explaining naturally why the universe has entered an epoch of accelerated expansion at a late stage of its evolution. Instead, K-essence is based on the idea of a dynamical attractor solution which causes it to act as a cosmological constant only at the onset of matter domination. Consequently, K-essence overtakes the matter density and induces cosmic acceleration at about the present epoch. Usually K-essence models are

restricted to the Lagrangian density of the form

$$S = \int d^4x \sqrt{-g}\, f(\phi)\, (\nabla\phi)^2 . \tag{1}$$

One of the motivations to consider this type of Lagrangian originates from string theory (Armendariz et al., 1999). For more details for K-essence applied to dark energy, you can see in (Copeland et al., 2006) and reference therein. Many works in SB formalism in the classical regime have been done, where the Einstein field equation is solved in a direct way, using a particular ansatz for the main scalar factor of the universe (Singh & Agrawal, 1991; Ram & Singh, 1995; Mohanty & Pattanaik, 2001; Singh & Ram, 2003), yet a study of the anisotropy behaviour trough the form introduced in the line element has been conected (Reddy & Rao, 2001; Mohanty & Sahu, 2003; 2004; Adhav et al., 2007; Rao et al., 2007; 2008a-2008b; Shri et al., 2009; Tripathy et al., 2009; Singh, 2009; Pradhan & Singh, 2010).

On another front, the quantization program of this theory has not been constructed. The main complication can be traced to the lack of an ADM type formalism. We can transform this theory to conventional one where the dimensionless scalar field is obtained from energy-momentum tensor as an exotic matter contribution, and in this sense we can use this formalism for the quantization program, where the ADM formalism is well known (Ryan, 1972).

In this work, we use this formulation to obtain classical and quantum exact solutions to anisotropic Bianchi Class A cosmological models with stiff matter. The first step is to write SB formalism in the usual manner, that is, we calculate the corresponding energy-momentum tensor to the scalar field and give the equivalent Lagrangian density. Next, we proceed to obtain the corresponding canonical Lagrangian $\mathcal{L}_{can}$ to Bianchi Class A cosmological models through the Legendre transformation, we calculate the classical Hamiltonian $\mathcal{H}$, from which we find the Wheeler-DeWitt (WDW) equation of the corresponding cosmological model under study. We employ in this work the Misner parametrization due that a natural way appear the anisotropy parameters to the scale factors.

The simpler generalization to Lagrangian density for the SB theory (Saez & Ballester, 1986) with the cosmological term, is

$$\mathcal{L}_{\rm geo} = \left(R - 2\Lambda - F(\phi)\phi_{,\gamma}\phi^{,\gamma}\right), \tag{2}$$

where $\phi^{,\gamma} = g^{\gamma\alpha}\phi_{,\alpha}$, R the scalar curvature, $F(\phi)$ a dimensionless function of the scalar field. In classical field theory with scalar field, this formalism corresponds to null potencial in the field ϕ, but the kinetic term is exotic by the factor $F(\phi)$.

From the Lagrangian (2) we can build the complete action

$$I = \int_\Sigma \sqrt{-g}(\mathcal{L}_{\rm geo} + \mathcal{L}_{\rm mat})d^4x, \tag{3}$$

where $\mathcal{L}_{mat}$ is the matter Lagrangian, g is the determinant of metric tensor. The field equations for this theory are

$$G_{\alpha\beta} + g_{\alpha\beta}\Lambda - F(\phi)\left(\phi_{,\alpha}\phi_{,\beta} - \frac{1}{2}g_{\alpha\beta}\phi_{,\gamma}\phi^{,\gamma}\right) = -8\pi G T_{\alpha\beta}, \tag{4a}$$

$$2F(\phi)\phi^{,\alpha}_{;\alpha} + \frac{dF}{d\phi}\phi_{,\gamma}\phi^{,\gamma} = 0, \tag{4b}$$

where G is the gravitational constant and as usual the semicolon means a covariant derivative. The equation (4b) take the following form for all cosmological Bianchi Class A models, assuming that the scalar field is only time dependent (here $\prime = \frac{d}{d\tau} = \frac{d}{Ndt}$)

$$3\Omega'\phi'\mathrm{F} + \phi''\mathrm{F} + \frac{1}{2}\frac{\mathrm{dF}}{\mathrm{d}\phi}\phi'^2 = 0,$$

which can be put in quadrature form as

$$\frac{1}{2}\mathrm{F}\phi'^2 = \mathrm{F}_0 e^{-6\Omega}, \tag{5}$$

this equation is seen as corresponding to a stiff matter content contribution.
The same set of equations(4a,4b) is obtained if we consider the scalar field ϕ as part of the matter budget, i.e. say $\mathcal{L}_\phi = -\mathrm{F}(\phi)g^{\alpha\beta}\phi_{,\alpha}\phi_{,\beta}$ with the corresponding energy-momentum tensor

$$\mathrm{T}_{\alpha\beta} = \mathrm{F}(\phi)\left(\phi_{,\alpha}\phi_{,\beta} - \frac{1}{2}g_{\alpha\beta}\phi_{,\gamma}\phi^{,\gamma}\right), \tag{6}$$

which is conserved and equivalent to a stiff (see appendix section 7). In this new line of reasoning, action (3) can be rewritten as a geometrical part (Hilbert-Einstein with Λ) and matter content (usual matter plus a term that corresponds to the exotic scalar field component of SB theory).
In this way, we write the action (3) in the usual form

$$\mathrm{I} = \int_\Sigma \sqrt{-g}\left(\mathrm{R} - 2\Lambda + \mathcal{L}_{\mathrm{mat}} + \mathcal{L}_\phi\right) d^4x, \tag{7}$$

and consequently, the classical equivalence between the two theories. We can infer that this correspondence also is satisfied in the quantum regime, so we can use this structure for the quantization program, where the ADM formalism is well known for different classes of matter (Ryan, 1972). Using this action we obtain the Hamiltonian for SB. We find that the WDW equation is solved when we choose one ansatz similar to this employed in the Bohmian formalism of quantum mechanics and the gravitational part in the solutions are the same that these found in the literature, years ago (Obregón & Socorro, 1996).
This work is arranged as follow. In section 2 we present the method used, employing the FRW cosmological model with barotropic perfect fluid and cosmological constant. In section 3 we construct the Lagrangian and Hamiltonian densities for the anisotropic Bianchi Class A cosmological model. In section 4 the classical solutions using the Jacobi formalism are found. Here we present partial results in the solutions for some Bianchi's cosmological models. Classical solution to Bianchi I is complete in any gauge, but the Bianchi II and $\mathrm{VI}_{h=-1}$, the solutions are found in particular gauge. Other Biachi's, only the master equation are presented. In Section 5 the complete cuantization scheme is presented, obtaining the corresponding Wheeler-DeWitt equation and its solutions are presented in unified way using the classification scheme of Ellis and MacCallum (Ellis & MacCallum, 1969) and Ryan and Shepley, (Ryan & Shepley, 1975).

2. The method

Let us start with the line element for a homogeneous and isotropic FRW universe

$$ds^2 = -N^2(t)dt^2 + a^2(t)\left[\frac{dr^2}{1-\kappa r^2} + r^2 d\Omega^2\right], \tag{8}$$

where $a(t)$ is the scale factor, $N(t)$ is the lapse function, and κ is the curvature constant that can take the values 0, 1 and -1, for flat, closed and open universe, respectively. The total Lagrangian density then reads

$$\mathcal{L} = \frac{6\dot{a}^2 a}{N} - 6\kappa N a + \frac{F(\phi)a^3}{N}\dot{\phi}^2 + 16\pi G N a^3 \rho - 2Na^3\Lambda, \tag{9}$$

where ρ is the energy density of matter, we will assume that it complies with a barotropic equation of state of the form $p = \gamma\rho$, where γ is a constant. The matter content is assumed as a perfect fluid $T_{\mu\nu} = (\rho + p)u_\mu u_\nu + g_{\mu\nu}p$ where u_μ is the fluid four-velocity satisfying $u_\mu u^\mu = -1$. Taking the covariant derivative we obtain the relation

$$3\dot{\Omega}\rho + 3\dot{\Omega}p + \dot{\rho} = 0,$$

whose solution becomes

$$\rho = \rho_\gamma e^{-3\Omega(1+\gamma)}, \tag{10}$$

where ρ_γ is an integration constant.

From the canonical form of the Lagrangian density (9), and the solution for the barotropic fluid equation of motion, we find the Hamiltonian density for this theory, where the momenta are defined in the usual way $\Pi_{q^i} = \frac{\partial\mathcal{L}}{\partial\dot{q}^i}$, where $q^i = (a, \phi)$ are the field coordinates for this system,

$$\begin{aligned} \Pi_a &= \frac{\partial\mathcal{L}}{\partial\dot{a}} = \frac{12a\dot{a}}{N}, &\rightarrow\quad \dot{a} &= \frac{N\Pi_a}{12a}, \\ \Pi_\phi &= \frac{\partial\mathcal{L}}{\partial\dot{\phi}} = \frac{2Fa^3\dot{\phi}}{N}, &\rightarrow\quad \dot{\phi} &= \frac{N\Pi_\phi}{2Fa^3}, \end{aligned} \tag{11}$$

so, the Hamiltonian density become

$$\mathcal{H} = \frac{a^{-3}}{24}\left[a^2\Pi_a^2 + \frac{6}{F(\phi)}\Pi_\phi^2 + 144\kappa a^4 + 48a^6\Lambda - 384\pi G\rho_\gamma a^{3(1-\gamma)}\right]. \tag{12}$$

Using the transformation $\Pi_q = \frac{dS_q}{dq}$, the Einstein-Hamilton-Jacobi (EHJ) associated to Eq. (12) is

$$a^2\left(\frac{dS_a}{da}\right)^2 + \frac{6}{F(\phi)}\left(\frac{dS_\phi}{d\phi}\right)^2 + 48a^6\Lambda - 384\pi G\rho_\gamma a^{3(1-\gamma)} = 0. \tag{13}$$

The EHJ equation can be further separated in the equations

$$\frac{6}{F(\phi)}\left(\frac{dS_\phi}{d\phi}\right)^2 = \mu^2, \tag{14}$$

$$a^2\left(\frac{dS_a}{da}\right)^2 + 48a^6\Lambda - 384\pi G\rho_\gamma a^{3(1-\gamma)} = -\mu^2, \tag{15}$$

where μ is a separation constant. With the help of Eqs. (11), we can obtain the solution up to quadratures of Eqs. (14) and (15),

$$\int \sqrt{F(\phi)}\, d\phi = \frac{\mu}{2\sqrt{6}} \int a^{-3}(\tau)\, d\tau\,, \tag{16a}$$

$$\Delta\tau = \int \frac{a^2 da}{\sqrt{\frac{8}{3}\pi G \rho_\gamma a^{3(1-\gamma)} - \frac{\Lambda}{3} a^6 - \nu^2}}\,, \tag{16b}$$

with $\nu = \frac{\mu}{12}$. Eq. (16a) readily indicates that

$$F(\phi)\dot{\phi}^2 = 6\nu^2 a^{-6}(\tau)\,. \tag{17}$$

Also, this equation could be obtained by solving equation (4b). Moreover, the matter contribution of the SB scalar field to the r.h.s. of the Einstein equations would be

$$\rho_\phi = \frac{1}{2} F(\phi)\dot{\phi}^2 \propto a^{-6}\,, \tag{18}$$

this energy density of a scalar field has the range of scaling behaviors (Andrew & Scherrer, 1998; Ferreira & Joyce, 1998), is say, scales exactly as a power of the scale factor like, $\rho_\phi \propto a^{-m}$, when the dominant component has an energy density which scales as similar way. So, the contribution of the scalar field is the same as that of stiff matter with a barotropic equation of state $\gamma = 1$. This is an interesting result, since the original SB theory was thought of as a way to solve the missing matter problem now generically called the dark matter problem. To solve the latter, one needs a fluid behaving as dust with $\gamma = 0$, it is surprising that such a general result remains unnoticed until now in the literature about SB. This is an instance of the results of the analysis of the energy momentum tensor of a scalar field by Marden (Marden, 1988) for General Relativity with scalar matter and by Pimentel (Pimentel, 1989) for the general scalar tensor theory. In both works a free scalar field is equivalent to a stiff matter fluid.
Furthermore, having identified the general evolution of the scalar field with that of a stiff fluid means that the Eq. (16b) can be integrated separately without a complete solution for the scalar field. In (Socorro et al., 2011) appear a compilation of exact solutions in the case of the original SB theory to FRW cosmological model and in (Socorro et al., 2010) were presented the classical and quantum solution to Bianchi type I.

3. The master Hamiltonian to Bianchi Class A cosmological models

Let us recall here the canonical formulation in the ADM formalism of the diagonal Bianchi Class A cosmological models. The metric has the form

$$\mathrm{ds}^2 = -\mathrm{dt}^2 + \mathrm{e}^{2\Omega(\mathrm{t})}\,(\mathrm{e}^{2\beta(\mathrm{t})})_{\mathrm{ij}}\,\omega^{\mathrm{i}}\,\omega^{\mathrm{j}}, \tag{19}$$

where $\beta_{\mathrm{ij}}(\mathrm{t})$ is a 3x3 diagonal matrix, $\beta_{\mathrm{ij}} = \mathrm{diag}(\beta_+ + \sqrt{3}\beta_-, \beta_+ - \sqrt{3}\beta_-, -2\beta_+)$, $\Omega(t)$ is a scalar and ω^{i} are one-forms that characterize each cosmological Bianchi type model, and that obey $\mathrm{d}\omega^{\mathrm{i}} = \frac{1}{2}\mathrm{C}^{\mathrm{i}}_{\mathrm{jk}}\omega^{\mathrm{j}} \wedge \omega^{\mathrm{k}}$, $\mathrm{C}^{\mathrm{i}}_{\mathrm{jk}}$ the structure constants of the corresponding invariance group, these are included in table 1.

Bianchi type	1-forms ω^i
I	$\omega^1 = dx^1, \quad \omega^2 = dx^2, \quad \omega^3 = dx^3$
II	$\omega^1 = dx^2 - x^1 dx^3, \quad \omega^2 = dx^3, \quad \omega^3 = dx^1$
$VI_{h=-1}$	$\omega^1 = e^{-x^1} dx^2, \quad \omega^2 = e^{x^1} dx^3, \quad \omega^3 = dx^1$
VII_0	$\omega^1 = dx^2 + dx^3, \quad \omega^2 = -dx^2 + dx^3, \quad \omega^3 = dx^1$
VIII	$\omega^1 = dx^1 + [1 + (x^1)^2] dx^2 + [x^1 - x^2 - (x^1)^2 x^2] dx^3,$ $\omega^2 = 2x^1 dx^2 + (1 - 2x^1 x^2) dx^3,$ $\omega^3 = dx^1 + [-1 + (x^1)^2] dx^2 + [x^1 + x^2 - (x^1)^2 x^2] dx^3$
IX	$\omega^1 = -\sin(x^3) dx^1 + \sin(x^1)\cos(x^3) dx^2,$ $\omega^2 = \cos(x^3) dx^1 + \sin(x^1)\sin(x^3) dx^2, \quad \omega^3 = \cos(x^1) dx^2 + dx^3$

Table 1. One-forms for the Bianchi Class A models.

We use the Bianchi type IX cosmological model as toy model to apply the method discussed in the previous section. The total Lagrangian density then reads

$$
\begin{aligned}
\mathcal{L}_{IX} = e^{3\Omega} \Bigg[& 6\frac{\dot{\Omega}^2}{N} - 6\frac{\dot{\beta}_+^2}{N} - 6\frac{\dot{\beta}_-^2}{N} + \frac{F(\phi)}{N}\dot{\phi}^2 + 16\pi G N \rho - 2N\Lambda \\
& + N e^{-2\Omega} \left\{ \frac{1}{2}\left(e^{4\beta_+ + 4\sqrt{3}\beta_-} + e^{4\beta_+ - 4\sqrt{3}\beta_-} + e^{-8\beta_+} \right) \right. \\
& \left. - \left(e^{-2\beta_+ + 2\sqrt{3}\beta_-} + e^{-2\beta_+ - 2\sqrt{3}\beta_-} + e^{4\beta_+} \right) \right\} \Bigg], \qquad (20)
\end{aligned}
$$

making the calculation of momenta in the usual way, $\Pi_{q^\mu} = \frac{\partial \mathcal{L}}{\partial \dot{q}^\mu}$, where $q^\mu = (\Omega, \beta_+, \beta_-, \phi)$

$$
\begin{aligned}
\Pi_\Omega &= \frac{12}{N} e^{3\Omega}\dot{\Omega}, \quad \rightarrow \quad \dot{\Omega} = \frac{N}{12} e^{-3\Omega}\Pi_\Omega, \\
\Pi_+ &= -\frac{12}{N} e^{3\Omega}\dot{\beta}_+, \quad \rightarrow \quad \dot{\beta}_+ = -\frac{N}{12} e^{-3\Omega}\Pi_+, \\
\Pi_- &= -\frac{12}{N} e^{3\Omega}\dot{\beta}_-, \quad \rightarrow \quad \dot{\beta}_- = -\frac{N}{12} e^{-3\Omega}\Pi_+, \\
\Pi_\phi &= \frac{2F}{N} e^{3\Omega}\dot{\phi}, \quad \rightarrow \quad \dot{\phi} = \frac{N}{2F} e^{-3\Omega}\Pi_\phi,
\end{aligned}
$$

and introducing into the Lagrangian density, we obtain the canonical Lagrangian as

$$
\mathcal{L}_{IX} = \Pi_{q^\mu}\dot{q}^\mu - N\mathcal{H}_\perp,
$$

with the Hamiltonian density

$$
\mathcal{H}_\perp = \frac{e^{-3\Omega}}{24}\left(-\Pi_\Omega^2 - \frac{6}{F(\phi)}\Pi_\phi^2 + \Pi_+^2 + \Pi_-^2 + U(\Omega, \beta_\pm) + C_1 \right), \qquad (21)
$$

where the gravitational potential becomes,

$$
U(\Omega, \beta_\pm) = 12 e^{4\Omega}\left(e^{4\beta_+ + 4\sqrt{3}\beta_-} + e^{4\beta_+ - 4\sqrt{3}\beta_-} + e^{4\beta_+} - 2\{ e^{4\beta_+} + e^{2\beta_+ - 2\sqrt{3}\beta_-} + e^{-2\beta_+ + 2\sqrt{3}\beta_-} \} \right),
$$

with $C_1 = 384\pi G\rho_1$ corresponding to stiff matter epoch, $\gamma = 1$.

The equation (21) can be considered as a master equation for all Bianchi Class A cosmological model in the stiff epoch in the Sáez-Ballester theory, with $U(\Omega, \beta_\pm)$ is the potential term of the cosmological model under consideration, that can read it to table II.

Bianchi type	Hamiltonian density $\mathcal{H}$
I	$\frac{e^{-3\Omega}}{24}\left[-\Pi_\Omega^2 - \frac{6}{F}\Pi_\phi^2 + \Pi_+^2 + \Pi_-^2 - 48\Lambda e^{6\Omega} + 384\pi G\rho_\gamma e^{-3(\gamma-1)\Omega}\right]$
II	$\frac{e^{-3\Omega}}{24}\left[-\Pi_\Omega^2 - \frac{6}{F}\Pi_\phi^2 + \Pi_+^2 + \Pi_-^2 - 48\Lambda e^{6\Omega} + 384\pi G\rho_\gamma e^{-3(\gamma-1)\Omega} + 12e^{4\Omega}e^{4\beta_+ + 4\sqrt{3}\beta_-}\right]$
VI_{-1}	$\frac{e^{-3\Omega}}{24}\left[-\Pi_\Omega^2 - \frac{6}{F}\Pi_\phi^2 + \Pi_+^2 + \Pi_-^2 - 48\Lambda e^{6\Omega} + 384\pi G\rho_\gamma e^{-3(\gamma-1)\Omega} + 48e^{4\Omega}e^{4\beta_+}\right]$
VII_0	$\frac{e^{-3\Omega}}{24}\left[-\Pi_\Omega^2 - \frac{6}{F}\Pi_\phi^2 + \Pi_+^2 + \Pi_-^2 - 48\Lambda e^{6\Omega} + 384\pi G\rho_\gamma e^{-3(\gamma-1)\Omega} + 12e^{4\Omega}\left(e^{4\beta_+ + 4\sqrt{3}\beta_-} - e^{4\beta_+} + e^{4\beta_+ - 4\sqrt{3}\beta_-}\right)\right]$
VIII	$\frac{e^{-3\Omega}}{24}\left[-\Pi_\Omega^2 - \frac{6}{F}\Pi_\phi^2 + \Pi_+^2 + \Pi_-^2 - 48\Lambda e^{6\Omega} + 384\pi G\rho_\gamma e^{-3(\gamma-1)\Omega} + 12e^{4\Omega}\left(e^{4\beta_+ + 4\sqrt{3}\beta_-} + e^{4\beta_+ - 4\sqrt{3}\beta_-} + e^{-8\beta_+} - 2\left\{e^{4\beta_+} - e^{-2\beta_+ - 2\sqrt{3}\beta_-} - e^{-2\beta_+ + 2\sqrt{3}\beta_-}\right\}\right)\right]$
IX	$\frac{e^{-3\Omega}}{24}\left[-\Pi_\Omega^2 - \frac{6}{F}\Pi_\phi^2 + \Pi_+^2 + \Pi_-^2 - 48\Lambda e^{6\Omega} + 384\pi G\rho_\gamma e^{-3(\gamma-1)\Omega} + 12e^{4\Omega}\left(e^{4\beta_+ + 4\sqrt{3}\beta_-} + e^{4\beta_+ - 4\sqrt{3}\beta_-} + e^{-8\beta_+} - 2\left\{e^{4\beta_+} + e^{2\beta_+ - 2\sqrt{3}\beta_-} + e^{-2\beta_+ + 2\sqrt{3}\beta_-}\right\}\right)\right]$

Table 2. Hamiltonian density for the Bianchi Class A models.

4. Classical scheme

In this section, we present the classical solutions to all Bianchi Class A cosmological models using the appropriate set of variables,

$$\begin{aligned}\beta_1 &= \Omega + \beta_+ + \sqrt{3}\beta_-,\\ \beta_2 &= \Omega + \beta_+ - \sqrt{3}\beta_-,\\ \beta_3 &= \Omega - 2\beta_+.\end{aligned} \tag{22}$$

4.1 Bianchi I

For building one master equation for all Bianchi Class A models, we begin with the simplest model give by the Bianchi I, and give the general treatment. The corresponding Lagrangian for this cosmological model is written as

$$\mathcal{L}_I = e^{\beta_1+\beta_2+\beta_3}\left[\frac{2\dot\beta_1\dot\beta_2}{N} + \frac{2\dot\beta_1\dot\beta_3}{N} + \frac{2\dot\beta_2\dot\beta_3}{N} + \frac{F(\phi)\dot\phi^2}{N} + 16N\pi G\rho_\gamma\, e^{-(1+\gamma)(\beta_1+\beta_2+\beta_3)} - 2N\Lambda\right], \tag{23}$$

the momenta associated to the variables (β_i, ϕ) are

$$
\begin{aligned}
\Pi_1 &= \frac{2}{\mathrm{N}}(\dot\beta_2+\dot\beta_3)e^{\beta_1+\beta_2+\beta_3}, \qquad \dot\beta_1 = \frac{\mathrm{N}}{4}e^{-(\beta_1+\beta_2+\beta_3)}(\Pi_2+\Pi_3-\Pi_1),\\
\Pi_2 &= \frac{2}{\mathrm{N}}(\dot\beta_1+\dot\beta_3)e^{\beta_1+\beta_2+\beta_3}, \qquad \dot\beta_2 = \frac{\mathrm{N}}{4}e^{-(\beta_1+\beta_2+\beta_3)}(\Pi_1+\Pi_3-\Pi_2),\\
\Pi_3 &= \frac{2}{\mathrm{N}}(\dot\beta_1+\dot\beta_2)e^{\beta_1+\beta_2+\beta_3}, \qquad \dot\beta_3 = \frac{\mathrm{N}}{4}e^{-(\beta_1+\beta_2+\beta_3)}(\Pi_1+\Pi_2-\Pi_3),\\
\Pi_\phi &= \frac{2\mathrm{F}\dot\phi}{\mathrm{N}}e^{\beta_1+\beta_2+\beta_3}, \qquad \dot\phi = \frac{\mathrm{N}}{2\mathrm{F}}e^{-(\beta_1+\beta_2+\beta_3)}\Pi_\phi,
\end{aligned}
\tag{24}
$$

so, the Hamiltonian is

$$
\begin{aligned}
\mathcal{H}_I = \frac{1}{8}e^{-(\beta_1+\beta_2+\beta_3)}\Big[&-\Pi_1^2-\Pi_2^2-\Pi_3^2+\frac{2}{\mathrm{F}}\Pi_\phi^2+2\Pi_1\Pi_2+2\Pi_1\Pi_3+2\Pi_2\Pi_3\\
&+16\Lambda e^{2(\beta_1+\beta_2+\beta_3)}-128\pi G\rho_\gamma e^{(1-\gamma)(\beta_1+\beta_2+\beta_3)}\Big],
\end{aligned}
\tag{25}
$$

using the hamilton equation, where $\prime = \frac{d}{d\tau} = \frac{d}{Ndt}$, we have

$$\Pi_1' = -4\Lambda e^{\beta_1+\beta_2+\beta_3} + 16\pi G(1-\gamma)\rho_\gamma e^{-\gamma(\beta_1+\beta_2+\beta_3)}, \tag{26}$$

$$\Pi_2' = -4\Lambda e^{\beta_1+\beta_2+\beta_3} + 16\pi G(1-\gamma)\rho_\gamma e^{-\gamma(\beta_1+\beta_2+\beta_3)}, \tag{27}$$

$$\Pi_3' = -4\Lambda e^{\beta_1+\beta_2+\beta_3} + 16\pi G(1-\gamma)\rho_\gamma e^{-\gamma(\beta_1+\beta_2+\beta_3)}, \tag{28}$$

$$\Pi_\phi' = \frac{1}{4}e^{-(\beta_1+\beta_2+\beta_3)}\frac{\mathrm{F}'}{\mathrm{F}^2\phi'}\Pi_\phi^2, \tag{29}$$

$$\beta_1' = \frac{1}{4}e^{-(\beta_1+\beta_2+\beta_3)}\left[-\Pi_1+\Pi_2+\Pi_3\right], \tag{30}$$

$$\beta_2' = \frac{1}{4}e^{-(\beta_1+\beta_2+\beta_3)}\left[-\Pi_2+\Pi_1+\Pi_3\right], \tag{31}$$

$$\beta_3' = \frac{1}{4}e^{-(\beta_1+\beta_2+\beta_3)}\left[-\Pi_3+\Pi_1+\Pi_2\right], \tag{32}$$

$$\phi' = \frac{1}{2\mathrm{F}}e^{-(\beta_1+\beta_2+\beta_3)}\Pi_\phi, \tag{33}$$

equations (26,27,28) implies

$$\Pi_1 = \Pi_2 + k_1 = \Pi_3 + k_2. \tag{34}$$

Also, the differential equation for field ϕ can be reduced to quadratures when we use equations (29) and (33), as

$$\frac{1}{2}\mathrm{F}(\phi)\phi'^2 = \phi_0 e^{-2(\beta_1+\beta_2+\beta_3)}, \qquad \Rightarrow \qquad \sqrt{\mathrm{F}(\phi)}d\phi = \sqrt{2\phi_0}\,e^{-(\beta_1+\beta_2+\beta_3)}d\tau, \tag{35}$$

which correspond to equation (5) obtained in direct way from the original Einstein field equation. The corresponding classical solutions for the field ϕ for this cosmological model can be seen in ref. (Socorro et al., 2010).

Using this result and the equation for the field ϕ given in (24) we can find that $2\frac{\Pi_\phi^2}{\mathrm{F}} = 16\phi_0$. From the hamilton equation for the momenta Π_1 can be written for the two equations of state

$\gamma = \pm 1$, introducing the generic parameter

$$\lambda = \begin{cases} -4\Lambda, & \gamma = 1 \\ -4\Lambda + 32\pi G\rho_1, & \gamma = -1 \end{cases} \tag{36}$$

as $\Pi_1' = \lambda e^{\beta_1+\beta_2+\beta_3}$, then re-introducing into the Hamiltonian equation (25) we find one differential equation for the momenta Π_1 as

$$\frac{4}{\lambda}\Pi_1'^2 + 2\Pi_1^2 - \kappa\Pi_1 - k_3 = 0, \tag{37}$$

where the corresponding constants are

$$\kappa = 2(k_1 + k_2), \quad k_3 = \begin{cases} k_1^2 + k_2^2 - 16\phi_0, & \gamma = -1 \\ k_1^2 + k_2^2 - 16\phi_0 + 128\pi G\rho_1, & \gamma = 1 \end{cases} \tag{38}$$

and whose solution is

$$\Pi_1 = \frac{\kappa}{6} \pm \frac{\sqrt{\kappa^2 + 12k_3}}{6} \sin\left[\frac{\sqrt{3\lambda}}{2}\Delta\tau\right]. \tag{39}$$

On the other hand, using this result in the sum of equation (52,53,54), we obtain that

$$\beta_1 + \beta_2 + \beta_3 = \mathrm{Ln}\left[\frac{\alpha}{\sqrt{12\lambda}} \cos\left[\frac{\sqrt{3\lambda}}{2}\Delta\tau\right]\right], \qquad \alpha = 2\sqrt{\kappa^2 + 12k_3}, \tag{40}$$

solution previously found in ref. (Socorro et al., 2010) using the Hamilton-Jacobi approach.

4.2 Bianchi's Class A cosmological models

The corresponding Lagrangian for these cosmological model are written using the Lagrangian to Bianchi I, as

$$\mathcal{L}_{II} = \mathcal{L}_I + N e^{\beta_1+\beta_2+\beta_3}\left[\frac{1}{2}e^{2(\beta_1-\beta_2-\beta_3)}\right], \tag{41}$$

$$\mathcal{L}_{VI_{h=-1}} = \mathcal{L}_I + N e^{\beta_1+\beta_2+\beta_3}\left[2e^{-2\beta_3}\right], \tag{42}$$

$$\mathcal{L}_{VII_{h=0}} = \mathcal{L}_I + N e^{\beta_1+\beta_2+\beta_3}\left[\frac{1}{2}e^{2(\beta_1-\beta_2-\beta_3)} + \frac{1}{2}e^{2(-\beta_1+\beta_2-\beta_3)} - e^{-2\beta_3}\right], \tag{43}$$

$$\begin{aligned}\mathcal{L}_{VIII} = \mathcal{L}_I + \frac{N}{2}e^{\beta_1+\beta_2+\beta_3}\Big[&e^{2(\beta_1-\beta_2-\beta_3)} + e^{2(-\beta_1+\beta_2-\beta_3)} + e^{2(-\beta_1-\beta_2+\beta_3)} \\ &-2\left(-e^{-2\beta_1} + e^{-2\beta_2} + e^{-2\beta_3}\right)\Big],\end{aligned} \tag{44}$$

$$\begin{aligned}\mathcal{L}_{IX} = \mathcal{L}_I + \frac{N}{2}e^{\beta_1+\beta_2+\beta_3}\Big[&e^{2(\beta_1-\beta_2-\beta_3)} + e^{2(-\beta_1+\beta_2-\beta_3)} + e^{2(-\beta_1-\beta_2+\beta_3)} \\ &-2\left(e^{-2\beta_1} + e^{-2\beta_2} + e^{-2\beta_3}\right)\Big],\end{aligned} \tag{45}$$

the momenta associated to the variables (β_i, ϕ) are the same as in equation (24), so, the generic Hamiltonian is

$$\mathcal{H}_A = \mathcal{H}_I - \frac{1}{2}e^{-(\beta_1+\beta_2+\beta_3)}\left[U_A(\beta_1, \beta_2, \beta_3)\right], \tag{46}$$

where the potential term $U_A(\beta_1, \beta_2, \beta_3)$ is given in table III, where A corresponds to particular Bianchi Class A models (I,II, $VI_{h=-1}$,$VII_{h=0}$,VIII,IX). If we choose the particular gauge to the lapse function $N = e^{(\beta_1+\beta_2+\beta_3)}$, the equation (46) is much simpler,

$$\mathcal{H}_A = \mathcal{H}_I - \frac{1}{2}\left[U_A(\beta_1, \beta_2, \beta_3)\right], \tag{47}$$

where $\mathcal{H}_I$ is as in equation (25) but without the factor $e^{-(\beta_1+\beta_2+\beta_3)}$

Bianchi type	**Potential** $U_A(\beta_1, \beta_2, \beta_3)$
I	0
II	$e^{4\beta_1}$
$VI_{h=-1}$	$4e^{2(\beta_1+\beta_2)}$
$VII_{h=0}$	$e^{4\beta_1} + e^{4\beta_2} - 2e^{2(\beta_1+\beta_2)}$
VIII	$e^{4\beta_1} + e^{4\beta_2} + e^{4\beta_3} - 2e^{2(\beta_1+\beta_2)} + 2e^{2(\beta_1+\beta_3)} + 2e^{2(\beta_2+\beta_3)}$
IX	$e^{4\beta_1} + e^{4\beta_2} + e^{4\beta_3} - 2e^{2(\beta_1+\beta_2)} - 2e^{2(\beta_1+\beta_3)} - 2e^{2(\beta_2+\beta_3)}$

Table 3. Potential $U_A(\beta_1, \beta_2, \beta_3)$ for the Bianchi Class A Models.

The Hamilton equations, for all Bianchi Class A cosmological models are as follows

$$\Pi_1' = -4\Lambda e^{\beta_1+\beta_2+\beta_3} + 16\pi G(1-\gamma)\rho_\gamma e^{-\gamma(\beta_1+\beta_2+\beta_3)} + \frac{\partial}{\partial\beta_1}\left(\frac{1}{2}e^{-(\beta_1+\beta_2+\beta_3)}\left[U_A(\beta_1, \beta_2, \beta_3)\right]\right), \tag{48}$$

$$\Pi_2' = -4\Lambda e^{\beta_1+\beta_2+\beta_3} + 16\pi G(1-\gamma)\rho_\gamma e^{-\gamma(\beta_1+\beta_2+\beta_3)} + \frac{\partial}{\partial\beta_2}\left(\frac{1}{2}e^{-(\beta_1+\beta_2+\beta_3)}\left[U_A(\beta_1, \beta_2, \beta_3)\right]\right), \tag{49}$$

$$\Pi_3' = -4\Lambda e^{\beta_1+\beta_2+\beta_3} + 16\pi G(1-\gamma)\rho_\gamma e^{-\gamma(\beta_1+\beta_2+\beta_3)} + \frac{\partial}{\partial\beta_3}\left(\frac{1}{2}e^{-(\beta_1+\beta_2+\beta_3)}\left[U_A(\beta_1, \beta_2, \beta_3)\right]\right), \tag{50}$$

$$\Pi_\phi' = \frac{1}{4}e^{-(\beta_1+\beta_2+\beta_3)}\frac{F'}{F^2\phi'}\Pi_\phi^2, \tag{51}$$

$$\beta_1' = \frac{1}{4}e^{-(\beta_1+\beta_2+\beta_3)}\left[-\Pi_1 + \Pi_2 + \Pi_3\right], \tag{52}$$

$$\beta_2' = \frac{1}{4}e^{-(\beta_1+\beta_2+\beta_3)}\left[-\Pi_2 + \Pi_1 + \Pi_3\right], \tag{53}$$

$$\beta_3' = \frac{1}{4}e^{-(\beta_1+\beta_2+\beta_3)}\left[-\Pi_3 + \Pi_1 + \Pi_2\right], \tag{54}$$

$$\phi' = \frac{1}{2F}e^{-(\beta_1+\beta_2+\beta_3)}\Pi_\phi. \tag{55}$$

In this cosmological models, it is remarkable that the equation for the field ϕ (35) is mantained for all Bianchi Class A models, and in particular, when we use the gauge $N = e^{\beta_1+\beta_2+\beta_3}$, the solutions for this field are independent of the cosmological models.

4.3 Classical solution in the gauge $N = e^{\beta_1+\beta_2+\beta_3}$, $\Lambda = 0$ and $\gamma = 1$

With these initial choices, the main equations are written for this gauge as (now a dot means $\frac{d}{dt}$)

$$\mathcal{H}_A = \frac{1}{8}\left[-\Pi_1^2 - \Pi_2^2 - \Pi_3^2 + \frac{2}{F}\Pi_\phi^2 + 2\Pi_1\Pi_2 + 2\Pi_1\Pi_3 + 2\Pi_2\Pi_3 - C_1\right] \\ -\frac{1}{2}\left[U_A(\beta_1, \beta_2, \beta_3)\right], \tag{56}$$

with $C_1 = 128\pi G\rho_1$.

The hamilton equation, for all Bianchi Class A cosmological models are

$$\dot{\Pi}_1 = +\frac{\partial}{\partial\beta_1}\left(\frac{1}{2}\left[U_A(\beta_1, \beta_2, \beta_3)\right]\right), \tag{57}$$

$$\dot{\Pi}_2 = +\frac{\partial}{\partial\beta_2}\left(\frac{1}{2}\left[U_A(\beta_1, \beta_2, \beta_3)\right]\right), \tag{58}$$

$$\dot{\Pi}_3 = +\frac{\partial}{\partial\beta_3}\left(\frac{1}{2}\left[U_A(\beta_1, \beta_2, \beta_3)\right]\right), \tag{59}$$

$$\dot{\Pi}_\phi = \frac{1}{4}\frac{\dot{F}}{F^2\dot{\phi}}\Pi_\phi^2, \tag{60}$$

$$\dot{\beta}_1 = \frac{1}{4}\left[-\Pi_1 + \Pi_2 + \Pi_3\right], \tag{61}$$

$$\dot{\beta}_2 = \frac{1}{4}\left[-\Pi_2 + \Pi_1 + \Pi_3\right], \tag{62}$$

$$\dot{\beta}_3 = \frac{1}{4}\left[-\Pi_3 + \Pi_1 + \Pi_2\right], \tag{63}$$

$$\dot{\phi} = \frac{1}{2F}\Pi_\phi. \tag{64}$$

4.3.1 Bianchi II

$$\dot{\Pi}_1 = 2e^{4\beta_1}, \tag{65}$$

$$\dot{\Pi}_2 = 0, \quad \rightarrow \quad \Pi_2 = p_2 = cte, \tag{66}$$

$$\dot{\Pi}_3 = 0, \quad \rightarrow \quad \Pi_3 = p_3 = cte, \tag{67}$$

$$\dot{\Pi}_\phi = \frac{1}{4}\frac{\dot{F}}{F^2\dot{\phi}}\Pi_\phi^2, \tag{68}$$

$$\dot{\beta}_1 = \frac{1}{4}\left[-\Pi_1 + p_2 + p_3\right], \tag{69}$$

$$\dot{\beta}_2 = \frac{1}{4}\left[-p_2 + \Pi_1 + p_3\right], \tag{70}$$

$$\dot{\beta}_3 = \frac{1}{4}\left[-p_3 + \Pi_1 + p_2\right], \tag{71}$$

$$\dot{\phi} = \frac{1}{2F}\Pi_\phi, \tag{72}$$

introducing (65) into (56) we find the differential equation for Π_1 as $\dot{\Pi}_1 = -\frac{1}{2}\Pi_1^2 + b\Pi_1 + c$ where the constants are defined as $b = p_2 + p_3$ and $c = 8\phi_0 - \frac{1}{2}\left(p_2^2 + p_3^2 + C_1\right)$. The solution for Π_1 is

$$\Pi_1 = b + \sqrt{-b^2 - 2c}\mathrm{Tan}\left[-\frac{1}{2}\sqrt{-b^2 - 2c}\Delta t\right], \tag{73}$$

and the solutions for β_i then are

$$\Delta\beta_1 = -\frac{1}{2}\mathrm{Ln}\left[\mathrm{Cos}\left(\frac{1}{2}\sqrt{-b^2 - 2c}\Delta t\right)\right], \tag{74}$$

$$\Delta\beta_2 = \frac{1}{2}p_3\Delta t + \frac{1}{2}\mathrm{Ln}\left[\mathrm{Cos}\left(\frac{1}{2}\sqrt{-b^2 - 2c}\Delta t\right)\right], \tag{75}$$

$$\Delta\beta_3 = \frac{1}{2}p_2\Delta t + \frac{1}{2}\mathrm{Ln}\left[\mathrm{Cos}\left(\frac{1}{2}\sqrt{-b^2 - 2c}\Delta t\right)\right], \tag{76}$$

$$\tag{77}$$

and the solution for the ϕ field is similar to (35)

$$\frac{1}{2}F(\phi)\dot{\phi}^2 = \phi_0, \qquad \Rightarrow \qquad \sqrt{F(\phi)}d\phi = \sqrt{2\phi_0}\,dt. \tag{78}$$

So, the solutions in the original variables are

$$\begin{aligned}
\Omega &= \frac{1}{6}\left[(p_2 + p_3)\Delta t + \mathrm{Ln}\left[\mathrm{Cos}\left(\frac{1}{2}\sqrt{-b^2 - 2c}\Delta t\right)\right]\right], \\
\beta_- &= \frac{\sqrt{3}}{6}\left[-\frac{1}{2}p_3\Delta t - \mathrm{Ln}\left[\mathrm{Cos}\left(\frac{1}{2}\sqrt{-b^2 - 2c}\Delta t\right)\right]\right], \\
\beta_+ &= \frac{1}{12}\left[(p3 - 2p_2)\Delta t - 2\mathrm{Ln}\left[\mathrm{Cos}\left(\frac{1}{2}\sqrt{-b^2 - 2c}\Delta t\right)\right]\right].
\end{aligned} \tag{79}$$

4.3.2 Bianchi $VI_{h=-1}$

$$\dot{\Pi}_1 = 4e^{2(\beta_1+\beta_2)}, \tag{80}$$

$$\dot{\Pi}_2 = 4e^{2(\beta_1+\beta_2)}, \qquad \rightarrow \qquad \Pi_2 = \Pi_1 + a_1, \tag{81}$$

$$\dot{\Pi}_3 = 0, \qquad \rightarrow \qquad \Pi_3 = p_3 = cte, \tag{82}$$

$$\dot{\Pi}_\phi = \frac{1}{4}\frac{\dot{F}}{F^2\dot{\phi}}\Pi_\phi^2, \tag{83}$$

$$\dot{\beta}_1 = \frac{1}{4}\left[-\Pi_1 + \Pi_2 + p_3\right], \tag{84}$$

$$\dot{\beta}_2 = \frac{1}{4}\left[-\Pi_2 + \Pi_1 + p_3\right], \tag{85}$$

$$\dot{\beta}_3 = \frac{1}{4}\left[-p_3 + \Pi_1 + \Pi_2\right], \tag{86}$$

$$\dot{\phi} = \frac{1}{2F}\Pi_\phi. \tag{87}$$

introducing (81) into (56) we find the differential equation for Π_1 as $\dot{\Pi}_1 - p_3\Pi_1 + k_1 = 0$ where $k_1 = \frac{1}{4}\left(p_3^2 + a_1^2 - 16\phi_0 + C_1 - 2a_1p_3\right)$ who solution become as

$$\Pi_1 = \frac{1}{p_3}\left[e^{p_3\Delta t} + k_1\right], \tag{88}$$

then the solutions for β_i become

$$\Delta\beta_1 = \frac{1}{4}(a_1 + p_3)\Delta t, \tag{89}$$

$$\Delta\beta_2 = \frac{1}{4}(p_3 - a_1)\Delta t, \tag{90}$$

$$\Delta\beta_3 = \frac{1}{4}(a_1 - p_3)\Delta t + \frac{1}{2p_3}\left[e^{p_3\Delta t} + k_1\right], \tag{91}$$

$$\tag{92}$$

and the solutions in the original variables are

$$\Omega = \frac{1}{12p_3}\left[2k_1 + p_3\left(a_1 + p_3\right)\Delta t + 2e^{p_3\Delta t}\right],$$

$$\beta_- = \frac{a_1}{4\sqrt{3}}\Delta t,$$

$$\beta_+ = -\frac{1}{12p_3}\left[2k_1 + p_3\left(a_1 - 2p_3\right)\Delta t + 2e^{p_3\Delta t}\right]. \tag{93}$$

5. Quantum scheme

The WDW equation for these models is achived by replacing $\Pi_{q^\mu} = -i\partial_{q^\mu}$ in (21). The factor $e^{-3\Omega}$ may be factor ordered with $\hat{\Pi}_\Omega$ in many ways. Hartle and Hawking (Hartle & Hawking, 1983) have suggested what might be called a semi-general factor ordering which in this case would order $e^{-3\Omega}\hat{\Pi}^2_\Omega$ as

$$-e^{-(3-Q)\Omega}\,\partial_\Omega e^{-Q\Omega}\partial_\Omega = -e^{-3\Omega}\,\partial^2_\Omega + Q\,e^{-3\Omega}\partial_\Omega\,,$$

$$-\frac{6}{F}\phi^s\frac{\partial}{\partial\phi}\phi^{-s}\frac{\partial}{\partial\phi} = -\frac{6}{F}\frac{\partial^2}{\partial\phi^2} + \frac{6s}{F}\phi^{-1}\frac{\partial}{\partial\phi}, \tag{94}$$

where Q and s are any real constants that measure the ambiguity in the factor ordering in the variables Ω and ϕ. We will assume in the following this factor ordering for the Wheeler-DeWitt equation, which becomes

$$\Box\Psi - \frac{6}{F(\phi)}\frac{\partial^2\Psi}{\partial\phi^2} + \frac{6s}{F}\phi^{-1}\frac{\partial\Psi}{\partial\phi} + Q\frac{\partial\Psi}{\partial\Omega} - U(\Omega,\beta_\pm)\,\Psi - C_1\Psi = 0, \tag{95}$$

where $\Box$ is the three dimensional d'Lambertian in the $\ell^\mu = (\Omega,\beta_+,\beta_-)$ coordinates, with signature (- + +).

When we introduce the Ansatz $\Psi = \chi(\phi)\psi(\Omega,\beta_\pm)$ in (95), we obtain the general set of differential equations (under the assumed factor ordering) for the Bianchi type IX cosmological model

$$\Box\psi + Q\frac{\partial\psi}{\partial\Omega} - \left[U(\Omega,\beta_\pm) + C_1 - \mu^2\right]\psi = 0, \tag{96}$$

$$\frac{6}{\mathrm{F}(\phi)}\frac{\partial^2\chi}{\partial\phi^2}-\frac{6\mathrm{s}}{\mathrm{F}}\phi^{-1}\frac{\partial\chi}{\partial\phi}+\mu^2\chi=0. \tag{97}$$

When we calculate the solution to equation (97), we find interesting properties on this, as

1. This equation is a master equation for the field ϕ for any cosmological model, implying that this field ϕ is an universal field as cosmic ground, having the best presence in the stiff matter era as an ingredient in the formation the structure galaxies and when we consider two types of functions, $\mathrm{F}(\phi)=\omega\phi^{\mathrm{m}}$ and $\mathrm{F}(\phi)=\omega e^{\mathrm{m}\phi}$, we have the following exact solutions (Polyanin & Zaitsev, 2003)
   (a) $F(\phi)=\omega\phi^m$
   the differential equation to solver is

$$\frac{d^2\chi}{d\phi^2}-s\phi^{-1}\frac{d\chi}{d\phi}+\alpha\phi^m\chi=0, \tag{98}$$

   with $\alpha=\frac{\omega\mu^2}{6}$. The solutions depend on the value to m and s,
   i. General solution for any $m\neq-2$ and s, are written in terms of ordinary and modify Bessel function,

$$\chi=c_1\phi^{\frac{1+s}{2}}Z_\nu\left(\frac{2\sqrt{\alpha}}{m+2}\phi^{\frac{m+2}{2}}\right), \tag{99}$$

   with c_1 an integration constant, Z_ν is a generic Bessel function, $\nu=\frac{1+s}{m+2}$ is the order. When $\alpha>0$ imply $\omega>0$, Z_ν become the ordinary Bessel function, (J_ν,Y_ν). If $\alpha<0,\rightarrow w<0, Z_\nu\rightarrow(I_\nu,K_\nu)$.
   ii. $m=-2$ and any s,

$$\chi=\phi^{\frac{1+s}{2}}\begin{cases} c_1\,\phi^{\mu}+c_2\phi^{-\mu}, & \text{si} \quad \mu>0\\ c_1\,+c_2 Ln\phi, & \text{si} \quad \mu=0\,,\\ c_1\,sin\,(\mu Ln\phi)+c_2\,cos\,(\mu Ln\phi)\,, & \text{if} \quad \mu<0\end{cases} \tag{100}$$

   where $\mu=\frac{1}{2}\sqrt{|(1+s)^2-4\alpha|}$.
   iii. $m=-6$ and $s=1$

$$\chi(\phi)=\phi^2\begin{cases} c_1\,sinh\left(\frac{\sqrt{|\alpha|}}{2\phi^2}\right)+c_2\,cosh\left(\frac{\sqrt{|\alpha|}}{2\phi^2}\right), & \alpha<0\rightarrow\omega<0\\ c_1\,sin\left(\frac{\sqrt{|\alpha|}}{2\phi^2}\right)+c_2\,cos\left(\frac{\sqrt{|\alpha|}}{2\phi^2}\right), & \alpha>0\rightarrow\omega>0\end{cases} \tag{101}$$

   (b) $F(\phi)=\omega e^{m\phi}$, for this case we consider the case $s=0$,

$$\frac{d^2\chi}{d\phi^2}+\alpha e^{m\phi}\chi=0, \tag{102}$$

   i. $m\neq0$

$$\chi=CZ_0\left(\frac{2\sqrt{\alpha}}{m}e^{\frac{m\phi}{2}}\right), \tag{103}$$

   with C is a integration constant and Z_0 is the generic Bessel function to zero order. So, if $\alpha>0$ then $\omega>0$, Z_0 is the ordinary Bessel function (J_0,Y_0). When $\alpha<0,\rightarrow$ $\omega<0, Z_0\rightarrow(I_0,K_0)$.

ii. for $m = 0$,

$$\chi = \begin{cases} c_1 \sinh\left(\sqrt{|\alpha|}\phi\right) + c_2 \cosh\left(\sqrt{|\alpha|}\phi\right), & \text{if } \alpha < 0 \rightarrow \omega < 0 \\ c_1 \sin\left(\sqrt{|\alpha|}\phi\right) + c_2 \cos\left(\sqrt{|\alpha|}\phi\right), & \text{if } \alpha > 0 \rightarrow \omega > 0 \end{cases} \tag{104}$$

2. If we have the solution for the parameter s=0 for arbitrary function F(ϕ), say χ_0, then we have also the solution for s=-2, as $\chi(s=-2) = \frac{\chi_0}{\phi}$.

To obtain the solution of the other factor of Ψ we use the particular value for the constants $C_1 = \mu^2$, and make the following Ansatz for the wave function

$$\psi(\ell^\mu) = W(\ell^\mu) e^{-S(\ell^\mu)}, \tag{105}$$

where $S(\ell^\mu)$ is known as the superpotential function, and W is the amplitude of probability to that employed in Bohmian formalism (Bohm, 1986), those found in the literature, years ago (Obregón & Socorro, 1996). So (96) is transformed into

$$\Box W - W\Box S - 2\nabla W \cdot \nabla S + Q\frac{\partial W}{\partial \Omega} - QW\frac{\partial S}{\partial \Omega} + W\left[(\nabla S)^2 - U\right] = 0, \tag{106}$$

where $\Box = G^{\mu\nu}\frac{\partial^2}{\partial \ell^\mu \partial \ell^\nu}$, $\nabla W \cdot \nabla \Phi = G^{\mu\nu}\frac{\partial W}{\partial \ell^\mu}\frac{\partial \Phi}{\partial \ell^\nu}$, $(\nabla)^2 = G^{\mu\nu}\frac{\partial}{\partial \ell^\mu}\frac{\partial}{\partial \ell^\nu} = -(\frac{\partial}{\partial \Omega})^2 + (\frac{\partial}{\partial \beta_+})^2 + (\frac{\partial}{\partial \beta_-})^2$, with $G^{\mu\nu} = \mathrm{diag}(-1,1,1)$, U is the potential term of the cosmological model under consideration.
Eq (106) can be written as the following set of partial differential equations

$$(\nabla S)^2 - U = 0, \tag{107a}$$

$$W\left(\Box S + Q\frac{\partial S}{\partial \Omega}\right) + 2\nabla W \cdot \nabla S = 0, \tag{107b}$$

$$\Box W + Q\frac{\partial W}{\partial \Omega} = 0. \tag{107c}$$

Following reference (Guzmán et al., 2007), first we shall choose to solve Eqs. (107a) and (107b), whose solutions at the end will have to fulfill Eq. (107c), which play the role of a constraint equation.

5.1 Transformation of the Wheeler-DeWitt equation

We were able to solve (107a), by doing the change of coordinates (22) and rewrite (107a) in these new coordinates. With this change, the function S is obtained as follow, with the ansatz (105),
In this section, we obtain the solutions to the equations that appear in the decomposition of the WDW equation, (107a), (107b) and (107c), using the Bianchi type IX Cosmological model. So, the equation $[\nabla]^2 = -(\frac{\partial}{\partial \Omega})^2 + (\frac{\partial}{\partial \beta_+})^2 + (\frac{\partial}{\partial \beta_-})^2$ can be written in the following way (see appendix section 8)

$$\begin{aligned} [\nabla]^2 &= 3\left[\left(\frac{\partial}{\partial \beta_1}\right)^2 + \left(\frac{\partial}{\partial \beta_2}\right)^2 + \left(\frac{\partial}{\partial \beta_3}\right)^2\right] - 6\left[\frac{\partial}{\partial \beta_1}\frac{\partial}{\partial \beta_2} + \frac{\partial}{\partial \beta_1}\frac{\partial}{\partial \beta_3} + \frac{\partial}{\partial \beta_2}\frac{\partial}{\partial \beta_3}\right] \\ &= 3\left(\frac{\partial}{\partial \beta_1} + \frac{\partial}{\partial \beta_2} + \frac{\partial}{\partial \beta_3}\right)^2 - 12\left[\frac{\partial}{\partial \beta_1}\frac{\partial}{\partial \beta_2} + \frac{\partial}{\partial \beta_1}\frac{\partial}{\partial \beta_3} + \frac{\partial}{\partial \beta_2}\frac{\partial}{\partial \beta_3}\right]. \end{aligned} \tag{108}$$

The potencial term of the Bianchi type IX is transformed in the new variables into

$$U = 12\left[\left(e^{2\beta_1} + e^{2\beta_2} + e^{2\beta_3}\right)^2 - 2e^{2(\beta_1+\beta_2)} - 2e^{2(\beta_1+\beta_3)} - 2e^{2(\beta_2+\beta_3)}\right]. \tag{109}$$

Then (107a) for this models is rewritten in the new variables as

$$3\left(\frac{\partial S}{\partial\beta_1} + \frac{\partial S}{\partial\beta_2} + \frac{\partial S}{\partial\beta_3}\right)^2 - 12\left[\frac{\partial S}{\partial\beta_1}\frac{\partial S}{\partial\beta_2} + \frac{\partial S}{\partial\beta_1}\frac{\partial S}{\partial\beta_3} + \frac{\partial S}{\partial\beta_2}\frac{\partial S}{\partial\beta_3}\right]$$
$$- 12\left[\left(e^{2\beta_1} + e^{2\beta_2} + e^{2\beta_3}\right)^2 - 4e^{2(\beta_1+\beta_2)} - 4e^{2(\beta_1+\beta_3)} - 4e^{2(\beta_2+\beta_3)}\right] = 0. \tag{110}$$

Now, we can use the separation of variables method to get solutions to the last equation for the S function, obtaining for the Bianchi type IX model

$$S_{IX} = \pm\left(e^{2\beta_1} + e^{2\beta_2} + e^{2\beta_3}\right). \tag{111}$$

In table 4 we present the corresponding superpotential function S and amplitude W for all Bianchi Class A models.
With this result, and using for the solution to (107b) in the new coordinates β_i, we have for W function as

$$W_{IX} = W_0\, e^{\left[(1+\frac{Q}{6})(\beta_1+\beta_2+\beta_3)\right]}, \tag{112}$$

and re-introducing this result into Eq. (107c) we find that $Q = \pm 6$. Therefore we have two wave functions

$$\begin{aligned}\Psi_{IX}(\beta_i) &= W_{IX}(\beta_i)\,\mathrm{Exp}\left[\pm\left(e^{2\beta_1} + e^{2\beta_2} + e^{2\beta_3}\right)\right]\\ &= \mathrm{Exp}\left[\pm\left(e^{2\beta_1} + e^{2\beta_2} + e^{2\beta_3}\right)\right]\begin{cases} W_0, & Q=-6\\ W_0\mathrm{Exp}\left[2\left(\beta_1+\beta_2+\beta_3\right)\right], & Q=6\end{cases}\end{aligned} \tag{113}$$

similar solutions were given by Moncrief and Ryan (Moncrief & Ryan, 1991) in standard quantum cosmology in general relativity. In table 4 we present the superpotential function S, the amplitude of probability W and the relations between the parameters for the corresponding Bianchi Class A models.
If one looks at the expressions for the functions S given in table 4, one notes that there is a general form to write them using the 3x3 matrix m^{ij} that appear in the classification scheme of Ellis and MacCallum (Ellis & MacCallum, 1969) and Ryan and Shepley (Ryan & Shepley, 1975), the structure constants are written in the form

$$C^i_{jk} = \epsilon_{jks}\, m^{si} + \delta^i_{[k} a_{j]}, \tag{114}$$

where $a_i = 0$ for the Class A models.
If we define $g_i(\beta_i) = (e^{\beta_1}, e^{\beta_2}, e^{\beta_3})$, with β_i given in (22), the solution to (107a) can be written as

$$S(\beta_i) = \pm[g_i\, M^{ij}\, (g_j)^T], \tag{115}$$

where $M^{ij} = m^{ij}$ for the Bianchi Class A, excepting the Bianchi type $VI_{h=-1}$ for which we redefine the matrix to be consistent with (115)

$$M^{ij} = (\beta_1 - \beta_2)\begin{pmatrix} 0 & 1 & 0\\ 1 & 0 & 0\\ 0 & 0 & 0\end{pmatrix}.$$

Bianchi type	Superpotential **S**	Amplitude of probability W	Constraint
I	constant	$e^{(\frac{r}{3}+\frac{b}{6}+\frac{\sqrt{3}c}{6})\beta_1+(\frac{r}{3}+\frac{b}{6}-\frac{\sqrt{3}c}{6})\beta_2+(\frac{r}{3}-\frac{b}{3})\beta_3}$	$r^2 - Qr - a^2 = 0$, $a^2 = b^2 + c^2$
II	$e^{2\beta_1}$	$e^{(a-1-\frac{Q}{6})\beta_1+a\beta_2+(a-b)\beta_3}$	$144b^2 - 144ab + 36 - Q^2 + 24aQ = 0$
$VI_{h=-1}$	$2(\beta_1 - \beta_2)\, e^{(\beta_1+\beta_2)}$	$e^{a(\beta_1+\beta_2)}$	$Q = 0$
$VII_{h=0}$	$e^{2\beta_1} + e^{2\beta_2}$	$e^{(1+\frac{Q}{6})(\beta_1+\beta_2+\beta_3)+a(\beta_1+\beta_2)}$	$Q^2 - 48a - 36 = 0$
VIII	$e^{2\beta_1} + e^{2\beta_2} - e^{2\beta_3}$	$W_0\, e^{[(1+\frac{Q}{6})(\beta_1+\beta_2+\beta_3)]}$	$Q = \pm 6$
IX	$e^{2\beta_1} + e^{2\beta_2} + e^{2\beta_3}$	$W_0\, e^{[(1+\frac{Q}{6})(\beta_1+\beta_2+\beta_3)]}$	$Q = \pm 6$

Table 4. Superpotential S, the amplitude of probability W and the relations between the parameters for the corresponding Bianchi Class A models.

Then, for the Bianchi Class A models, the wave function Ψ can be written in the general form

$$\Psi = \chi(\phi)\, W(\beta_i) \exp\left[\pm[g_i\, M^{ij}\, (g_j)^T]\right]. \tag{116}$$

6. Final remarks

Using the analytical procedure of hamilton equation of classical mechanics, in appropriate coordinates, we found a master equation for all Bianchi Class A cosmological models, we present partial result in the classical regime for three models of them, but the general equation are shown for all them. In particular, the Bianchi type I is complete solved without using a particular gauge. The Bianchi type II and $VI_{h=-1}$ are solved introducing a particular gauge. An important results yields when we use the gauge $N = e^{\beta_1+\beta_2+\beta_3}$, we find that the solutions for the ϕ field are independent of the cosmological models, and we find that the energy density associated has a scaling behaviors under the analysis of standard field theory to scalar fields (Andrew & Scherrer, 1998; Ferreira & Joyce, 1998), is say, scales exactly as a power of the scale factor like, $\rho_\phi \propto a^{-m}$. More of this can be seen to references cited before. On the other hand, in the quantum regime, wave functions of the form $\Psi = W e^{\pm S}$ are the only known exact solutions for the Bianchi type IX model in standard quantum cosmology. In the SB formalism, these solutions are modified only for the function χ, $\Psi = \chi(\phi)\, W(\ell^\mu)\, e^{\pm S(\ell^\mu)}$ when we include the particular ansatz $C_1 = \mu^2$. This kind of solutions already have been found in supersymmetric quantum cosmology (Asano et al., 1993) and also for the WDW equation defined in the bosonic sector of the heterotic strings (Lidsey., 1994). Recently, in the books (Paulo, 2010) appears all solutions in the supersymmetric scheme similar at our formalism. We have shown that they are also exact solutions to the rest of the Bianchi Class A models in SB quantum cosmology, under the assumed semi-general factor ordering (94). Different procedures seem to produce this particular quantum state, where S is a solution to the corresponding classical Hamilton-Jacobi equation (107a).

7. Appendix: Energy momentum tensor

From Eq. (6) we see that the effective energy momentum tensor of the scalar field is

$$T_{\alpha\,\beta} = F(\phi)\left(\phi_{,\alpha}\phi_{,\beta} - \frac{1}{2} g_{\alpha\beta}\phi_{,\gamma}\phi^{,\gamma}\right), \tag{117}$$

this energy momentum tensor is conserved, as follows from the equation of motion for the scalar field

$$
\begin{aligned}
\nabla^{\beta} T_{\alpha\,\beta} = \nabla^{\beta}\left[F(\phi)\left(\phi_{,\alpha}\phi_{,\beta} - \frac{1}{2}g_{\alpha\beta}\phi_{,\gamma}\phi^{,\gamma}\right)\right] &= F'(\phi)\phi^{,\beta}\left(\phi_{,\alpha}\phi_{,\beta} - \frac{1}{2}g_{\alpha\beta}\phi_{,\gamma}\phi^{,\gamma}\right) \\
&+F(\phi)\left(\phi_{,\alpha}{}^{;\beta}\phi_{,\beta} + \phi_{,\alpha}\phi_{,\beta}{}^{;\beta} - \frac{1}{2}g_{\alpha\beta}\phi_{,\gamma}{}^{;\beta}\phi^{,\gamma} - \frac{1}{2}g_{\alpha\beta}\phi_{,\gamma}\phi^{,\gamma;\beta}\right) \\
&= F'(\phi)\left(\frac{1}{2}\phi_{,\gamma}\phi^{,\gamma}\phi_{,\alpha}\right) + F(\phi)\left(\phi_{,\alpha}{}^{;\beta}\phi_{,\beta} + \phi_{,\alpha}\phi_{,\beta}{}^{;\beta} - g_{\alpha\beta}\phi_{,\gamma}{}^{;\beta}\phi^{,\gamma}\right) \\
&= \frac{1}{2}\phi_{,\alpha}\left(F'(\phi)\phi_{,\gamma}\phi^{,\gamma} + 2F(\phi)\phi_{,\beta}{}^{;\beta}\right) = 0. \qquad (118)
\end{aligned}
$$

Now we proceed to show that the energy momentum tensor has the structure of an imperfect stiff fluid,

$$
T_{\alpha\,\beta} = (\rho + p)U_{\alpha}U_{\beta} + p g_{\alpha\,\beta} = (2\rho)[U_{\alpha}U_{\beta} + \frac{1}{2}g_{\alpha\,\beta}], \qquad (119)
$$

here ρ is the energy density, p the pressure, and U_{α} the velocity If we choose for the velocity the normalized derivative of the scalar field, assuming that it is a timelike vector, as is often the case in cosmology, where the scalar field is only time dependent

$$
U_{\alpha} = S^{-1/2}\phi_{,\alpha}, \qquad S = -\phi_{,\sigma}\phi^{,\sigma}. \qquad (120)
$$

It is evident that the energy momentum tensor of the SB theory is equivalent to a stiff fluid with the energy density given by

$$
\rho = \frac{S\,F(\phi)}{2} = -\frac{\phi_{,\sigma}\phi^{,\sigma}\,F(\phi)}{2}. \qquad (121)
$$

Therefore the most important contribution of the scalar field occurs during a stiff matter phase that is previous to the dust phase.

8. Appendix: Operators in the β_i variables

The operators who appear in eqn (95) are calculated in the original variables $(\Omega, \beta_+, \beta_-)$; however the structure of the cosmological potential term gives us an idea to implement new variables, considering the Bianchi type IX cosmological model, these one given by eqn (22). The main calculations are based in the following

$$
\begin{aligned}
\frac{\partial}{\partial\Omega} &= \frac{\partial}{\partial\beta_1} + \frac{\partial}{\partial\beta_2} + \frac{\partial}{\partial\beta_3}, \\
\frac{\partial^2}{\partial\Omega^2} &= \frac{\partial^2}{\partial\beta_1^2} + \frac{\partial^2}{\partial\beta_2^2} + \frac{\partial^2}{\partial\beta_3^2} + 2\left[\frac{\partial^2}{\partial\beta_1\partial\beta_2} + \frac{\partial^2}{\partial\beta_1\partial\beta_3} + \frac{\partial^2}{\partial\beta_2\partial\beta_3}\right], \\
\frac{\partial}{\partial\beta_+} &= \frac{\partial}{\partial\beta_1} + \frac{\partial}{\partial\beta_2} - 2\frac{\partial}{\partial\beta_3}, \\
\frac{\partial^2}{\partial\beta_+^2} &= \frac{\partial^2}{\partial\beta_1^2} + \frac{\partial^2}{\partial\beta_2^2} + 4\frac{\partial^2}{\partial\beta_3^2} + 2\left[\frac{\partial^2}{\partial\beta_1\partial\beta_2} - 2\frac{\partial^2}{\partial\beta_1\partial\beta_3} - 2\frac{\partial^2}{\partial\beta_2\partial\beta_3}\right], \\
\frac{\partial}{\partial\beta_-} &= \sqrt{3}\left(\frac{\partial}{\partial\beta_1} - \frac{\partial}{\partial\beta_2}\right),
\end{aligned}
$$

$$\frac{\partial^2}{\partial \beta_-^2} = 3\left(\frac{\partial^2}{\partial \beta_1^2} + \frac{\partial^2}{\partial \beta_2^2} - 2\frac{\partial^2}{\partial \beta_1 \partial \beta_2}\right). \tag{122}$$

So, the operator $(\nabla)^2, \Box, \nabla S \nabla W$ are written as

$$\begin{aligned}
(\nabla)^2 &= G^{\mu\nu}\frac{\partial}{\partial \ell^\mu}\frac{\partial}{\partial \ell^\nu}, \qquad G^{\mu\nu} = \mathrm{diag}(-1,1,1), \qquad \ell^\mu = (\Omega, \beta_+, \beta_1)\\
&= 3\left\{\left(\frac{\partial}{\partial \beta_1}\right)^2 + \left(\frac{\partial}{\partial \beta_2}\right)^2 + \left(\frac{\partial}{\partial \beta_3}\right)^2 - 2\left[\frac{\partial}{\partial \beta_1}\frac{\partial}{\partial \beta_2} + \frac{\partial}{\partial \beta_1}\frac{\partial}{\partial \beta_3} + \frac{\partial}{\partial \beta_2}\frac{\partial}{\partial \beta_3}\right]\right\}\\
&= 3\left\{\left[\frac{\partial}{\partial \beta_1} + \frac{\partial}{\partial \beta_2} + \frac{\partial}{\partial \beta_3}\right]^2 - 4\left[\frac{\partial}{\partial \beta_1}\frac{\partial}{\partial \beta_2} + \frac{\partial}{\partial \beta_1}\frac{\partial}{\partial \beta_3} + \frac{\partial}{\partial \beta_2}\frac{\partial}{\partial \beta_3}\right]\right\},\\
\Box &= G^{\mu\nu}\frac{\partial^2}{\partial \ell^\mu \partial \ell^\nu} = 3\left(\frac{\partial^2}{\partial \beta_1^2} + \frac{\partial^2}{\partial \beta_2^2} + \frac{\partial^2}{\partial \beta_3^2}\right) - 6\left(\frac{\partial^2}{\partial \beta_1 \partial \beta_2} + \frac{\partial^2}{\partial \beta_1 \partial \beta_3} + \frac{\partial^2}{\partial \beta_2 \partial \beta_3}\right),\\
\nabla S \cdot \nabla W &= G^{\mu\nu}\frac{\partial S}{\partial \ell^\mu}\frac{\partial W}{\partial \ell^\nu}\\
&= 3\left(\frac{\partial S}{\partial \beta_1}\frac{\partial W}{\partial \beta_1} + \frac{\partial S}{\partial \beta_2}\frac{\partial W}{\partial \beta_2} + \frac{\partial S}{\partial \beta_3}\frac{\partial W}{\partial \beta_3}\right)\\
&\quad -3\left(\frac{\partial S}{\partial \beta_1}\frac{\partial W}{\partial \beta_2} + \frac{\partial S}{\partial \beta_1}\frac{\partial W}{\partial \beta_3} + \frac{\partial S}{\partial \beta_2}\frac{\partial W}{\partial \beta_3} + \frac{\partial S}{\partial \beta_2}\frac{\partial W}{\partial \beta_1} + \frac{\partial S}{\partial \beta_3}\frac{\partial W}{\partial \beta_1} + \frac{\partial S}{\partial \beta_3}\frac{\partial W}{\partial \beta_2}\right).
\end{aligned} \tag{123}$$

9. Acknowledgments

This work was partially supported by CONACYT grant 56946. DAIP (2010-2011) and PROMEP grants UGTO-CA-3. This work is part of the collaboration within the Instituto Avanzado de Cosmología. Many calculations were done by Symbolic Program REDUCE 3.8.

10. References

Saez, D. & Ballester, V.J. (1986). *Physics Letters A*, 113, 467.

Socorro, J.; Sabido, M.; Sánchez G. M.A. & Frías Palos M.G. (2010). Anisotropic cosmology in Sáez-Ballester theory: classical and quantum solutions, *Revista Mexicana de Física*, 56(2), 166-171.

Sabido, M.; Socorro, J. & Luis Arturo Ureña, (2011). Classical and quantum Cosmology of the Sáez-Ballester theory, *Fizika B*, in press.

Armendariz-Picon, C.; Mukhanov, V. & Steinbardt, P.J. (2000) *Phys. Lett*, 85, 4438; (2001), *Phys. Rev. D* 63, 103510.

Armendariz-Picon, C., Damour, T. & Mukhanov, V., (1999). *Phys. Lett. B* 458 209 (1999); Garriga, J. & Mukhanov, V. (1999), *Phys. Lett. B* 458, 219.

Copeland, E.J.; Sami, M. & Tsujikawa S. (2006). *Dynamics of dark energy, Int. J. Mod. Phys. D,* 15 1753, [arXiv:hep-th 0603057].

Singh, T. & Agrawal, A.K. (1991). *Some Bianchi-type cosmological models in a new scalar-tensor theory, Astrophys. and Space Sci.,* 182, 289.

Shri, R. & Singh, J.K. (1995). *Cosmological model in certain scalar-tensor theories, Astrophys. Space Sci,* 234, 325.

Reddy, D.R.K. & Rao, N.V. (2001), *Some cosmological models in scalar-tensor theory of gravitation, Astrophys. Space Sci,* 277, 461.

Mohanty, G. & Pattanaik, S.K. (2001). *Theor. Appl. Mech,* 26, 59.
Singh, C.P. & Ram Shri, (2003). *Astrophys. Space Sci.,* 284, 1199.
Mohanty, G. & Sahu, S.K. (2003). *Bianchi* VI_0 *cosmological model in Saez and Ballester theory. Astrophys. Space Sci. ,* 288, 611.
Mohanty, G. & Sahu, S.K. (2004). *Bianchi type-I cosmological effective stiff fluid model in Saez and Ballester theory. Astrophys. Space Sci.,* 291, 75.
Adhav, K.S.; Ugale, M.R.; Kale, C.B. & Bhende, M.P. (2007). *Bianchi Type VI String Cosmological Model in Saez-BallesterŠs Scalar-Tensor Theory of Gravitation, Int. J. Theor. Phys.,* 46 3122.
Rao, V.U.M.; Vinutha, T. & Vihaya Shanthi, M. (2007). *An exact Bianchi type-V cosmological model in Saez-Ballester theory of gravitation, Astrophys. Spa. Sci.,* 312, 189.
Rao, V.U.M.; Vijaya Santhi, M. & Vinutha, T. (2008). *Exact Bianchi type II, VIII and IX string cosmological models in Saez-Ballester theory of gravitation, Astrophys. Spa. Sci.,* 314, 73.; *Exact Bianchi type II, VIII and IX perfect fluid cosmological models in Saez-Ballester theory of gravitation, Astrophys. Spa. Sci.,* 317, 27.
Shri, R.; Zeyauddin, M. & and Singh, C.P. (2009). *Bianchi type-V cosmological models with perfect fluid and heat flow in Saez-Ballester theory, Pramana,* 72 (2), 415.
Tripathy, S.K.; Nayak, S.K.; Sahu, S.K. & Routray, T.R. (2009). *Massive String Cloud Cosmologies in Saez-Ballester Theory of Gravitation, Int. J. Theor. Phys.,* 48, 213.
Singh, C.P. (2009). *LRS Bianchi Type-V Cosmology with Heat Flow in Scalar-Tensor Theory, Brazilian Journal of Physics,* 39 (4), 619.
Pradhan, A. & Singh, S.K. (2010). *Some Exact Bianchi Type-V Cosmological Models in Scalar Tensor Theory: Kinematic Tests, EJTP,* 7 (24), 407.
Ryan, M.P.(1972). *Hamiltonian cosmology,* (Springer, Berlin,).
Obregón, O. & Socorro, J, (1996). $\Psi = \mathrm{W}\,e^{\pm\Phi}$ *Quantum Cosmological solutions for Class A Bianchi Models, Int. J. Theor. Phys.,* 35, 1381. [gr-qc/9506021].
Liddle, A.R. & Scherrer, R.J. (1998). *Classification of scalar field potential with cosmological scaling solutions, Phys. Rev. D,* 59, 023509.
Ferreira, P.G. & Joyce, M. (1998). *Cosmology with a primordial scaling field, Phys. Rev. D,* 58, 023503.
Marden, M. (1988). *Class. and Quantum Grav,* 5, 627.
Pimentel, L.O. (1989). *Class. and Quantum Grav,* 6, L263.
Hartle, J. & Hawking, S.W. (1983). *Phys. Rev. D,* 28, 2960.
Polyanin, Andrei C. & Zaitsev Valentin F. (2003). *Handbook of Exact solutions for ordinary differential equations, Second edition,* Chapman & Hall/CRC.
Bohm, D. (1986). *Phys. Rev.,* 85, 166.
Guzmán, W.; Sabido, M.; Socorro, J. & Arturo L. Ureña-López, (2007). *Int. J. Mod. Phys. D,* 16 (4), 641, [gr-qc/0506041]
Moncrief, V. & Ryan, M.P., (1991). *Phys. Rev. D,* 44, 2375.
Ellis G.F.R. & MacCallum M.A.H. (1969). *Comm. Math. Phys,* 12, 108.
Ryan, M.P. & Shepley, L.C. (1975). *Homogeneous Relativistic Cosmologies* (Princenton)
Asano, M.; Tanimoto, M. & Yoshino N. (1993). *Phys. Lett. B,* 314, 303.
Lidsey, J.E. (1994). *Phys. Rev. D,* 49, R599.
Moniz, P.V, (2010). *Quantum cosmology -the supersymmetric perspective- Vol. 1 & 2,* Lecture Notes in Physics 803 & 804, (Springer, Berlin).

11

A New Cosmological Model

J.-M. Vigoureux[1], B. Vigoureux[1] and M. Langlois[2]
[1]*Institut UTINAM, UMR CNRS 6213,*
Université de Franche-Comté, Besançon Cedex
[2]*Passavant, 25360*
France

1. Introduction

The constant c was first introduced as the speed of light. However, with the development of physics, it came to be understood as playing a more fundamental role, its significance being not directly that of a usual velocity (even though its dimensions are) and one might thus think of c as being a fundamental constant of the universe (for a discussion on the speed of light, see, for example, (Ellis & Uzan, 2005)). Moreover, the advent of Einsteinian relativity, the fact that c appears in phenomena where there is neither light nor any motion (for example in $E = mc^2$ which shows that c can in principle be measured with a weighing scale and a thermometer (Braunbeck, 1937) or in the relation $(\epsilon_0\mu_0)^{-1/2} = c$ showing that c can be obtained from electro*static* and magneto*static* experiments (Maxwell, 1954)) and its dual-interpretation in terms of "speed" of light and of "speed" of gravitation [1] forces everybody to associate c with the theoretical description of space-time itself rather than that of some of its specific contents. We could not in fact be satisfied by such results and we may think that these different aspects of "c" reflect an underlying structure we do not yet comprehend.
All this invites us to connect c to the geometry of the universe. Noting then that both c and the expansion of the universe *provide a universal relation between space and time* which both have the physical dimension of a velocity, we consider that these two facts cannot be a fortuitous coincidence and that they consequently are two different aspects of a same phenomenon. We thus consider that c must be related to the expansion of the universe and we postulate as a fundamental law of nature (Vigoureux et al., 1988) that

$$c = \alpha\dot{a} = Cst \tag{1}$$

where α is a positive constant and where $a(t)$ is the cosmic scale factor which can be assimilated to the radius of the universe in the case of a spherical geometry (of course, all results also holds when taking $c = 1$). Equation (1) of course means that the scale factor increases at a constant expanding rate. Such a case is usually expected to describe an empty expanding universe (as is for example the Milne universe) or, at the least, an universe in which the density of matter and radiation are so small that they have negligible effect on the flat spacetime geometry. *However, as we shall see, in our model where appears a cosmological constant*

[1] Answering to the question by saying that light and gravitation correspond to zero rest-mass particules does not change the problem.

term, a constant velocity of expansion does not need such an empty universe. [2] Let us also note that eq.(1) verifies the condition $\dot{H} + (1+q)H^2 = 0$ where $q = -\ddot{a}a/\dot{a}^2$ is the deceleration parameter and where H is the Hubble parameter. In our case that equation in fact reduces to $\dot{H} + H^2 = 0$ the solution of which is $H = 1/t$ and consequently $a \sim t$ as expected from eq.(1). Eq.(1) permits to define c *from the knowledge of the geometry of space-time only*, that is from its size and its age. It thus really gives c the statute of a true geometrical fundamental magnitude of the universe, whereas its value 299,792,458 metres per second not only has no geometrical meaning, but also has no meaning at all in the early universe when metres and seconds cannot be defined [3]. On the contrary, it is in fact to be underlined that defining c from the size and the age of the universe has a meaning at all times.

Our aim in this chapter is to show that solving Friedmann's equations with eq.(1), which thus appears as an additional constraint, can explain unnatural features of the standard cosmology without needing any other hypothesis such as those of the inflationary universe or of varying speed of light cosmologies. We thus show that using eq.(1) can solve

- the flatness problem: in our model, the universe dispays the same evolution as a flat universe and *must appear to be flat whatever it may be* (spherical or not);
- the horizon problem: there is no particle horizon;
- the uniformity of the cosmic microwaves background radiation and the small-scale inhomogeneity problem: we show that it is the same tiny part of the early universe that we can observe in any direction around us so that it is quite normal to find the observed background homogeneity. Moreover, it becomes obvious that the universe at time t_{CMB} of the cosmic microwave background radiation can be quite inhomogeneous so that its inhomogeneities can be understood as the seeds of cosmological structures (galaxies and clusters of galaxies).
- We also show that it permits to fit observational data of type Ia supernovae without having to consider an accelerating expansion of the universe: in the standard cosmology, the interpretation of such observations need to use for q a value close to -0.5 for today and a value of 0.5 for very high redshits. On the contrary, our calculations show that all observations can be explained by using $q = 0$ at all times. So, provide we use eq.(1), the linear approach for the cosmological scale factor is well supported by observations;
- Studying then the cosmological term problem which is to understand why ρ_Λ is not only small but also of the same order of magnitude as the present mass density ρ_M of the universe, we finally show how our model also answers that problem.

In each part, we begin by introducing briefly the problem we consider. We then present our results. Some of them have been published (Viennot & Vigoureux, 2009; Vigoureux et al., 1988; 2001; 2003; 2008). However they have not been presented in details. Moreover we also need them for a coherent presentation of our model. We thus present them for clarity and for their subsequent uses in this chapter. In any case, all results are discussed in a detailed way.

In concluding, we first discuss the originality of eq.(1) which has the advantage of giving unity to number of results which, for some of them, have been found by various authors

[2] Usually, such a linear variation of the scale factor leads to at least two special cases. One is an empty universe ($T_{\mu\nu} = 0$) with $k = -1$. The other is a flat universe with the equation of state $p = -\rho c^2/3$. It is consequently concluded that such a variation of the scale factor cannot describe the universe in which we live. However, it would be to conclude too quickly to deduce that any flat-spacetime metric must describe an empty universe: we shall see that in our model, the metric of a spherical universe, *for example*, can be reduced to that of a flat space-time metric.

[3] *For example*, in its 1960 definition, the meter is defined as "the length equal to 1,650,763.73 wavelengths in vacuum of the radiation corresponding to the transition between the levels 2p10 and 5d5 of the krypton 86 atom." Such a definition has obviously no meaning when atoms did not exist.

from number of different (and sometimes ad hoc) hypotheses. We also open our subject to some of its consequences in other fields of physics. In fact, we consider eq.(1) as *a general law of nature* (Vigoureux et al., 1988) which also concerns other fields of physics such as special relativity, quantum theory or electromagnetism. Some of these ideas will be shortly open in our conclusion.

2. Friedmann equations

We briefly summarize here some well-known results for clarity and for their subsequent uses in this chapter.
Einstein's field equation which relates the geometry of space-time to the energy content of the universe can be written

$$R_{ij} - \frac{1}{2}R\,g_{ij} = 8\pi G\left(T_{ij} - \frac{\Lambda}{8\pi G}\,g_{ij}\right) \tag{2}$$

As is usual now, the cosmological term Λ has been moved from the left-hand side (curvature side) to the right-hand side of the Einstein equation and has thus been included inside the energy-momentum tensor term. This permits to interpret Λ as a part of the matter content of the universe rather than as a purely geometrical entity.
Taking into account the fact that on very large scale the universe is spatially homogeneous and isotropic to an excellent approximation (which implies that its metric takes the Robertson-Walker form) Einstein's equations reduce to the two Friedmann equations (a dot refers to a derivative with respect to the cosmic time t)

$$\frac{\dot{a}^2}{a^2} = \frac{8\pi G\rho}{3} - \frac{kc^2}{a^2} + \frac{\Lambda}{3} \tag{3}$$

and

$$\frac{\ddot{a}}{a} = -\frac{4\pi G}{3}(\rho + 3\frac{p}{c^2}) + \frac{\Lambda}{3} \tag{4}$$

where G, ρ and p are the gravitationnal constant, matter-energy density and fluid pressure respectively ; $a(t)$ is the cosmic scale factor characterizing the relative size of the spatial sections as a function of time. As usual, the curvature parameter k takes on values $-1, 0, +1$ for negatively curved, flat, and positive curved spatial sections (open, flat or closed universes) respectively. Note that the cosmological constant Λ will appear in what follows as a time-dependant function.
The energy conservation can be found by differentiation of eq.(3) and by using eq.(4). It can also be found by introducing Λ in the energy-momentum tensor and then using Einstein's field equation. We get

$$\frac{\dot{\Lambda}}{8\pi G} + \dot{\rho} = -3\left(\rho + \frac{p}{c^2}\right)\frac{\dot{a}}{a} \tag{5}$$

3. The solutions of the Friedmann equations

We solve here Friedmann's equations with the additional constraint (1) which expresses a restriction on usual variables characterizing the problem.
Using eq.(1), Friedmann equations (3) and eq.(4) become

$$\frac{\dot{a}^2}{a^2}(1 + k\alpha^2) = \frac{8\pi G\rho}{3} + \frac{\Lambda}{3} \tag{6}$$

$$0 = -\left(\rho + 3\frac{p}{c^2}\right) + \frac{\Lambda}{4\pi G} \tag{7}$$

These two above eqs.(6, 7) show that when taking $\Lambda \neq 0$, the linear variation of the scale factor $a(t) = ct/\alpha$ obtained from eq.(1), does not lead to an empty universe. Moreover, the fact that $\ddot{a}(t) = 0$ in the second one could appear inconsistant with observations. It will however be shown that observations which need the condition $\ddot{a}(t) \neq 0$ in the standard model can be explained without it when using eq.(1).
These equations can be solved in the most general case by using the equation of state parameter w of a perfect fluid:

$$p(t) = w\,\rho(t)\,c^2 \tag{8}$$

with w a constant ($w = \frac{1}{3}$ for the radiation dominated epoch and $w = 0$ in the case of an universe dominated by cold matter). Solving eq.(6,7) with (8) we obtain

$$\rho(t) = \frac{(1+k\alpha^2)\,c^2}{4\pi G(1+w)\alpha^2}\,\frac{1}{a(t)^2} \tag{9}$$

showing that the cosmic mass density varies with the reciprocal of the squared cosmic scale, and

$$\Lambda(t) = (1+3w)4\pi G\,\rho(t) = \frac{(1+3w)\,(1+k\alpha^2)\,c^2}{(1+w)\alpha^2}\,\frac{1}{a(t)^2} \tag{10}$$

Such a variation of $\rho(t)$ and of $\Lambda(t)$ with $a(t)^{-2}$ will be discussed at the end of this part. It comes from the presence of the term $\dot{\Lambda}$ in eq.(5). This can be seen by introducing eq.(10) into the left-hand side of eq.(5) which becomes

$$\frac{\dot{\Lambda}}{8\pi G} + \dot{\rho} = \frac{(1+3w)}{2}\dot{\rho} + \dot{\rho} = \frac{3}{2}(1+w)\dot{\rho} \tag{11}$$

so that the energy conservation becomes

$$\dot{\rho} = -2\rho\frac{\dot{a}}{a} \tag{12}$$

where the multiplicating factor 2 appears instead of 3.
Eq.(9) also gives (for a spherical universe):

$$M = \frac{4\pi}{3}a^3\rho = \frac{(1+k\alpha^2)c^2}{3G(1+w)\alpha^2}a(t) \overset{k=1,w=0}{=} \frac{c^2(1+\alpha^2)}{3G\,\alpha^2}a(t) \tag{13}$$

showing that the total mass of the universe scales with its cosmic radius (that unexpected result is discussed at the end of that part). Using that last equation, we note that

$$\frac{GM}{Rc^2} = \frac{(1+k\alpha^2)}{3(1+w)\alpha^2} \overset{k=1,w=0}{=} \frac{(1+\alpha^2)}{3\alpha^2} \tag{14}$$

which is a general expression of Mach's principle (Assis, 1994; Brans & Dicke, 1961) showing that our model can fulfil the principle of equivalence of rotation (Fahr & Heyl, 2006).
It is often useful to introduce the critical density ρ_c:

$$\rho_c = \frac{3H^2}{8\pi G} \overset{\text{eq.(1)}}{=} \frac{3c^2}{8\pi G\alpha^2 a^2} \tag{15}$$

and the density parameter Ω (we take the effects of a cosmological constant into account by including the vacuum energy density $\rho_\Lambda = \Lambda/8\pi G$ into the total density). We thus find, whatever may be the value of w

$$\Omega = \frac{\rho_{total}}{\rho_c} = \frac{\rho + \rho_\Lambda}{\rho_c} = (1 + k\alpha^2) \tag{16}$$

We thus find that the density ρ of the universe may be, as expected on the basis of number of recent observations, of the same order of the critical density ρ_c.
The expressions for Ω and Ω_Λ are

$$\Omega = \frac{\rho}{\rho_c} = \frac{(1 + k\alpha^2)}{3} \frac{2}{1+w} \tag{17}$$

$$\Omega_\Lambda = \frac{\rho_\Lambda}{\rho_c} = \frac{(1 + k\alpha^2)}{3} \frac{1+3w}{1+w} \tag{18}$$

Solving the three above results for Ω_Λ and Ω we obtain in the case of an universe dominated by cold matter ($w = 0$) and vacuum energy

$$\Omega = \frac{2}{3}(1 + k\alpha^2) \qquad \Omega_\Lambda = \frac{1}{3}(1 + k\alpha^2) \tag{19}$$

so that we get $(\Omega, \Omega_\Lambda) = (0.66(1 + k\alpha^2), 0.33(1 + k\alpha^2))$. This result gives $\Omega/\Omega_\Lambda = 2$ instead of the value $\Omega/\Omega_\Lambda = 1/2$ usually obtained from recent observations. However, it is to be emphasized, firstly, that this latter *numerical* result has not be obtained from direct measurements but from interpretations using explicitely the standard model, and secondly that it comes from explaining recent observations of type Ia supernovae in terms of an accelerating expansion of the universe which will appear as unnecessary in our model. It is worth recalling (an example will be given in the next part) that the same observations can lead to different numerical results when interpreted with different theories.
Discussion : the above results call two remarks:
- The first one concerns the variation of Λ with respect to time and, more precisely, its $a(t)^{-2}$ variation in eq. (10). In this connection, let us note that cosmologies with a time variable cosmological "constant" have been extensively discussed in the litterature (Dolgov, 1983; Ford, 1985; Ratra & Peebles, 1988) and that it has been shown that they not only lead to no conflict with existing observations (Riess et al., 2004) but also that they are suggested by recent observations (Axenides & Perivolaropoulos, 2002; Baryshev et al., 2001; Chernin et al., 2000; Overduin & Cooperstock, 1998) for example to solve the so-called coincidence problem. More precisely, the $a(t)^{-2}$ variation of Λ has been shown to be in conformity with quantum gravity by Chen and Wu (Chen & Wu, 1990) and consistent with the result of Özer (Özer & Taha, 1987) and other authors (Khadekar & Butey, 2009; Mukhopadhyay et al., 2011; Ray et al., 2011) who obtained it in different contexts (S. Ray, for example, consider $\Lambda \sim H^2$ leading thus, in our case (i.e. when using eq.(1)), to $\Lambda \sim a^{-2}$).
- The second remark deals with the variation of masses with $a(t)$. That result could appear surprising, but, as explained in (Fahr & Heyl, 2007), it has yet been emphasized as possibly true from completely different reasonings by many physicists (Dirac, 1937; Einstein, 1917; Fahr & Heyl, 2006; Fahr & Zoennchen, 2006; Hoyle, 1990; 1992; Whitrow, 1946). It moreover appears, on one hand, that a scaling of masses with the cosmic scale factor is the most natural scale required to make the theory of general relativity conformally scale-invariant (H. Weyl's

requirement) and, on the other hand, that it expresses a necessary condition to extend the equivalence principle with respect to rotating reference systems to the whole universe (Mach's principle). We do not discuss here the possible explanations for such a variation of masses with the scale factor. They are discussed in (Fahr & Heyl, 2007). It still remains that here is an important point to explore more deeply.

4. The flatness problem

The observable universe is close to a flat Friedmann universe in which the energy density ρ_M takes the critical value ρ_c ($\Omega_0 \sim 1$) and the homogeneous spatial surfaces are euclidean. That result is all the more surprising that the flat Friedmann model is unstable. In fact, small deviations from $\Omega = 1$ must quickly grow as time increases. The observation of $\Omega_0 \sim 1$ now therefore requires extreme fine-tuning of the cosmological initial conditions at the beginning of the universe. The question has thus been asked to know how Ω could have been so highly fine-tuned in the past.

A solution of this problem has been proposed in the context of inflationary scenarios. In these scenarios, k has not to vanish and ρ may not start out close to ρ_c, but there is an early period of rapid growth of the universe in which Ω rapidly approachs unity. In few words, the flatness problem is thus resolved from the fact that when a geometry is scale up by a great factor then it appears locally flat.

In this part, we show that when using eq.(1), the universe dispays the same evolution as a flat universe and *must appear to be flat whatever it may be* (spherical or not). Within our model, it is consequently not surprising to find it to be flat:

In the conventional cosmology, the Friedmann eq.(3) gives in the case of a flat universe ($k = 0$)

$$\frac{\dot{a}^2}{a^2} = \frac{8\pi G\rho}{3} + \frac{\Lambda}{3} \tag{20}$$

That equation is to be compared with eq.(6) we have obtained by using eq.(1) in eq.(3):

$$\frac{\dot{a}^2}{a^2} = \frac{8\pi G}{3}\frac{\rho}{1+k\alpha^2} + \frac{\Lambda}{3(1+k\alpha^2)}. \tag{21}$$

That eq.(21) *which describes both flat, closed or open universes following the value of* k is quite similar to eq.(20) which characterizes *a flat universe* in the standard cosmology. A comparison between these two equations thus shows that if eq.(1) is valid, the universe *must appear to be flat* whatever may be its geometric form (*whatever may be the value of* k) but with more or less matter than expected in the standard model following the value -1 or $+1$ of k since the density $\rho/(1+k\alpha^2)$ appears in eq.(21) instead of ρ in eq.(20).

These unexpected results can easily be verified: let us consider a flat universe with energy matter density ρ' and cosmological constant Λ'. Whatever may be ρ' and Λ', we may write their values $\rho' = \rho/(1+k\alpha^2)$ and $\Lambda' = \Lambda/(1+k\alpha^2)$ with $k = \pm 1$. Using then the Robertson-Walker metric of a *flat universe*

$$ds^2 = -c^2dt^2 + a(t)^2\left(dr^2 + r^2(d\theta^2 + sin^2\theta\, d\phi^2)\right) \tag{22}$$

and using $\rho/(1+k\alpha^2)$ and $\Lambda/(1+k\alpha^2)$ instead of ρ' and of Λ' respectively in the energy-momentum tensor of the Einstein equation, directly lead to eq.(21) which, using eq.(1), becomes:

$$\frac{\dot{a}^2}{a^2} = \frac{8\pi G\rho}{3} - \frac{kc^2}{a^2} + \frac{\Lambda}{3} \tag{23}$$

Although it has been obtained from equations describing a flat universe in usual cosmologies, we thus find the Friedmann equation which corresponds (when $k = \pm 1$) to *non flat universes*

5. The horizon and the smoothness problems

The horizon and the smoothness problems were identified in the 1970s. They point out that different regions of the universe which cannot have "contacted" each other due to the great distances between them, have nevertheless the same temperature and the same density to a high degree of accuracy (one part in one hundred thousand). Given the fact that the exchange of information or energy cannot take place at velocities greater than that of light such a result, which underlines the uncanny homogeneity of the universe across apparently causally disconnected regions, should not be possible. In the standard cosmology the problem is consequently to understand how the universe can be so smooth at large angular sizes, if different parts of it were never in contact or in communication [4]. That problem may have been answered by inflationary theory or by variable speed of light theory.

- Inflation provides the following explanation: before the inflationary area, the part of the universe that we can observe would have occupied a very tiny space and there would have been plenty of time for everything in this space to be homogeneized. However, it gives no clear explanations of why the universe would have then exponentially grown.
- The idea of varying speed of light cosmologies, as originally proposed by Moffat (Moffat, 1993) is that a higher propagation velocity for light in the cosmological past would have increased the propagation of causality so that all or most of the universe could thus have been causally connected.

In this part, we first show that, using eq.(1), the space-time of any observer is closed on itself so that there is no horizon problem. We then show that it is the same tiny part of the early universe that we see in every directions around us, so that it is quite logic to find the observed uniformity in terms of temperature and density of the cosmological microwaves background (CMB).

In the standard isotropic and homogeneous model of the universe, the Robertson-Waker metric may be written

$$ds^2 = -c^2dt^2 + a(t)^2\left(\frac{dr^2}{1-kr^2} + r^2d\Omega^2\right) = -c^2dt^2 + a(t)^2dl^2 \tag{24}$$

where t is the co-moving proper time and where $d\Omega^2 = d\theta^2 + sin^2\theta\, d\phi^2$ is the metric on a two-sphere. More generally, that equation can also be written

$$ds^2 = -c^2dt^2 + a(t)^2\left(d\chi^2 + \sigma^2(\chi)\, d\Omega^2\right) \tag{25}$$

where χ is the standard radial coordinate. In that equation, the three possible elementary topologies are defined by $\sigma(\chi) = \chi$ for a flat universe, $\sigma(\chi) = \sin\chi$ for a closed universe and $\sigma(\chi) = \sinh\chi$ for an open universe. Using the line element (25) the coordinate of the particle horizon is obtained by writing that the light we detect now at $t = t_0$ must have been emitted at

[4] In conventional cosmologies, the horizon at time of last scattering ($z \sim 1100 - 1500$) now substends an angle of order 1.5 degree. Therefore no physical influence could have smoothed out initial inhomogeneities and brought points at a redshif $z = 1100 - 1500$ that are separated by more than a few degrees to the same temperature and density.

the beginning of the universe ($t = 0$). Noting that the path of light is given by setting $ds^2 = 0$ and taking light rays travelling in the radial direction, eq.(25) gives for the particle horizon

$$ds^2 = -c^2dt^2 + a(t)^2d\chi^2 = 0 \overset{\text{eq.(1)}}{\Longrightarrow} \chi_H = \pm \int_0^{t_0} \frac{cdt}{a(t)} \overset{\text{eq.(1)}}{=} \pm\alpha \int_0^{t_0} \frac{dt}{t} = \pm\infty \qquad (26)$$

The integral does not converge and it can easily be shown that there is no particle horizon whatever may be the geometry of the universe ($k = -1, 0, +1$). Our model is thus horizon-free and allows the interactions to eventually homogenize the whole universe. Moreover, in the case of a spherical universe, it implies that the our "antipodes" can be seen by us now.
Our model could thus explain the observed uniformity in terms of temperature and density of the cosmological microwaves background radiation (CMB) without needing an inflationary expansion or a varying speed of light hypothesis. However, although it has no particle horizon so that all space points could have undergone physical interactions with each others, it shows that *the observed homogeneity does not come from such causal interactions, but from the fact that it is the same "tiny part" of the primitive universe that we see in any direction around us*:
Let us consider the case of *a spherical universe* ($k = +1$). Because of the symmetry, the rays that correspond to photons' world lines can be chosen so that $d\phi = d\theta = 0$. Solving then eq.(25) for light ($ds^2 = 0$) with these conditions and using eq.(1) give the radial coordinate χ as a function of time

$$ds^2 = -c^2dt^2 + a(t)^2d\chi^2 = 0 \overset{\text{eq.(1)}}{\Longrightarrow} \chi(t) = -\alpha \ln \frac{a(t)}{a(t_0)} = -\alpha \ln \frac{t}{t_0} \qquad (27)$$

where we take for the initial condition $\chi = 0$ at the present value $t = t_0$ of the cosmic time and where χ increases toward the past. Eq.(27) shows that when using eq.(1), the space-time of any observer is closed on itself at early times defined by $\chi(t) = n\pi$. The first of these (our spatio-temporal antipode which is defined as the point where the radial coordinate $\chi(t)$ takes the value π), is denoted A on fig. 1.
- Since it can be seen identically in any direction around us, it can reasonably be identified to the source of the cosmic microwaves background radiation (CMB).
- Since it is then *the same "tiny part"* of the early universe that we can observe *in any directions around us* (the cosmic microwave background radiation arriving at the earth from all directions in the sky *does come* from the *same* tiny part of the early universe), *it is not surprising to observe a very high uniformity in terms of temperature and of density of the CMB.* [5]
- Neither inflation nor other hypotheses are consequently required to explain the high isotropy of the *CMB*.
All these results are shown on fig. 1, on which the logarithmic spiral (eq.27) corresponds to our past light cone (present observers are at point O). The point A represents our spatiotemporal antipode and thus corresponds to that "tiny part" of the universe that we observe in any directions around us (the "source" of the CMB).

[5] To give a simple example, consider we are on the north Pole of the Earth and that light must propagate by following Earth's surface. Looking at the farthest point of us, we would see the same point of the south Pole all around us and our background would then appear surprisingly homogeneous.

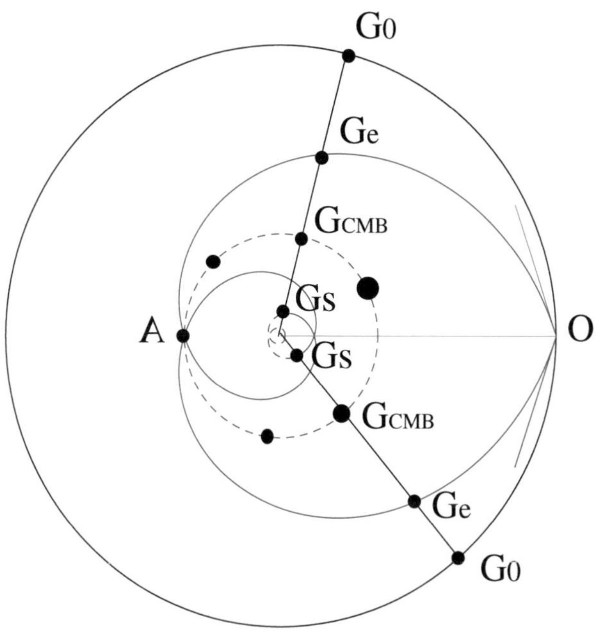

Fig. 1. We consider a spherical universe. The circle of radius $R(t_0)$ represents the universe at time t_0. The logarithmic spiral corresponds to the past light cone of the observer O, that is, to trajectories of all the light rays that he/she receives at $t = t_0$. The point A can be seen in any direction around O. It can thus be identified to the "source" of the CMB. The dashed circle corresponds to the universe at time t_{CMB} when the CMB was formed. A represents only a very tiny part of the universe at that time so that, *at that time*, the seeds of galaxies we observe now (points G_e) were not at A, but here and there on that dashed circle. They are *symbolically* illustrated by grey circles on the dashed circle. Note that they have not the same size. In fact at t_{CMB} the universe did not need to be homogeneous (and was certainly not) so that the seeds of these galaxies at that time could be quite different the ones from the others. The two radius are the world lines of two galaxies: G_{CMB} are galaxies (or their seeds) at time t_{CMB}; G_e gives their positions at the time t_e they emitted the light we receive now at t_0; G_0 are their current (and unknown) positions now. Because of the spiral form of the light cone, it could *theoretically* be possible (if universe was not opaque before t_{CMB}) to see *behind galaxies* G_e we see, their earlier seeds (the points G_s near the big-bang, on the galaxy world lines and on the light cone of O). However, no *radiation* coming from "before the last scattering surface" can be "visible" now *by definition*. May be other "isotropic points" $\chi = n\pi$ with $n > 1$ could be the "source" of isotropic cosmic particles backgrounds.

To be clear, and to show that we do see the same "tiny part A" of the universe in every direction we look, let us note that the spatial volume enclosed between the coordinate hyperspheres of radius $\chi_0 - \Delta\chi_0$ and χ_0 is

$$\Delta V_{\chi_0} = \int_{\chi_0-\Delta\chi_0}^{\chi_0} \int_0^{\pi} \int_0^{2\pi} (a_0 e^{-\frac{\chi}{\alpha}})^3 \sin^2\chi \, \sin\theta \, d\chi \, d\theta \, d\phi \tag{28}$$

Making the change of variable $\chi \rightarrow -\alpha \ln \frac{t}{t_0}$ (eq.(27)), that expression becomes

$$\Delta V_t = \int_{t-\Delta t}^{t} \int_0^{\pi} \int_0^{2\pi} \frac{c^3 t^2}{\alpha^2} \sin^2(-\alpha \ln \frac{t}{t_0}) \sin\theta \, dt \, d\theta \, d\phi \qquad (29)$$

It expresses the value of the spatial volume of the observed universe corresponding to past times between $t - \Delta t$ and t. Its expression being not simple, we only present its variation with respect to t on fig. 2. That figure shows that the farther back we look in the past, the smaller ΔV is, or, in other words, that the volume of the universe we progressively add to our observed universe when looking farther and farther tends toward 0 when t tends toward t_{CMB}.
Integrating eq.(29) over the past history of the universe, from t_{CMB} up to the present, we find the apparent volume V_{app} of the universe (the volume which is seen). Taking then $\alpha = 1$ or 0.3 (see at the end of that paragraph) this volume is only few percents of the universe at the present time t_0.

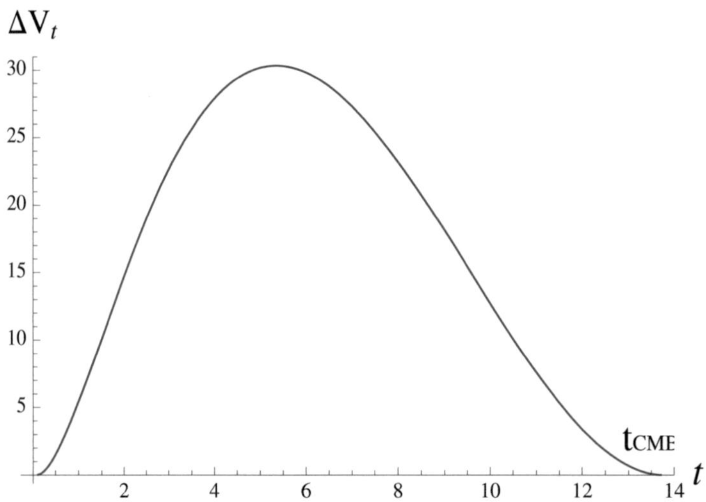

Fig. 2. ΔV_t (arbitrary units) versus time t (in billion years) : ΔV_t represents the volume of the universe we can observe now corresponding to past cosmic times between $t - \Delta t$ and t. We see that when time tends toward t_{CMB} (at about 13.7 Gy) that volume tends toward 0. In other words, as we look back in time, the spatial part of the observed volume of the universe that corresponds to times between $t - \Delta t$ and t, spreads out, then reaches a maximum and then starts to decrease to be all the more small that we approach t_{CMB}. That figure has been drawn by taking $c = 1$, $\alpha = 0.35$ and $\Delta t = 100$ million years. Time increases from $t_0 = 0$ (present time) to 14 billion years in the past.

It can be added that the identification of A with the source of the CMB *could* permit to calculate the value of α in eq.(1). Using both the right-hand part of eq.(27) with $\chi = \pi$ and taking the usual value given by nucleosynthesis for t_{CMB} (the radiation was created when atoms formed at around 360 000 years after the big-bang) thus would give $\alpha \sim 0.3$. With such a value, the theoretical value of Ω (16) would be $\Omega \sim 1.1$.

6. The cosmic microwaves background radiation and the small scale homogeneity

A related comment concerns the problem of the small-scale inhomogeneities needed to produce astronomical structures that are now observed. Cosmologists are usually searching in fluctuations of the *CMB* the density fluctuations that led to galaxies clusters and giant voids. In this context, the uniformity of the CMB leads to another problem of the standard cosmology: if the universe was so smooth, then how did anything form ? There must have been some bumps in the early universe that could grow to create the structures (galaxies and clusters of galaxies) we see locally. This problem no more exists when using eq.(1).

In our model, the small fluctuations that we observe now in the CMB *are not those which gave birth to the structures of the universe we can observe*. In fact, as shown on fig. 1, the galaxies which emitted the light we receive at $t = t_0$ were not at A at time $t_A = t_{CMB}$ (and consequently *their seeds were not in the CMB we observe*) but on the circle of radius $ct_A = ct_{CMB}$ which represents the universe at time t_{CMB}. In that light, the uniformity of the CMB not only is obvious (since it is the same tiny part of the universe that we see in any direction we look), but also it does not pose any problem to understand the cosmic structures we observe now. In fact nothing imposes that inhomogeneities of the universe at that time (that is on the dashed circle in fig. 1) be so small as thoses observed in its very tiny part A (that is to say in the CMB). We cannot know others regions (other than A) of the circle of radius $ct_A = ct_{CMB}$ and they, in fact, may be have overdense parts. Of course, it remains that studying the small inhomogeneities of the microwaves background may be useful to understand the past history of the universe.

We can note that it could *theoretically* be possible (if the universe was not opaque before t_{CMB}) to observe the seeds G_s (G_s for G_{seed}) which gave birth to galaxies and cosmic structures. The two images would then be observed the one behind the other (see fig. 1: behind galaxy G_e, and beyond the point A, the black points G_s are simultaneously on the world line of G_e and in our light cone).

We can also note that others points defined by $\chi = n\pi$ with $n > 1$ (n integer) are also "isotropic points" which could be "seen" as a homogeneous background in all directions around us (as does the *CMB*). Of course, they cannot correspond to light sources since the universe was *by definition* opaque before the "last scattering time". However they may correspond to sources of isotropic cosmic particles backgrounds.

Remark: These above results can be illustrated by mapping the 3-spatial sphere onto a 3-dimensional hyperplane by a 3-dimensional stereographic projection. Restricting ourselves to the spherical case (k=1) and using $\sigma(\chi) = \sin\chi$ in eq.(25) gives

$$ds^2 = -c^2dt^2 + a(t)^2(d\chi^2 + \sin^2\chi\, d\Omega^2) \tag{30}$$

Making then the change of variable $\mathcal{R} = 2\tan(\chi/2)$ we get the metric on the 3-hyperplane

$$ds^2 = -c^2dt^2 + \frac{a(t)^2}{(1+\frac{\mathcal{R}^2}{4})^2}(d\mathcal{R}^2 + \mathcal{R}^2 d\Omega^2) \tag{31}$$

Using it, it is straigthforward to show that all points at infinity are the image of the same antipodal point on S^3 so that we can understand that it is really the same point we see in all directions around us when looking at the CMB. Such a stereographic projection sends meridians of the 3-sphere (light world lines that do pass through the place of the observer) to straight lines on the hyperplane making their way toward the observer. Apart from a change of scale when looking increasingly far, the 3-hyperplane consequently corresponds more closely to the universe which is seen by each of us.

Also note that, whatever may be the value of k, using eq.(1) transforms the above metric (25) into a flat spacetime metric admitting Minkowski coordinates: writing $a = ct/\alpha$ from eq.(1) and $t = t_0 e^{u/\alpha}$ as suggested by eq.(27) gives $dt/t = du/\alpha$ so that eq.(25) gives

$$ds^2 = \frac{c^2 t_0^2}{\alpha^2} e^{2u/\alpha}(-du^2 + d\chi^2 + \sigma^2(\chi)\, d\Omega^2) = a(t)^2(-du^2 + d\chi^2 + \sigma^2(\chi)\, d\Omega^2) \qquad (32)$$

where the term $a(t)^2 = c^2 t^2/\alpha^2$ represents the factor by which the scale changes in different locations. Using the conformal time u ($u = \int \frac{cdt}{a} = \alpha \int \frac{dt}{t}$) has thus the advantage of leading to a "conformally flat" metric.

7. Apparent luminosity and observation of type Ia supernovae

Following pioneering works related in (Norgaard-Nielsen et al., 1989), recent observations of type Ia supernovae (Perlmutter et al., 1999; Riess et al., 1998; 2004; Schwarzschild, 2004; Tonry et al., 2003; Wang et al., 2003) have provided a robust extension of the Hubble diagram to $1 < z < 1.8$. These results have shown that observations *cannot be fitted by using the usual distance modulus expression with* $\Lambda = 0$ *both for* $z < 1$ *and for* $z > 1$. To fit new data points at redshift 1.755 the standard model thus needs to consider that the expansion of the universe is accelerating, an effect that is generally attributed to the existence of an hypothetic "dark energy".

In that part, we show that eq.(1) leads to another expression for the distance-moduli which can fit all the data without needing for an acceleration of the expansion (fig. 3).

Distances are measured in terms of the "distance modulus" $\mu = m - M$ where m is the apparent magnitude of the source and M its absolute magnitude. The standard expression for the distance-moduli with respect to z can be found in (Tonry et al., 2003; Weinberg, 1972). Our aim here is to calculate μ in our model:

let an object be at cosmic radial coordinate χ and consider that the light that it emitted at cosmic time t_e is just reaching us at time t_0. The luminosity distance d_L of the object can be expressed as (Weinberg, 1972)

$$d_L = (\frac{a^2(t_o)}{a(t_e)})\chi = a(t_o)(1+z)\chi \qquad (33)$$

Using eq.(1) and noting H_0 the Hubble constant at the present time, that expression becomes

$$d_L = (\frac{c}{\alpha H_0})(1+z)\chi. \qquad (34)$$

χ can be obtained from calculations similar to that of eq.(26):

$$\chi = \int_{t_e}^{t_o} c\frac{dt}{a(t)} \overset{\text{eq.(1)}}{=} \alpha \ln(\frac{a(t_o)}{a(t_e)}) = \alpha \ln(1+z) \qquad (35)$$

so that

$$d_L = \frac{c}{H_0}(1+z)\ln(1+z) \qquad (36)$$

Expressing the distance modulus μ in terms of d_L then gives

$$\mu = 25 + 5\log d_L = 25 + 5\log(\frac{c}{H(t_0)}) + 5\log(1+z) + 5\log\ln(1+z) \qquad (37)$$

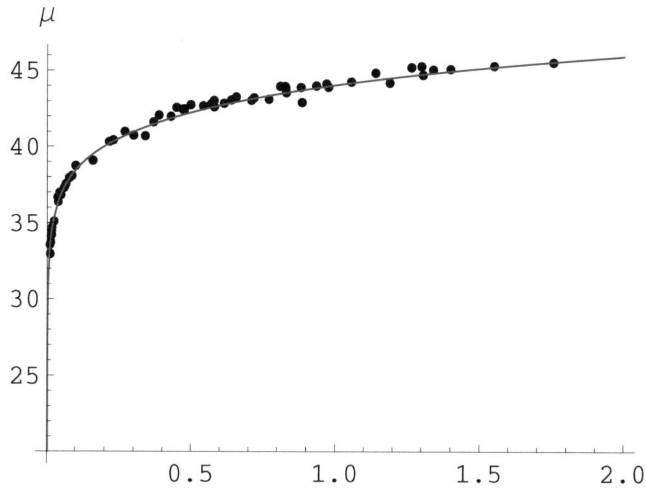

Fig. 3. Distance modulus μ vs redshift z in our model. The data points are taken from table 5 of the High-z Supernova Search Team (Riess et al., 2004). Whereas conventional cosmologies fail to fit all experimental data *both for* $z < 1$ *and for* $z > 1$, this is possible when using eq.(1). The full line, which represents predictions of the present model (eq.37), has been drawn by using $H_0 = 68\ km.s^{-1}.Mpc^{-1}$ (note a typewriting error in (Vigoureux et al., 2008) where we wrote $H_0 = 58\ km.s^{-1}.Mpc^{-1}$).

where c is in km.s^{-1} and H in km.s^{-1}.Mpc^{-1}.
The variation of μ with respect to z is shown on fig 3 . Fig. 3 has been obtained by using the value $H(t_0) = 68\,km/sec/Mpc$ which agrees well with usual determinations of the Hubble constant ($H(t_0) = 73 \pm 4km/sec/Mpc$). It shows that eq.(37) can permit to fit all experimental values in the whole range $z < 1$ and for $z > 1$ without any other hypothesis. The use of eq.(1) thus succeeds in explaining all the data without having to consider an acceleration of the expansion of the universe. To be clear, whereas in the standard model observations of type Ia supernovae lead to give the deceleration parameter q a value close to -0.5 for today and close to $+0.5$ for very high redshifts, we are able to explain all these observations by taking $q = 0$ at all times, as required by eq.(1).
Noting that different fitting of experimental points gives 63 < H < 70 at the present time and that eq.(1) leads to a scale factor proportional to time (and thus to $H(t) = 1/t$) the age of the universe in our model is about 14 billion years.
Remark: the above calculation uses the usual relation $a(t_0)/a(t_e) = 1 + z$ where $a(t_e)$ is the scale factor at the time of emission and where $a(t_0)$ is the scale factor at the time of observation. The redshift z undergone by radiation from a comoving object as it travels to us today is thus related to the scale factor at which it was emitted. It can easily be shown that this relation is still valid in our model and that it may consequently be used in calculations leading to eq.(37): using the Robertson-Walker metric (25), consider ligth reaching us (at $\chi = 0$) at the present time t_0 and emitted by a galaxy at a distant position $\chi = \chi_e$ and at a time t_e. Two crests arriving at t_0 and $t_0 + \Delta t_0$ were emitted at $t_e + \Delta t_e$. Since light has travelled radially inwards along a null geodesic, we get

$$c \int_{t_e}^{t_0} \frac{dt}{a(t)} = -\int_0^{\chi_e} d\chi = c \int_{t_e+\Delta t_e}^{t_0+\Delta t_0} \frac{dt}{a(t)} \tag{38}$$

Over the period of one cycle of a light wave, the scale factor is essentially a constant. This yields $\Delta t_e/a(t_e) = \Delta t_0/a(t_0)$. Now, the observed and emitted wavelength λ_0 and λ_e are related to Δt_0 and Δt_e by $\lambda_{e,0} = c\Delta t_{e,0}$ so that the cosmological redshift $z = (\lambda_0 - \lambda_e)/\lambda_e = a_0/a_e - 1$ takes it usual expression and its use is consequently valid in the above calculation.

8. The cosmological constant and the cosmic coincidence problem

In the standard model, the cosmological constant has been introduced to account for anomalies observed in cosmological data and especially for explaining supernovae observations (Carroll, 2001). That introduction rises a new cosmological problem which is to explain the so-called *cosmic coincidence problem*, that is to understand why ρ_Λ (the dark energy density) is not only small but also, as current type Ia supernovae observations indicate, of the same order of magnitude as the present mass density ρ_M of the universe.

In fact, in usual models, the mass density ρ_M changes with time whereas the vacuum energy is constant. These two energy densities have thus evolved differently throughout the history of the universe and it is consequently very hard to explain why ρ_M and ρ_Λ would coincide today. Such a coincidence would require that the early universe had been very fine-tuned (Henttunen et al., 2006) but the underlaying models of particle physics cannot provide a natural explanation to the necessity of a so carefully fine-tuning.

That problem can be solved, arguably at least, by the anthropic principle argument. There are however other potential solutions based on physical arguments alone.

The most common is to consider that ρ_Λ really is not a constant. Peebles and Ratra, for example, (Peebles & Ratra, 1988; Ratra & Peebles, 1988) have thus considered a model in which the vacuum energy depends on a scalar field that changes as the universe expands. The vacuum is then treated as a form of matter and the cosmological constant thus turns out to be a measure of the energy density of the vacuum.

The quintessence model has then been proposed. It consists in a slowly varying energy component with a negative equation of state. That "dark energy" associated with the scalar field slowly evolves down its potential according to an attractor-like solution of the equation of motion, *regardless of the initial conditions* and can thus resolve the coincidence problem.

However the proposed solutions cannot satisfy *exactly* the *necessary* conditions $p_\Lambda = -\rho_\Lambda c^2$ and $\rho_\Lambda(t) \sim a(t)^{-n}$ with $n \neq 0$. They consequently cannot exactly generate the cosmological constant.

In the above part, we have shown that we do not need introducing a cosmological constant in order to explain type Ia supernovae observations. As explained just under eq.(7), we however need it to satisfy the Friedmann equations when adding them the additionnal constraint (1). Our aim in that part is to show that in the model we propose, we find not only that vacuum can *exactly* verify the condition $p_\Lambda = -\rho_\Lambda c^2$ but also that ρ_Λ and ρ_M have the same order of magnitude *at all times*.

To explain the origin of the cosmological constant, let us consider a quintessence fluid the density and the pressure of which (denoted ρ_Λ and p_Λ) being thus to be included in the Friedmann's equations. Assuming, as is usual, that the equation of state of that fluid has the form

$$p_\Lambda = \gamma \rho_\Lambda c^2 \tag{39}$$

where the constant γ, which has to be determined, must be negative to get an anti-gravity. The cosmological constant can thus be written

$$\Lambda = 8\pi G \rho_\Lambda \tag{40}$$

in eq.(3), and

$$\Lambda = -4\pi G \left(\rho_\Lambda + 3\frac{p_\Lambda}{c^2}\right) = -4\pi G \rho_\Lambda (1+3\gamma) \tag{41}$$

in eq.(4). Of course, these two expressions for Λ must be equal so that the two Friedmann equations are coherent (and consequently the quintessence fluid can generate *exactly* the cosmological constant Λ) *if and only if*

$$8\pi G \rho_\Lambda = -4\pi G \rho_\Lambda (1+3\gamma) \quad \Rightarrow \quad \gamma = -1 \tag{42}$$

or, by inserting this result inside eq.(39), *if and only if*

$$p_\Lambda = -\rho_\Lambda c^2 \tag{43}$$

So, the value $\gamma = -1$ is that which must be found. Apart from that "coherence reason", two other reasons can be considered in support of it: first, observations of supernovae indicate that $\gamma = -1.02^{+0.13}_{-0.19}$ (Riess et al., 2004); second, the value $\gamma = -1$ is a necessary and sufficient condition for the energy-momentum tensor of the vacuum to be Lorentz invariant [6] (see for example (Jordan, 2005)).
In that part we first show that the standard model cannot satisfy *exactly* that value and consequently that it cannot exactly generate the cosmological constant. We then show that the present model can generate it:
Let us first consider the conventional model ($\dot{a} \neq Cst$). Introducing eq.(40) and eq.(41) into Friedmann equations (3) and eq.(4) respectively gives

$$\frac{\dot{a}^2}{a^2} = \frac{8\pi G(\rho + \rho_\Lambda)}{3} - \frac{kc^2}{a^2} \tag{44}$$

$$\frac{\ddot{a}}{a} = -\frac{4\pi G}{3}\left(\rho(1+3w) + \rho_\Lambda(1+3\gamma)\right) \tag{45}$$

Derivating eq.(44) and inserting eq.(45) into the result then leads to

$$-3\frac{\dot{a}}{a}\left(\rho(1+w) + \rho_\Lambda(1+\gamma)\right) = \dot{\rho} + \dot{\rho}_\Lambda \tag{46}$$

the solution of which for ρ_Λ is

$$\rho_\Lambda \propto \frac{1}{a^{3(\gamma+1)}} \tag{47}$$

Introducing the coherence condition $\gamma = -1$ (eq.42) into eq.(47) then leads to $\rho_\Lambda = Cst$, and to $\Lambda = Cst$. These results make Λ to be a pure constant but in that case the quintessence fluid does not dilute when the universe expands. The key problem then remains to explain the cosmic coincidence: if ρ_Λ is constant whereas ρ_M varies, why these two quantities should be comparable today ? This shows, that, in the usual cosmology
- if the cosmic fluid can generate *exactly* the cosmological constant ($\gamma = -1$ *exactly*), then $\rho_\Lambda = Cst$ and consequently the standard model cannot explain the cosmic coincidence, and
- if the standard model want to explain the cosmic coincidence (ρ_Λ does vary with respect

[6] The vacuum must be Lorentz invariant or one would have a preferred frame. The stress-energy tensor of the vacuum is diagonal and this tensor must be invariant. The only Lorentz invariant nonzero rank tensor is the metric diag$(-1,1,1,1)$ in a local inertial frame so if the vacuum energy density is non-zero the pressure has to be $-\rho c^2$.

to $a(t)$), then, eq.(47) shows that ($\gamma \neq -1$) and consequently it cannot *exactly* generate the cosmological constant.
In other words, in usual theories, the condition $\gamma = -1$ is not compatible with the other condition $\rho_\Lambda \sim a(t)^{-n}$ ($n \neq 0$) and thus provides no answer to the fine-tuning problem.
Contrarily to what is found with these theories, the two conditions $\gamma = -1$ (or $p_\Lambda = -\rho_\Lambda c^2$) and $\rho_\Lambda \propto R^{-n}$ with $n \neq 0$ can be *simultaneously* fulfilled when using eq.(1):
When using eq.(1) and eqs.(40, 41), the two Friedmann's equations (6) and (7) can be written:

$$\left(\frac{c}{\alpha a}\right)^2 = \frac{8\pi G}{3}\frac{\rho_M + \rho_\Lambda}{1+k\alpha^2} \tag{48}$$

$$0 = \rho_M(1+3w) + \rho_\Lambda(1+3\gamma) \tag{49}$$

It is obvious that these two equations do have solutions even when $\gamma = -1$. They are

$$\rho_M = \frac{c^2}{8\pi G}\frac{(1+k\alpha^2)}{\alpha^2}\frac{2}{1+w}\frac{1}{a^2} \tag{50}$$

$$\rho_\Lambda = \frac{c^2}{8\pi G}\frac{(1+k\alpha^2)}{\alpha^2}\frac{1+3w}{1+w}\frac{1}{a^2} \tag{51}$$

As discussed in section 3, such a variation of ρ_Λ and of the cosmological "constant" term as a^{-2} has been shown to lead to no conflict with existing observations (Riess et al., 2004) and to be in conformity with quantum cosmology (Chen & Wu, 1990).
We thus have

$$\rho \propto \frac{1}{a^2} \qquad \rho_\Lambda \propto \frac{1}{a^2} \qquad \Lambda \propto \frac{1}{a^2} \tag{52}$$

whatever may be the equation of state of the cosmic fluid. Contrarily to what is obtained in the standard cosmology, the present model thus do fulfil the two conditions $\gamma = -1$ and $\rho_\Lambda \propto a^{-n}$ (with $n \neq 0$) simultaneously. It can consequently explain the origin of the cosmological constant with a quintessence fluid which dilutes when the universe expands. It can also solve the problem of the "cosmic coincidence": in this model, the "cosmological constant" in fact varies in the same way as ρ_M and has always been comparable to it. Since the two fluids dilute in the same way and evolve together, it is not suprising to find that they can coincide now.
Moreover, the above equations (50, 51) also show
- that the two energy densities ρ_M and ρ_Λ are exactly equal when $w = \frac{1}{3}$ that is to say in the radiation dominated epoch.
- that $\rho_M = 2\rho_\Lambda$ in the matter epoch.
Let us also recall that eq.(1) can also explain why the mass density of the cosmic fluid is so near the critical density ρ_c: using eq.(15) in fact gives

$$\rho_\Lambda = \rho_c\frac{(1+k\alpha^2)}{3}\frac{1+3w}{1+w} \overset{w=0}{=} \rho_c\frac{(1+k\alpha^2)}{3} \tag{53}$$

$$\rho_M = \rho_c\frac{(1+k\alpha^2)}{3}\frac{2}{1+w} \overset{w=0}{=} \rho_c\frac{2(1+k\alpha^2)}{3} \tag{54}$$

so that

$$\rho_\Lambda \sim \rho_c \qquad \rho_{total} = \rho_c(1+k\alpha^2) \tag{55}$$

9. Conclusions

We advocate the possibility that the universal relations existing between space and time in the so-called "speed of light" and in the expansion of the universe are two aspects of a same phenomenon:
Introducing eq.(1) as an additionnal constraint to solve the Friedmann equations leads to interesting ways to explain number of unanswered problems of the standard cosmology without needing usual hypotheses as, for example, the present accelerating expansion of the universe or the inflation scenario which assumes that the universe went through an early period of exponential growth without worrying about how this came about.
We have shown how eq.(1) can solve the flatness and the horizon problems, the problem of the observed uniformity in term of temperature and density of the cosmological background radiation, the small-scale inhomogeneity problem (with the one of the seeds of galaxies and of cosmic structures) and the cosmic coincidence problem. Reconsidering the Hubble diagram of distance moduli and redshifts as obtained from recent observations of type Ia supernovae, we have also shown that all the new data can be understood without needing an accelerating universe.
Whereas a cosmological constant is useless in the present model to explain such observations, we however need it for coherence in Friedmann's equations. Concerning that point, one appealing feature of our results is that eq.(1) permits to accommodate simultaneously the equation of state $p_\Lambda = -\rho_\Lambda c^2$ of the quintessence fluid which generates the cosmological constant Λ (so that it can *perfectly generate* the cosmological constant), with a varying density $\rho_\Lambda \propto a^{-n}$ (with $n = 2$ in our case) which appears to be a necessary condition to avoid the cosmic coincidence problem.
The present model also explains why ρ, ρ_c and ρ_Λ are comparable today. At this point, let us recall (Vigoureux et al., 2008) that, with eq.(1), a spherical universe (for example) displays the same evolution as a flat universe in the standard model (section 4).
One of our results may however appear unnatural: the total mass M of the universe would scale with $a(t)$. Although such a variation has been shown to be the most natural one to extend the equivalence principle with respect to rotating reference frame to the whole universe (Mach's principle); although it appears to be the most natural scale to fulfil the Weyl's requirement of conformally scale invariance; although it has also been emphasized as possibly true by physicists as Dirac, Einstein or Hoyle as discussed in (Fahr & Heyl, 2007), it however remains to be carefully studied.
Eq.(1) may provide an alternative way to solve the standard cosmological problems and our results appear compatible with astronomical observations. It leads however to some numerical values which may seem contradict with some of these (for example, concerning the proportion of ordinary matter and of black matter, we find $(\Omega_M, \Omega_\Lambda) = (0.66, 0.33)$ when usual experiments would rather give $(\Omega_M, \Omega_\Lambda) = (0.3, 0.7)$). However, one has to be careful before concluding such a question: as liked to recall Einstein, theory and observations are interdependent and there are no observation which can be directly interpretable without referring to a given theory. To be able to construct a picture of the world, we must interpret the observational data *within a given theory* and we may occasionally forget that we use theories all the time while we may think of us as giving observational results independently of any theory. Because of this, our results cannot be too quickly compared with numerical values *deduced from* the standard big-bang cosmology. An example of this is given by looking at eqs.(20, 21) showing that a flat universe corresponding to a given value of the energy density of matter ρ in usual cosmology, may correspond to a spherical universe with another density $\rho/(1 + k\alpha^2)$

in our model. Another example is given by the interpretations of observations of type Ia supernovae: the values $q \sim -0.5$ for today and $q \sim +0.5$ obtained in the standard model for very high redshifts are not independant of any theory. On the contrary, they correspond to the values that must be used *inside the usual theory* to explain observations. As explained above, *in our model*, these same observations lead to the quite different value $q = 0$ at all times.

Eq.(1) can thus solve usual problems of cosmology. An important remark about this is that these latter have been solved by using one single hypothesis. It is in fact to be emphasized that *all our results have been obtained from the only hypothesis that the speed of light is related to the expansion of the universe.* An important feature of eq.(1) is thus *its unifying power*. Eq.(1) gives unity to number of results which, for some of them, have yet been obtained by other authors by introducing many quite different, and sometimes *ad hoc*, hypotheses.

In order to illustrate the importance of such an unifying power of our proposition, let us present a brief outline of some of the wide variety of hypotheses which have yet been used to solve one or the other problem:

The variation of ρ and Λ as a^{-2} in our equations (9, 10), has yet been obtained from some very general arguments in line with quantum cosmology and with dimensional considerations (Chen & Wu, 1990) or by postulating the invariance of equations under a change of scale (Canuto et al., 1977). It has also been directly postulated to explore its consequences as did, for example, Berman (Berman, 1991) who made the hypothesis that $\Lambda(t) = Bt^{-2}$ and $\rho(t) = At^{-2}$ (leading then to some of our results). Others (Lima & Carvalho, 1994; Mukhopadhyay et al., 2011) consider the phenomenological assumption $\Lambda \sim H^2$ (Overduin & Cooperstock, 1998). Fahr and Heyl (Fahr & Heyl, 2007) also make the assumption that the total mass density of the universe (matter and vacuum) scales with a^{-2} and find the relation $c = \dot{a}(t)$ in the particular case $k = 0$. They then show that such a scaling abolishes the horizon problem and that the cosmic vacuum energy density can then be reconcilied with its theoretical expected value. Others postulated the Mach's principle or, as did Özer (Özer & Taha, 1987), make the assumption that the equality $\rho_M = \rho_c$ is a time-independant feature of the universe from which they deduce $\Lambda \sim a(t)^{-2}$. Similarly it has also been postulated the ratio ρ_Λ/ρ_M to be constant in time (Freese et al., 1987)... In a similar way Bacinich and Kriz (Bacinich & Kriz, 1995) found the same logarithmic spiral form of the light cone from the quite different consideration of a local conservation of the CMB flux...

Eq.(1) may not only unify different results which can have been proposed from number of different hypotheses, but it may also illustrate and unify different questions about light (see the introduction). It may thus interest other fields of physics such as special relativity, quantum theory or electromagnetism.

In its light

- the energy $E = m_0c^2$ of a given rest mass m_0 can be seen as originating from the expansion of the universe: it would in fact correspond to a form of "comoving kinetic energy" of any comoving object carried away by the expansion of the universe ($E = m_0c^2 = \alpha^2 m_0 \dot{a}^2$);
- by connecting the light phenomenon (and more generally electromagnetic radiations) to the expansion of the universe, eq.(1) also *illustrates* the assumption that the speed of electromagnetic radiations is indifferent to both its emitter and its absorber and that it can be neither compounded with that of an object nor transformed away by the choice of a suitable reference frame. This independance of place (homogeneity), direction (isotropy), source and detector motions can be understood when connecting c to the expansion of the universe. It can thus be *illustrated* by imagining an insect moving on an expanding balloon: the velocity of the insect is obviously independant of that of the ballon expansion and it is not because the

insect would go faster or slower that the balloon would expand differently.
In these views, an essential feature of eq.(1) is perhaps to suggest a cosmic interpretation of light phenomena which would thus *essentially* appear essentially as a consequence of the expansion of the universe rather than as a propagating phenomenon.
The expansion of the universe in fact induces two kinds of change in the universe : a growth of its radius (cdt) and a growth of its circumference ($dx = ad\chi$), the second being a consequence of the first. Both are equivalent so that $cdt = ad\chi = dx$. That equivalence makes the expansion to appear in space although it is *essentially* a time phenomenon. In the same way, light appears to propagate into space although its 4-velocity (c, 0, 0, 0) clearly expresses *its temporal nature*. In other words, light, and electromagnetic phenomena, are carried by the time axis (the radius of the universe) but, because of the expansion, they *appear* to propagate *into* space (and so they appear "diagonally" in space-time diagrams). To be clear, consider a comoving point A in the expanding universe. Because of the expansion, although it has no dynamical motion, its relation to us *in our ligth cone* is expressed by the time extension of the distance $D = \int a(t)d\chi = \int cdt = c\Delta t$ so that its instant relation to us *appears* to propagate at velocity $dD/dt = c$ whatever may be its comoving coordinates.
This may perhaps throw light on current and fondamental problems that are the time symmetry of Maxwell's equations, the emission theory or other problems in quantum theory where considering light as a *propagating* phenomenon often leads to paradoxes.
The complete time symmetry of Maxwell's equations (whereas the observed electromagnetic phenomena are *asymmetric* with respect to time) in fact tells us that electromagnetic interaction proceeds not only forward in time (from the emitter to the detector), but also backwards in time (from the detector to the emitter). In practice, retarded fields are selected because they appear to correspond to reality, whereas advanced fields are discarded on the grounds that they are contrary to experiments. However, it seems we need it on a theoretical ground: purely retarded solutions of Maxwell's equations embodies an electrodynamical arrow of time not recognized by the equations themselves. That question has been asked for a long time: it is generally assumed that a radiating body emits light in every direction, quite regardless of whether there are near or distant objects which may ultimately absorb that light (in other words, it radiates "into space"). However, Tetrode, yet in 1922, (Tetrode, 1922) made the assumption that an atom never emits light except to another atom so that the emitter and the absorber both act in the emission process, the first one to emit light and the other one "to tell" the emitter that it is ready to absorb. He thus proposed to eliminate the idea of a mere emission of light and substituted the idea of a process of exchange of energy between two definite atoms or molecules. Such propositions were reconsidered by G. N. Lewis in 1926, and then, in 1927, by Bridgman *who held that it is wrong to speak of light as something travelling*. Their paper gave birth to the Wheeler-Feynman absorber theory of radiation (Wheeler & Feynman, 1945) in which there is no radiation proper (see also (Hoyle & Narlikar, 1995)). They thus anticipated a quantitative theory of electrodynamics using both retarded and advanced potential the interest of which is perhaps to try to give both to the emitter and the absorber the same importance.
Such a use of advanced waves may be somewhat provoking. In fact, it is. However it seems possible to consider such a dual interaction between an emitter and a detector as the translation in the langage of classical waves physics of what may reallycorrespond to an elongation (a dilation) phenomenon (as in the stretching of an elastic band where the "interaction" between the two ends cannot be accredited to one or to the other end). As written above, because of the expansion of the universe, the relation of two comoving objects (the two

ends of the elastic band in our example) in fact appears to us *as if* a signal was propagating between them at velocity c. Such a description would suppress the provoking acausal action from an absorber to an emitter.

As expected by all the above authors, our aim is thus to note that connecting the light phenomenon to the expansion of the universe may perhaps permit to consider light as an effect of the stretching of the spacetime rather than as a propagating phenomenon. May be eq.(1), could thus also open a way to reconsider the origin of electromagnetism.

10. References

Assis, A. K. T. (1994). *Weber's electrodynamics*, Kluwer Academic Publishers, Collection : Fundamental Theories of Physics, ISBN-10: 0792331370.

Axenides, M. & Perivolaropoulos L. (2002), Dark energy and the quietness of the local Hubble flow, *Phys. Rev.D*, 65, 127301.

Bacinich E.J. & Kriz T.A., 1995, Photon trajectory attributes of an expanding hypersphere, *Phys Essays* 8(4) 506-517, 1995.

Baryshev Yu, Chernin A. & Teerikorpi, P. (2001), The cold local Hubble flow as a signature of dark energy, *Astronomy and Astrophysics*, 378, 3, 729-734.

Benabed K. & Bernardeau F. (2001), Testing quintessence models with large-scale structure growth, *Phys. Rev. D*, 64, 083501.

Berman M.S. (1991), Cosmological models with a variable cosmological term, *Phys Rev D*, 43(4), 1075-1078.

Brans C. & Dicke R.H., Mach's principle and a relativistic theory of gravitation, *Phys Rev*, 124(3), 925-935.

Braunbeck W. 1937, Die empirische Genauigkeit des Masse-Energy-Verhältnisses, *Z. Phys*, 107, 1-11.

Bridgman P.W. (1927) *The logic of modern physics*, New York, MacMillan and co.

Canuto V. Hsieh S.H. & Adams P.J. (1977), Scale-covariant theory of gravitation and astrophysical applications, *Phys Rev D*, 16, 1643-1663.

Carroll S. M. (2001), *Living Rev. Relativity, 3, 1,* www.livingreviews.org/lrr-2001-1, The cosmological constant.

Chen W. & Wu Y.S. (1990), Implications of a cosmological constant varying as R^{-2}, *Phys. Rev. D*, 41, 695-698.

Chernin A., Teerikorpi, P. & Baryshev Yu. (2000), Why is the Hubble flow so quiet ? astro-ph/0012021.

Dirac P.A.M., (1937), The Cosmological Constants, *Nature*, 139, 323-323.

Dirac P.A.M., (1938), A New Basis for Cosmology, *Proc. Roy. Soc. A*, 165, 199-208.

Dolgov A.D. (1983), in *The very early universe*, Gibbons G.W., Hawking S.W. & Siklos S.T.C. eds. (Cambridge University Press, Cambridge, New York).

Einstein A. (1917), *Über die Spezielle und Allgemeine Relativitätstheorie*, Vieweg, Braunschweig.

Ellis G. & Uzan J.-P., (2005), c is the speed of light isn't it ? *Am. J. Phys.* , 73 (3), 240-247.

Fahr H.J. & Heyl M., (2006), *Astron. Naschr.*, 327(4), 383-386.

Fahr H.J. & Zoennchen J.H. (2006), Cosmological implications of the Machian principle, *Naturwissenschaften*, 93, 577-587.

Fahr H.J. & Heyl, M., (2007), Cosmic vacuum energy decay and creation of cosmic matter, *Naturwissenschaften* 94, 709-724.

Fahr H.J. & Heyl M. (2007), About universes with scale-related total masses and their abolition of presently outstanding cosmological problems, Astron. Naschr./astron Notes, 238, n 2, 192-199.

Ford L. (1985), Quantum instability of de Sitter spacetime, *Phys. Rev. D*, 31, 710-717.

Freese, K., Adams, F.C., Frieman J.A., Mottola (1987), *E., Nucl Phys B*, 287, 797.

Henttunen K, Multamäki T. & Vilja I., (2006), Complex supergravity quintessence models confronted with Sn Ia data, *Phys. Lett. B*, 634, 5-8.

Hogarth J.E. (1962), Cosmological considerations of the absorber theory of radiation, *Proc. Roy. Soc A* 267, 365-383.

Hoyle F. , (1990), On the relation of the large numbers problem to the nature of mass, *Astrophys. Space Sci*, 168, 59-88.

Hoyle F. , (1992), Mathematical theory of the origin of matter, *Astrophys. Space Sci*, 198, 195-230.

Hoyle F. & Narlikar J.V. (1990), Cosmology and action-at-a-distance electrodynamics, *Reviews of Modern Physics*, 67, 113-155.

Jordan T.F. (2005), Cosmology calculations almost without general relativity, *Am. J. Phys*, 73(7), 653-662.

Khadekar G.S. & Butey B.P., (2009), Higher dimensional cosmological model of the universe with decaying Λ cosmology with varying *G*, *Int J Theor Phys*, 48: 2618-2624.

Lewis G. N. (1926), The nature of light, *Physics*, 12 1926.

Lima A.S. & Carvalho, Dirac's cosmology with varying cosmological constant (1994), *Gen. Relativ. Grav.* 26, 909-916.

Mach E. (1904), *La Mecanique*, Hermann, Paris.

Maxwell J. K., *A treatise in electromagnetism and magnetism*, Dover, New York, vol II, pp. 434-436.

McCrea W.H. & Milne E.A., (1934), Newtonian universes and the curvature of space, *Q. J. Math*, 5, 73-80.

Milne, E. A. (1934), A newtonian expanding universe, *Q. J. Math*, 5, 64-72.

Moffat J.W., (1993), Superluminary universe: a possible solution of the initial value problem in cosmology, *Int J. Mod Phys*, D2, 351-366.

Mukhopadhyay U., Ray S.,Usmani A.A. & Ghosh P.P., (2011), Time variable Λ and the accelerating universe, *Int. J. Theor. Phys.*, 50, p. 752-759.

Norgaard-Nielsen H. U., Hansen, L., Jørgensen H. E., Aragon Salamanca A. & Ellis R. S. (1989), The discovery of a type Ia supernova at a redshift of 0.31, *Nature*, 339, 523-525.

Overduin, J. M. & Cooperstock, F. I., (1998), Evolution of the scale factor with a variable cosmological term, *Phys. Rev. D*, 58, 043506, 1-23. [astro-phys/9805260].

Özer M. & Taha M.O., (1987), A model of the universe free of cosmological problems, *Nucl. Phys.*, B 287, 776-796.

Peebles P.J.E. & Ratra B., (1988), Cosmology with a time-variable cosmological "constant", *Astrophys. J.*, 325, L17-L20.

Perlmutter, S, Aldering G., Goldhaber G. et al., (1999), Measurements of Ω and Λ from 42 High-Redshift Supernovae, *Astrophys. J.*, 517, 2, 565-586. [astro-ph/9812133].

Ratra B. & Peebles P.J.E., (1988), Cosmological consequences of a rolling homogeneous scalar field, *Phys. Rev. D*, 37(12), 3406-3427.

Ray S., Khlopov M., Ghosh P.P. & Mukhopadhyay U., (2011), Phenomenology of ΛCDM model: a possibility of accelerating universe with positive pressure, *Int. J. Theor. Phys.*, 50, 939-951.

Riess A.G., Philipenko A.V., Challis P. *et* al. (1998), Observational evidence from supernovae for an accelerating universe and a cosmological constant, *Astron. J.*, 116, 1009-1038. [astro-ph/9805201].

Riess A.G. *et* al. (2004), Type Ia Supernova Discoveries at z > 1 from the Hubble Space Telescope: Evidence for Past Deceleration and Constraints on Dark Energy Evolution, *Astrophys. J.*, 607, 665-687.

Schwarzschild B. , (2004), High-Redshift Supernovae Reveal an Epoch When Cosmic Expansion Was Slowing Down, *Phys Today*,19-21.

Tetrode H. (1922), Über den Wirkungszammenhang des Welt. Eine Erweiterung der klassichen Dynamik, *Zeitschrift für Physik* 10, 317.

Thirring H. (1918), Über die Wirkung rotierender ferner Massen in der Einsteinschen Gravitationstheorie, *Phys Zeitschrift* 19, 33-39.

Tonry J. L. et *al*, (2003), Cosmological Results from High-z Supernovae, *Astrophys. J.*, 594, 1-24.

Viennot D., Vigoureux J.-M., (2007), The Cosmological constant and the coincidence problem in a new cosmological interpretation of the universal constant *c*. *Int. J. Theor. Phys.*, 48(8), 2246-2252.

Vigoureux B., Vigoureux J. M. & Vigoureux P. (1988), Essai sur une théorie géométrique de l'Univers reliant la vitesse de la lumière à l'expansion de l'Univers, Ann. Sc. Univ. F. Comté, 4, 19-30.

Vigoureux B., Vigoureux J.M. & Vigoureux P. (2001), Connecting c to the Expansion of the Universe: Cosmological Consequences, *Physics Essays*, 14, 4, 314-319.

Vigoureux J. M. , Vigoureux P. & Vigoureux B. (2003), The Einstein Constant c in Light of Mach's Principle. Cosmological Applications, *Foundations of Physics Letters*, 16, 2,183-193.

Vigoureux J.-M., Vigoureux P. & Vigoureux B. (2008), Cosmological Applications of a Geometrical Interpretation of c, *Int. J. Theor. Phys.*, 47, 4, 928-935.

Wang L. Goldhaber G., Aldering G. & Perlmutter S., (2003), Multicolor Light Curves of Type Ia Supernovae on the Color-Magnitude Diagram: a Novel Step Toward More Precise Distance and Extinction Estimates, *Astrophys. J.*, 590, 944-970 (astro-ph/0302341).

Weinberg S, *Gravitation and cosmology*, Wiley and Sons. New York 1972.

Wheeler J.A. & Feynman R.P. (1945), Interaction with the Absorber as the Mechanism of Radiation, *Rev. Mod. Phys.*, 17, 157-181.

Whitrow G.J. (1946), The Mass of the Universe, *Nature*, 158, 165-166.

12

C-Field Cosmological Model for Barotropic Fluid Distribution with Variable Gravitational Constant

Raj Bali
Department of Mathematics, University of Rajasthan, Jaipur
India

1. Introduction

The importance of gravitation on the large scale is due to the short range of strong and weak forces and also to the fact that electromagnetic force becomes weak because of the global neutrality of matter as pointed by Dicke and Peebles [1965]. Motivated by the occurrence of large number hypothesis, Dirac [1963] proposed a theory with a variable gravitational constant (G). Barrow [1978] assumed that G α t^{-n} and obtained from helium abundance for – $5.9 \times 10^{-13} < n < 7 \times 10^{-13}$, $\left|\frac{\dot{G}}{G}\right| < (2 \pm .93)x10^{-12} yr^{-1}$ by assuming a flat universe.

Demarque et al. [1994] considered an ansatz in which G α t^{-n} and showed that $|n| < 0.1$ corresponds to $\left|\frac{\dot{G}}{G}\right| < 2x10^{-1} yr^{-1}$. Gaztanga et al. [2002] considered the effect of variation of gravitational constant on the cooling of white dwarf and their luminosity function and concluded that $\left|\frac{\dot{G}}{G}\right| < 3x10^{-1} yr^{-1}$.

To achieve possible verification of gravitation and elementary particle physics or to incorporate Mach's principle in General Relativity, many atempts (Brans and Dicke [1961], Hoyle and Narlikar [1964]) have been made for possible extension of Einstein's General Relativity with time dependent G.

In the early universe, all the investigations dealing with physical process use a model of the universe, usually called a big-bang model. However, the big-bang model is known to have the short comings in the following aspects.

i. The model has singularity in the past and possibly one in future.
ii. The conservation of energy is violated in the big-bang model.
iii. The big-bang models based on reasonable equations of state lead to a very small particle horizon in the early epochs of the universe. This fact gives rise to the 'Horizon problem'.
iv. No consistent scenario exists within the frame work of big-bang model that explains the origin, evolution and characteristic of structures in the universe at small scales.
v. Flatness problem.

Thus alternative theories were proposed from time to time. The most well known theory is the 'Steady State Theory' by Bondi and Gold [1948]. In this theory, the universe does not have any singular beginning nor an end on the cosmic time scale. For the maintenance of uniformity of mass density, they envisaged a very slow but continuous creation of matter in contrast to the explosive creation at t = 0 of the standard FRW model. However, it suffers the serious disqualifications for not giving any physical justification in the form of any dynamical theory for continuous creation of matter. Hoyle and Narlikar [1966] adopted a field theoretic approach introducing a massless and chargeless scaler field to account for creation of matter. In C-field theory, there is no big-bang type singularity as in the steady state theory of Bondi and Gold [1948]. Narlikar [1973] has explained that matter creation is a accomplished at the expense of negative energy C-field. He also explained that if overall energy conservation is to be maintained then the primary creation of matter must be accompanied by the release of negative energy and the repulsive nature of this negative reservoir will be sufficient to prevent the singularity. Narlikar and Padmanabhan [1985] investigated the solution of modified Einstein's field equation which admits radiation and negative energy massless scalar creation field as a source. Recently Bali and Kumawat [2008] have investigated C-field cosmological model for dust distribution in FRW space-time with variable gravitational constant.

In this chapter, we have investigated C-field cosmological model for barotropic fluid distribution with variable gravitational constant. The different cases for $\gamma = 0$ (dust distribution), $\gamma = 1$ (stiff fluid distribution), $\gamma = 1/3$ (radiation dominated universe) are also discussed.

Now we discuss Creation-field theory (C-field theory) originated by Hoyle and Narlikar [1963] so that it may be helpful to readers to understand Creation-field cosmological model for barotropic fluid distribution with variable gravitational constant.

2. Hoyle-Narlikar creation-field theory

Hoyle's approach (1948) to the steady state theory was via the phenomena of creation of matter. In any cosmological theory, the most fundamental question is "where did the matter (and energy) we see around us originate?" by origination, we mean coming into existence by primary creation, not transmutation from existing matter to energy or vice-versa. The Perfect Cosmological Principle (PCP) deduces continuous creation of matter. In the big-bang cosmologies, the singularity at t = 0 is interpreted as the primary creation event. Hoyle's aim was to formulate a simple theory within the framework of General Relativity to describe such a mechanism.

Now I discuss this method since it illustrates the power of the Action-principle in a rather simple way.

The action principle

The creation mechanism is supposed to operate through the interaction of a zero rest mass scalar field C of negative energy with matter. The action is given by

$$A = \frac{1}{16\pi G}\int R\sqrt{-g}d^4x - \sum_a ma\int da - \frac{1}{2}f\int C_iC^i\sqrt{-g}d^4x + \sum_a\int C_i da^i \tag{2.1}$$

where $C_i = \dfrac{\partial C}{\partial x^i}$ and f > 0, is a coupling constant between matter and creation field.

The variation of a stretch of the world line of a typical particle 'a' between the world points A_1 and A_2 gives

$$\delta A = \int_{A_1}^{A_2} ma \left\{ \frac{d^2 a^i}{da^2} + \Gamma^i_{kl} \frac{dx^k}{da} \frac{dx^l}{da} \right\} g_{ik} \delta a^k da - \left[\left\{ ma \frac{da^i}{da} g_{ik} - C_k \right\} \cdot \delta a^k \right]_{A_1}^{A_2} \tag{2.2}$$

Now suppose that the world-line is not endless as it is usually assumed but it begins at A_1 and the variation of the world line is such that $\delta a^k \neq 0$ at A_1. Thus for arbitrary δa^k which vanish at A_2, we have

$$\frac{d^2 a^i}{da^2} + \Gamma^i_{kl} \frac{da^k}{da} \frac{da^l}{da} = 0 \tag{2.3}$$

along A_2A_2 while at A_1,

$$ma \frac{da^i}{da} g_{ik} = C_k \tag{2.4}$$

The equation (2.3) tells us that C-field does not alter the geodesic equation of a material particle. The effect of C-field is felt only at A_1 where the particle comes into existence. The equation (2.4) tells us that the 4-momentum of the created particle is balanced by that of the C-field. Thus, there is no violation of the matter and energy-momentum conservations law as required by the action principle. However, this is achieved because of the negative energy of the C-field. The variation of C-field gives from $\delta A = 0$,

$$C^i_{;i} = \frac{1}{f} n \tag{2.5}$$

where n = number of creation events per unit proper 4-volume. By creation event, we mean points like A_1, if the word line had ended at A_2 above, we would have called A_2 an annihilation event. In n, we sum algebraically (i.e. with negative sign for annihilation events) over all world-line ends in a unit proper 4-volume. Thus the C-field has its sources only in the end-points of the world-lines.
Finally, the variation of g_{ik} gives the Einstein's field equation

$$R^{ik} - \frac{1}{2} R g^{ik} = -8\pi G \left[\underset{(m)}{T^{ik}} + \underset{(C)}{T^{ik}} \right] \tag{2.6}$$

Here $\underset{(m)}{T^{ik}}$ is the energy-momentum of particles a, b, ... while

$$\underset{(C)}{T^{ik}} = -f \left\{ C^i C^K - \frac{1}{2} g^{ik} C^l C_l \right\} \tag{2.7}$$

is due to Hoyle and Narlikar (1964).
A comparison with the standard energy-momentum tensors of scalar fields shows that the C-field has negative energy. Thus, when a new particle is created then its creation is accompanied by the creation of the C-field quanta of energy and momentum. Since the C-

field energy is negative, it is possible to have energy momentum conservation in the entire process as shown in (2.4).

3. The metric and field equations

We consider FRW space time in the form

$$ds^2 = dt^2 - R^2(t)\left[\frac{dr^2}{1-kr^2} + r^2 d\theta^2 + r^2 \sin^2\theta d\phi^2\right] \tag{3.1}$$

where k = 0, -1, 1

The modified Einstein's field equation in the presence of C-field is due to Hoyle and Narlikar [1964]) is given by

$$R_i^j - \frac{1}{2}\overline{R}g_i^j = -8\pi G\left[\underset{(m)}{T_i^j} + \underset{(C)}{T_i^j}\right] \tag{3.2}$$

where $\overline{R} = g^{ij}R_{ij}$, is the scalar curvature, $\underset{(m)}{T_i^j}$ is the energy-momentum tensor for matter and $\underset{(C)}{T_i^j}$ the energy-momentum tensor for C-field are given by

$$\underset{(m)}{T_i^j} = (\rho + p)v_1 v^j - p g_i^j \tag{3.3}$$

and

$$\underset{(C)}{T_i^j} = -f\left[C_i C^j - \frac{1}{2}g_i^j C^l C_l\right] \tag{3.4}$$

p being isotropic pressure, ρ the matter density, f > 0. We assume that flow vector to be comoving so that $v^1 = 0 = v^2 = v^3$, $v^4 = 1$ and $C_i = \frac{\partial C}{\partial x^i}$.

The non-vanishing components of energy-momentum tensor for matter are given by

$$\underset{(m)}{T_1^1} = (\rho + p)\cdot 0 - p = -p \tag{3.5}$$

Similarly

$$\underset{(m)}{T_2^2} = -p = \underset{(m)}{T_3^3} \tag{3.6}$$

$$\underset{(m)}{T_4^4} = (\rho + p)\cdot 1 - p = \rho \tag{3.7}$$

The non-vanishing components of energy-momentum tensor for Creation field are given by

$$\underset{(C)}{T_1^1} = -f\left[0 - \frac{1}{2}\cdot 1\cdot g^{44}C_4^2\right] = \frac{1}{2}f\dot{C}^2 \tag{3.8}$$

Similarly

$$\underset{(C)}{T_2^2} = \frac{1}{2} f \dot{C}^2 = \underset{(C)}{T_3^3} \tag{3.9}$$

$$\underset{(C)}{T_4^4} = -f\left[g^{44}C_4^2 - \frac{1}{2}\cdot 1 \cdot g^{44}C_4^2\right] = -f\left[C_4^2 - \frac{1}{2}C_4^2\right] = -\frac{1}{2} f \dot{C}^2 \tag{3.10}$$

where $C_4 = \dot{C}$

The modified Einstein field equation (3.10) in the presence of C-field for the metric (3.1) for variable G(t) leads to

$$\frac{3\dot{R}^2}{R^2} + \frac{3k}{R^2} = 8\pi G(t)\left[\rho - \frac{1}{2} f \dot{C}^2\right] \tag{3.11}$$

$$\frac{2\ddot{R}}{R} + \frac{\dot{R}^L}{R^2} + \frac{k}{R^2} = -8\pi G(t)\left[p - \frac{1}{2} f \dot{C}^2\right] \tag{3.12}$$

4. Solution of field equations

The conservation equation

$$\left(8\pi G T_i^j\right)_{;j} = 0 \tag{4.1}$$

leads to

$$\frac{\partial}{\partial x^j}\left(8\pi G T_i^j\right) + 8\pi G T_i^l \Gamma_{lj}^j - 8\pi G T_l^j \Gamma_{ij}^l = 0$$

which gives

$$\frac{\partial}{\partial t}\left(8\pi G T_4^4\right) + 8\pi G\left[T_1^1\left(\Gamma_{11}^1 + \Gamma_{12}^2 + \Gamma_{13}^3\right) + T_2^2\left(T_{23}^3\right) + T_3^3(0)\right.$$

$$\left. T_4^4\left(\Gamma_{14}^1 + \Gamma_{24}^2 + \Gamma_{34}^3\right)\ \right] - 8\pi G\left[T_1^1\left(\Gamma_{14}^1 + \Gamma_{11}^1\right) + T_2^2\left(\Gamma_{24}^2 + \Gamma_{12}^2\right)\right.$$

$$\left. T_3^3\left(\Gamma_{34}^3 + \Gamma_{13}^3 + \Gamma_{23}^3\right) + T_4^4(0)\ \right] \tag{4.2}$$

which leads to

$$\frac{\partial}{\partial t}\left[8\pi G\left(\rho - \frac{1}{2} f \dot{C}^2\right)\right] + 8\pi G\left[\left(\frac{1}{2} f \dot{C}^2 - p\right)\left(\frac{kr}{1-kr^2} + \frac{1}{r} + \frac{1}{r}\right)\right.$$

$$\left. + \left(\frac{1}{2} f \dot{C}^2 - p\right)\cot\theta + \left(\rho - \frac{1}{2} f \dot{C}^2\right)\frac{3\dot{R}}{R}\ \right] - 8\pi G\left[\left(\frac{1}{2} f \dot{C}^2 - p\right)\right.$$

$$\left(\frac{\dot{R}}{R}+\frac{kr}{1-kr^2}\right)+\left(\frac{1}{2}f\dot{C}^2-p\right)\left(\frac{\dot{R}}{R}+\frac{1}{r}\right)+\left(\frac{1}{2}f\dot{C}^2-p\right)\left(\frac{\dot{R}}{R}+\frac{1}{r}+\cot\theta\right)=0 \tag{4.3}$$

which gives

$$8\pi\dot{G}\left(\rho-\frac{1}{2}f\dot{C}^2\right)+8\pi G\left(\dot{\rho}-f\dot{C}\ddot{C}\right)+8\pi\dot{G}\left(\frac{3\dot{R}}{R}\rho+\frac{3\dot{R}}{R}\rho-\frac{3\dot{R}}{R}f\dot{C}^2\right)=0 \tag{4.4}$$

which yields $\dot{C}=1$ when used in source equation. Using $\dot{C}=1$ in (3.11), we have

$$8\pi G\rho=\frac{3\dot{R}^2}{R^2}+\frac{3k}{R^2}+4\pi Gf \tag{4.5}$$

We assume that universe is filled with barotropic fluid i.e. $p = \gamma\rho$ $(0 \le \gamma \le 1)$, p being the isotropic pressure, ρ the matter density. Now using $p = \gamma\rho$ and $\dot{C}=1$ in (3.12), we have

$$\frac{2\ddot{R}}{R}+\frac{\dot{R}^2}{R^2}+\frac{k}{R^2}=-8\pi G(t)\left[\gamma\rho-\frac{1}{2}f\right] \tag{4.6}$$

Equations (4.5) and (4.6) lead to

$$\frac{2\ddot{R}}{R}+(1+3\gamma)\frac{\dot{R}^2}{R^2}=(1-\gamma)4\pi Gf-(1+3\gamma)\frac{k}{R^2} \tag{4.7}$$

To obtain the deterministic solution, we assume that

$$G = R^n \tag{4.8}$$

where R is scale factor and n is a constant. From equations (4.7) and (4.8), we have

$$2\ddot{R}+(3\gamma+1)\frac{\dot{R}^2}{R}=(1-\gamma)AR^{n+1}-\frac{k}{R}(3\gamma+1) \tag{4.9}$$

where

$$A = 4\pi f \tag{4.10}$$

To find the solution of (4.9), we assume that

$$\dot{R}=F(R) \tag{4.11}$$

Thus

$$\ddot{R}=\frac{d\dot{R}}{dt}=\frac{dF}{dt}=\frac{dF}{dR}\frac{dR}{dt}=FF' \tag{4.12}$$

where

$$F'=\frac{dF}{dR}$$

Using (4.11 and (4.12) in (4.9), we have

$$\frac{dF^2}{dR}+\frac{(3\gamma+1)F^2}{R}=(1-\gamma)AR^{n+2}-\frac{k(3\gamma+1)}{R} \tag{4.13}$$

Equation (4.13) leads to

$$F^2=\left(\frac{dR}{dt}\right)^2=\frac{A(1-\gamma)R^{n+2}}{(n+3\gamma+3)}-k \tag{4.14}$$

which leads to

$$\frac{dR}{\sqrt{R^{n+2}-\frac{k(n+3\gamma+3)}{A(1-\gamma)}}}=\sqrt{\frac{A(1-\gamma)}{(n+3\gamma+3)}}dt \tag{4.15}$$

To get determinate value of R in terms of cosmic time t, we assume n = -1. Thus equation (4.15) leads to

$$\frac{dR}{\sqrt{R-\frac{k(3\gamma+2)}{A(1-\gamma)}}}=\sqrt{\frac{A(1-\gamma)}{(3\gamma+2)}}dt \tag{4.16}$$

From equation (4.16), we have

$$R=(at+b)^2+\frac{k(3\gamma+2)}{A(1-\gamma)} \tag{4.17}$$

where

$$a=\frac{1}{2}\sqrt{\frac{A(1-\gamma)}{(3\gamma+2)}}\ ,\quad b=\frac{N}{2} \tag{4.18}$$

where N is the constant of integration
Therefore, the metric (3.1) leads to

$$ds^2=dt^2-\left[(at+b)^2+\frac{k(3\gamma+2)}{A(1-\gamma)}\right]^2\left[\frac{dr^2}{1-kr^2}+r^2d\theta^2+r^2\sin^2\theta d\phi^2\right] \tag{4.19}$$

where

$$\gamma\neq 1.$$

Taking a = 1, b = 0, the metric (4.19) reduces to

$$ds^2=dt^2-\left[t^2+\frac{k(3\gamma+2)}{A(1-\gamma)}\right]^2\left[\frac{dr^2}{1-kr^2}+r^2d\theta^2+r^2\sin^2\theta d\phi^2\right] \tag{4.20}$$

5. Physical and geometric features

The homogeneous mass density (ρ), the isotropic pressure (p) for the model (4.19) are given by

$$8\pi\rho = \frac{12a^2(at+b)^2 + 3k}{\left[(at+b)^2 + \frac{k(3\gamma+2)}{A(1-\gamma)}\right]} + A \tag{5.1}$$

$$8\pi p = 8\pi\gamma\rho = \frac{12a^2\gamma(at+b)^2 + 3k\gamma}{\left[(at+b)^2 + \frac{k(3\gamma+2)}{A(1-\gamma)}\right]} + A\gamma \tag{5.2}$$

$$G = R^{-1} = \frac{1}{\left[(at+b)^2 + \frac{k(3\gamma+2)}{A(1-\gamma)}\right]} \tag{5.3}$$

q = Deceleration parameter

$$= -\frac{\frac{\ddot{R}}{R}}{\frac{\dot{R}^2}{R^2}}$$

where R is scale factor given by (4.17). Thus

$$q = -\frac{\left[2a^2(at+b)^2 + \frac{2ka^2(3\gamma+2)}{A(1-\gamma)}\right]}{4a^2(at+b)^2} + A \tag{5.4}$$

To find C (creation field)

Using p = γρ, (5.1), (5.3) and (4.17) in (4.4), we have

$$\frac{d\dot{C}^2}{dt} + \frac{10t}{\left[t^2 + \frac{k(3\gamma+2)}{A(1-\gamma)}\right]}\dot{C}^2 = \frac{4}{A}\left[\frac{6t^3(3\gamma+2) + \frac{3k(3\gamma+1)}{2}t + \frac{6kt(3\gamma+2)}{A(1-\gamma)}}{\left\{t^2 + \frac{k(3\gamma+2)}{A(1-\gamma)}\right\}}\right]$$

$$+\frac{A(3\gamma+2)t}{2\left\{t^2 + \frac{k(3\gamma+2)}{A(1-\gamma)}\right\}} \tag{5.5}$$

Equation (5.5) is linear in $\dot{C}^2$. The solution of (5.5) is given by

$$\dot{C}^2 = \frac{4(3\gamma+2)}{A(1-\gamma)} \tag{5.6}$$

which gives

$$\dot{C} = 1 \tag{5.7}$$

which agrees with the value used in source equation. Here $\frac{(3\gamma+2)}{\pi f(1-\gamma)} = 1$ which gives $\gamma = \frac{\pi f - 2}{\pi f + 3}$. Equation (5.7) leads to

$$\mathrm{C} = \mathrm{t} \tag{5.8}$$

Thus creation field increases with time.
Taking a = 1, b = 0 in equations (5.1) – (5.4), we have

$$8\pi\rho = 12 + 4\pi f \tag{5.9}$$

$$8\pi p = 8\pi\gamma f = 12\gamma + 4\pi\gamma f \tag{5.10}$$

$$G = \frac{1}{t^2 + \frac{k}{4}} \tag{5.11}$$

$$q = -\left(\frac{1}{2} + \frac{k}{8t^2}\right) \tag{5.12}$$

6. Discussion

The matter density (ρ) is constant for the model (4.20). The scale factor (R) increases with time. Thus inflationary scenario exists in the model (4.20). $\left|\frac{\dot{G}}{G}\right| \cong \frac{1}{t} = H$ where H is Hubble constant. $G \to \infty$ when $t \to 0$ and $G \to 0$ when $t \to \infty$. The deceleration parameter (q) < 0 which indicates that the model (4.20) represents an accelerating universe. The creation field C increases with time and $\dot{C} = 1$ which agrees with the value taken in source equation. The matter density ρ = constant as given by (5.9). This result may be explained as : Referring to Hoyle and Narlikar [2002], Hawking and Ellis [1973], the matter is supposed to move along the geodesic normal to the surface t = constant. As the matter moves further apart, it is assumed that more matter is continuously created to maintain the matter density at constant value. For k = 0, γ = 0 and for k = ±1 γ = 0, we get the same results as obtained by Bali and Tikekar [2007], Bali and Kumawat [2008] respectively.
The coordinate distance to the horizon r_H is the maximum distance a null ray could have travelled at a time t starting from the infinite past i.e.

$$r_H(t) = \int_{\infty}^{t} \frac{dt}{R^3(t)} \tag{5.13}$$

We could extend the proper time t to (-∞) in the past because of non-singular nature of the space-time. Now

$$r_H(t) = \int_0^t \frac{dt}{\alpha t^3} \tag{5.14}$$

where $\alpha = \sqrt{\frac{4\pi f(1-\gamma) - k(3\gamma+1)}{3\gamma+1}}$

This integral diverges at lower time showing that the model (4.20) is free from horizon.

Special Cases:

i. Dust filled universe i.e. $\gamma = 0$, the metric (4.20) leads to

$$ds^2 = dt^2 - \left(t^2 + \frac{k}{2\pi f}\right)^2 \left[\frac{dr^2}{1-kr^2} + r^2 d\theta^2 + r^2 \sin^2\theta d\phi^2\right] \tag{5.15}$$

For k = 0, the metric (5.15) leads to the model obtained by Bali and Tikekar (2007).

ii. For k = ±1, $\gamma = 0$, the model (5.15) leads to the model obtained by Bali and Kumawat (2008).

iii. For $\gamma = 1/3$ (Radiation dominated universe), the model (4.20) leads to

$$ds^2 = dt^2 - \left(t^2 + \frac{9k}{8\pi f}\right)\left[\frac{dr^2}{1-kr^2} + r^2 d\theta^2 + r^2 \sin^2\theta d\phi^2\right] \tag{5.16}$$

For $\gamma = 1$ (stiff fluid universe), the model (4.20) does not exist.

7. References

[1] Dicke, R.H. and Peebles, P.J.E. : Space-Science Review 4, 419 (1965).

[2] Dicke, R.H. : Relativity, Groups and Topology, Lectures delivered at Les Houches 1963 edited by C. De Witt and B. De Witt (Gordon and Breach, New York).

[3] Barrow, J.D. : Mon. Not. R. Astron. Soc. 184, 677 (1978).

[4] Demarque, P., Krause, L.M., Guenther, D.B. and Nydam, D. : Astrophys. J. 437, 870 (1994).

[5] Gaztanaga, E., Gracia-Berro, E., Isern, J., Bravo, E. and Dominguez, I. : Phys. Rev. D65, 023506 (2002).

[6] Brans, C. and Dicke, R.H. : Phys. Rev. 124, 925 (1961).

[7] Hoyle, F. and Narlikar, J.V. : Proc. Roy. Soc. A282, 191 (1964).

[8] Bondi, H. and Gold, T. : Mon. Not. R. Astron. Soc. 108, 252 (1948).

[9] Hoyle, F. and Narlikar, J.V. : Proc. Roy. Soc. A290, 162 (1966).

[10] Narlikar, J.V. and Padmanabham, T. : Phys. Rev. D32, 1928 (1985).

[11] Bali, R. and Kumawat, M. (2008) : Int. J. Theor. Phys., DOI10.1007/s10773-009-0146-3.

[12] Hoyle, F. and Narlikar, J.V. : Proc. Roy. Soc. A278, 465 (1964).

[13] Narlikar, J.V. : Introduction to Cosmology, Cambridge University Press, p.140 (2002).

[14] Hawking, S.W. and Ellis, G.F.R. : The large scale structure of space-time, Cambridge University Press, p.126 (1973).

[15] Bali, R. and Tikekar, R. : Chin. Phys. Lett. 24, 3290 (2007).

Part 5

More Mathematical Approaches

13

Separation and Solution of Spin 1 Field Equation and Particle Production in Lemaître-Tolman-Bondi Cosmologies

Antonio Zecca
Dipartimento di Fisica dell' Universita' - Via Celoria, Milano
GNFM, Gruppo Nazionale per la Fisica Matematica, Sesto Fiorentino (Fi)
Italy

1. Introduction

An attractive issue in general relativity is the separation, and possibly the solution, of field equation of arbitrary spin in space-time of physical relevance, especially from the cosmological point of view. The knowledge of the normal mode solutions is a basic tool in view of a quantization of the field that in turns can lead to a further adjustement of the theoretical formulation of the cosmological model.
In case of the Robertson-Walker (RW) space-time metric, that is the base of spherically symmetric homogeneous standard cosmology (Weinberg, 1972), the problem has been widely considered (Penrose and Rindler, 1984; Fulling, 1989; Parker and Toms, 2009). Recently that goal can be found solved, for arbitrary spin value, in RW metric by the Newmann-Penrose formalism (Zecca, 2009). The separation method employed to that end has been developed in the line of Chandrashekar's separation of Dirac equation in Kerr metric (Chandrasekhar, 1983). In the specific case of spin 0, 1/2, 1 it has been pointed out (Zecca, 2009a; 2010a; 2010b) that particle creation (annihilation) in expanding universe is possible. (Particle production by universe expansion was originally discussed by Parker (1969; 1971); see also Parker and Toms, 2009). The presence of this effect modifies the gravitational dynamics of the Universe. An extension of the Standard Cosmology has also been proposed that includes the back reaction due to particle production (Zecca, 2010).
The separation of field equation of arbitrary spin has been obtained also in Schwarzschild metric (Zecca, 2006b). This metric is interesting because it represents the gravitational field outside a spherical central non rotating mass such as stars, planets, black holes, .. . In this metric however the separated radial equation are much more difficult to disentagle.
Another situation of relevance concerns the spherically symmetric non homogeneous metrics, and in particular the one that is the base of the Lemaître-Tolman-Bondi (LTB) cosmological model. This metric represents a spherically symmetric inhomogeneous universe filled with freely falling dust matter without pressure. The model can be completely integrated and the general solution of the Einstein equation depends on three arbitrary functions of the radial coordinate. (For a comprehensive study of the model see Krasinski, 1997). The separation of the field equation for spins 0, 1/2 has been shown to be possible also in this model under a special choice of the mentioned integration functions. The surviving configuration remains

however sufficiently general because the cosmological model still depends on an arbitrary function of the radial coordinate (Zecca, 2000; 2001).
In the line of the above considerations, it would be desirable to extend the solution of the field equation to higher spin values. This seems a difficult task in the LTB metrics. Indeed in curved space-time the spinor formulation of field equation of spin value greater than 1, in general involves the knowledge of the Weyl spinor (e.g., Illge, 1993 and references therein). Contrarily to what happens in the Robertson-Walker (RW) metric, the Weyl spinor does not vanish in the LTB metrics (e. g., Zecca, 2000a) and makes the solution of the field equation much more complex.

Therefore, in the present Chapter, we study the spin 1 field equation in LTB models. This is a case that, as far as the author knows, has not yet been considered. Moreover it is the case of the higher spin values where the field equation is insensible to the presence of the Weyl spinor (Illge, 1993). On physical grounds the interest of the spin 1 field case lies in that in the massless case it can be interpreted, in a standard way, in terms of electromagnetic field and in the massive case in terms of Proca fields (Illge, 1993; Penrose and Rindler, 1984; Zecca, 2006). For what concerns the separation of the equation, it is performed for a general LTB metric by using the Newmann-Penrose formalism based on a previously determined null tetrad frame. At this general level of the metric, the angular dependence separates. The separated angular equations coincide with those relative to spin 1 field in Robertson-Walker and Schwarzschild metric that have been previously integrated (Zecca, 1996; 2005a; 2006b). The complete variable separation can be then achived for a class of LTD cosmological models. This is obtained under a factorization assumption $Y = Z(r)T(t)$ on the time and radial dependence of the physical radius $Y(r,t)$, the same assumption under which the spin 0 and spin $1/2$ field equations have been previously separated. There results that the separated radial dependence can be reduced to the solution of two independent disentangled ordinary differential equations. These equations still depend on an arbitrary radial function that is an integration function of the cosmological model. For what concerns the separated time dependence, it can be reduced to the solution of two coupled time equations. These equations do not depend on any arbitrary function and have therefore an absolute character in the class of LTB model satisfying the factorization assumption. In turn the time equations can be decoupled and reduced to ordinary differential equations of known form. However due to the special dependence on the physical parameters, an integration by series, that is explicitly performed in every case, results unavoidable.
Finally a quantization of the scheme is performed by mimicking the procedure previously developed for spin 1 field equation in the RW metric (Zecca, 2009a). In that case, the number of one mode particle production per unit of time at time t was found to be proportional to the Hubble "constant" $\dot{R}(t)/R(t)$. Here the quantization procedure again leads to preview particle creation (annihilation) in expanding universe for the LTB models admitting a factorization assumption of the physical radius Y. Moreover it is coherent with the generally admitted big bang origin assumption of the universe because it avoids considering "in states" with underlying Minkowskian space-time at time $t = -\infty$ as often assumed in different examples (Birrell and Davies, 1982; Moradi, 2008; Parker and Toms, 2009). There results a generalization of the RW case. Here the number of one mode particle creation per unit of time, at a given time, is proportional to $\dot{Y}(r,t)/Y(r,t) = \dot{T}(t)/T(t)$. The quantity of particles produced by universe expansion, does not seem of relevance at a generic time of the cosmological evolution, especially at the present time. Instead, for a cosmological model

admitting a big bang origin, an enormous number of particles is foreseen to be produced near the big bang.

2. Spin 1 field equation in a class of spherically symmetric comoving system

The spin 1 field equation for particles of mass m_0 can be formulated in a general curved space-time by the spinor equation (Penrose and Rindler, 1984) in terms of the spinors Φ_{AB}, $\Theta_{AX'}$

$$\begin{aligned} \nabla^{A}_{X'}\Phi_{AB} &= -i\mu_* \Theta_{BX'} \\ \nabla^{X'}_{A}\Theta_{BX'} &= i\mu_* \Phi_{AB} \end{aligned} \tag{1}$$

with $\Phi_{AB} = \Phi_{BA}$, $\sqrt{2}\mu_*$ the mass of the particle, $\nabla_{AX'}$ the covariant spinor derivative. The formulation (1) holds in a general curved space-time (see e.g., Illge, 1993, and references therein). The object is to solve the system of equations (1) in the general comoving spherically symmetric Lemaître-Tolman-Bondi (LTB) metric whose line element is given by

$$ds^2 = g_{\mu\nu}dx^\mu dx^\nu = dt^2 - e^\Gamma dr^2 - Y^2(d\theta^2 + \sin^2\theta d\varphi^2) \tag{2}$$

with $\Gamma = \Gamma(r,t)$, $Y = Y(r,t)$. (See e.g., Krasinski, 1997). The Newmann-Penrose (1962) formalism is a powerfull tool to that end. Accordingly we consider the null tetrad frame $\{l^i, n^i, m^i, m^{\star i}\}$ that was considered in Zecca 1993, for which the directional derivatives and the non trivial spin coefficients, that we report for reader's convenience, are

$$\begin{aligned}
D &\equiv \partial_{00'} = l^i\partial_i = \frac{1}{\sqrt{2}}(\partial_t + e^{-\Gamma/2}\partial_r), \\
\Delta &\equiv \partial_{11'} = n^i\partial_i = \frac{1}{\sqrt{2}}(\partial_t - e^{-\Gamma/2}\partial_r), \\
\delta &\equiv \partial_{01'} = m^i\partial_i = \frac{1}{Y\sqrt{2}}(\partial_\theta + i\csc\theta\,\partial_\varphi), \\
\delta^\star &\equiv \partial_{10'} = m^{\star i}\partial_i = \frac{1}{Y\sqrt{2}}(\partial_\theta - i\csc\theta\,\partial_\varphi). \\
\rho &= -\frac{1}{Y\sqrt{2}}(\dot{Y} + Y'e^{-\Gamma/2}), \\
\mu &= \frac{1}{Y\sqrt{2}}(\dot{Y} - Y'e^{-\Gamma/2}) \\
\beta &= -\alpha = \frac{\cot\theta}{2Y\sqrt{2}}, \\
\epsilon &= -\gamma = \frac{\dot{\Gamma}}{4\sqrt{2}}
\end{aligned} \tag{3}$$

where $\dot{Y} = \partial Y/\partial t$, $Y' = \partial Y/\partial r$. For the definitions see e. g., Chandrasekhar, 1983 and Penrose and Rindler, 1984.

By expliciting the covariant spinor derivatives in terms of the directional derivatives and spin coefficients (3) the equation (1) reduces to the system of coupled differential equations

$$
\begin{aligned}
&(D-2\rho)\Phi_{10}-(\delta^{\star}-2\alpha)\Phi_{00}=i\mu_{\star}\Theta_{00'}\\
&(D-\rho+2\epsilon)\Phi_{11}-\delta^{\star}\Phi_{10}=i\mu_{\star}\Theta_{10'}\\
&(\Delta+\mu-2\gamma)\Phi_{00}-\delta\Phi_{01}=-i\mu_{\star}\Theta_{01'}\\
&(\Delta+2\mu)\Phi_{10}-(\delta+2\beta)\Phi_{11}=-i\mu_{\star}\Theta_{11'}\\
&(D-\rho)\Theta_{01'}-\delta\Theta_{00'}=-i\mu_{\star}\Phi_{00}\\
&(D-\rho+2\epsilon)\Theta_{11'}-(\delta+2\beta)\Theta_{10'}+\mu\Theta_{00'}=-i\mu_{\star}\Phi_{10}\\
&(\delta^{\star}+2\beta)\Theta_{01'}-(\Delta+\mu-2\gamma)\Theta_{00'}+\rho\Theta_{11'}=-i\mu_{\star}\Phi_{01}\\
&\delta^{\star}\Theta_{11'}-(\Delta+\mu)\Theta_{10'}=-i\mu_{\star}\Phi_{11}
\end{aligned}
\tag{4}
$$

(Note that the situation is similar to the general case of arbitrary spin field equation in RW space-time (Zecca, 2009) when specialized to spin $s = 1$). To separate the system (4) it is useful to put

$$
\begin{aligned}
&\Phi_{AB}(r,\theta,\varphi,t)=\alpha(t)\phi_k(r)S_k(\theta)e^{im\varphi}, \quad k=A+B=0,1,2\\
&\Theta_{00'}(r,\theta,\varphi,t)=A(t)\phi_1(r)S_1(\theta)e^{im\varphi}\\
&\Theta_{10'}(r,\theta,\varphi,t)=A(t)\phi_2(r)S_2(\theta)e^{im\varphi}\\
&\Theta_{01'}(r,\theta,\varphi,t)=-A(t)\phi_0(r)S_0(\theta)e^{im\varphi},\\
&\Theta_{11'}=-\Theta_{00'}
\end{aligned}
\tag{5}
$$

where, for convenience, we we assume $m = 0, \pm 1, \pm 2, \ldots$. By using (5) into equation (4) the angular dependence factors out and one is left with the equations in the $r,\ t$ variables

$$
\begin{aligned}
&(D-2\rho)(\alpha\phi_1)-\frac{\lambda_1}{Y\sqrt{2}}\alpha\phi_0=i\mu_* A\phi_1\\
&(D-\rho+2\epsilon)(\alpha\phi_2)-\frac{\lambda_2}{Y\sqrt{2}}\alpha\phi_1=i\mu_* A\phi_2\\
&(\Delta+\mu+2\epsilon)(\alpha\phi_0)-\frac{\lambda_3}{Y\sqrt{2}}\alpha\phi_1=i\mu_* A\phi_0\\
&(\Delta+2\mu)(\alpha\phi_1)-\frac{\lambda_4}{Y\sqrt{2}}\alpha\phi_2=i\mu_* A\phi_1\\
&(D-\rho)(A\phi_0)+\frac{\lambda_3}{Y\sqrt{2}}A\phi_1=i\mu_*\alpha\phi_1\\
&(D-\rho+2\epsilon)(A\phi_1)-\mu A\phi_1+\frac{\lambda_4}{Y\sqrt{2}}A\phi_2=i\mu_*\alpha\phi_2\\
&(\Delta+\mu+2\epsilon)(A\phi_1)+\rho A\phi_1+\frac{\lambda_1}{Y\sqrt{2}}A\phi_0=i\mu_*\alpha\phi_1\\
&(\Delta+\mu)(A\phi_2)+\frac{\lambda_2}{Y\sqrt{2}}A\phi_1=i\mu_*\alpha\phi_2
\end{aligned}
\tag{6}
$$

Instead the angular functions satisfy the equations

$$
\begin{aligned}
L_1^- S_0 &= \lambda_1 S_1, \\
L_0^- S_1 &= \lambda_2 S_2, \\
L_0^+ S_1 &= \lambda_3 S_0, \\
L_1^+ S_2 &= \lambda_4 S_1,
\end{aligned}
\tag{7}
$$

where it has been set $L_n^{\pm} = \partial_\theta \mp m \csc\theta + n \cot\theta$. λ_i $(i = 1,2,3,4)$ are the corresponding separation constants. These equations are the same of those relative to the separation of spin 1 field in RW space-time (cfr. Zecca 2005; 2009). By setting $\lambda_1\lambda_3 = \lambda_2\lambda_4 = -\lambda^2$ the angular equations can be reduced to an eigenvalue problem (Zecca, 1996) whose solutions are expressible (Zecca, 2005) in terms of Legendre functions and Jacobi polynomials (For the definitions see e.g., Abramovitz and Stegun, 1970):

$$
\begin{aligned}
& S_{1lm} = (1-\xi^2)^{\frac{m}{2}} P_l^m(\xi), \quad l = |m|, |m|+1, .. \\
& S_{2lm} = (1-\xi)^{\frac{m-1}{2}} (1+\xi)^{\frac{m+1}{2}} P_{l-m}^{(m+1,m-1)}(\xi), \qquad m \geq 1,\ l = m, m+1, .. \\
& S_{2lm} = (1+\xi)^{\frac{|m|-1}{2}} (1-\xi)^{\frac{|m|+1}{2}} P_{l-m}^{(|m|-1,|m|+1)}(\xi),\ m \leq 1,\ l = |m|, |m|+1, .. \\
& S_{2l0} = \sin\theta\, P_{l+2}^{(1,1)}(\cos\theta), \quad l = 0,1,2, .. \\
& S_{0lm}(\theta) = S_{2l-m}(\theta), \qquad (\xi = \cos\theta),
\end{aligned}
\tag{8}
$$

with λ that takes the values $\lambda^2 = l(l+1)$, $l = 0,1,2,..$ By possibly considering a normalization factor, the angular functions satisfy

$$
\int d\Omega\ S_{ilm}(\theta) e^{im\varphi} \left(S_{il'm'}(\theta) e^{im'\varphi} \right)^* = \delta_{ll'}\delta_{mm'} \quad (i = 0,1,2)
\tag{9}
$$

a relation usefull in view of an ortho-normalization of the complete solution of (1).

For what concerns the separation of the r and t dependence in (6), it does not seem to be obtainable in general even by using the explicit expression of the spin coefficients. In the following we confine within a class of LTB model for which Γ is related to the function Y and Y itself can be given in an explicit parametric factorized form.

3. Variable separation in Lemaître - Tolman - Bondi cosmological models

The system (6) can be further separated in its r, t dependence in a sufficiently large class of cosmological models. Suppose to that end that the universe is filled with freely falling dust like matter without pressure, as seen in the comoving spherically symmetric space-time coordinates (2). If the proper energy momentum tensor is considered, the corresponding Einstein equation can be integrated exactly in parametric form and gives rise to what is widely known as the Lemaître-Tolman-Bondi (LTB) cosmological model. (For a comprehensive study of the model see Krasinski, 1997; in the Newman-Penrose formalism see e.g., Zecca, 1993).

The explicit solution is the following (Demianski and Lasota, 1973)

$$
\begin{aligned}
&Y = G\frac{m(r)}{2E(r)}(\cosh\eta - 1); \;\; t = t_0(r) + G\frac{m(r)}{(2E(r))^{\frac{3}{2}}}(\sinh\eta - \eta), \quad \eta > 0, \quad E > 0 \\
&Y = G\frac{m(r)}{-2E(r)}(1 - \cos\eta); \;\; t = t_0(r) + G\frac{m(r)}{(-2E(r))^{\frac{3}{2}}}(\eta - \sin\eta), \;\; 0 \le \eta \le 2\pi, \;\; E < 0 \\
&Y = \left[\frac{3}{2}(2m(r))^{\frac{1}{2}}(t - t_0(r))\right]^{\frac{2}{3}}, \qquad E = 0
\end{aligned} \tag{10}
$$

$m(r)$, $E(r)$, $t_0(r)$ are arbitrary integration functions that depend only on the radial coordinate and G the gravitational constant. In particular $m(r)$ can be interpreted as the mass contained in a sphere of radius Y, $m(r) = 4\pi G \int_0^r \sigma(r,t) Y^2(r,t) Y'(r,t) dr$, $\sigma(r,t)$ being the matter density. Moreover Γ and Y are no more independent but

$$
\exp\Gamma \;=\; \frac{Y'^2(r,t)}{1 + 2E(r)} \tag{11}
$$

a relation usefull for the following purposes.
Suppose now to choose $t_0(r) = 0$ in every case and, in case $E \neq 0$,

$$
G\, m(r) = \; (2|E|)^{\frac{3}{2}} \tag{12}
$$

With this choices the physical radius in (10) reads

$$
\begin{aligned}
&Y = E^{\frac{1}{2}}(\cosh\eta - 1); \quad t = \sinh\eta - \eta, \quad \eta > 0, \quad E > 0 \\
&Y = |E|^{\frac{1}{2}}(1 - \cos\eta); \quad t = \eta - \sin\eta, \quad 0 \le \eta \le 2\pi, \quad E < 0 \\
&Y = \left(\frac{9}{2}\right)^{\frac{1}{3}} m^{\frac{1}{3}}\, t^{\frac{2}{3}}, \qquad E = 0
\end{aligned} \tag{13}
$$

These assumptions are sufficient to separate the system (6). Indeed from (13), Y is in every case of the form $Y = Z(r)T(t)$. By using this factorization and relation (11) in the expression of the directional derivatives and spin coefficients, one is able to separate the time dependence from eq. (6). The result is expressed in terms of the coupled time equation

$$
\begin{aligned}
&\dot{\alpha}T + 2\dot{T}\alpha - im_0 AT = -ik\alpha \\
&A\dot{T} + \dot{A}T - im_0\alpha T = ikA
\end{aligned} \tag{14}
$$

These equations are formally those of the separation of the spin 1 field equation in RW metric. Therefore the solutions $\alpha_k(t)$, $A_k(t)$ satisfy the constraint

$$
T^3(t)\left[A_k(t)\alpha^*_{-k}(t) + A^*_{-k}(t)\alpha_k(t)\right] = const \tag{15}
$$

The result follows from Zecca (2006a) after the substitution $R(t) \;\rightarrow\; T(t)$. Also this property is an usefull tool for the normalization of the complete solution of (1).

Instead, for what concerns the radial dependence, one obtains

$$
\begin{aligned}
ik &= \frac{\sqrt{1+2E}}{Z'}\frac{\phi_1'}{\phi_1} + \frac{2}{Z}\sqrt{1+2E} - \frac{\lambda_1}{Z}\frac{\phi_0}{\phi_1} \\
ik &= \frac{\sqrt{1+2E}}{Z'}\frac{\phi_2'}{\phi_2} + \frac{1}{Z}\sqrt{1+2E} - \frac{\lambda_2}{Z}\frac{\phi_1}{\phi_2} \\
-ik &= \frac{\sqrt{1+2E}}{Z'}\frac{\phi_0'}{\phi_0} + \frac{1}{Z}\sqrt{1+2E} + \frac{\lambda_3}{Z}\frac{\phi_1}{\phi_0} \\
-ik &= \frac{\sqrt{1+2E}}{Z'}\frac{\phi_1'}{\phi_1} + \frac{2}{Z}\sqrt{1+2E} + \frac{\lambda_4}{Z}\frac{\phi_2}{\phi_1}
\end{aligned} \tag{16}
$$

k is a separation constant, the same in all equations, to ensure consistency in the separation procedure.

4. Decoupling and properties of the radial solutions

The equations (16) are similar to the corresponding ones of the RW metric (Zecca, 2005) and can therefore be disentangled in a similar way. By defining the operator

$$
A_b = \sqrt{1+2E}\left(\frac{1}{Z'}\frac{d}{dr} + \frac{b}{Z}\right) - ik, \quad b \in \mathbf{C} \tag{17}
$$

eqs. (16) reads

$$
\begin{aligned}
A_2\phi_1 &= \frac{\lambda_1}{Z}\phi_0 & \qquad A_1\phi_2 &= \frac{\lambda_2}{Z}\phi_1 \\
A_1^*\phi_0 &= -\frac{\lambda_3}{Z}\phi_1 & \qquad A_2^*\phi_1 &= -\frac{\lambda_4}{Z}\phi_2
\end{aligned} \tag{18}
$$

and can be easily reduced to equations in a single function

$$
\begin{aligned}
ZA_2ZA_1^*\phi_0 &= -\lambda_1\lambda_3\phi_0 \\
ZA_1ZA_2^*\phi_1 &= -\lambda_2\lambda_4\phi_1 \\
ZA_1^*ZA_2\phi_1 &= -\lambda_1\lambda_3\phi_1 \\
ZA_2^*ZA_1\phi_2 &= -\lambda_2\lambda_4\phi_2
\end{aligned} \tag{19}
$$

By taking into account that $\lambda_1\lambda_3 = \lambda_2\lambda_4 = -\lambda^2$, one has further that the radial solutions satisfy $\phi_1 \equiv \phi_1^*$, $\phi_0 \equiv \phi_2^*$. Therefore it suffices to solve two independent ordinary differential equations. By expliciting the equations for ϕ_0, ϕ_1 one obtains respectively

$$
\begin{aligned}
&\frac{Z}{Z'^2}(1+2E)\phi_0'' + \left[(1+2E)\left(\frac{4}{Z'} - \frac{ZZ''}{Z'^3}\right) + \frac{E'Z}{Z'}\right]\phi_0' + \\
&\qquad + \left[\frac{E'}{Z'} + \frac{2-\lambda^2+4E}{Z} + k^2 Z + 2ik\sqrt{1+2E}\right]\phi_0 = 0
\end{aligned} \tag{20}
$$

$$
\begin{aligned}
&\frac{Z}{Z'^2}(1+2E)\phi_1'' + \left[(1+2E)\left(\frac{4}{Z'} - \frac{ZZ''}{Z'^3}\right) + \frac{E'Z}{Z'}\right]\phi_1' + \\
&\qquad + \left[\frac{2E'}{Z'} + k^2 Z + \frac{2-\lambda^2+4E}{Z}\right]\phi_1 = 0
\end{aligned} \tag{21}
$$

Note that the Robertson-Walker metric is a special case of the LTB metric with $Y = rR(t)$, $Z(r) = r$, $2E(r) = -ar^2, (a = 0, \pm 1)$. One can check that with this choice, eqs. (20), (21) become exactly the separated radial equation of spin 1 field in RW metric that were derived in (Zecca, 2005). In RW flat case, normal modes of the field equation, have also been determined (Zecca, 2006a) and a quantization procedure developed leading to the possibility of particle production in expanding universe (Zecca, 2009a). Consequently a simple extension of the Standard Cosmological model has been proposed to include particle production (Zecca, 2010). Instead in the curved cases of the RW metric the eqs. (20), (21) have been solved by reduction to Heun's equation (Zecca, 2009a) without however succeding in determining the normal modes.

In the LTB case, the solution of the radial equations seems quite difficult for a general $E(r)$. In particular this is due to the presence of the square root term in (20). One could try to reduce the equations by expliciting, as assumed in (13), $Z(r) = |E(r)|^{1/2}$ for $E \neq 0$ and $Z(r) = \left(9m(r)/2\right)^{1/3}$ for $E = 0$. However, even with these specifications into the radial equations, the solution does not become easier.

5. Solution of the separated time equations

In the previous Sections the spin 1 field equation has been separated in the three classes of LTB cosmological models, each of them depending on an arbitrary radial function. The resulting time equations (14) are, contrarily to the radial equations, independent of any model integration function. Therefore it seems usefull to give the explicit solution of the time equations in each case. By setting $B(t) = \alpha(t)T^2(t)$, $\gamma(t) = A(t)T(t)$ the equations (14) can be easily reported to the form

$$\begin{aligned} im_0 B &= \dot{\gamma}T - ik\gamma, \\ \ddot{\gamma}T + \dot{\gamma}\dot{T} + \gamma\Big(m_0^2 T + \frac{k^2}{T}\Big) &= 0 \end{aligned} \tag{22}$$

In this way it suffices to solve the equation for $\gamma(t)$ to obtain $\alpha(t)$ and $A(t)$. The object is now of integrating the equation (22) for γ by distinguishing according to the different situations of E in (13).

5.1 Time equation for $E = 0$

Here $T(t) = t^{2/3}$. When substituted into the equation for γ in (22) and then by setting $s = t^{1/3}$ one obtains

$$\frac{d\gamma}{ds^2} + (9m_0^2 s^4 + 9k^2)\gamma = 0 \tag{23}$$

The solution of (23) can be given by both odd and even regular functions that can be determined by series. By setting $\gamma = \sum_0^\infty c_n s^n$ into (23) one has the recurrence relation

$$\begin{aligned} &(n+1)(n+2)c_{n+2} + 9k^2 c_n = 0, \quad n = 0,1,2,3, \\ &(n+1)(n+2)c_{n+2} + 9k^2 c_n + 9m_0^2 c_{n-4} = 0, \quad n = 4,5,\ldots \end{aligned} \tag{24}$$

Two independent integral γ_0, γ_1 can be obtaind by setting respectively $c_1 \neq 0$, $c_0 = 0$ and $c_0 \neq 0$, $c_1 = 0$. As a consequence of the recurrence relation (24), the general solution is of the form $\gamma(s) = a_0\gamma_0(s) + a_1\gamma_1(s)$, γ_0, γ_1 being respectively an odd and an even function.

The radius of convergence of the series is different from 0, on account of general results (e.g., Moon and Spancer, 1961; Magnus and Winkler, 1979). One has therefore the t dependence

$$\begin{aligned} \gamma_0(t) &= c_1 t^{\frac{1}{3}} + c_3 t + c_5 t^{\frac{5}{3}} + c_7 t^{\frac{7}{3}} + \dots \\ A_0(t) &= \gamma_0 T^{-1} = \frac{c_1}{t^{\frac{1}{3}}} + c_3 t^{\frac{1}{3}} + c_5 t + c_7 t^{\frac{5}{3}} + \dots \end{aligned} \tag{25}$$

and $\alpha_0(t) = B_0(t)T^{-2}(t)$ where $B_0(t)$ follows from (25), the first equation (22) and the expression of $T(t)$. Similarly for $\alpha_1(t)$, $A_1(t)$.

5.2 Time equation for $E < 0$

Since in the present case $T(\eta) = 1 - \cos\eta$, $t = \eta - \sin\eta$, the eq. (22) can be reported to a differential equation in the variable η

$$\begin{aligned} &\frac{d^2\gamma}{d\eta^2} + \left[\nu_0 + \nu_1\cos\eta + \nu_2\cos 2\eta\right]\gamma = 0, \qquad 0 \le \eta \le 2\pi \\ &\nu_0 = \frac{3}{2}m_0^2 + k^2, \qquad \nu_1 = -2m_0^2, \qquad \nu_2 = \frac{m_0^2}{2} \end{aligned} \tag{26}$$

Note that, by setting $\chi = \eta/2$, the equation (26) assumes the form of a Wittaker-Hill equation (Magnus and Winkler 1979) of period π;

$$\begin{aligned} &\frac{d^2\gamma}{d\chi^2} + \left[\lambda_0 + 4\bar{m}q\cos(2\chi) + 2q^2\cos(4\chi)\right]\gamma = 0 \\ &\lambda_0 = 4k^2 + 3m_0^2, \qquad q = \pm m_0, \qquad \bar{m} = \pm 2m_0 \end{aligned} \tag{27}$$

The interest in this form of the equation lies in that it may have periodic solutions of period π or 2π. However this possibility is prevented in the present case because the parameter $\bar{m} = \pm 2m_0$ is not, as required, an integer number (see e.g., Magnus and Winkler 1979, Theorem 7.9), m_0 being the mass of the particle. Therefore it is convenient to solve directly eq. (26) by series. It appears that a solution of (26) can be an odd or an even function, We consider separately the cases. By setting $\gamma(\eta) = \sum_0^\infty c_{2n}\eta^{2n}$ into the equation for γ in (26), one obtains for the coefficients the recurrence relation

$$(2n+2)(2n+1)c_{2n+2} + \nu_0 c_{2n} + \sum_{j=0}^{n} \frac{(-1)^j}{(2j)!}\left[\nu_1 + \nu_2 2^{2j}\right] c_{2n-2j} = 0, \quad n = 0,1,2,\dots \tag{28}$$

If instead one looks for odd solutions, $\gamma(\eta) = \sum_0^\infty c_{2n+1}\eta^{2n+1}$, one finds from (26) the recurrence relation

$$(2n+3)(2n+2)c_{2n+3} + \nu_0 c_{2n+1} + \sum_{j=0}^{n} \frac{(-1)^j}{(2j)!}\left[\nu_1 + \nu_2 2^{2j}\right] c_{2n+1-2j} = 0, \quad n = 0,1,2,\dots \tag{29}$$

In both cases the coefficients are completely determined by the first one. To obtain $\gamma(t)$ one has to reverte the expression $t = \eta - \sin\eta$ to have $\eta = \eta(t)$ to be substituted in the series expression of the solution.

5.3 Time equation for $E > 0$

By expressing now the unknown function γ in terms of η with $T(\eta) = \cosh\eta - 1,\ t = \sinh\eta - \eta,\ (\eta > 0)$, the γ-equation in (22) becomes

$$\frac{d^2\gamma}{d\eta^2} + \left[m_0^2(\cosh\eta - 1)^2 + k^2\right]\gamma = 0 \tag{30}$$

that can be put into the form

$$\begin{gathered}\frac{d^2\gamma}{d\eta^2} + \left[\sigma_0 + \sigma_1\cosh\eta + \sigma_2\cosh 2\eta\right]\gamma = 0\\ \sigma_0 = k^2 + \frac{3}{2}m_0^2, \qquad \sigma_1 = -2m_0^2, \qquad \sigma_2 = \frac{m_0^2}{2}\end{gathered} \tag{31}$$

The last equation can be integrated by series by distinguishing again between even and odd solutions. By setting $\gamma_1(\eta) = \sum_0^\infty c_{2n}\eta^{2n}$ into (31) one has the recurrence relation for the coefficients c_n's

$$(2n+2)(2n+1)c_{2n+2} + (\sigma_0+\sigma_1+\sigma_2)c_{2n} + \sum_{j=1}^{n}\frac{c_{2n-2j}}{(2j)!}\left(\sigma_1 + \sigma_2 2^{2j}\right) = 0, \quad n = 0,1,2,\ldots \tag{32}$$

Instead by setting $\gamma_1(\eta) = \sum_0^\infty c_{2n+1}\eta^{2n+1}$ into (31) one has

$$(2n+3)(2n+2)c_{2n+3} + (\sigma_0+\sigma_1+\sigma_2)c_{2n+1} + \sum_{j=1}^{n}\frac{c_{2n+1-2j}}{(2j)!}\left(\sigma_1 + \sigma_2 2^{2j}\right) = 0, \quad n = 0,1,.. \tag{33}$$

Here the general solution, $\gamma(t) = a_1\gamma_1(t) + a_2\gamma_2(t)$, follows again by expressing $\eta = \eta(t)$ into $\gamma_1(\eta),\ \gamma_2(\eta)$.

5.3.1 Time equation for $E > 0$ and large t

In the present case one can also determine the behaviour of the situation for large t (large η). To that end, by setting $y = \exp\eta$, the equation (30) becomes

$$\frac{d^2\gamma}{dy^2} + \frac{1}{y}\frac{d\gamma}{dy} + \left[\frac{m_0^2}{4} - m_0^2\frac{1}{y} + \frac{k^2 + 3m_0^2/2}{y^2} - \frac{m_0^2}{y^3} + \frac{m_0^2}{4}\frac{1}{y^4}\right]\gamma = 0 \tag{34}$$

that is in a suitable form for the mentioned purpose. By looking for asymptotic solutions of the form

$$\gamma(\eta) = y^\delta e^\chi \sum_{n=0}^{\infty}\frac{c_{-n}}{y^n} \tag{35}$$

one finds, by inserting into eq. (34),

$$\chi = \frac{-1 \pm \sqrt{1 - m_0^2}}{2}, \qquad \delta = \pm\frac{m_0^2}{\sqrt{1 - m_0^2}} \tag{36}$$

Therefore by considering the dominant term in (35), one has, for $y \to \infty$

$$\gamma(y) \sim y^{\pm\frac{m_0^2}{\sqrt{1-m_0^2}}} e^{\frac{-1\pm\sqrt{1-m_0^2}}{2}y} \tag{37}$$

that is a decaying behaviour, except for $m_0 = 1$ in which case the approximation is not valid. Note that for large t, $t \sim e^{\eta}/2 = y/2$ so that the behaviour (37) is also the same of that of $\gamma(t)$ for large t.

6. Remarks and comments

In the previous Sections the spin 1 field equation has been separated in LTB space-times and reduced to ordinary differential equations in one variable. The angular dependence of the wave spinor factors out in a general LTB metric. Due to spherical symmetry it is the same that the corresponding one in Robertson-Walker and Schwarzschild metric. The further separation of the time and radial coordinates has been possible in LTB cosmologies for which the physical radius has the factorised form $Y = Z(r)T(t)$. This assumption still let the LTB comological model depend on an arbitrary function $E(r)$ (or $m(r)$). As a consequence the separated time dependence is essentially unique in the sense that it depends only on the sign of E or on its vanishing. The time equations have been separated and integrated in all cases.
Instead the radial dependence is reported to the solution of two independent ordinary differential equations that explicitly depend on E. The choice $E(r) = 0$, $Z(r) = r$, $T(t) = R(t)$ ($R(t)$ the radius on the universe in the RW metric) reduces the scheme to a special case of the RW space-time. In this case the radial equations can be explicitly solved (Zecca, 2005). Moreover if one considers toghether with (1) also its complex conjugate equation, a scalar product, induced by a conserved current, can be defined between solutions of (1). Correspondingly normal modes can be defined, that are the base for a quantization of the scheme. In turn this implies that particle creation is possible and that the number of one mode created particles per unit time in expanding universe is proportional to $\dot{R}(t)/R(t)$ (Zecca, 2009a). These results, applied to the present LTB scheme with $E = 0$, $R(t) = T(t) = t^{2/3}$, give that the number of one mode created particles per unit time is proportional to $\dot{T}/T = 2/(3t)$.
Suppose now $E \neq 0$. The procedure of the mentioned RW case, can be applied to define a scalar product between solution of (1), as induced by the conserved current (Zecca, 2006a; 2009a). This product finally factorizes in a product of reduced scalar products in a single variable as a consequence of the assumption $Y = Z(r)T(t)$. By taking into account the orthogonality relation (9) for the angular solutions, the relation (15) for the time dependence and by proceeding as in Zecca, 2006a, one is finally left with a one dimensional scalar product for the solutions of the radial equations (20), (21). If the assumptions on $E(r)$ are such that the solutions of (20), (21) result ortho-normal in the reduced scalar product, then one recover a set of normal mode for the solutions of (1). Accordingly, a quantization procedure can be devoloped as in the flat RW case (Zecca, 2009a). On account of the complete analogy of the two schemes, again one obtains the results of Zecca (2009a) with the substitution $R(t) \to T(t)$. Therefore (with the mentioned suitable choice of E) the balance $n(t)$ of one mode created and

annihilated particles per unit of time is

$$n(t) \propto \frac{\dot{T}}{T} = \frac{\sinh\eta}{(\cosh\eta - 1)^2}; \quad t = \sinh\eta - \eta, \quad \eta > 0, \qquad E > 0 \tag{38}$$

$$n(t) \propto \frac{\dot{T}}{T} = \frac{\sin\eta}{(1 - \cos\eta)^2}; \quad t = \eta - \sin\eta, \quad 0 \le \eta \le 2\pi, \quad E < 0 \tag{39}$$

Therefore, for an LTB cosmology for which $\dot{Y} = Z(r)\dot{T}(t) \neq 0$ particle production is non trivial. Note that for these models one has

$$Y \propto Z(r)\, t^{\frac{2}{3}}, \qquad t \to 0 \tag{40}$$

$$n(t) \propto \frac{\dot{T}}{T} \propto \frac{2}{3}\frac{1}{t}, \qquad t \to 0 \tag{41}$$

for both $E > 0$ and $E < 0$. Hence the cosmological model admits a big bang origin at time $t = 0$ and, if particle production is taken for grant, there is, near the big bang origin, an enormous production of particles that does not depend on the sign of E. This is in some way the converse of what happens in the flat RW metric where particle production is possible for different cosmological dynamics, but with a well defined spatial configuration.

We now briefly comment the separation method employed here. The complete separation of (6) has been done under the special condition (12) for which the physical radius results to be factorized in the time and radial dependence. It would be interesting to know whether the mentioned condition is also in some sense necessary to obtain separated time and radial equations. This would throw also light in the separation of scalar and Dirac field equations that can be separated in LTB models under the same condition (Zecca, 2009; 2001). Solutions of (6) not involving Y-factorizations would be as well of interest.

Another point is the problem of the separation of field equations of spin values higher than 1 in LTB models. This is attractive because the explicit recursive structure of (4) is the same that in the Robertson-Walker metric that in turn is a special case of the general recursive structure for field equations of arbitrary spin (Zecca, 2009). However, as mentioned in the introduction, the presence of a non vanishing Weyl spinor as it happens in LTB metric (e. g., Penrose and Rindler, 1984; Zecca, 2000a) requires a more complex formulation of the field equation for spin greater than 1 (see e. g., Illge, 1993 and references therein). Also in this case it would be interesting to know whether the condition (11) still plays a central role for the separation of the equation, at least in the simplest case of spin $s = 3/2$. The problem is currently under investigation.

As final comment, if particle production is taken for grant, its effect is of modifying the gravitational dynamics of the universe. Therefore it should be taken into account in the formulation of a cosmological model. A precise formulation of the gravitational modification seems problematic. The previous quantization scheme does indeed foresee particle production but it does not specify where and with what density the particles are produced. However, by mediating over possible spatial distributions, a simple modification of the Standar Cosmological model has been proposed by an ansatz on the definition of energy density and of the pressure of the universe (Zecca, 2010).

7. References

Abramovitz, W.; Stegun, I. E. (1970). *Handbook of Mathematical Functions*. Dover, New York.

Birrell, N. D.; Davies, P. C. W. (1982). *Quantum fields in Curved Space*. Cambridge University Press, Cambridge.

Chandrasekhar, S. (1983) *The Mathematical Theory of Black Holes*. Oxford University Press, New York, 1983.

Demianski, H.; Lasota, J. P. (1973). Black Holes in Expanding Universe. *Nature Physical Science*, Vol 242, pp. 53-55.

Fulling, S. A. (1989) *Aspects of Quantum Field Theory in Curved Space-time*. Cambridge University Press. Cambridge.

Illge, R. (1993). Massive Fields of Arbitrary spin in Curved Space-times. *Communications in Mathematical Physics*, Vol. 158, pp. 433-457.

Krasinski, A. (1997). *Inhomogeneous Cosmological Models*. Cambridge University Press. Cambridge.

Magnus, W. and Winkler, S. (1979). *Hill's Equation*. Dover Publication, New York.

Moon, P.; Spencer, D. E. (1961). *Field Theory Handbook*. Springer Verlag, Berlin.

Moradi, S. (2008). Creation of scalar and Dirac Particles in Asymptotically flat Robertson Walker Space-times. *International Journal of Theoretical Physics*. Vol. 47, pp. 2807-2818.

Newman, E.; Penrose, R. (1962). An Approach to Gravitational Radiation by a Method of Spin Coefficients. *Journal of Mathematical Physics*, Vol. 3, pp. 566-578.

Parker, L. (1969). Quantized Fields and Particle Creation in Expanding Universes. I. *Physical Review*, Vol. 183, pp. 1057-1068.

Parker, L. (1971). Quantized Fields and Particle Creation in Expanding Universes . II. *Physical Review*, Vol. 3, pp. 346-356.

Parker, L.; Toms, D. (2009). *Quantum Field Theory in Curved Space - Time*. Cambridge University Press. Cambridge

Penrose, R.; Rindler, W. (1984). *Spinors and Space-time*. Cambridge University Press, Cambridge.

Weinberg, S. (1972). *Gravitation and Cosmology*, John Wiley Inc., New York.

Zecca, A. (1993). Some Remarks on Dirac's Equation in the Tolman-Bondi Geometry, *International Journal of Theoretical Physics*. Vol. 32, pp. 615-624.

Zecca, A. (1996). Separation of the Massless Spin-1 Equation in Robertson-Walker Space-Time. *International Journal of Theoretical Physics*. Vol. 35, pp. 323-331.

Zecca, A. (2000). Dirac Equation in Lemaître-Tolman-Bondi Models. *General Relativity and Gravitation*. Vol. 32, pp. 1197-1206.

Zecca, A. (2000a). Weyl Spinor and Solutions of Massless Free Field Equations. *International Journal of Theoretical Physics*. Vol. 39, pp, 377-387.

Zecca, A. (2001). Scalar Field Equation in Lemaître-Tolman-Bondi Cosmological Models. *Il Nuovo Cimento*. Vol. 116B. pp. 341-350.

Zecca, A. (2005). Solution of the Massive Spin 1 Equation in expanding Universe. *Il Nuovo Cimento B*. Vol. 120. pp. 513-520 .

Zecca, A. (2005a). Massive Spin 1 Equation in Schwarzschild Geometry. *Il Nuovo Cimento*, Vol. 120B, pp, 1017-1020.

Zecca, A. (2006). Proca fields interpretation of spin 1 equation in Robertson - Walker space - time. *General Relativity and Gravitation*. Vol. 38, pp. 837-843.

Zecca, A. (2006a). Normal modes for massive spin 1 equation in Robertson - Walker space - time. *International Journal of Theoretical Physics*. Vol. 45, pp. 1958-1964.

Zecca, A. (2006b). Massive field equations of arbitrary spin in Schwarzschild geometry: separation induced by spin-3/2 case. *International Journal of Theoretical Physics*, Vol. 45, pp. 2208-2214.

Zecca, A. (2009). Variable separation and solutions of massive field equations of arbitrary spin in Robertson-Walker space-time, *Advanced Studies in Theoretical Physics*, Vol. 3, pp. 239-250.

Zecca, A. (2009a). Spin 1 Mode Particles Production in Models of Expanding Universe. *Advanced Studies in Theoretical Physics*, Vol. 3, pp. 493-502.

Zecca, A. (2010). The Standard Cosmological Model Extended to Include Spin 1 Mode Particle Production. *Advanced Studies in Theoretical Physics*, Vol. 4, pp. 91-100.

Zecca, A. (2010a) Instantaneous Creation of scalar particles in expanding universe. *Advanced Studies in Theoretical Physics*, Vol. 4. pp, 797-804.

Zecca, A. (2010b). Quantization of Dirac field and particle production in expanding universe. *Advanced Studies in Theoretical Physics*, Vol. 4, pp. 951-961.

14

On the Dilaton Stabilization by Matter

Alejandro Cabo Montes de Oca
Departamento de Física Teórica, Instituto de Cibernética, Matemática y Física, La Habana Cuba

1. Introduction

The Dilaton field is a scalar partner of the graviton in the context of superstring theory (1). Then, the background fields in the vacuum state of this theory should involve its component in common with the metric ones in the basic action (2; 3). To the simplest approximation the Dilaton is a massless scalar field showing a special sort of interaction with the matter modes. This type of coupling, determines that a time varying Dilaton induces time-dependent coupling constants. Therefore, to overcome this difficulty this field should remain constant at the current stage of cosmological evolution. In addition, unless it becomes very massive, its existence can imply an observable kind of "Fifth force", being similar to the ones which are currently associated to the observations of the Dark Matter. The constraints posed by current experimental observations determine the lower bound on the mass of the Dilaton to be of the order $m < 10^{-12}$GeV (4) . However, there are attempts to make a time dependent Dilaton consistent with late time cosmology (see (5)).
Therefore, the Dilaton stabilization problem has been the objective of an intense research activity in recent times due to its physical relevance. It can be emphasized that the Dilaton is one of various scalar fields emerging from the formulation of superstring theory in its low-energy limit. Scalar fields describing the sizes and shapes of the extra spatial dimensions associated in this theory are also arising, and are called "moduli fields". The stabilization of such moduli fields has also been the object of recent attention, specially in connection with Type IIB superstring theory. The introduction of fluxes within the compactification spaces has made it possible to stabilize various moduli fields (7). Moreover, gaugino condensation effects (8) has been argued to stabilize the Dilaton field in the framework of heterotic superstring theory (9) and also in string gas cosmology (10).
It can be underlined that, since Dilaton stabilization has special relevance for late time cosmology, there is motivation for finding mechanisms which do not directly rest on the concrete assumptions defining the nature of the extra dimensions. Further motivation to search for alternative Dilaton stabilization mechanisms appears in connection with String Gas Cosmology (SGC). The SGC (11; 12) is a model of early universe cosmology which employs new degrees of freedom and symmetries of string theory, and couples these elements with gravity and Dilaton fields into a classical action model. The Universe is assumed to start as a compact space filled with a gas of strings. Then, because in string theory there is a maximal temperature for a gas of closed strings, the cosmological evolution in SGC starts from a phase of almost constant temperature, called the "Hagedorn phase". The SGC allows to define a non-singular cosmology in which there is no initial Big Bang explosion. Also, it has been identified that the thermal fluctuations in a gas of closed strings in the Hagedorn phase gives

justification to the observed scale-invariant spectrum of cosmological fluctuations in Nature (13; 14), by adding a particular prediction of a slight blue tilt for gravitational waves (15). In this, the consistency of the picture also requires that the Dilaton field be stabilized during the Hagedorn phase. Therefore, in the SGC theory the Dilaton is also required to be fixed at very early times as well as at very late times.
In the present review chapter I will resume the conclusions of two studies previous done in common with various collaborators, in connection with the Dilaton vacuum field. They were presented in Refs. (32; 33). Each of these works assumes different properties for the Dilaton field as described below in the following two subsections:

1.1 a) Small mass Dilaton

In the discussion done in (32), which will be resumed in the section 2 of this chapter, the Dilaton field was assumed to show a small mass. Therefore a static solution of the KG equation for the Dilaton in interaction with gravity and dust matter was searched in that work. The configuration found showed a large region of homogeneity close to a central symmetry point, which becomes increasingly spatially varying at large distances. The existence of this static solution essentially rests on the presence of an interaction of the Dilaton field with pressureless matter. The solution obtained was a generalization of one formerly investigated in Ref. (18; 19) in the absence of matter. The special behavior of the scalar field in such works led to the proposal made in Ref. (18) about considering it as representing the Dilaton of the string theory (20). The idea came from the arising circumstance that when you fix the value of the scalar field (which have dimension of mass) at the central symmetry point to be at the Planck scale, by also requiring an amount of Hubble effect similar to the experimental one, the radius of existence of the solution gets a value $R = 10^{28} cm$ which is near the radius of the Universe. Also very much curious is that the values of KG mass of the scalar field obtained by fixing the above parameters, results to be of the order of $1/R$. That is, a very small value which seems compatible with a very tiny mass acquired by the Dilaton due to boundary conditions or non perturbative effects, which could deviate its mass from its vanishing first approximation.
It should be remarked that the assumption about the isotropic and homogeneous nature of our Universe, that is the Cosmological Principle, is central to modern Cosmology (16). However, recent experimental observations suggest the possibility for the break down of the validity of the principle at large scales (17). Accepting such a breaking will become necessary if the obtained solution result to be realized in Nature. Various static models of the Universe have been considered. Among them are the ones of Einstein, Le Maitre and de'Sitter, respectively. Originally, Einstein (16) examined a Universe filled of uniformly distributed matter but obtained a non-static metric. This result motivated him to introduce in his equations the Cosmological Constant term λ, with the objective of allowing the obtaining of a static solution. In connection with the solution discussed in (32) it follows that the centrally symmetric static scalar field which satisfies the Einstein-Klein-Gordon equations (EKG), curves the space time in a form resembling the one in the de'Sitter space in a large neighborhood of the origin of coordinates (19). The fact that the scalar field is more weakly varying along the radial distances when its value at the origin is lower is an interesting arising property to underline. The associated densities of energy and pressure are positive and negative respectively and weakly varying, approximating the presence of a positive Cosmological Constant. These properties suggested the idea proposed in (18) about considering the Dark Energy (DE) as described by a scalar field in this approximately homogeneous field configuration studied in (19). As mentioned before, this assumption will determine the abandoning of the

Cosmological Principle in favor of what could be imagined as a kind of "Matryoshka" model of the Universe. In this conception, proposed in (18; 19), we could be living inside of a particular configuration in which the Dilaton field has a definite value resulted from the collapse of string matter in fermionic states. Then, the idea comes to the mind about that the Dilaton field could be radiated by the string matter in fermionic states under the extreme conditions of the collapse of fermion matter. The effective realization of this picture in Nature, could lead to the possibility that the astrophysical black-holes (by example the ones which are expected to exist near the centers of the Galaxies) could be no other things that small Universes in which the Dilaton field gets a different value to the external one. This change could be produced again by the collapse of fermion matter in falling to the collapsed configuration, upon the possible radiation of zero angular momentum modes, that is of the Dilaton to furnish the variation of the internal Dilaton field. We find this picture as an interesting one and think that its exploration is worth considering. One point to note, is that the proposed collapsed structures would resemble the so called "gravastars" in Refs. (35; 36). At this point it might be helpful to underline that given that the above recurrent picture is realized in Nature, supports the interest of the ideas argued in Ref. (37), about the connections between the cosmological constant and the quantum behavior of matter in such internal universes.
An important outcome emerged in the examination of the problem, is that the coexistence of the scalar field as described by the EKG equations including also the dust energy momentum tensor does not allow the existence of static solutions, at least in centrally symmetric configurations, in the absence of Dilaton - matter interaction (26). However, when the interaction is allowed a solution appears. The introduction of the coupling does not damage the almost homogenous character of the solution in a relative large region around the origin of the central symmetry, being far away form the limits of the Universe. Another interesting outcome is that the distributions of matter and Dilaton field both show a very close behavior. That is, the scalar field is able to sustain an amount of matter being almost proportional between them.

1.2 a) Large mass stabilized by matter Dilaton

In Ref. (33), which results will be reviewed in section 3 of this chapter, in an alternative way as in Ref. (32), the possibility for the Dilaton to acquire an appreciable mass due to its generic interaction with the matter fields was investigated. In other words, the idea which motivated this study was the universal type of coupling of the Dilaton to the matter fields. This property, could not only lead to an unwanted effect as the mentioned time-dependence of the coupling constants, but it also can give the possibility that quantum effects due to the interaction of the Dilaton with matter, could generate interesting contributions to the Dilaton effective potential. This question was started to be explored in Ref. (28). That work considered the cosmological periods when the additional spatial dimensions of superstring theory were already stabilized and the study was done in the framework of a four-dimensional field theory. The main objective of study was then the interaction of the Dilaton with massive fermions. These masses can be defined by fluxes through internal manifolds. Also, in late time cosmology, the masses could had been generated after supersymmetry breaking. In an alternative early universe cosmology, one may also take into account fermion masses generated by thermal effects. Ref. (28) considered a simple form for the Dilaton gravity action in which a massive Dirac fermion term was included (29). The Einstein frame, was chosen which does not show Dilaton field dependence in the kinetic terms for the fermions. Alternatively, the fermion mass becomes a function of the Dilaton through an universal exponential factor in Dilaton gravity (2; 3). The chosen action described the low energy effective interaction of Super-Yang-Mills fermions with the Dilaton field in superstring theory (28). The effective potential for the Dilaton field was evaluated up to two loop corrections in the small Dilaton radiative quantum field limit. That leads to a Yukawa like interaction term which allows standard QFT calculations. A fixed value of the cosmological scale factor was assumed. The outcome of the work was, thanks to the appearing of logarithms in the loop calculations, that the Dilaton field appeared in the result in powers multiplied by the exponential factors of the field. This structure, in the one loop approximation clearly indicated the spontaneous generation of vacuum mean value of the Dilaton field.

Motivated by the dynamical generation of the Dilaton result in Ref. (28), in Ref. (33) we addressed the evaluation of next corrections 3-loop terms to the 2-loop evaluation of the effective potential for the Dilaton field. The main issue explored in this work was the possibility of the appearance in the improved potential of stabilizing effects which were in fact absent in the second order correction, and which are suspected to be created by the existence of massive matter upon the mean value of the Dilaton.

The results obtained indicated, for the fermion mass being selected at the GUT or the top quark mass scales, that the mean value of the Dilaton field tends to be stabilized at a high value being close to the Planck mass or the GUT scale, respectively. Therefore, it was suggested that the appearance of mass for matter in the course of the evolution of the Universe can generate a stabilizing action on the vacuum expectation value of the Dilaton field making it unobservable. This effect will tend to stop the time evolution of the mean value, as it is convenient for String Theory consistency.

It should be remarked that in in Ref. (33), in the process of extending the work to include higher loop corrections, we have noticed that in Ref. (28), the kinetic term of the Dilaton Lagrangian was chosen with a negative sign. This selection, although not changing the one loop correction, led to a sign modification of the 2-loop terms, which suggested the existence of minima in the effective action argued in Ref. (28). However, in spite of this non physical adopted assumption, the indication about the dynamical generation in Ref.(28) remained a valid one, because the change in the metric did not affected the one-loop correction, the basic quantity indicating the dynamical generation effect. The results of the work in Ref. (33) and reviewed in this chapter, corrected the evaluation of the two loop term, and indicated that its place in the stabilizing effect over the Dilaton field is played by higher order contributions.

The exposition of section 3 will proceed as follows: In subection 3.1, the notation and basic formulation are given. Subsection 3.2 presents the elements of the one, two and three loops evaluation of the effective potential. Subsection 3.3 discuss the results of the evaluation. In the concluding subsection 3.4 the conclusions of the work are resumed and commented.

2. A cosmological model with a nearly massless Dilaton field

As remarked in the introduction this section 2 will resume the discussion of the work (32) in which the Dilaton field was assumed as a scalar field obeying the Einstein-Klein-Gordon equations in which the mass is assumed to be small. It should be underlined that this previous assumption resulted in radical contrast with the outcome of the later work reported in (33), which will be also reviewed in this chapter. However, the appearance of a large mass suggested by the discussion done in (33), as it will commented at the last section of the chapter devoted to the conclusions, could not result to be excluding some of the most motivating suggestions advanced in Ref. (32).

Given the isotropic and stationary character of the solution which was searched in Ref. (32), the structure of the metric was proposed in the standard form

$$ds^2 = v(\rho)dx^{o2} - u(\rho)^{-1}d\rho^2 - \rho^2(sin^2\theta d\varphi^2 + d\theta^2),$$

$$x^o = ct, \quad x^1 = \rho, \tag{1}$$

$$x^2 = \varphi, \; x^3 \equiv \theta, \tag{2}$$

from which the components of the Einstein tensor $G_{\mu\nu}$ were computed. Since the metric tensor is diagonal and only depending on ρ, the only non vanishing components of $G_{\mu\nu}$ resulted in

$$G_0^0 = \frac{u'}{\rho} - \frac{1-u}{\rho^2},$$

$$G_1^1 = \frac{u}{v}\frac{v'}{\rho} - \frac{1-u}{\rho^2},$$

$$G_2^2 = G_3^3 = \frac{u}{2v}v'' + \frac{uv'}{4v}(\frac{u'}{u} - \frac{v'}{v}) + \frac{u}{2\rho}(\frac{u'}{u} + \frac{v'}{v}).$$

The components G_2^2 and G_3^3 generated second order equations in the temporal component of the metric, which explicitly did not played an important role thanks to the Bianchi identities (16)

$$G^{\nu}_{\mu;\nu} = 0, \tag{3}$$

which were employed in the discussion. Assumed the satisfaction of the Einstein equations the G^{ν}_{μ} tensor was substituted by the energy momentum tensor T^{ν}_{μ}. Equation (3) was interpreted as a set of dynamical equations for the variables of the problem e, p and ϕ.

2.1 Matter and Dilaton dark energy

In this subsection let us sketch the way followed in (32) for obtaining two of the necessary equations needed to show the existence of the mentioned static model for the Universe: the Bianchi relations (3) and the static equation for the scalar field coupled to matter.
The action for the scalar field-matter in the given space time was written in the form

$$S_{mat-\phi} = \int L\sqrt{-g}d^4x, \tag{4}$$

where g is the determinant of the metric tensor, and it was considered that the Lagrangian density was taking the form:

$$L = \frac{1}{2}(g^{\alpha\beta}\phi_{,\alpha}\phi_{,\beta} + m^2\phi^2) + j\phi + L_{e,p}. \tag{5}$$

The first and the third terms of the right member of (5) are the Lagrangian densities of the KG scalar field and the dust-like matter respectively, while the second term was an interaction term between both quantities which was assumed to exist. The strength of the interaction was represented by the constant source j. Note that, the existing coupling of the Dilaton to matter fields made this supposition a natural one in our case in which the scalar field was considered as representing the Dilaton.
As it was previously mentioned we assumed for the matter, the perfect fluid expression (16):

$$(T_{e,p})^{\nu}_{\mu} = p\,\delta^{\nu}_{\mu} + u^{\nu}u_{\mu}(p+e), \tag{6}$$

where p was the pressure of the matter. Note that in the work it was assumed a pressureless matter $p = 0$. However, for bookkeeping purposes, it was employed the expression for a general pressure p up to the end when the limit $p = 0$ was fixed.
As usual u^ν denoted the contra-variant components of the 4-velocity of the fluid in the system of reference under consideration. In addition since the search for static configurations was undertaken, the 4-velocity took the simple form $u^\nu = \delta_0^\nu$.
From the Lagrangian L in (5) and the above remarks the energy momentum tensor of the scalar field coupled with the matter got the form

$$T_\mu^\nu = -\frac{\delta_\mu^\nu}{2}(g^{\alpha\beta}\phi_{,\alpha}\phi_{,\beta} + m^2\phi^2 + 2j\phi) + g^{\alpha\nu}\phi_{,\alpha}\phi_{,\mu} + p\delta_\mu^\nu + \delta_0^\nu\delta_\mu^0(p+e). \tag{7}$$

From equation (7), the Bianchi relation for $\mu = 1$ in (3) transformed in

$$-\phi j' + p' + \frac{v'}{2v}(p+e) = 0.$$

In case under consideration this is the only one of the four Bianchi relations which became different from zero.
The dynamical equation for the scalar field determining the extremum of the action $S_{mat-\phi}$, resulted in the form

$$\frac{\delta S_{mat-\phi}}{\delta\phi} \equiv \frac{d}{dx^\mu}\frac{\partial L}{\partial\phi_{,\mu}} - \frac{\partial L}{\partial\phi} \equiv \frac{1}{\sqrt{-g}}\frac{\partial}{\partial x^\mu}(\sqrt{-g}g^{\mu\nu}\phi_{,\nu}) - m^2\phi - j = 0, \tag{8}$$

which after introducing the components of the metric tensor was simplified to become

$$u\phi'' + u\phi'(\frac{1}{2}\frac{v'}{v} + \frac{1}{2}\frac{u'}{u} + \frac{2}{\rho}) - m^2\phi - j = 0.$$

Note that if $u = v = 1$, that is, in Minkowski space, relation (9) reduces to the static KG equation for scalar field interacting with an external source j. It might be helpful to notice that natural units

$$[e] = [p] = cm^{-4}, [m] = cm^{-1}, [\phi] = cm^{-1},$$

were employed.

2.2 Einstein equations

The extremum of the action $S_{mat-\phi}$with respect to the metric led to the Einstein equations in the absence of a Cosmological Constant

$$G_\mu^\nu = G\,T_\mu^\nu, \tag{9}$$

where in natural units $G = 8\pi \times l_p^2$ and $l_p = 1.61 \times 10^{-33}cm$ is the Planck length.

From relation (7), the Einstein equations (9) were expressed in the form

$$\frac{u'}{\rho}-\frac{1-u}{\rho^2}=-G[\frac{1}{2}(u\phi_{,\rho}^2+m^2\phi^2+2j\phi)+e], \tag{10}$$

$$\frac{u}{v}\frac{v'}{\rho}-\frac{1-u}{\rho^2}=G[\frac{1}{2}(u\phi_{,\rho}^2-m^2\phi^2-2j\phi)+p]. \tag{11}$$

As it was mentioned above, the third Einstein equation was not needed for determining the solution, because its satisfaction was implied by the other equations. This expression only imposed the continuity of the derivative of v with respect to the radial variable since it is a second order differential equation.
It was assumed that j, which gives the form of the interaction term between the dark energy and matter is of the form:

$$j=g\sqrt{e},$$

where g is a coupling constant for the interaction matter-scalar field. In the natural system of units $[g]=cm^{-1}$.
With the aim of working with dimensionless forms of the equations (10) and (11), we defined the new variables and parameters

$$r\equiv m\rho,\quad \Phi\equiv\sqrt{8\pi}l_p\phi,$$

$$J\equiv\frac{\sqrt{8\pi}l_p}{m^2}j,\quad \epsilon\equiv\frac{8\pi l_p^2}{m^2}e,\quad \gamma\equiv\frac{g}{m}.$$

Also, it was fixed the small mass of the Dilaton field to the value estimated in Ref. (18) for assuring the observed strength of the Hubble effect in the regions near the origin. Interestingly, this value resulted in the very small quantity, $m=4\times10^{-29}cm^{-1}$. This mass is compatible with the zero mass Dilaton in the lowest approximation. In addition the mass was of the order of the inverse of the estimated radius of the Universe, as it was observed in Ref. (18).
Therefore, the set of working equations resulted in the form

$$\frac{u_{,r}}{r}-\frac{1-u}{r^2}=-\frac{1}{2}(u\Phi_{,r}{}^2+\Phi^2)-J\Phi-\epsilon, \tag{12}$$

$$\frac{u}{v}\frac{v_{,r}}{r}-\frac{1-u}{r^2}=-\frac{1}{2}(-u\Phi_{,r}{}^2+\Phi^2+2J\Phi), \tag{13}$$

$$\epsilon\frac{v_{,r}}{2v}-\Phi J_{,r}=0, \tag{14}$$

$$u\Phi_{,rr}-\Phi-J=-u\Phi_{,r}(\frac{1}{2}\frac{v_{,r}}{v}+\frac{1}{2}\frac{u_{,r}}{u}+\frac{2}{r}). \tag{15}$$

2.3 The solutions near the center of symmetry

In Ref. (32) it was searched for smooth solutions around the origin. Thus, the continuity of the derivatives v and ϕ, in all places including the origin, was required. Then, after considering the equations in a neighborhood of the origin, the asymptotic field values were written in the form

$$u = 1 + u_1 r^2 ...,$$
$$v = 1 + v_1 r^2 ...,$$
$$\Phi = \Phi_0 + \Phi_1 r^2 ...,$$
$$\epsilon = \epsilon_0 + \epsilon_1 r^2 ...,$$

where u_1, v_1, Φ_1, ϵ_1 after substitution of the asymptotic solution in the equations were determined in the form

$$u_1 = -\frac{1}{3}(\frac{\Phi_0^2}{2} + J_0\Phi_0 + \epsilon_0), \tag{16}$$
$$v_1 = -\frac{1}{3}(\frac{\Phi_0^2}{2} + J_0\Phi_0 - \frac{\epsilon_0}{2}), \tag{17}$$
$$\Phi_1 = -\frac{1}{6}(\Phi_0 + J_0), \tag{18}$$
$$\epsilon_1 = -\frac{\epsilon_0^{\frac{3}{2}}}{3\gamma\Phi_0}(\frac{\Phi_0^2}{2} + J_0\Phi_0 - \frac{\epsilon_0}{2}), \tag{19}$$
$$J_0 = \gamma\epsilon_0^{\frac{1}{2}}.$$

Note that the spacial dependence of the metric tends to have an homogeneous structure near the center of symmetry. The quantities Φ_0, ϵ_0 and the dimensionless coupling constant γ remained as free parameters. Extensions of this work, could be considered to optimize the parameters, aiming to compare the predictions of the model with redshift vs. stelar magnitude in the supernovae obervations. In the next subsection we resume the study done about the behavior of the solution at all radial distances for given physically motivated values of the parameters.

2.4 The solutions at a arbitrary radial values

The numerical solutions of the equations (12)-(15) were considered, by selecting the parameter values $\gamma = -0.75$, $\Phi_0 = 2.2$ and $\epsilon_0 = 1$. These specific choosing corresponded to a coupling constant $g = 2.9 \times 10^{-29} cm^{-1}$, a value of the scalar field at the origin $\phi_0 = 2.7 \times 10^{32} cm^{-1}$ (that is, laying at the Planck scale) and a matter energy density of $e = 2.3 \times 10^7 cm^{-4}$. The determined numerical solutions of the equations (12)-(15) are illustrated in the figures (1)-(4).

These parameters were a priori selected with the aim of fixing the estimated value of 0.7/0.3 for the ratio of the Dark Energy to the matter energy content in the Universe (27) and an approximate value of the Hubble effect.
From Fig. (1) the global similarity between the space-time being examined and the de Sitter static solution can be observed. Moreover, due to the chosen value of the Dilaton mass suggested in Ref. (18), the size of the Universe (defined as the radial distance at which the singularity of the structure appears) is of the order of the estimated value $10^{29} cm$. In Fig.(2) the dependence of the temporal metric is shown, it evidenced that the observer near the origin measures a redshift which was imposed to show a value being near to the one currently observed.

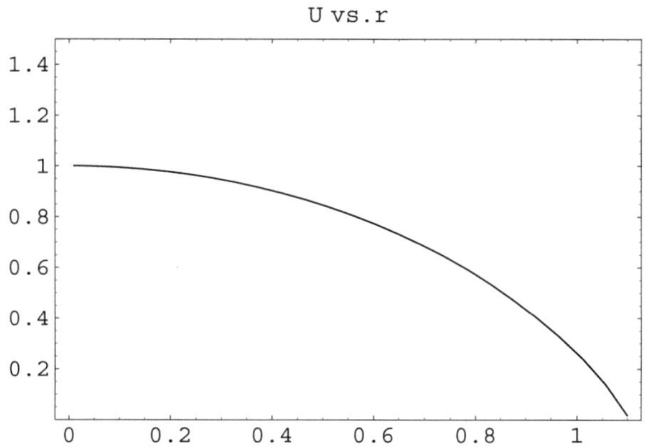

Fig. 1. The radial contraviant component of the metric $g^{11} \equiv u(r)$ behaved basically as in the deSitter Universe having the size $R \equiv 0.25 \times 10^{29}$cm.

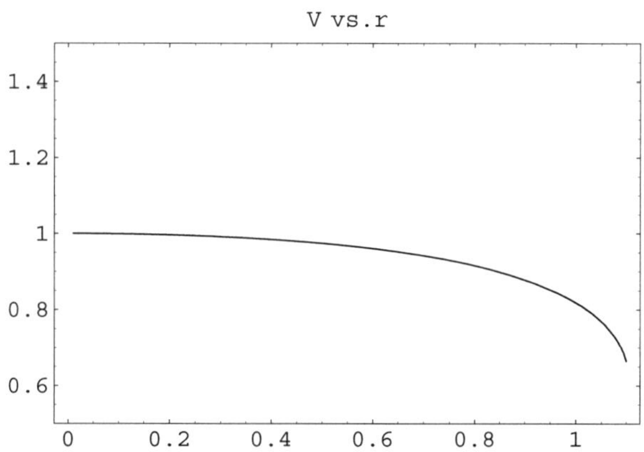

Fig. 2. Temporal component of the metric $g_{00} \equiv$v(r). Its decreasing behavior shows the redshift of the light arriving to the central zone regions from the outside regions. The radius of the singularity at the far away regions is $R \equiv 0.25 \times 10^{29}$cm.

Figures (3) and (4) illustrate the obtained distribution of energy and scalar field respectively. Note the similarity between both quantities. That is, the presence of the Dilaton-Matter coupling not only allowed the existence of the static solution, but in addition it also produced a configuration in which the proportion of matter and dark energy became approximately constant over large regions of the space time.

3. Large mass Dilaton stabilization by matter

As it had been mentioned in the Introduction, this section will review the results of the work presented in Ref. (33) . In this study it was investigated the possibility that the Dilaton could be stabilized at large values and masses as a direct consequence its universal type of interaction with matter. The review will be ordered as follows: In subsection 3.1, the notation and basic formulation employed in Ref. (33) will be given. Subsection 3.2 will review the

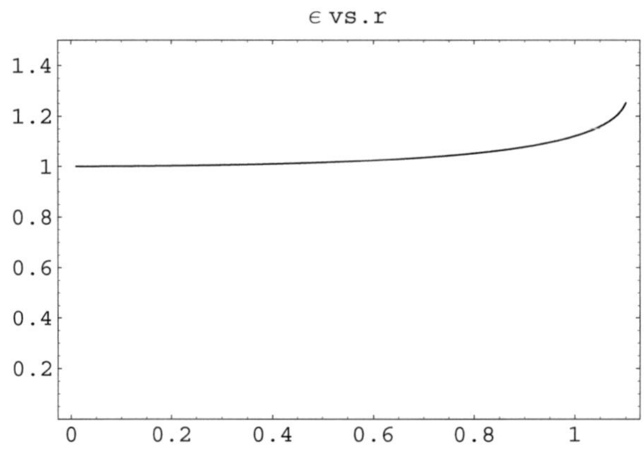

Fig. 3. The matter distribution $e(r)$ resulted as slowly varying with the radial distance. The coupling between the scalar field and the matter $J\,\Phi$ was central in allowing the existence of the static solution, in which also the matter to Dark energy content ratio resulted a slowly varying. The radial singularity defining the end of the space time at $R = 0.25 \times 10^{29}$cm.

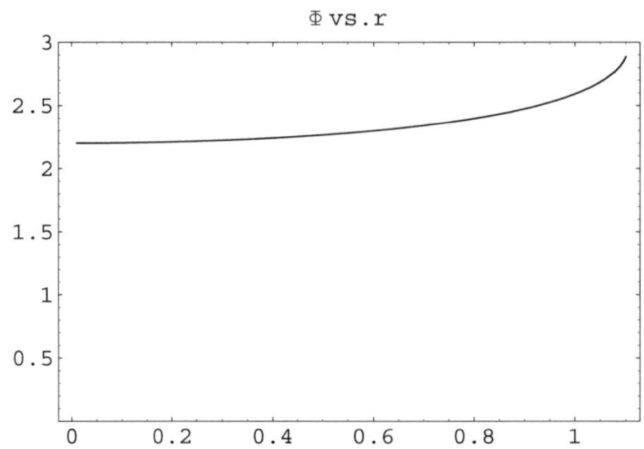

Fig. 4. The scalar field slowly varied with the radial component and behaved very closely with the matter density $e(r)$; The radial singularity defining the end of the space time is at $R = 0.25 \times 10^{29}$cm. There is no static metric with Dilaton and matter in coexistence without interaction.

elements of the one, two and three loops evaluation of the effective potential. The subsection 3.3 discuss the results of the evaluations done.

3.1 The Dilaton action and generating functional

In Ref. (33) it was considered a model of the Dilaton field interacting with fermion matter in the form

$$\begin{aligned} S = & \int d^4x\sqrt{-g(x)}\Big(\frac{1}{2\kappa^2}g^{\mu\nu}(x)\partial_\mu\phi^r(x)\partial_\nu\phi^r(x) + \overline{\Psi}(x)\big(i\frac{g^{\mu\nu}\gamma_\mu\overleftrightarrow{\partial}_\nu}{2} - m\big)\Psi(x) \\ & -\overline{\Psi}(x)g_Y^*\,\phi^r(x)\Psi(x) + j(x)\phi^r(x) + \overline{\Psi}(x)\eta(x) \;+\overline{\eta}(x)\Psi(x)\Big), \end{aligned} \tag{20}$$

$$m = \exp(\alpha^* \, \phi) m_f, \tag{21}$$
$$g_Y^* = \alpha^* \, m, \tag{22}$$
$$\alpha^* = -\frac{3}{4}, \tag{23}$$
$$x^\mu = (x^0, x^1, x^2, x^3), \quad \overleftrightarrow{\partial} = \overrightarrow{\partial} - \overleftarrow{\partial}, \quad \{\gamma_\mu, \gamma_\nu\} = 2g_{\mu\nu}(x), \tag{24}$$
$$g_{\mu\nu}(x) = \begin{pmatrix} 1 & 0 & 0 & 0 \\ 0 & -1 & 0 & 0 \\ 0 & 0 & -1 & 0 \\ 0 & 0 & 0 & -1 \end{pmatrix}, \quad \sqrt{-g(x)} = 1. \tag{25}$$

That is, we considered the Dilaton field interacting with a massive fermion in the Einstein frame, in which the metric $g_{\mu\nu}$ was approximated by the Minkowski one in order to simplify the evaluations. The gravitational constant was explicitly introduced, and natural units were employed for the distances and mass. The vacuum value of the Dilaton field is named as ϕ and its radiative part is called ϕ^r. Note that is was assumed that the radiative part is small in order to retain only the first term in the expansion of the exponential. This was the Yukawa approximation which was employed in Ref. (33). All the results are functions of the vacuum field ϕ.

The parameter defining the Dilaton field dependent exponential, the Planck length $\kappa = l_P$ and mass m_P were defined by the expressions

$$\kappa^2 = \frac{8\pi G\hbar}{c^3}, \tag{26}$$
$$\kappa = l_P = \frac{1}{m_P} = 8.10009 \times 10^{-33} \text{ cm}, \tag{27}$$
$$G = 6.67 \times 10^{-8} \text{ cm}^3 \text{ g}^{-1} \text{ s}^{-2}, \tag{28}$$
$$\hbar = 1.05457 \times 10^{-27} \text{ cm}^2 \text{ g s}^{-1}, \tag{29}$$
$$c = 2.9979245800 \times 10^{10} \text{ cm s}^{-1}. \tag{30}$$

In the above formula for the action, the coordinates and times are measured in cm, the masses m in the natural unit cm^{-1} and the Dilaton field is dimensionless.

Starting from the classical action, the work considered corrections up to 3-loops for the effective action, assuming a homogenous and time independent value of the Dilaton mean field ϕ as

$$\frac{\Gamma[\phi]}{V^{(4)}} = -V^{eff}(\phi), \tag{31}$$

where $V^{(4)}$ is the four dimensional volume. In order to eliminate the explicit appearance of the gravitational constant from the diagram technique for evaluating the effective action, its appearance was eliminated from the equations by redefining the Dilaton field value, the α^* constant and the coupling as

$$\varphi = \phi/\kappa, \tag{32}$$
$$\alpha = \alpha^* \kappa = -\frac{3}{4}\kappa, \tag{33}$$
$$g_Y = g_Y^* \kappa. \tag{34}$$

After these changes, the above written classical action S, to be used for generating the Feynman expansion, was expressed as follows

$$S\,[\overline{\Psi},\Psi,\varphi^r,\varphi] = \int d^4x\,(\frac{1}{2}g^{\mu\nu}(x)\partial_\mu\varphi^r(x)\partial_\nu\varphi^r(x) + \overline{\Psi}(x)(i\frac{g^{\mu\nu}\gamma_\mu\overset{\leftrightarrow}{\partial}_\nu}{2} - m)\Psi(x)$$
$$-\overline{\Psi}(x)g_Y\,\varphi^r(x)\Psi(x) + j(x)(\,\varphi + \varphi^r(x)) + \overline{\Psi}(x)\,\eta(x)\; + \overline{\eta}(x)\,\Psi(x)). \quad (35)$$

The expansion was considered in $d = 4 - 2\epsilon$ dimensions for implementing dimensional regularization scheme. Accordingly, the coupling constant g_Y was modified by the introduction of the regularization scale parameter μ as follows

$$g_Y^2 = \mu^{2\epsilon}(g_Y^0)^2,$$

where g_Y^0 is the usual coupling constant in four dimensions.

3.1.1 The generating functional and the effective action

In this subsection, we will sketch the main expressions defining the perturbative calculation which was considered in Ref. (33). The generating functional of the Green functions Z , its connected part W and the mean field values were defined by the formulae

$$Z[\overline{\eta},\eta,j] = \int \mathcal{D}\overline{\Psi}\mathcal{D}\Psi\mathcal{D}\varphi^r \exp(i\,S\,[\overline{\Psi},\Psi,\varphi^r,\varphi]), \quad (36)$$

$$W[\overline{\eta},\eta,j] = \frac{1}{i}\log Z[\overline{\eta},\eta,j], \quad (37)$$

$$\frac{\delta\,W}{i\,\delta j(x)} = \varphi + \langle\varphi^r(x)\rangle, \quad (38)$$

$$\frac{\delta\,W}{i\,\delta\overline{\eta}(x)} = \langle\Psi(x)\rangle, \quad (39)$$

$$\frac{\delta\,W}{-i\,\delta\eta(x)} = \langle\overline{\Psi}(x)\rangle. \quad (40)$$

Note that the mean Dilaton field φ was considered as homogeneous and the mean value of the radiative part $\langle\varphi^r(x)\rangle$ was assumed to vanish when the sources are zero. The effective action was defined as the Legendre transform of Z depending on the mean field values as:

$$\Gamma[\langle\Psi\rangle,\langle\overline{\Psi}\rangle,\varphi+\langle\varphi^r\rangle] = \frac{1}{i}\log Z[\overline{\eta},\eta,j] - \int dx[j(x)(\,\varphi + \langle\varphi^r(x)\rangle) +$$
$$\langle\overline{\Psi}(x)\rangle\,\eta(x)\; + \overline{\eta}(x)\,\langle\Psi(x)\rangle\,], \quad (41)$$

$$\frac{\delta\,\Gamma}{\delta\langle\varphi^r(x)\rangle} = -j(x), \quad (42)$$

$$\frac{\delta\,\Gamma}{\delta\langle\overline{\Psi}(x)\rangle} = -\eta(x), \quad (43)$$

$$\frac{\delta\,\Gamma}{\delta\langle\Psi(x)\rangle} = \overline{\eta}(x). \quad (44)$$

The expression for Z, after writing the Yukawa vertex part of the Lagrangian in terms of the functional derivatives over the sources and integrating the gaussian functional integral that remains, led to the Wick expansion formula:

$$Z[\overline{\eta},\eta,j] = \exp\left[i\int dx\, g_Y \frac{\delta}{i\delta j(x)}\frac{\delta}{-i\delta\eta(x)}\frac{\delta}{i\delta\overline{\eta}(x)}\right]\times$$
$$\exp\left[\left[\int dx\, dy\, (\overline{\eta}(x)S(x-y)\eta(y)+\frac{1}{2}j(x)D(x-y)j(y))\right]\right], \qquad (45)$$

$$S(x-y) = \int \frac{dp^d}{(2\pi)^d}\frac{\exp(-i\, p.(x-y))}{m-\gamma^\mu p_\mu}, \qquad (46)$$

$$D(x-y) = \int \frac{dk^d}{(2\pi)^d}\frac{\exp(-i\, k.(x-y))}{-(k^2-i\epsilon)}, \qquad (47)$$

in which S and D are the fermion and Dilaton free propagators, respectively. The notation for fermions and scalar field related quantities, and the definition of the Feynman rules for the generation of the analytic expressions for the various contributions, were exactly the ones described in Ref. (30), for the cases of scalar and fermion fields. Specifically, for the momentum space rules, the propagators and the only existing vertex are graphically illustrated in figure 5.

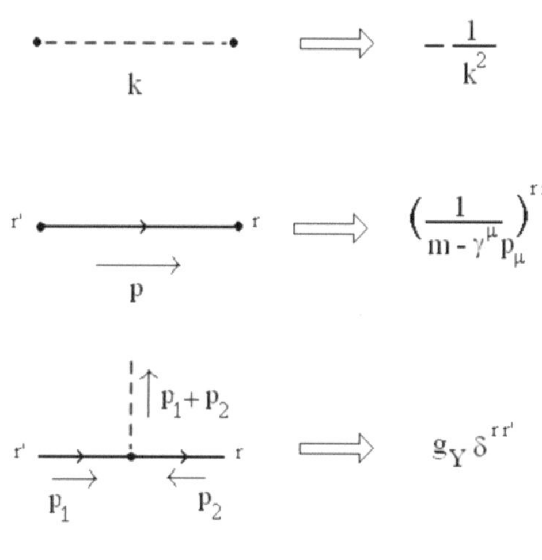

Fig. 5. The figure illustrates the Feynman rules for the particular Yukawa model approximation adopted for the Dilaton action in Ref. (33)

3.2 Effective potential evaluation

Let us resume in this section the evaluations of the effective potential for the Dilaton field done in Ref. (33). They followed after employing the perturbative expansion described in the past section. The diagrams which were considered are depicted in Fig. 6. They included up to three loops corrections. The contributions were exactly evaluated for the one and two loops terms. In addition, the three loop term D_{32} also was analytically calculated in terms of Master integrals. However, the three loop diagrams D_{31} and D_{33} were determined only in their leading terms of order $\log\left(\frac{m}{\mu}\right)^3$. We expect to be able in evaluating the non leading corrections (lower powers of $log\left(\frac{m}{\mu}\right)$) in extending the work done in Ref.(33). The results for each diagram are reviewed in various subsections below.

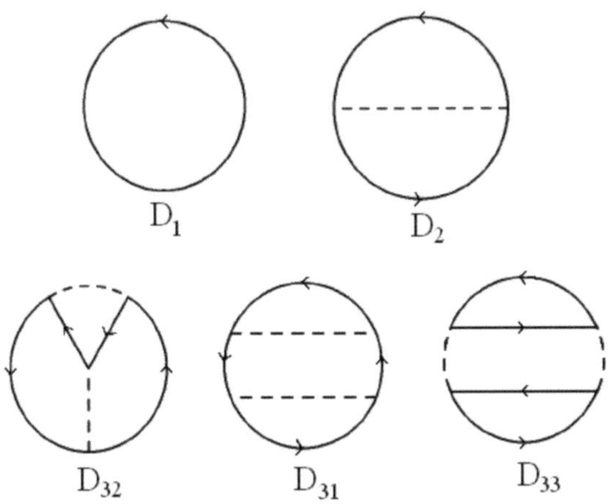

Fig. 6. The one, two and three loops Feynman diagrams considered in Ref. (33) . The one and two loop corrections D_1 and D_2 were exactly calculated. In the case of the three loops terms, the D_{32} was completely evaluated in terms of the listed Master integrals in Ref. (31). The D_{31} and D_{33} were determined only in their leading logarithm correction.

3.2.1 One loop term D_1

The analytic expression for the one loop diagram D_1 and its derivative over m^2 had the forms

$$\Gamma^{(1)} = V^{(d)} \int \frac{dp^d}{(2\pi)^d i} Tr \log(m^2 - p^2), \tag{48}$$

$$\frac{d}{d\,m^2}\Gamma^{(1)} = 4V^{(d)} \int \frac{dp^d}{(2\pi)^d i} \frac{1}{m^2 - p^2}. \tag{49}$$

The result for the momentum integral entering in the derivate of $\Gamma^{(1)}$ over m^2, after divided by $\mu^{2\epsilon}\, V^{(d)}$ (in order to define a 4-dimensional energy density) and integrated over m^2, allowed to write for the one loop effective action density the expression (See Ref. (31))

$$\gamma_1(m,\epsilon,\mu) \equiv \frac{\Gamma^{(1)}}{\mu^{2\epsilon} V^{(d)}} = m^4 (\frac{m}{\mu})^{-2\epsilon} \frac{8\pi^{2-\epsilon}}{(2\pi)^{4-2\epsilon}} \Gamma\left(-1+\epsilon\right). \tag{50}$$

After employing the minimal substraction (MS) scheme, that is, getting the finite part by eliminating the pure pole part in ϵ the Laurent expansion of $\gamma(m,\epsilon)$ and taking the limit $\epsilon \to 0$, the one loop contribution to the effective action density as a function of m and μ becomes written in the form

$$\gamma_1(m,\mu) = 0.0506606 m^4 \left(2. \log\left(\frac{m}{\mu}\right) - 2.95381\right). \tag{51}$$

Note that the negative of this term, which defines the one loop effective potential led to a the dynamical generation of the Dilaton field for positive values of $\alpha^* \ \phi$ as follows from $log(m) = log(m_f) + \alpha^* \ \phi$. This was the effect which motivated the study started in Ref. (28).

3.2.2 Two loop term D_2

For the two loop contribution D_2 the analytic expression was

$$\begin{aligned}\gamma_2(m,\epsilon,\mu) \equiv \frac{\Gamma^{(2)}}{\mu^{2\epsilon}V^{(d)}} &= \frac{1}{2}(g_Y^0)^2 \int \frac{dp_1^d}{(2\pi)^d i}\frac{dp_2^d}{(2\pi)^d i}\frac{4(m^2+p_1.p_2)}{(m^2-p_1^2)(m^2-p_2^2)(p_1-p_2)^2} \\ &= \frac{1}{2}(g_Y^0)^2 m^{2d-4}\int \frac{dq_1^d}{(2\pi)^d i}\frac{dq_2^d}{(2\pi)^d i}\frac{4(1+q_1.q_2)}{(1-q_1^2)(1-q_2^2)(q_1-q_2)^2} \\ &= 2(g_Y^0)^2 m^4 m^{-4\epsilon}(2\int \frac{dq_1^d}{(2\pi)^d i}\frac{dq_2^d}{(2\pi)^d i}\frac{1}{(1-q_1^2)(1-q_2^2)(q_1-q_2)^2} - \\ &\quad -\frac{1}{2}(\int \frac{dq_1^d}{(2\pi)^d i}\frac{1}{(1-q_1^2)})^2),\end{aligned} \tag{52}$$

where the identity $q_1.q_2 = \frac{1}{2}(q_1^2 - 1 + q_2{}^2 - 1) + 1 - \frac{1}{2}(q_1 - q_2)^2$ was employed. The two momentum integrals appearing in the last line are the simplest Master integrals for scalar fields as listed in Ref. (31). The results for them in that reference are:

$$\int \frac{dq_1^d}{(2\pi)^d i}\frac{dq_2^d}{(2\pi)^d i}\frac{1}{(1-q_1^2)(1-q_2^2)(q_1-q_2)^2} = \frac{(d-2)(\pi)^d\Gamma\left(1-\frac{d}{2}\right)^2}{2(d-3)(2\pi)^{2d}}, \tag{53}$$

$$\int \frac{dq_1^d}{(2\pi)^d i}\frac{1}{(1-q_1^2)} = \frac{(\pi)^{\frac{d}{2}}\Gamma\left(1-\frac{d}{2}\right)}{(2\pi)^d}. \tag{54}$$

They allowed to write for the regularized two loop effective action density the expression

$$\gamma_2(m,\epsilon,\mu) = -m^4(\frac{m}{\mu})^{-4\epsilon}\frac{2(g_Y^0)^2(\pi)^d}{(2\pi)^{2d}}(-\frac{d-2}{d-3}+\frac{1}{2})\Gamma\left(1-\frac{d}{2}\right)^2. \tag{55}$$

Expanding in Laurent series in ϵ and disregarding the pole part in the limit $\epsilon \to 0$, led in Ref. (33) to the two loop perturbative contribution to the effective action

$$\gamma_2(m,\mu) = 0.0000200507(g_Y^0)^2 m^4\,(48.\log^2\left(\frac{m}{\mu}\right) - 173.783\log\frac{m}{\mu} + 183.83\,). \tag{56}$$

As it was noted in the Introduction, in Ref. (28) it was employed an inappropriate negative kinetic term for the Dilaton field. This change, although not affecting the one fermion loop contribution, which is not altered by the sign of the boson propagator, drastically modified the sign of the two loop term which linearly depends on the Dilaton propagator. In the previous evaluation, the two loop terms determined the existence of minima for the Dilaton potential. Therefore, the consequence of the change in sign fixed by the consideration in Ref. (33) of the correct positive kinetic energy term, should be further investigated in connection with the existence of stabilizing minima for the scalar field. This circumstance determined the motivation for the new three loop corrections considered in Ref. (33) and reviewed in this chapter.

3.2.3 Three loops terms

Let us resume the evaluation of the three loop terms in Ref. (33).

3.2.4 Diagram D_{32}

The D_{32} term is the only of the 3-loops diagrams which is not composed of two fermion or boson self energy insertions connected in series. For the D_{31} and D_{33} cases we had difficulties in reducing their contributions to a linear combination of tabulated Master integrals. This obstacle only allowed us to calculate their leading term in the expansion in $\log(\frac{m}{\mu})$. However, for D_{32} it was possible to express it as a sum over the Master integrals given in Ref. (31). The analytic expression of the diagram was

$$\Gamma^{(32)} = -V^{(d)}\frac{1}{4}(g_Y)^4 \int \frac{dp_1^d}{(2\pi)^d i}\frac{dp_2^d}{(2\pi)^d i}\frac{dp_3^d}{(2\pi)^d i} \times$$

$$\frac{\mathrm{Tr}\left[(m+p_2^\mu\gamma_\mu)(m+(p_2^\mu+p_3^\mu-p_1^\mu)\gamma_\mu)(m+p_3^\mu\gamma_\mu)(m+p_1^\mu\gamma_\mu)\right]}{(m^2-p_1^2)(m^2-p_2^2)(m^2-p_3^2)(m^2-(p_2+p_3-p_1)^2)(p_1-p_3)^2(p_1-p_2)^2}$$

$$= -V^{(d)}\frac{1}{4}(g_Y)^4 \int \frac{dp_1^d}{(2\pi)^d i}\frac{dp_2^d}{(2\pi)^d i}\frac{dp_3^d}{(2\pi)^d i} \times \tag{57}$$

$$\frac{m^4+c_1(p_1,p_2,p_3)m^2+c_2(p_1,p_2,p_3)}{(m^2-p_1^2)(m^2-p_2^2)(m^2-p_3^2)(m^2-(p_2+p_3-p_1)^2)(p_1-p_3)^2(p_1-p_2)^2},$$

$$c_1(p_1,p_2,p_3) = 3p_2.p_3+p_1.p_2+p_1.p_3+p_2^2+p_3^2-p_1^2 \tag{58}$$

$$c_2(p_1,p_2,p_3) = p_1^2\, p_2.p_3+p_2^2\, p_1.p_3+p_3^2\, p_1.p_2-2\, p_1.p_2\, p_1.p_3. \tag{59}$$

After defining

$$\begin{aligned} z_1 &= p_1^2-m^2,\\ z_2 &= p_2^2-m^2,\\ z_3 &= p_3^2-m^2,\\ z_4 &= (p_1-p_2)^2,\\ z_5 &= (p_1-p_3)^2,\\ z_6 &= (p_2-p_1+p_3)^2-m^2, \end{aligned} \tag{60}$$

and employing various vectorial identities expressing the squares of the differences between any two momenta in terms of the scalar product between them and the squares of the considered momenta, the integral defining Γ^{32} was written as follows

$$\Gamma^{32} = -V^{(d)}\frac{1}{4}(g_Y)^4 \int \frac{dp_1^d}{(2\pi)^d i}\frac{dp_2^d}{(2\pi)^d i}\frac{dp_3^d}{(2\pi)^d i} \times$$

$$\frac{m^4+c_1(z)\,m^2+c_2(z)}{z_1\, z_2\, z_3\, z_4\, z_5\, z_6},$$

$$z = (z_1, z_2, z_3, z_4, z_5, z_6), \tag{61}$$

$$c_1(z) = \frac{3}{2}(z_1+z_2+z_3+z_6)-2(z_4+z_5)+6m^2, \tag{62}$$

$$c_2(z) = \frac{1}{2}(z_1z_6+z_2z_3-z_4z_5+m^2(z_1+z_2+z_3+z_6)+2m^4). \tag{63}$$

Therefore, there exist one or two z factors in the denominator that can be canceled by the terms of the quadratic polynomial in these quantities. This property allowed the integral to be decomposed in a linear combination of the Master integrals listed in Ref. (31). The result for the action density

$$\gamma_{32}(m,\mu,\epsilon) = \frac{\Gamma^{(32)}}{\mu^{2\epsilon}V^{(d)}} \tag{64}$$

was expressed in terms of only five of them as follows

$$\gamma_{32}(m,\mu,\epsilon) = -(g_Y^0)^4\, m^4 \left(\frac{m}{\mu}\right)^{-6\epsilon} \left(8I_1(\epsilon)+8I_2(\epsilon)-4I_3(\epsilon)+I_5(\epsilon)-\frac{I_7(\epsilon)}{2}\right),$$

where the functions $I_1(\epsilon)$, $I_2(\epsilon)$, $I_3(\epsilon)$, $I_5(\epsilon)$ and $I_7(\epsilon)$ resulted to be given by

$$\begin{aligned} I_1(\epsilon) &= \frac{2^{-3(4-2\epsilon)-9}\pi^{-\frac{3}{2}(4-2\epsilon)}(5(4-2\epsilon)-18)M_1(\epsilon)^3}{1-2\epsilon} + \\ &\quad \frac{2^{-3(4-2\epsilon)-6}\pi^{-3(4-2\epsilon)}(3(4-2\epsilon)-10)(3(4-2\epsilon)-8)\left(M_5(\epsilon)-\frac{8\epsilon}{2(4-2\epsilon)-7}M_4(\epsilon)\right)}{\epsilon^2}, \end{aligned} \tag{65}$$

$$I_2(\epsilon) = -\frac{2^{-3(4-2\epsilon)-2}\pi^{-3(4-2\epsilon)}}{1-2\epsilon}\left(\frac{M_1(\epsilon)^3(2-2\epsilon)^2}{1-2\epsilon}+(3(4-2\epsilon)-8)M_4(\epsilon)\right), \tag{66}$$

$$I_3(\epsilon) = -\frac{2^{-3(4-2\epsilon)-3}\pi^{-3(4-2\epsilon)}}{\epsilon}\left(\frac{2(2-2\epsilon)^2M_1(\epsilon)^3}{1-2\epsilon}+(3(4-2\epsilon)-8)M_5(\epsilon)\right), \tag{67}$$

$$\begin{aligned} I_5(\epsilon) &= (2\pi)^{-3(4-2\epsilon)}M_4(\epsilon), \\ I_7(\epsilon) &= (2\pi)^{-3(4-2\epsilon)}M_5(\epsilon), \end{aligned} \tag{68}$$

in terms of the Master integrals (See Ref. (31)):

$$M_1(\epsilon) = \pi^{\frac{1}{2}(4-2\epsilon)}\Gamma\left(\frac{1}{2}(2\epsilon-4)+1\right), \tag{69}$$

$$M_2(\epsilon) = -\frac{(2-2\epsilon)\,M_1(\epsilon)^2}{2(1-2\epsilon)}, \tag{70}$$

$$M_3(\epsilon) = 2^{\frac{1}{2}(2\epsilon-4)}\Gamma\left(\frac{1}{2}(4-2\epsilon)\right)\Gamma\left(\frac{1}{2}(2\epsilon-1)\right)M_1(\epsilon)^2, \tag{71}$$

$$M_4(\epsilon) = \frac{2^{1-2\epsilon}\Gamma\left(\frac{1}{2}(8-3(4-2\epsilon))\right)\Gamma\left(\frac{1}{2}(2\epsilon-1)\right)}{\Gamma\left(\frac{1}{2}(7-2(4-2\epsilon))\right)\Gamma\left(\frac{1}{2}(2\epsilon-2)\right)}M_1(\epsilon)^3, \tag{72}$$

$$\begin{aligned} M_5(\epsilon) &= (-2-\frac{5}{3}\epsilon - \frac{1}{2}\epsilon^2 + \frac{103}{12}\epsilon^3 + \frac{7}{24}(163-128\zeta(3))\epsilon^4 + \\ &\quad (\frac{9055}{48}+\frac{136\pi^4}{45}+\frac{1}{3}(\pi^2-\log(2)^2)(32\log(2)^2)-168\zeta(3) \\ &\quad -256\mathrm{Li}_4(\frac{1}{2})\ )\epsilon^5\ )\ M_1(\epsilon)^3, \end{aligned} \tag{73}$$

where the special functions $\mathrm{Li}_n(\frac{1}{2})$ and $\zeta(n)$ are defined as

$$\mathrm{Li}_n(x) = \sum_{k=1}^{\infty} \frac{1}{2^k k^n}, \tag{74}$$

$$\zeta(n) = \sum_{k=1}^{\infty} \frac{1}{k^n}. \tag{75}$$

Finally, the application of the before described MS procedure led to the following formula for the contribution to the vacuum effective action density of the diagram D_{32}

$$\begin{aligned}\gamma_{32}(m,\mu) = \; & (g_Y^0)^4 m^4 \, (0.0000329114 \log^5\left(\frac{m}{\mu}\right) - 0.000105904 \log^4\left(\frac{m}{\mu}\right) + \\ & 0.0000165851 \log^3\left(\frac{m}{\mu}\right) + 0.000441159 \log^2\left(\frac{m}{\mu}\right) \\ & -0.00074347 \log\left(\frac{m}{\mu}\right) + 0.000388237).\end{aligned} \tag{76}$$

It can be noted that this term has a high quintic power of $\log^5(\frac{m}{\mu})$ which was determined by the also high pole of the ϵ expansion present in the function I_1. This represents the highest power of the $\log\left(\frac{m}{\mu}\right)$ expansion appearing in the results. The next higher power, the fourth one, also is arising in this term.

3.2.5 Diagram D_{31}

We were not able to exactly evaluate this contribution (and also the one associated to D_{33}) in terms of Master integrals. Therefore, for both of these terms we limited ourself to evaluate their leading terms in the expansion in powers of $\log\left(\frac{m}{\mu}\right)$. For this purpose, use was made of the circumstance that (at variance with D_{32}, but in coincidence with the case of D_{33}) this term corresponds to a loop formed by two one loop self-energy insertions. Since these self-energy terms are explicitly calculable in terms of hypergeometric functions, both terms were expressed as single momentum integral in d dimensions. The diagram had the original analytic expression

$$\begin{aligned}\Gamma^{31} &= -V^{(d)}\frac{1}{2}(g_Y)^4 \int \frac{dp_1^d}{(2\pi)^d i}\frac{dp_2^d}{(2\pi)^d i}\frac{dp_3^d}{(2\pi)^d i} \times \\ & \frac{Tr\left[(m+p_2^\mu\gamma_\mu)(m+p_1^\mu\gamma_\mu)(m+p_3^\mu\gamma_\mu)(m+p_1^\mu\gamma_\mu)\right]}{(m^2-p_1^2)^2(m^2-p_2^2)(m^2-p_3^2)(p_1-p_3)^2(p_1-p_2)^2} \\ &= -V^{(d)}\frac{1}{2}(g_Y)^4 \int \frac{dp_1^d}{(2\pi)^d i}\frac{dp_2^d}{(2\pi)^d i}\frac{dp_3^d}{(2\pi)^d i} \times \\ & \frac{m^4 + d_1(p_1,p_2,p_3)m^2 + d_2(p_1,p_2,p_3)}{(m^2-p_1^2)^2(m^2-p_2^2)(m^2-p_3^2)(p_1-p_3)^2(p_1-p_2)^2},\end{aligned} \tag{77}$$

$$d_1(p_1,p_2,p_3) = p_1^2 + 2p_1.p_2 + 2p_1.p_3 + p_2.p_3 \;, \tag{78}$$

$$d_2(p_1,p_2,p_3) = 2\, p_1.p_2\; p_1.p_3 - \; p_1^2\; p_2.p_3. \tag{79}$$

Then, it was defined the fermion self-energy integral and its related vector as follows

$$
\begin{aligned}
s_{31}(p^2) &= \int \frac{dp_1^d}{(2\pi)^d i} \frac{1}{(m^2 - p_1^2)(p_1 - p)^2} \\
&= -\frac{\pi^{\frac{d}{2}}}{(2\pi)^d} \Gamma(\epsilon) \int_0^1 dx\, x^{-\epsilon} (m^2 - p^2(1-x) - i\delta)^{-\epsilon} \\
&= \frac{\pi^{\frac{d}{2}}}{(2\pi)^d} \Gamma(\epsilon) (m^2)^{-\epsilon} \frac{{}_2F_1(1-\epsilon, \epsilon, 2-\epsilon, -\frac{(\frac{p}{m})^2}{1-(\frac{p}{m})^2})}{\epsilon - 1}, && (80) \\
v_\mu(p^2) &= \int \frac{dp_1^d}{(2\pi)^d i} \frac{p_{1\mu}}{(m^2 - p_1^2)(p_1 - p)^2} \\
&= a(p^2)\ p_\mu, && (81) \\
a(p^2) &= \frac{p^2 + m^2}{2p^2} s_{31}(p^2) - \frac{L(m,\epsilon)}{2p^2}, && (82) \\
L(m,\epsilon) &= \int \frac{dp_1^d}{(2\pi)^d i} \frac{1}{(m^2 - p_1^2)} = \frac{\pi^{\frac{d}{2}} m^{d-2}}{(2\pi)^d} \Gamma(1 - \frac{d}{2}). && (83)
\end{aligned}
$$

In the above expressions, the Feynman parametric integral was explicitly evaluated by employing the algebraic calculation program Mathematica. After again defining the action density contribution as

$$
\gamma_{31}(m, \mu, \epsilon) = \frac{\Gamma^{(31)}}{\mu^{2\epsilon} V^{(d)}}, \quad (84)
$$

performing the Wick rotation in the momenta and extracting the d-dimensional solid angle arising form the angular integrals, this quantity is expressed as the integral

$$
\begin{aligned}
\gamma_{31}(m, \mu, \epsilon) &= -\frac{2(g_Y^0)^4 \left(\frac{m}{\mu}\right)^{-6\epsilon} c(m,\epsilon)}{\epsilon^2} \int_0^\infty \frac{p^{3-2\epsilon}}{(p^2+1)^2} f(p,\epsilon) dp, && (85) \\
f(p,\epsilon) &= \epsilon^2 f_1(p,\epsilon) \Gamma(\epsilon)^2 + f_2(p,\epsilon)\, \epsilon\, \Gamma(\epsilon) + f_3(p,\epsilon), && (86) \\
f_1(p,\epsilon) &= (1-p^2)(3 - \frac{(1-p^2)^2}{4p^2})(s_{31}^*(p^2,\epsilon))^2, && (87) \\
f_2(p,\epsilon) &= (2 - \frac{(1-p^2)^2}{2p^2}) s_{31}^*(p^2,\epsilon) L^*(\epsilon), && (88) \\
f_3(p,\epsilon) &= -\frac{(1-p^2)^2 (L^*(\epsilon))^2}{4p^2}, && (89) \\
s_{31}^*(p,\epsilon) &= -\frac{2^{2\epsilon-4} \pi^{\frac{1}{2}(4-2\epsilon)+2\epsilon-4} {}_2F_1\left(1-\epsilon, \epsilon; 2-\epsilon; \frac{p^2}{p^2+1}\right)}{\epsilon - 1}, && (90) \\
c(m,\epsilon) &= \frac{2^{2\epsilon-3} m^4 \pi^{\frac{1}{2}(4-2\epsilon)+2\epsilon-4}}{\Gamma\left(\frac{1}{2}(4-2\epsilon)\right)}, && (91) \\
L^*(\epsilon) &= \epsilon L(1,\epsilon). && (92)
\end{aligned}
$$

As it was mentioned before, we were not able yet to find an epsilon expansion (rigorous or sufficiently approximated numerical one) allowing to exactly evaluate this integral after removing the regularization. Therefore, in order to determine an approximation for γ_{31} we have made use of an assumption suggested by an exploration done about the asymptotic power expansion at infinity of the integrand as a function of the momentum integration variable p. It followed that all the terms of the expansion after integrated, show a single pole structure in their Laurent expansion in ϵ. Then, it suggests that the full divergence of the integral at $d=4$ is defined by a single pole in ϵ. Assuming this property, the extraction of the leading correction in $\log(\frac{m}{\mu})$ should be defined by the maximal power of $\log(\frac{m}{\mu})$ appearing in the coefficient of the zero order term in the expansion of the modified integral $\gamma_{31}(m,\mu,\epsilon)$

$$\gamma^{*}_{31}(m,\mu,\epsilon) = -\frac{2(g^0_Y)^4\left(\frac{m}{\mu}\right)^{-6\,\epsilon}c(m,0)}{\epsilon^2}\int_0^\infty \frac{p^{3-2\epsilon}}{(p^2+1)^2}f(p,0)dp. \tag{93}$$

Note that any other power of ϵ in the expansions of $c(m,\epsilon)$ and $f(p,\epsilon)$ will reduce the maximal order of the negative powers of epsilon in the full expansion of $\gamma_{31}(m,\mu,\epsilon)$, which determines the leading correction in the expansion. For $f(p,0)$ it followed

$$f(p,0) = \frac{p^4}{1024\pi^4} - \frac{17p^2}{1024\pi^4} + \frac{7}{256\pi^4} - \frac{1}{256\pi^4 p^2}. \tag{94}$$

Then, the use of the formula

$$\int_0^\infty \frac{p^{3-2\epsilon+m}}{(p^2+1)^2}dp = -\frac{\pi}{4}(m-2\epsilon+2)\csc(\frac{\pi}{2}(m-2\epsilon)), \tag{95}$$

which shows the $\frac{1}{\epsilon}$ singularity, allowed to write for γ_{31} the leading logarithm correction to its finite part

$$\gamma_{31}(m,\mu) = -0.0000228551(g^0_Y)^4\, m^4 \log^3\left(\frac{m}{\mu}\right). \tag{96}$$

3.2.6 Diagram D_{33}

As it was remarked before, this term was treated in a similar way as it was D_{31}. Now, the corresponding self-energy insertions were boson ones. Again, the two self-energy loops were explicitly calculable in terms of hypergeometric functions. The starting analytic expression of the diagram was

$$\begin{aligned}
\Gamma^{33} &= V^{(d)}\frac{1}{4}(g_Y)^4\int \frac{dp^d}{(2\pi)^d i}\frac{dp_1^d}{(2\pi)^d i}\frac{dp_2^d}{(2\pi)^d i}\times\frac{1}{(p^2)^2} \\
&\frac{\mathrm{Tr}[(m+p_1^\mu\gamma_\mu)(m+(p+p_1)^\nu\gamma_\nu)]\mathrm{Tr}[(m+p_2^\mu\gamma_\mu)(m+(p_2+p)^\nu\gamma_\nu)]}{(m^2-p_1^2)^2(m^2-(p_1+p)^2)(m^2-p_2^2)^2(m^2-(p_2+p)^2)}, \\
&= V^{(d)}4(g_Y)^4\int \frac{dp^d}{(2\pi)^d i}\frac{dp_1^d}{(2\pi)^d i}\frac{dp_2^d}{(2\pi)^d i}\times\frac{1}{(p^2)^2}\times \\
&\frac{(m^2+p_1.(p_1+p))(m^2+p_2.(p_2+p))}{(m^2-p_1^2)^2(m^2-(p_1+p)^2)(m^2-p_2^2)^2(m^2-(p_2+p)^2)},
\end{aligned} \tag{97}$$

where the fermion traces were evaluated for writing the second form of the integral. The last expression evidenced the decomposition in two serial self-energy terms.
After rotating to Euclidean space the momenta variables of the integration regions and the external momentum, the fermion selfenergy integral and its related vector integral were written as follows (See Ref. (30))

$$\begin{aligned}
s_{33}(q^2,\epsilon) &= \int \frac{dq_1^d}{(2\pi)^d i} \frac{1}{(m^2+q_1^2)(m^2+(q+q_1)^2} \\
&= \frac{(m)^{-2\epsilon}}{(4\pi)^{\frac{d}{2}}}\Gamma(\epsilon)\int_0^1 dx\,(1+(\frac{q}{m})^2 x(1-x))^{-\epsilon} \\
&= \frac{(m)^{-2\epsilon}}{(4\pi)^{\frac{d}{2}}}\Gamma(\epsilon)\mathcal{F}(\frac{q^2}{m^2}), \\
\mathcal{F}(q^2) &= \int_0^1 dx\,(1+q^2 x(1-x))^{-\epsilon} \\
&= -\frac{2^{-\epsilon-1}(q+\sqrt{q^2+1})(1-\frac{q}{\sqrt{q^2+1}})^{\epsilon} F_1(1-\epsilon,\epsilon,2-\epsilon,\frac{1}{2}(\frac{q}{\sqrt{q^2+1}}+1))}{q(\epsilon-1)} - \\
&\quad \frac{2^{-\epsilon-1}(q-\sqrt{q^2+1})(1+\frac{q}{\sqrt{q^2+1}})^{\epsilon} F_1(1-\epsilon,\epsilon,2-\epsilon,\frac{1}{2}(-\frac{q}{\sqrt{q^2+1}}+1))}{q(\epsilon-1)}, \\
v_{33\mu}(p^2) &= \int \frac{dp_1^d}{(2\pi)^d i} \frac{p_{1\mu}}{(m^2-p_1^2)(m^2-(p+p_1)^2} \\
&= a(p^2)\ p_\mu, \\
a(p^2) &= -\frac{1}{2}s_{33}(p^2,\epsilon).
\end{aligned}$$

Again the result for parametric Feynman integral was analytically evaluated thanks to the use of the algebraic calculation program Mathematica. Thus, after extracting the Euclidean angular integrals and performing some transformations, the effective action density contribution

$$\gamma_{33}(m,\mu,\epsilon) = \frac{\Gamma^{(33)}}{\mu^{2\epsilon}V^{(d)}} \tag{98}$$

was expressed as single momentum integral in the range (0,∞) as follows

$$\gamma_{33}(m,\mu,\epsilon) = \frac{4c(m,\epsilon)(g_Y^0)^4\left(\frac{m}{\mu}\right)^{-6\epsilon}}{\epsilon^2}\int_0^\infty dp\,\frac{p^{3-2\epsilon}}{(p^2+r^2)^2}g(p,\epsilon), \tag{99}$$

$$g(p,\epsilon) = \epsilon^2 g_1(p,\epsilon)\Gamma(\epsilon)^2 + g_2(p,\epsilon)\,\epsilon\,\Gamma(\epsilon) + g_3(p,\epsilon), \tag{100}$$

$$g_1(m,\epsilon) = \left(\frac{p^2}{2}+2\right)^2 s_{33}^*(p,\epsilon)^2, \tag{101}$$

$$g_2(m,\epsilon) = -2\left(\frac{p^2}{2}+2\right) L(\epsilon)\, s_{33}^*(p,\epsilon), \tag{102}$$

$$g_3(m,\epsilon) = L(\epsilon)^2, \tag{103}$$

$$s^*_{33}(p,\epsilon) = \frac{s_{33}(p,\epsilon)}{\Gamma(\epsilon)}, \tag{104}$$

$$c(m,\epsilon) = \frac{2^{2\epsilon-3} m^4 \pi^{\frac{1}{2}(4-2\epsilon)+2\epsilon-4}}{\Gamma\left(\frac{1}{2}(4-2\epsilon)\right)}, \tag{105}$$

$$L(\epsilon) = \frac{\pi^{2-\epsilon}}{(2\pi)^{4-2\epsilon}}\Gamma(-1+\epsilon). \tag{106}$$

Finally, by employing a similar procedure for extracting the leading logarithmic correction in $\log\left(\frac{m}{\mu}\right)$ for D_{31}, the analogous contribution for D_{33} followed in the form

$$\gamma_{33}(m,\mu) = -0.000329114\ (g_Y^0)^4\ m^4 \log^3\left(\frac{m}{\mu}\right).$$

3.3 Discussion of the results

This subsection resume the results obtained in Ref. (33) for the effective action density. The total effective potential value $v(m,\mu)$, resulted to be given by the sum of all the evaluated terms after changing their sign. The total potential and its various contributions were written as follows

$$v(m,\mu) = v_1(m,\mu) + v_2(m,\mu) + v_{31}(m,\mu) + v_{33}(m,\mu) + v_{32}(m,\mu), \tag{107}$$

$$\begin{aligned}\frac{v_1(m,\mu)}{m^4} &= -\frac{\gamma_1(m,\mu)}{m^4} \\ &= -0.0506606\left(2.\log\left(\frac{m}{\mu}\right) - 2.95381\right),\end{aligned} \tag{108}$$

$$\begin{aligned}\frac{v_2(m,\mu)}{m^4} &= -\frac{\gamma_2(m,\mu)}{m^4} \\ &= -0.0000200507(g_Y^0)^2(183.83 - 173.783\log\left(\frac{m}{\mu}\right) + 48.\log^2\left(\frac{m}{\mu}\right)),\end{aligned} \tag{109}$$

$$\frac{v_{31}(m,\mu)}{m^4} = -\frac{\gamma_{31}(m,\mu)}{m^4} = 0.0000228551(g_Y^0)^4 m^4 \log^3\left(\frac{m}{\mu}\right), \tag{110}$$

$$\frac{v_{33}(m,\mu)}{m^4} = -\frac{\gamma_{33}(m,\mu)}{m^4} = 0.000329114(g_Y^0)^4 m^4 \log^3\left(\frac{m}{\mu}\right), \tag{111}$$

$$\begin{aligned}\frac{v_{32}(m,\mu)}{m^4} &= -\frac{\gamma_{32}(m,\mu)}{m^4} \\ &= -(g_Y^0)^4 m^4\, 10^{-3}(0.0329114\log^5\left(\frac{m}{\mu}\right) - 0.105904\log^4\left(\frac{m}{\mu}\right) \\ &+ 0.0165851\log^3\left(\frac{m}{\mu}\right) + 0.441159\log^2\left(\frac{m}{\mu}\right) - \\ &0.74347\log\left(\frac{m}{\mu}\right) + 0.388237).\end{aligned} \tag{112}$$

The renormalization point for μ was chosen at the same value of the fermion mass m_f, under consideration, that is $Log(\frac{m_f}{\mu}) -> 0$. Also, it was defined a new scaled scalar field Φ and interaction parameter g by mean of

$$\Phi = \alpha\varphi, \tag{113}$$

$$g_Y^0 = \alpha m = g\exp(\Phi), \tag{114}$$

$$g = \alpha m_f. \tag{115}$$

Then, the evaluated total contribution to the effective potential for the Dilaton $v(m,\mu)$ was expressed as a function $v(\Phi, g)$ as follows

$$\begin{aligned} \frac{v(\Phi,g)}{m_f^4} \equiv \frac{v(m,\mu)}{m_f^4} = & -0.0000329114e^{8\Phi}g^4\Phi^5 + 0.000105904e^{8\Phi}g^4\Phi^4 \\ & +0.000289673e^{8\Phi}g^4\Phi^3 + e^{4\Phi}\left(-0.000441159e^{4\Phi}g^4 - 0.000962436e^{2\Phi}g^2\right)\Phi^2 + \\ & e^{4\Phi}\left(0.00074347e^{4\Phi}g^4 + 0.00348448e^{2\Phi}g^2 - 0.101321\right)\Phi + \\ & e^{4\Phi}\left(-0.000388237e^{4\Phi}g^4 - 0.00368594e^{2\Phi}g^2 + 0.149642\right). \end{aligned} \tag{116}$$

New functions u_5, u_4 and u_3 representing approximations of the potential were defined in the form

$$\frac{u_5(\Phi,g)}{m_f^4} = \frac{v(\Phi,g)}{m_f^4}, \tag{117}$$

$$\begin{aligned} \frac{u_4(\Phi,g)}{m_f^4} = & \; 0.000105904e^{8\Phi}g^4\Phi^4 + 0.000289673e^{8\Phi}g^4\Phi^3 + \\ & e^{4\Phi}\left(-0.000441159e^{4\Phi}g^4 - 0.000962436e^{2\Phi}g^2\right)\Phi^2 + \\ & e^{4\Phi}\left(0.00074347e^{4\Phi}g^4 + 0.00348448e^{2\Phi}g^2 - 0.101321\right)\Phi + \\ & e^{4\Phi}\left(-0.000388237e^{4\Phi}g^4 - 0.00368594e^{2\Phi}g^2 + 0.149642\right), \end{aligned} \tag{118}$$

$$\begin{aligned} \frac{u_3(\Phi,g)}{m_f^4} = & +0.000289673e^{8\Phi}g^4\Phi^3 + \\ & e^{4\Phi}\left(-0.000441159e^{4\Phi}g^4 - 0.000962436e^{2\Phi}g^2\right)\Phi^2 + \\ & e^{4\Phi}\left(0.00074347e^{4\Phi}g^4 + 0.00348448e^{2\Phi}g^2 - 0.101321\right)\Phi + \\ & e^{4\Phi}\left(-0.000388237e^{4\Phi}g^4 - 0.00368594e^{2\Phi}g^2 + 0.149642\right). \end{aligned} \tag{119}$$

Note that u_5 coincided v and is of order five in the powers of Φ. The function u_4, u_3 were defined as retaining only all the terms up to order Φ^4 and Φ^3 respectively of the original function u_5 . Therefore, these functions basically correspond to the expansion of order five, four and three in powers of $\log\left(\frac{m}{\mu}\right)$. They were defined in order to study the influence of increasing the order of the perturbative expansion in powers of $\log\left(\frac{m}{\mu}\right)$.

To evidence the dependence on Φ and g of the three functions (after divided by the common factor m_f^4), they were plotted in figure 7. The range of values of $g = m_f\alpha$ was chosen $(0,1)$ as suggested by the fact that α is of the order of the Planck length and thus the physical values of the considered fermion mass are expected to determine g to be smaller than one. The plot of u_5 showed that there is a threshold value of g, below which the potential exhibits minima tending to stabilize the vacuum mean value of the Dilaton field. This behavior was also shown by the approximated potentials u_4 and u_3, a fact that indicated that after disregarding the higher quintic and quartic terms in the expansion in $log(\frac{m}{\mu})$, the existence of Dilaton stabilizing minima is not affected.
When considering the full evaluated potential curve u_5, illustrated at the top plot of figure 7, it can be observed that after lowering the g value below a critical threshold, the minimum as a function of Φ stops to exist at a critical value g_{min}. However, in the case of u_4 and u_3, the minimum exists for arbitrary values of $g- > 0$. That is, when the potential approximation is bounded from below, the potential shows stabilizing minima at any small value of g close to zero. The field value at the minima grow when the coupling tends to vanish. It can be noted, that the non bounded from below character of the approximated potential calculated, is determined by the fact that the quintic power of Φ correction turned to be negative. However, the physical system under consideration is one in which the total effective potential can be expected to show an exact bounded from below character. Thus, the next corrections are expected to exhibit a bounded from below behavior. In accordance with this expectation, in studying the g dependence at small values, we employed the approximated potential function u_4, assuming that it represents a reasonably good approximation of the exact potential.

3.4 Dilaton field and mass for m_f at the GUT scale

In Ref. (33) it was firstly considered that the highest fermion mass m_f is given by the GUT mass scale

$$\begin{aligned} m_f &= m_{GUT} = 5.06773 \times 10^{29}\text{cm}^{-1} \\ &\equiv 10^{16}\text{ GeV}, \end{aligned} \tag{120}$$

which produced for the coupling g the value

$$g = m_f\,\alpha = -\frac{3}{4}\kappa\, m_{GUT} = -0.0030789542773.$$

The potential u_4 as a function of the field Φ for this particular value of g is shown in figure 8. The minimum of the curve determined an estimate for the vacuum value of the Dilaton field given by

$$\Phi_{vac} = 5.8576156 \;= \alpha\;\varphi_{vac}, \tag{121}$$

$$\varphi_{vac} = -\frac{4}{3}5.8576156\frac{1}{\kappa}. \tag{122}$$

This result indicated that the vacuum mean value of the Dilaton field, after assuming that the fermion mass is in the GUT scale, became stabilized in the scale of the Planck mass.
For the mass of the field excitation it followed that its value was determined by the second derivative of the potential curve taken at the minimum, which is given by

$$\frac{d^2}{d\Phi^2}u_4(\Phi,-0.0030789542773)\bigg|_{\Phi=\Phi_{vac}^{(m_{GUT})}} = 1.28179 \times 10^{11} m_f^4. \tag{123}$$

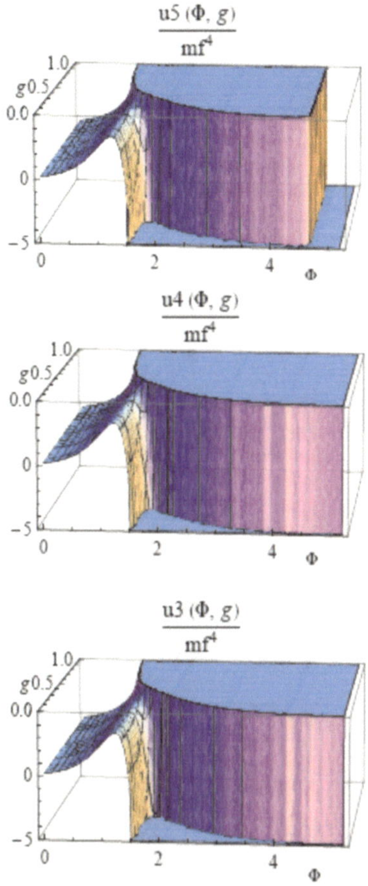

Fig. 7. The three figures show, from top to bottom, the potentials u_5, u_4 and u_3 dependence on the field Φ and the coupling g, respectively. The potential scale is chosen for a high magnification range (the minima of the surface at fixed g values are very far below the plotted range) in order to evidence the presence of a threshold for the appearing of the minima when the value of g decreases below $g = 1$. Note that for Φ smaller than some units and not to small values of g, the three plotted graphs are similar, indicating that the elimination of the highest fifth, and also the next to highest fourth, powers of the field (or, of the logarithm in the original expansion) in defining u_4 and u_3 respectively, are not affecting the results in the mentioned region. The circumstance that the exact evaluated contribution has a negative leading term of order five (which makes the result unbounded from below) explains that for the plot of u_5 the minima disappear for sufficiently small values of g. However, the fact that exact potential should be expected to be bounded from below, we consider that supports our assumption about employing the bounded from below approximations of the potential u_4 (or u_3) in evaluating the Dilaton properties at the small values of g defined by the GUT and m_{top} mass scales.

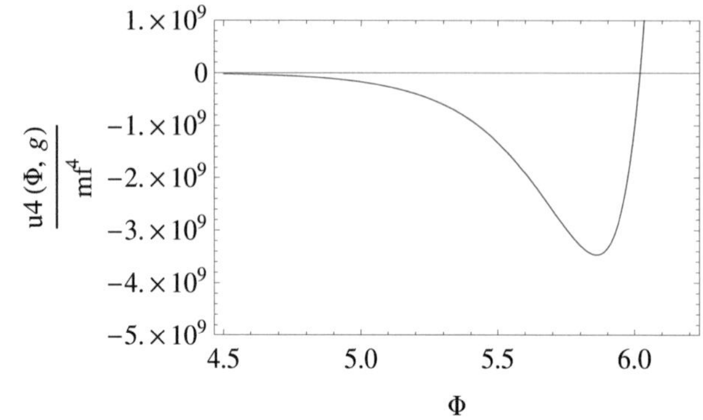

Fig. 8. The effective potential u_4 defined by Eq. 118 as a function of the Dilaton field Φ. The fermion mass was fixed to correspond to the GUT mass m_{GUT} and the renormalization scale μ was chosen to coincide with this mass. The minimum of the potential was near the value $\Phi = 5.7$, which indicates that the field is bound to a high value near the Planck scale.

In order to estimate the Dilaton mass it was considered the linearized equation of motion for the mean field

$$(\frac{1}{\alpha^2}\partial^2 + \frac{d^2}{d\Phi^2}u_4(\Phi, -0.00307895)\Big|_{\Phi=\Phi_{vac}^{(m_{GUT})}})\,\Phi = 0, \tag{124}$$

in which the factor $\frac{1}{\alpha^2}$ multiplying the D'Alembertian appeared due to the previously done change of field variables $\Phi = \alpha\,\varphi$.
The above wave equation led to the dispersion relation for the Dilaton modes

$$(-\frac{1}{\alpha^2}p^2 + \frac{d^2}{d\Phi^2}u_4(\Phi, -0.00307895)\Big|_{\Phi=\Phi_{vac}^{(m_{GUT})}}) = 0, \tag{125}$$

which for the case of the particle at rest $p = (m_D^{(m_{GUT})}, 0, 0, 0)$ determined for the Dilaton the mass estimate

$$\begin{aligned} m_D^{(m_{GUT})} &= \sqrt{\frac{d^2}{d\Phi^2}u_4(\Phi, -0.00307895)\Big|_{\Phi=\Phi_{vac}^{(m_{GUT})}} m_{GUT}^2} \mid \alpha \mid \\ &= 5.58626 \times 10^{32} \quad \text{cm}^{-1}. \end{aligned} \tag{126}$$

Therefore, the predicted order of the mass for the Dilaton also became an extremely high value which makes this field mode undetectable in a direct way.

3.5 Dilaton mean value and mass for m_f at the *top* quark mass scale

It was also of interest to take as m_f the highest currently known fermion mass: that is, the top quark one

$$m_{top} = 172.0 \pm 0.9 \quad \text{GeV} = 8.7164 \times 10^{15} \quad \text{cm}^{-1}. \tag{127}$$

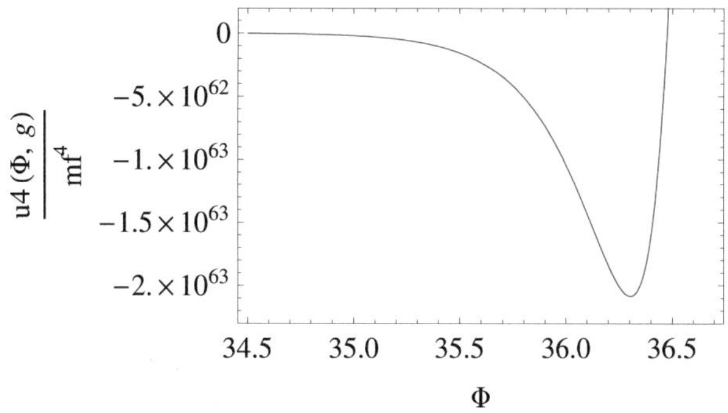

Fig. 9. The effective potential u_4 plotted as a function of the Dilaton field Φ. In this case the coupling was defined by a fermion mass correspond to the top quark one m_{top} and the renormalization scale μ was also chosen to coincide with this value. The minimum of the potential is now near the value $\Phi = 36.7765$, which indicated that the field is again staying at a high value.

Then, the coupling g in this case got the small value

$$g = m_f\,\alpha = -\frac{3}{4}\kappa\, m_{top} = -5.32659 \times 10^{-17}. \tag{128}$$

Figure 9 shows the dependence of the potential u_4 as a function of the field Φ at the above value of the coupling g. The minimum of the curve in this case gave for the mean Dilaton field at the vacuum

$$\Phi_{vac}^{(m_{top})} = 36.3020096 = \alpha\ \varphi_{vac}^{(m_{top})}, \tag{129}$$

$$\varphi_{vac}^{(m_{top})} = -\frac{4}{3}36.3020096\frac{1}{\kappa}. \tag{130}$$

This result predicts that, assuming that the maximal fermion mass in Nature is given by the top quark one, which means a lower bound for the physical masses, the vacuum field of the Dilaton, again becames stabilized in a scale, which although not being so high, is yet close to the Planck mass.

In this case the dispersion relation for the Dilaton modes resulted in the form

$$(-\frac{1}{\alpha^2}p^2 + \frac{d^2}{d\Phi^2}u_4(\Phi, -5.32659 \times 10^{-17})\Big|_{\Phi=\Phi_{vac}^{(m_{top})}}) = 0. \tag{131}$$

But, after evaluating for the second derivative of the potential at the minimum to be

$$\frac{d^2}{d\Phi^2}u_4(\Phi, -5.32659 \times 10^{-17})\Big|_{\Phi=\Phi_{vac}^{(m_{top})}} = 6.86404 \times 10^{64}, \tag{132}$$

and fixing again the rest frame momentum $p = (m_D^{(m_{top})},0,0,0)$ estimated for the Dilaton mass the value

$$m_D^{(m_{top})} = \sqrt{\frac{d^2}{d\Phi^2}u_4(\Phi,-5.32659\times 10^{-17})\Big|_{\Phi=\Phi_{vac}^{(m_{top})}}} \times m_{top}^2 \mid \alpha \mid$$

$$= 7.07209\times 10^{29} \qquad \text{cm}^{-1}. \tag{133}$$

Henceforth, also in this case the predicted mass for the Dilaton turned to be a high value being now close to the GUT scale. Thus, it can be expected that for maximal fermion masses in Nature ranging between the lower bound m_{top} and the GUT scale one, the Dilaton gets stabilized at a large field value as required by string phenomenology. In addition the resulting values of its mass, for the same range of m_f, became also out of the current observability range of particle detectors.

4. Conclusions

We had reviewed and commented some issues linked with the possible roles of the Dilaton in Cosmology and its stabilization properties under the existence of massive fermion matter, which were advanced in Refs. (32; 33).
In the work (33), the fermion field mass values were considered in two cases: the top quark mass representing the lower bound of all existing but yet unknown fermion masses in Nature, and the energy scale of the grand unification theories of order 10^{16} GeV. In both situations, the results indicated that the Dilaton mean field becomes stabilized at the very high values required by its role in allowing gravity to have its observed properties. Then, the same existence of matter seems to be a possible source of the dynamical fixation of the Dilaton field at the high values, required by String Theory to imply the observable Einstein theory of gravity. Furthermore, the evaluations pointed out that the Dilaton field also resulted to be strongly bound around its mean value, by showing a large mass being close to the GUT or Planck scales. Therefore, a possible explanation for the lack of observable consequences of the Dilaton scalar field in Nature was underlined. The discussion included contributions to the effective potential up to 3-loops. They allowed to consider the influence of the inclusion of different leading perturbative correction on the main conclusions. After, disregarding in the evaluated potential: a) the highest order term (quintic) in the expansion in powers of $Log(\frac{m}{\mu})$ (which determined the unbounded from below structure of the potential at large Φ values) or b) the two highest orders (the quintic and the quartic ones), the obtained modified potentials were both bounded from below at high field values. This procedure allowed that minima as functions of Φ exist for arbitrarily small values of the coupling g, which allowed to evaluate at the small coupling values fixed by the GUT and top quark masses. The fact that the Yukawa theory under consideration should exhibit a bounded from below potential, then supported the adopted procedure for estimating the vacuum mean values and mass of the Dilaton field. However, further higher loop evaluations are convenient to define more precise estimated values of the Dilaton vacuum field and mass and also for checking that they do not affect the picture. Moreover, it will be also helpful to perform an evaluation of the vacuum mean value of the square of the radiation Dilaton field (basically defined by a Dilaton propagator tadpole diagram). A result of $< 0|((\Phi^r(x))^2|0 >$ being much smaller than 1 will check a main assumption adopted in this work: that the QFT with exponential interaction associated to the

Dilaton field could be well approximated by the here employed Yukawa QFT description for each value of the mean field Φ.

On another hand the results of the work (32) presented a static solution of the EKG equations in which the Dilaton was represented by a scalar field showing a small mass of the order of the inverse of the estimated radius of the Universe. An interaction of the Dilaton with matter was also included. It arose that the existence of the interaction was central in allowing the arising of the static solution. The model parameters were able to be fixed for determining a relation matter-Dark Energy ratio close to the one observed, and which slowly changes with the increase of the radial distance. However, the assumed small mass of the Dilaton rises doubts about the validity of this picture for the Dilaton. Consequently, these doubts also translate to the possible feasibility of the picture speculated in the introduction in which the Universe could show a kind a "Matryoshka" structure, in which our Universe could result to be the interior of a kind of "gravastar" (See Refs. (35; 36)). However, the fixation of the Dilaton field to a high and rigid value induced by the validity of the results of Ref. (33) could perhaps not to exclude the realization of the mentioned picture. We would like to underline here an idea, which is already being discussed in the literature, and that in our view could support the considered picture, by also furnishing a concrete explanation for the origin and smallness of the Cosmological Constant. The first reference about this point of view we received from the work reported in Ref. (34). In this article it was pointed out that the one loop quantization of pure gravity determines deSitter spacetime as a natural solution. More importantly, it was also underlined that the effect is a consequence of a very much natural effect: the condensation of the massless and also gravitationally attracting between themselves gravitons. In other words that work proposed as the source of the Cosmological Constant (that is, the validity in Nature of the deSitter space time) the expected to occur condensation of gravitons once the gravity is assumed to be quantized. The smallness of the CC could be associated to the weakness of the attraction between gravitons. It should be also pointed out that the technical difficulty linked with the non renormalizability of pure gravity, should not be considered as a serious obstacle to the possibility of the relevance in Nature of this effect. This seems to be so, whenever we accept the relevance of string theory in Physics, because a quantized gravity should be described by string theory in an expected to be finite way. If such is the case, the just mentioned graviton condensation effect should be expected in the finite calculational framework of string theory for the quantum gravitational effects. Finally, the connection of this picture with our discussion comes form the possibility that the graviton condensation effect could allow a possible realization of the "Matryoshka" model of the Universe. In it, the collapse of usual matter could occur between regions showing a difference in the Cosmological Constant values. Such configurations can be imagined as being closely resembling the so called "gravastars" discussed in Refs. (35; 36). In ending, I would like to remark about the fact that the observed CC is considered as a surprisingly small quantity as compared with the energy density of the vacuum modes for the fields associated to the observed particles in Nature. However, it can be noted that in the framework of QFT in Minkowski spacetime, such vacuum densities are automatically canceled by the normal ordering rules in the field quantization procedure. Therefore, one could suspect that such vacuum contributions can be also efficiently canceled by consistent quantization procedures in curved spacetimes. The validity of this expectation could perhaps enforce the vanishing of the CC when QFT is considered in Minkowski space time and the gravitons are assumed as pure classical modes. However, just the gravitons are assumed to be quantum waves, their massless character in addition with the natural gravitational attraction existing among them, strongly suggest the appearance of the graviton condensate already underlined in Ref. (34), as being equivalent to the instability of the Minkowski spacetime to become a deSitter one. This last point also rises the idea about that the one loop effective action for gravity which should be generated by the graviton condensation effect, could also play a role in explaining the large scale effects currently attributed the existence of the Dark Matter. We expect to be able of considering some of these questions in further extensions of the work.

5. References

[1] Green, M. B.; Schwartz, H. H. & Witten, E.(1987). *Superstring theory*, Cambridge University Press, Cambridge.

[2] Veneziano, G. (1991). Scale factor duality for classical and quantum strings, *Phys. Lett. B*, 265, (287-294).

[3] Tseytlin, A. A. & Vafa, C. Elements of string cosmology, *Nucl. Phys. B* 372, (443-446), arXiv:hep-th/9109048.

[4] Adelberger, E. G.; Heckel, B. R. & Nelson, A. E. (2003). Tests of the gravitational inverse-square law, *Ann. Rev. Nucl. Part. Sci.* 53, (77-121), arXiv:hep-ph/0307284.

[5] Damour, T. & Polyakov, A. M. (1994). The string Dilaton and a least coupling principle, *Nucl. Phys. B* 423 (532-558), arXiv:hep-th/9401069.

[6] Dasgupta, K.; Rajesh, G. & Sethi, S. (1999). M theory, orientifolds and G-flux, *JHEP* 9908, 023, arXiv:hep-th/9908088.

[7] Giddings, S. B.; Kachru, S. & Polchinski, J. (2001). Hierarchies from fluxes in string compactifications, *Phys. Rev. D* 66, 106006, arXiv:hep-th/0105097.

[8] Ferrara, S.; Girardello, L. & Nilles, H. P. (1983). Breakdown of local supersymmetry through gauge fermion condensates, *Phys. Lett. B* 125 (457-460);
Affleck, I.; Dine, M. & Seiberg, N. (1983). Supersymmetry breaking by instantons, *Phys. Rev. Lett.* 51, (1026-1029);
Affleck, I.; Dine, M. & Seiberg, N. (1984) Dynamical supersymmetry breaking in supersymmetric QCD, *Nucl. Phys. B* 241, (493-534);
Affleck, I.; Dine, M. & Seiberg, N. (1985). Dynamical supersymmetry breaking in four-dimensions and its phenomenological implications, *Nucl. Phys. B* 256, (557-559);
Shifman, M. A. & Vainshtein, A. I. (1988). On gluino condensation in supersymmetric gauge theories. SU(N) and O(N) Groups, *Nucl. Phys. B* 296, (445-461) [*Sov. Phys. JETP* 66, 1100 (1987)].

[9] Dine, M.; Rohm, R.; Seiberg, N. & Witten, E. (1985). Gluino condensation in superstring models, *Phys. Lett. B* 156, (55-60).

[10] Danos, R. J.; Frey, A. R. & Brandenberger, R. H. (2008). Stabilizing moduli with thermal matter and nonperturbative effects, *Phys.Rev.D* 77, 126009, arXiv:hep-th/0802.1557.

[11] Brandenberger, R. H. & Vafa, C.(1989) Superstrings in the early Universe, *Nucl. Phys. B* 316, (391-410).

[12] Brandenberger, R. H. (2007). String gas cosmology and structure formation: A brief review, *Mod. Phys. Lett. A* 22, (1875-1885), arXiv:hep-th/0702001;
Brandenberger, R. H. (2006). Moduli stabilization in string gas cosmology, *Prog. Theor. Phys. Suppl.* 163, (358-372), arXiv:hep-th/0509159;
Battefeld. T. & Watson, S. (2006). String gas cosmology, *Rev. Mod. Phys.* 78, (435-454), arXiv:hep-th/0510022.

[13] Nayeri, A.; Brandenberger, R. H. & Vafa, C. (2005). Producing a scale-invariant spectrum of perturbations in a Hagedorn phase of string cosmology, arXiv:hep-th/0511140;

Nayeri, A. (2006). Inflation free, stringy generation of scale-invariant cosmological fluctuations in D = 3 + 1 dimensions, arXiv:hep-th/0607073.
[14] Brandenberger, R. H.; Nayeri, A.; Patil, S. P. & Vafa, C. (2007). String gas cosmology and structure formation, *Int. J. Mod. Phys. A* 22, (3621-3642), arXiv:hep-th/0608121.
[15] Brandenberger, R. H.; Nayeri, A.; Patil, S. P. & Vafa, C. (2007). Tensor modes from a primordial Hagedorn phase of string cosmology, *Phys. Rev. Lett.* 98, 231302, arXiv:hep-th/0604126.
[16] Weinberg, S. (1971) *Gravitation and Cosmology*, MIT Press.
[17] Alnes, H. & M. Amarzguioui, M. (2006). CMB anisotropies seen by an off-center observer in a spherically symmetric inhomogeneous universe, *Phys. Rev. D* 74, 103520, astro-ph/0607334.
[18] Cabo, A. & Amezaga, A. (1999). Can the Red Shift be a consequence of the Dilaton field?, arXiv:hep-th/9903265.
[19] Cabo, A. & Ayón-Beato, E. (1999). About black holes with non trapping interior, *Int. J. Mod. Phys. A*, Vol. A14, (2013-2022).
[20] Green, M.; Schwartz, J. & Witten E. (1887), *Superstring Theory*, 1, 2, Cambridge University Press.
[21] Polshinski, J. (1994). What is String Theory?, arXiv:hep-th/9411028v1.
[22] Quevedo, F. (1998). *Superstring phenomenoloy, an overview, Nucl. Phys. Proc. Suppl.*, 62, (134-143), hep-ph/9707434.
[23] Font, A.; Ibanez, L. E.; Quevedo F. & Sierra, A., (1990). The construction of "realistic" four dimensional strings through orbifolds, *Nucl. Phys. B* 331, (421-474), CERN-TH-5326/89.
[24] Susskind, L. & Lindesay, J. (2005). *An Introduction to Black Holes*, World Scientific, arxiv.org/pdf/0901.3724.
[25] Perlmutter, S.; et al.(1999). Measurements of omega and lambda from 42 high redshift supernovae. *Astrophysics. J.* 517, (565-586), astro-ph/9812133.
[26] 't Hooft, G. (2001). *Introduction to General Relativity*, Rinton Press, Inc. Princeton, New Jersey, e-Print: astro-ph/9812133
[27] Balantekin, A. Ed., (2006). Review of Particle Physics, *J. Phys. G*, 33.
[28] Cabo, A. & Brandenberger, R. H. (2009). Could fermion masses play a role in the stabilization of the dilaton in cosmology?, *JCAP* 02 015.
[29] Elizalde, E.; Naftulin, S. & Odintsov, S. D. (1994). One-loop divergence in dilaton gravitation with neutral fermions, *Phys. Rev. D* 49, (2852-2861).
[30] Muta, T. (1987). *Foundations of Quantum Chromodynamics*, World Scientific Lecture Notes Vol. 5, World Scientific Publishing Co. Pte. Ltd., Singapore.
[31] Schroder, Y. & Vuorinen, A. (2005). High-precision epsilon expansion of single-mass-scale four-loop vacuum bubbles, *JHEP* 0506, 051, arXiv:hep-ph/05033209v1.
[32] Cabo-Bizet, A & Cabo Montes de Oca, A. (2007). Static Universe model existing due to the Matter-Dark Energy coupling, arXiv:0709.0073v2 [gr-qc].
[33] Cabo, A.; Brandenberger, R. H.; Roos, M. & Erfani, E. (2010). Dilaton stabilization by massive fermion matter, arXiv:1011.4871v1 [hep-th].
[34] Marochnik, L.; Usikov, D. & Vereshkov, G. (2008). Cosmological acceleration from virtual gravitons, *Found. Phys.* 38, (546-555), arXiv:0709.2537v2 [gr-qc].
[35] Laughlin, R. B. (2003). Emergent Relativity, *Int. J. Mod. Phys. A*, 18,(831-853).
[36] Chapline, G. (2003). Quantum phase transitions and the failure of classical general relativity, *Int. J. Mod. Phys. A*,18,(3587-3590).

[37] Alfonso-Faus, A. (2009). Artificial contradiction between cosmology and particle physics: the lambda problem, *Astrophys.Space Sci.*,321,(69-72), arXiv:0811.3933.

15

A Polytropic Solution of the Expanding Universe – Constraining Relativistic and Non-Relativistic Matter Densities Using Astronomical Results

Ahmet M. Öztaş[1] and Michael L. Smith[2]
[1]*Engineering Physics, Hacettepe University, Ankara*
[2]*Jannek Fuel Inc., Phoenix*
[1]*Turkey*
[2]*USA*

1. Introduction

Important programs are being pursued currently to collect distance and recession velocity data from supernovae type Ia of several groups collecting and analyzing data (SNe Ia) (Astier et al., 2006; Hicken et al., 2009; Wood-Vasey et al., 2007) and from gamma ray bursts (GRB) (Schaefer, 2007) and most recently from other supernovae types. These events are our best hope of "standard candles" looking back in a quantitative manner towards the epochs of recombination and early galaxy formation, to nearly to the beginning of time (Leibundgut, 2008). While some optical data have been obtained from orbiting observatories, data of good quality have been collected from grounded telescopes with the hope to collect several thousand distance/velocity pairs of ever increasing quality over the next few years. Likewise, interpretation of the cosmic microwave background (CMB), which consists of signals remaining from the very primitive Universe after 13 billion years of cooling, may provide another independent source of data for estimating these parameters (Komatsu et al., 2009). Interpretation of the Sloan Digital Sky Survey (SDSS) data are being continually refined and the interpretations broadened to include analysis for the baryonic acoustic oscillations (BAO) remnant signals (Eisenstein et al., 2005). Another tool for estimation of gross Universe structure is the X-ray emissions from galaxy clusters (Vikhlinin et al., 2008) with future data collections planned to answer several cosmological questions (Vikhlinin et al., 2009). Data from dozens of studies have been used to estimate the Hubble constant, the Universe age and to support the new idea of Dark Energy (DE) assisted universe expansion (Carroll et al., 1992). Many decades ago Chandrasekhar successfully investigated the nature of white dwarf stars with gravitational fields large enough to quickly accrete considerable surrounding gas (Chandrasekhar, 1964). He predicted that as the object mass increases, Pauli instability of the dwarf star constituents is approached with a thermonuclear critical mass dependent both on the nature of the matter, the energy contained within and by the repulsive pressure exerted from the various nuclear species. When the instability limit is reached the object suffers collapse quickly followed by thermonuclear explosion which we gaze at in awe as a supernova (Chandrasekhar & Trooper, 1964). This very successful description, which has been confirmed many times by observation, is dependent on the polytropic generalizations of matter and energy. The useful relationships of Chandrasekhar demand unique constants which differ between common matter and relativistic matter or radiation. This property allows us to easily discriminate between the two matter types and manipulate both using modeling with astronomical data such as that from SNe Ia events. We present a brief derivation of the general polytropic model in Appendix 5.1.

People are currently investigating the physics of SNe Ia explosions and other cosmological data with aim to better use improved data for solving many problems of astronomy and physics (Linden et al., 2009), but reports of calculations of neutrino abundance using these data are rare. Systematic error remains in the SNE Ia data, however, with SNe Ia luminosity being correlated to the host galaxy size and mass density (Kelly et al., 2009). We and others have recently warned of even more systematic errors entering analyses of these data, both from the viewpoint of mathematical arguments (Oztas & Smith, 2006) and a call to use proper statistical analysis of the SNe Ia data (Hartnett & Oliveira, 2007; Oztas et al., 2008). A similar situation, though not identical in nature, has recently arisen regarding the analysis of the CMB data. The CMB temperature maps published by the Wilkinson Microwave Anisotropy Probe (WMAP) team are inconsistent with the differential, time-ordered data from which the descriptive maps are reconstructed (Liu et al., 2009). When a simple correction to the analytical routine is used the resulting maps become much "smoother" losing the vast majority of temperature details (Li & Liu, 2009; Liu & Li, 2010a). Indeed, these reinterpreted maps seem featureless; consistent with extreme homogeneity of the early Universe. Since interpretation of the analysis of the WMAP group is under question, we leave comparisons between CMB observations and our analysis of SNe Ia data until that time when these issues are resolved (Roukema, 2010) and concentrate on the SNe Ia and BAO analyses. We have also recently published results from an attempt to combine SNe Ia and GRB data for analysis of cosmological parameters (Smith et al., 2010). Unfortunately the GRB data are exceptionally noisy making firm conclusions impossible.

Here we extend the use of polytropic tools for the first time to estimate ranges for the average common matter and relativistic matter densities of the Universe, as well as better estimating spacetime curvature and DE. We analyze SNe Ia data using luminary distances and associated distance errors rather than log distances and log errors. Because we use actual distances rather than the logs (not a trivial difference), we also include a data pair for the earth with no error for the first time and shall show this important data pair should be included in all analyses utilizing SNe Ia data. Using actual estimated errors for distance rather than log errors accents the differences between models with and without the cosmic constant and between results from different reports. We examine data from four recent SNe Ia collections and can roughly estimate the current, non-relativistic matter density and the order of the relativistic matter density. We obtain these values using models combining the polytropic indexes with two variations of the Friedmann-Robertson-Walker (FRW), the *standard model.* Our results from the model admitting significant spacetime but without the cosmic constant are significantly different from the results using this term, Ω_Λ, but in a flat Universe. In general, we also find the matter densities for the models not invoking the cosmic constant to be much lower than those with that term. Our calculated low matter densities agree with several estimates from Big Bang Nucleosynthesis (BBN); calculations derived from first principles (Burles et al., 2001a;b). Not surprisingly, we also find the typical values for non-relativistic matter to be at least one order of magnitude larger than the relativistic matter density. Our results are driven by the great difference in distance errors between nearby and distant SNe Ia which are not evident when modeling using log errors. The matter densities from many solutions here are near the low end of the range as predicted by others while the estimated DE and/or spacetime portions of the Universe are often larger than previous findings. We must emphasize such results, varying from popular expectations, are not unexpected since we take care in weighing observational errors.

On the other hand our results derived from reported BAO parameters do agree with currently popular values for Ω_m and Ω_Λ. The suggestion was made over a decade ago that the SDSS data could be useful to estimate neutrino abundance (Hu et al., 1998) though reports of progress using that data have not come to our attention. Here we have expanded the range of data which can be used for such analysis by considering the BAO solutions. Our findings are extremely sensitive to total matter density and the ratio of polytropic matter species. We present a brief explanation of our analytical technique in Appendix 5.2.

We are also aware of reports suggesting the Universe may be modeled considering some components as a Chaplygin gas (Setare, 2009) but we think our approach significantly different and perhaps more rigorous. We have not performed calculations using SNe Ia data, that include the effects of Cold Dark Matter (CDM), because the properties are not understood not allowing us to estimate these polytropic constants. We suggest the SNe Ia and SDSS/BAO data are extremely useful observations, which in addition to constraining the Hubble constant, the Universe age and DE, can also be used to define the limits of non-relativistic and relativistic matter densities. This is because large spacetime is the important independent variable rather than the much more commonly tested energy, pressure, density, etc. Better constraints on these cosmological values can be made using larger SNe Ia (and SDSS) data sets, hopefully with smaller errors, which should be available in the near future after systematic errors have been corrected (Kelly et al., 2009). Likewise estimates of H_0 should await reconciliation between groups analyzing the CMB data (Li & Liu, 2009; Liu & Li, 2010a). The overall aim of all this mental anguish will hopefully lead to a much better understanding of the dynamics and parameters of our expanding Universe and hence our origin and eventual fate.

2. Expansion of the polytropic Universe

We begin with two equations of state we suggest describe the early Universe both during and after recombination, allowing $c = 1$ as usually presented

$$(\frac{\dot{a}}{a})^2 = \frac{8\pi G}{3}\rho + \frac{\Lambda}{3} + \frac{k}{a^2} \tag{1}$$

$$\frac{\ddot{a}}{a} = \frac{-4\pi G}{3}(\rho + 3p) + \frac{\Lambda}{3}. \tag{2}$$

Here ρ is the material content density, p the pressure, a the expansion factor, G the gravitational constant, Λ the cosmic constant and k the constant of integration; typically indicating spacetime curvature. For the general situation both variables include contributions from normal matter, relativistic matter and radiation. Cold dark matter (CDM) is sometimes considered to be described within ρ. We take the derivative of the first equation with respect to time and use the result to eliminate $\ddot{a}$ and arrive at the following relationship

$$\dot{\rho} = -3\frac{\dot{a}}{a}(\rho + p). \tag{3}$$

We presume adiabatic processes dominate Universe expansion and use the polytropic relationship between pressure and density from Chandrasekhar as $p = K\rho^\gamma$, where $\gamma = 1 + 1/n$, and n is the polytropic index. This relationship is a well-known equation of state with $n \to \infty$ for isothermal processes and $n = -1$ for isobaric processes. By substitution with Chandrasekhar's relationship we now have

$$\dot{\rho} = -3\frac{\dot{a}}{a}(\rho + K\rho^{\gamma}) \tag{4}$$

and with separation of variables we can then integrate both sides in a few steps

$$\int \frac{\dot{\rho}}{\rho + K\rho^{\gamma}} dt = -3 \int \frac{\dot{a}}{a} dt. \tag{5}$$

We separate the integrand on the left-hand side into parts and perform some algebra

$$\begin{aligned}
\frac{1}{\gamma - 1} \int \left(\frac{\gamma}{\rho} - \frac{1 + \gamma K \rho^{\gamma-1}}{\rho + K\rho^{\gamma}} \right) \dot{\rho} dt &= -3 \int \frac{\dot{a}}{a} dt \\
\frac{1}{\gamma - 1} (\ln(\rho^{\gamma}) - \ln(\rho + K\rho^{\gamma})) &= -3\ln(a) + \ln(C) \\
\ln \left(\frac{\rho^{\gamma}}{\rho + K\rho^{\gamma}} \right) &= (\gamma - 1)(-3\ln(a) + \ln(C)) \\
\frac{\rho^{\gamma}}{\rho + K\rho^{\gamma}} &= \left(\frac{C}{a^3} \right)^{\gamma - 1} \\
\frac{\rho^{\gamma-1}}{1 + K\rho^{\gamma-1}} &= \left(\frac{C}{a^3} \right)^{\gamma - 1}.
\end{aligned} \tag{6}$$

We can rearrange Eq. (6) in two steps

$$\rho^{\gamma-1} = \frac{C^{\gamma-1}}{a^{3(\gamma-1)} - KC^{\gamma-1}} \tag{7}$$

to arrive at a useful relationship

$$\rho = \frac{C}{(a^{3(\gamma-1)} - KC^{\gamma-1})^{\frac{1}{(\gamma-1)}}}. \tag{8}$$

By presuming an expansion factor of $a_0 = 1$ in Eq. (6), we can solve for C to eliminate this term

$$\frac{\rho_0^{\gamma-1}}{1 + K\rho_0^{\gamma-1}} = C^{\gamma-1} \tag{9}$$

and substituting Eq. (9) into Eq. (8) we arrive at our relationship of interest

$$\rho = \frac{\rho_0}{(a^{3(\gamma-1)}(1 + K\rho_0^{\gamma-1}) - K\rho_0^{\gamma-1})^{\frac{1}{(\gamma-1)}}}. \tag{10}$$

We presume the Universe consists of different matter species with unique values of K_i and γ_i describing each variety. We can eliminate each K_i species and use the present values for our parameters by adhering to the *cosmological principle* that expansion occurs isotropically for each species, since both non-relativistic and relativistic matter species have been dilute, except for stars, since near singularity. We also presume that energy, relativistic matter and

non-relativistic matter only weakly interact currently, except for the stars, and the contribution of radiant energy to the Universe has been tiny since recombination.
In the neo-Newtonian framework two phases of matter are important - non-relativistic matter and relativistic matter and the values we shall use for the associated constants for each are listed (Chandrasekhar, 1983) and we present brief derivations in Appendix 5.1. We estimate the relative values for the Chandrasekhar constants for relativistic and normal matter from considerations of cosmological parameters presented for solution of the SDSS data (Eisenstein et al., 2005) in our "Modeling and Results" section.

non-relativistic	relativistic
$n_{nr} = 3/2$	$n_r = 3$
$\gamma_{nr} = 5/3$	$\gamma_r = 4/3$
$K_{nr} = 0.645$	$K_r = 1.124$

With these generalizations we can separate the matter density into two different variables within the Friedmann relationship

$$\left(\frac{\dot{a}}{a}\right)^2 = \frac{8\pi G}{3}(\rho_{nr} + \rho_r) + \frac{\Lambda}{3} - \frac{k}{R_0^2 a^2} \tag{11}$$

where ρ_{nr} and ρ_r are the densities of non-relativistic and relativistic matter.
For use with SNe Ia data these are redefined as normalized parameters with

$$\Omega_r = \frac{\rho_{0,r}}{\rho_c}, \quad \Omega_{nr} = \frac{\rho_{0,nr}}{\rho_c}, \quad \Omega_\Lambda = \frac{\Lambda}{3H_0^2} \quad \text{and} \quad \Omega_k = -\frac{k}{R_0^2 H_0^2} \tag{12}$$

where Ω_r and Ω_{nr} can be calculated as averages from the present range for H_0, the Hubble constant. For this work we use the critical density parameter $\rho_c = \dfrac{3H_0^2}{8\pi G}$ for the case of a Universe without DE($\Lambda = 0$) and so do not require a flat Universe.
We present a more general variation of the common normalization condition for the parameters of interest as

$$1 = \Omega_{nr} + \Omega_r + \Omega_\Lambda + \Omega_k. \tag{13}$$

When we substitute τ for $H_0 t$ we can simplify these equations into a usable form as derived below

$$\left(\frac{da}{d\tau}\right)^2 = a^2 \left\{ \frac{\Omega_{nr}}{\left(a^2\left(1 + K_{nr}\rho_{0,nr}^{2/3}\right) - K_{nr}\rho_{0,nr}^{2/3}\right)^{3/2}} + \frac{\Omega_r}{\left(a\left(1 + K_r\rho_{0,r}^{1/3}\right) - K_r\rho_{0,r}^{1/3}\right)^3} + \Omega_\Lambda + \frac{\Omega_k}{a^2} \right\} \tag{14}$$

$$\left(\frac{da}{d\tau}\right)^2 = \frac{1}{a}\left\{ \frac{\Omega_{nr}}{A_{nr}^{3/2}} + \frac{\Omega_r}{A_r^3} + \Omega_\Lambda a^3 + \Omega_k a \right\} \tag{15}$$

and for brevity we use $A_{nr}(a) = \left(1 + K_{nr}\rho_{0,nr}^{2/3}\right) - \frac{K_{nr}\rho_{0,nr}^{2/3}}{a^2}$ and $A_r(a) = \left(1 + K_r\rho_{0,r}^{1/3}\right) - \frac{K_r\rho_{0,r}^{1/3}}{a}$. We collect the normalized terms in a familiar form

$$d\tau = \frac{\sqrt{a}da}{\sqrt{\frac{\Omega_{nr}}{A_{nr}^{3/2}} + \frac{\Omega_r}{A_r^3} + \Omega_\Lambda a^3 + \Omega_k a}} \tag{16}$$

and presuming the null geodesic of the FRW Universe and multiplying by R_0 we get

$$\frac{dR_0 r}{dt} = \frac{R_0}{R}(1 - kr^2)^{1/2} \tag{17}$$

$$R_0\frac{dr}{dt} = \frac{1}{a(t)}(1 - kr^2)^{1/2} \tag{18}$$

and by rearranging the differential equation for the separation of variables

$$R_0\frac{dr}{(1 - kr^2)^{1/2}} = \frac{dt}{a(t)} \tag{19}$$

and introducing H_0 we can obtain a relationship between which shall begin to allow us to calculate the two forms of interesting matter in relationship with Universe expansion

$$H_0 R_0\frac{dr}{(1 - kr^2)^{1/2}} = \frac{d\tau}{a(\tau)}$$

$$H_0 R_0\frac{dr}{(1 + \Omega_k R_0^2 H_0^2 r^2)^{1/2}} = \frac{d\tau}{a(\tau)}. \tag{20}$$

We now reintroduce Eq.(16) for $d\tau$

$$\frac{H_0 R_0 dr}{\sqrt{1 + \Omega_k R_0^2 H_0^2 r^2}} = \frac{1}{a}\frac{\sqrt{a}da}{\sqrt{\frac{\Omega_{nr}}{A_{nr}^{3/2}} + \frac{\Omega_r}{A_r^3} + \Omega_\Lambda a^3 + \Omega_k a}}. \tag{21}$$

We then integrate both sides to allow for the frequency drop to respond to the Universe expansion

$$\int_0^{r_1}\frac{H_0 R_0 dr}{\sqrt{1 + \Omega_k R_0^2 H_0^2 r^2}} = \int_{a_1}^{1}\frac{da}{\sqrt{a}\sqrt{\frac{\Omega_{nr}}{A_{nr}^{3/2}} + \frac{\Omega_r}{A_r^3} + \Omega_\Lambda a^3 + \Omega_k a}}. \tag{22}$$

By collecting several variables into a simpler term $\sqrt{\Omega_k}R_0 H_0 r = y$ and introducing the redshift relation in terms of frequency decline $a = 1/(1+z) = \xi$ on the right-hand side (we shall use ξ as the ratio of observed frequency to emitted frequency) and after a few steps we get

$$D_L = \frac{c}{\xi H_0 \sqrt{|\Omega_k|}} \text{sinn} \left\{ \sqrt{|\Omega_k|} \int_{\xi_1}^{1} \frac{d\xi}{\sqrt{\xi} \sqrt{\frac{\Omega_{nr}}{A_{nr}^{3/2}} + \frac{\Omega_r}{A_r^3} + \Omega_\Lambda \xi^3 + \Omega_k \xi}} \right\} \tag{23}$$

where the integration begins with the past ξ_1 to the present 1, sinn is either *sinh* or *sin* dependent on positive or negative spacetime curvature and the speed of light is in km/s.
For a universe without DE we simply drop the $\Omega_\Lambda \xi^3$ term and allow spacetime curvature and we designate this solution as Ω_r-ST. For a flat, relativistic universe with DE we can greatly simplify the above relationship as

$$D_L = \frac{c}{\xi H_0} \int_{\xi_1}^{1} \frac{d\xi}{\sqrt{\xi} \sqrt{\frac{\Omega_{nr}}{A_{nr}^{3/2}} + \frac{\Omega_r}{A_r^3} + \Omega_\Lambda \xi^3}} \tag{24}$$

which we denote as the Ω_r-DE model.
For purposes of comparison we also fit the now famous DE relationship for a flat universe without relativistic matter in terms of frequency decline

$$D_L = \frac{c}{\xi H_0} \int_{\xi_1}^{1} \frac{d\xi}{\xi \sqrt{\frac{\Omega_m}{\xi} + \Omega_\Lambda \xi^2}} \tag{25}$$

which we designate as the *Simple*-DE model. For evaluation of the FRW model without DE we simply replace the $\Omega_\Lambda \xi^2$ term from the denominator with the $\Omega_k \xi$ term leaving us with an integral which has been solved analytically (Oztas & Smith, 2006; Peebles, 1993)

$$D_L = \frac{c}{\xi H_0 \sqrt{|\Omega_k|}} \text{sinn}[2(\text{arctanh}(\sqrt{|\Omega_k|}) - \text{arctanh}(\frac{\sqrt{|\Omega_k|}}{\sqrt{\frac{\Omega_m}{\xi} + \Omega_k}}))] \tag{26}$$

which we designate as the *Analytic*-ST model and sometimes as the spacetime model.
Notice the normalized matter density terms for Eqs.(23,24) encompass, non-relativistic, relativistic and light-energy densities. We also need to emphasize the introduction of terms to account for the two natures of matter also places a dependence of matter density on the Hubble constant and hence spacetime expansion. This can be easily justified since the range of SNe Ia data from the present to z=1.55 covers the majority of the Universe age and very large changes of matter densities. By use of the above models, solidly based on the work of Chandrasehkar, we attempt to avoid introducing additional parameters as sometimes done by *ad hoc* modification of the equation of state, for instance as a Chaplygin gas.
We are able to estimate the Chandrasekhar constants for normal and relativistic matter with principles used to estimate parameter fits with SDSS data Eisenstein et al. (2005). The densities for relativistic and normal matter species, ρ_i, define the various terms of Ω_i as

$$\Omega_i = \frac{\rho_{0,i}}{\rho_c} \quad \text{for} \quad K_i \rho_{0,i}^{\gamma-1} = K_i \rho_c^{\gamma-1} \Omega_i^{\gamma-1}. \tag{27}$$

So we can use these terms as coefficients for both species of normal and relativistic matter

$$B_{nr} = K_{nr} \rho_c^{2/3}, \qquad B_r = K_r \rho_c^{1/3}.$$

We now define the current matter densities in terms of B_i, matter densities and the expansion factor as

$$\rho_{nr} = \frac{\rho_c \Omega_{nr}}{\left(a^2\left(1 + B_{nr}\Omega_{nr}^{2/3}\right) - B_{nr}\Omega_{nr}^{2/3}\right)^{3/2}}, \quad \rho_r = \frac{\rho_c \Omega_r}{\left(a(1 + B_{nr}\Omega_r^{2/3}) - B_r\Omega_{nr}^{1/3}\right)^3}. \tag{28}$$

We follow the lead of Eisenstein, using their A and R parameters and perform evaluations independent of H_0, to uncover the relative dependencies of B_i species on normalized matter density Ω_m. Details of our technique for evaluation are presented in Appendix 5.2. Results and relative errors with evaluations at several popular matter densities are presented below.
There is a recent report indicating systematic errors of up to 10% buried in the luminosities of many SNe Ia observations. For this reason we present results considering four independent reports of SNe Ia distances and redshifts, rather than a single or combined set. Hopefully, removal of more systematic error from these data, which shall soon be made public (Kelly et al., 2009), will allow better detailed investigation than the 3 parameters used here. A recent submission claiming serious problems with interpretation of the WMAP5 presentations dissuades us from investigating this area with our polytropic model until these difficulties have been resolved (Liu & Li, 2010a;b; Moss et al., 2010; Roukema, 2010).

3. Modeling and results

Calculations are made using only recently published SNe Ia sets. We first examine the 397 data pairs of Hicken and coworkers from their Table 1 as they prefer (Hicken et al., 2009). We next present fits of the combined 288 SNe Ia treated with the MLCS2K2 Light Curve Fits from data recently compiled from several sources as presented by (Kessler et al., 2009). We also present results from the list of Kowalski and coworkers (Kowalski et al., 2008), which are 307 SNe Ia culled from the Union compilation of 414 SN Ia including data from the Supernova Legacy Survey, the ESSENCE Survey, the Hubble Space Telescope and some older data and finally, we present results from modeling with 162 SNe Ia data presented by Wood-Vasey and coworkers (Wood-Vasey et al., 2007). In addition, we add the present frequency ratio of 1 at a distance of 0 with no error, for an exacting data pair at no financial cost to all data sets (Oztas et al., 2008).
The actual luminary distances and geometric errors are extracted from the published log distance and log errors rather than use log-log estimates and we best-fit these curves using robust minimization. The fitting routines for Tables 1 through 4 with solution trace examples presented in Figure 1, allow two or three free parameters, one of which is always the Hubble constant. We prefer H_0 as a free parameter having noticed the goodness of fit to be highly dependent on the freedom of this parameter (no surprise this). The other free parameters are either Ω_{nr} or Ω_{nr} with Ω_k while the remainder of the estimates is always the spacetime curvature or relativistic matter density, for instance, $\Omega_k = 1 - \Omega_{nr}$ or $\Omega_r = 1 - \Omega_{nr} - \Omega_k$. We do not think the data firm enough to report results from models containing 4 parameters; necessary for simultaneous solution of Ω_k, Ω_Λ, Ω_{nr} and Ω_r.
Figure 1 is an illustration of the data with the curves from the fits for *Simple*-DE and *Analytic*-ST models to the complete Hicken *et al.* data. Note the very large errors associated with ancient SNe Ia distances, to the left on the graph, compared to those of more nearby explosions, at the lower right-side. Rightly so - one should expect noisy data from signals more than half the Universe age. Precise distance determination is a problem which has plagued astronomers from time immemorial (Sharaf & Sendi, 2010). These very large errors mean the data from ancient SNe Ia play a much smaller role in determining the fit parameters

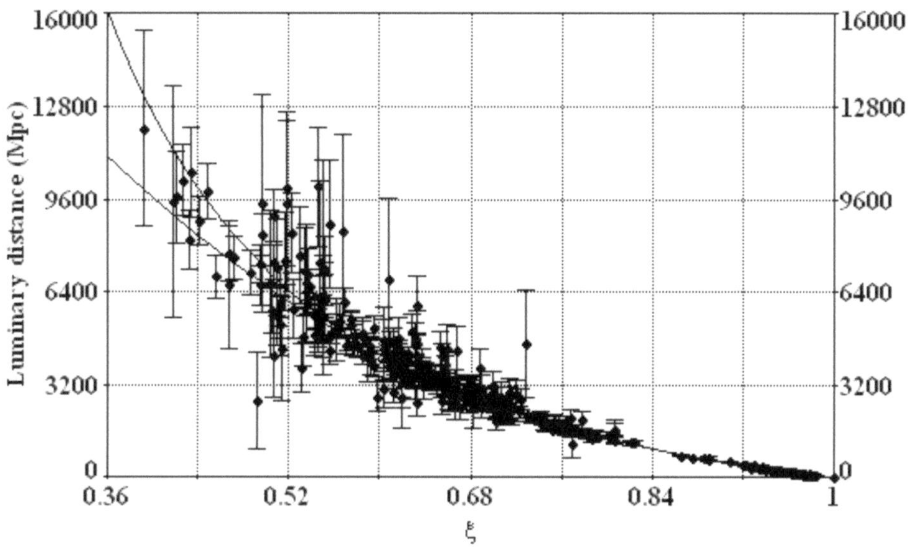

Fig. 1. Comparison of the two Standard Models using the abscissa of observed SNe Ia galaxy frequency ratios rather than redshifts with all 397 pairs plus today. The bottom curve represents the fit for the *Simple*-DE model, and the top line the fit for the *Analytic*-ST model.

Model	FP	H_0	Ω_{nr} or Ω_m	Ω_Λ or Ω_k	$\chi^2/$(N-FP)
Ω_r-DE	3	68.4 ± 0.5	0.05 ± 0.03	$0.95 \pm 0.01(\Omega_\Lambda)$	1.37
Simple-DE	2	67.9 ± 0.5	$0.27 \pm 0.03(\Omega_m)$	$0.73(\Omega_\Lambda)$	1.40
Ω_r-ST	3	73.6 ± 0.6	0.034 ± 0.30	$0.966 \pm 0.01(\Omega_k)$	1.79
Analytic-ST	2	67.5 ± 0.5	$0.0003 \pm 0.06(\Omega_m)$	$0.9997(\Omega_k)$	1.92

*With N the number of data pairs (Hicken et al., 2009) after culling 11 outliers(387) and FP the number of free parameters. H_0 in km s^{-1}Mpc^{-1}.

Table 1. Results with data from 386 SNe Ia and today

than nearby supernovae. This contrasts to the more usual fitting regimes which place value on the ancient distances almost as strongly as those of nearby SNe Ia. Rather than distance errors increasing somewhat marginally between nearby SNe Ia and distant explosions as usually presented in the typical log/log plots of luminary magnitude *vs.* redshift, we present a more realistic view of error estimates which are incredibly large from distant signals. The results of this can be seen as the obvious difference between curve traces for both *standard models* in the figure. It is more usual that the traces for these two models be nearly inseparable on displayed graphs - and they do overlap at the smaller distances - lower right-side.

There are 11 data pairs of the Hicken *et al.* set (397 pairs) which reside more than 2500 km s^{-1}Mpc^{-1} from the best fit *Analytic*-ST model; $> 3\sigma$. We remove these and calculate the values from 387 data pairs with results in Table 1; the goodness of fits for three models are nicely improved by discarding these 11 pairs. This improvement is surprising considering the 11 being outliers which should least impact the robust curve fitting process. None of the models however, return excellent fits with this data as judged by a χ^2/N-FP of less than 1.35 (Hartnett & Oliveira, 2007). Noteworthy of this data reduction is that the *Simple*-DE model now presents a better fit than the *Analytic*-ST, for the reverse is true when all 398 data pairs are

Model	FP	H_0	Ω_{nr} or Ω_m	Ω_Λ or Ω_k	$\chi^2/(\text{N-FP})$
Ω_r-DE	3	74.2 ± 0.5	0.04 ± 1	$0.96 \pm 1(\Omega_\Lambda)$	2.24
Ω_r-ST	3	79.8 ± 0.6	0.0003 ± 0.02	$0.9997 \pm 1(\Omega_k)$	2.34
Analytic-ST	2	67.8 ± 0.5	$0.42 \pm 0.09(\Omega_m)$	$0.58(\Omega_k)$	2.66
Simple-DE	2	69.9 ± 0.7	$0.76 \pm 0.07(\Omega_m)$	$0.24(\Omega_\Lambda)$	3.05

*With N the number of data pairs (283) and FP the number of free parameters (Kessler et al., 2009). H_0 in km $s^{-1}Mpc^{-1}$.

Table 2. Results with data from 282 SNe Ia and today

Model	FP	H_0	Ω_{nr} or Ω_m	Ω_Λ or Ω_k	$\chi^2/(\text{N-FP})$
Analytic−ST	2	68.0 ± 0.5	$0.0002 \pm 0.05(\Omega_m)$	$0.9996(\Omega_k)$	1.34
Ω_r-ST	3	77.4 ± 0.6	0.0002 ± 0.03	$0.9997 \pm 1(\Omega_k)$	1.44
Ω_r-DE	3	75.4 ± 0.6	0.045 ± 0.03	$0.955 \pm 0.02(\Omega_\Lambda)$	1.49
Simple−DE	2	75.3 ± 0.7	$0.14 \pm 0.02(\Omega_m)$	$0.86(\Omega_\Lambda)$	1.50

*With N the number of data pairs (308) and FP the number of free parameters (Kowalski et al., 2008). H_0 in km $s^{-1}Mpc^{-1}$.

Table 3. Results with data from 307 SNe Ia and today

used. While the matter density for the *Simple*-DE model is near that expected from previous publications, ≈0.27, for instance the BAO analysis (Eisenstein et al., 2005), the matter density returned by the *Analytic*-ST model is quite low. Low matter density values are also found from the fits of both polytropic models and the Hubble constant also tends on the low side of the usual expectation for 3 of the 4 models. When we consider the more detailed polytropic model which includes the cosmic constant, Ω_r-DE, we find it fits the data better than other models, being on the border of a good fit.

When we discard our exact data pair of today on earth and analyze with only the 386 SNe Ia data we find significant differences in the goodness of fit of the two *standard models*. Exclusion of this single data pair allows the models to "drift" from the origin at x, y of exactly 1,0 and the values for $\chi^2/(\text{N-FP})$ increase by nearly 0.1 for both the *Simple*-DE model and the *Analytic*-ST model (results not presented).

In our next analyses we use the 288 data pairs from Kessler and friends (Kessler et al., 2009) after culling 6 pair outside 2500 Mpc from the curve defined by the *Analytic*-ST fit. Even after discarding these outliers, the fits are not considered good as judged by the $\chi^2/(\text{N-FP})$, Table 2. The values for matter densities, Ω_m, for the two *standard models* are much greater than commonly reported. This might be evidence for systematic error which is revealed by our direct plot and "hidden" in the typical log/log plots. Here we find the polytropic models to be significant improvements over the two *standard models*, again with rather low values for Ω_m and with somewhat high values for the Hubble constant, compared with expectations.

We next extend our analysis to all 307 SNe Ia data of Kowalski (Kowalski et al., 2008) and we find the modeling to yield significantly better fits as judged by lower $\chi^2/(\text{N-FP})$ for all models, Table 3. The fit for the *Analytic*-ST model with a $\chi^2/(\text{N-FP})$ of 1.34 suggests a good fit but again the *Simple*-DE model does not fit the data very well. The matter densities of the better fitting models are much lower than those usually published from log/log plots, even the normalized matter density, Ω_m, of the *Simple*-DE model, at 0.14, is well below the oft presented 0.25 to 0.27.

Model	DF	H_0	Ω_{nr} or Ω_m	Ω_Λ or Ω_k	$\chi^2/(\text{N-FP})$
Simple-DE	2	66.6 ± 0.7	$0.21 \pm 0.04(\Omega_m)$	$0.79(\Omega_\Lambda)$	2.12
Ω_r-ST	3	69.0 ± 0.7	0.025 ± 0.02	$0.975 \pm 0.02(\Omega_k)$	2.35
Ω_r-DE	3	68.0 ± 0.8	0.065 ± 0.02	$0.935 \pm 0.05(\Omega_\Lambda)$	2.60
Analytic-ST	2	65.2 ± 0.7	$0.08 \pm 0.10(\Omega_m)$	$0.92(\Omega_k)$	2.61

*With N the number of data pairs (163) and FP the number of free parameters (Wood-Vasey et al., 2007). H_0 in km $s^{-1}Mpc^{-1}$.

Table 4. Results with data from 162 SNe Ia and today

We examine our final data set published by Wood-Vasey (Wood-Vasey et al., 2007) consisting of 60 distance SNe Ia from the ESSENCE Supernova Survey normalized with about 100 other observations. This is the only time the *standard model*, *Simple*-DE, is the the best fit, Table 4. The *Simple*-DE model also presents a value for Ω_m of 0.21, not much lower than currently popular values. Notice that all values for $\chi^2/(\text{N-FP})$ are much greater than two other analyses (Tables 1,3) and similar to the large values reported in Table 2. These, and the results from the Kessler *et al.* data may be caused by assigning consistently smaller errors to distance measurements, compared with other reports. This results in the curve fit being described primarily by the nearby SNe Ia events, with very low assigned errors, almost totally ignoring earlier SNe Ia. The curves from this data and the fits of the Kessler data are "flatter" than expected, reflecting the near linear alignments of nearby SNe Ia as graphed, resulting in large values of $\chi^2/(\text{N-FP})$. The Ω_r-ST models typically return a value for H_0 of 69 to 80 km $s^{-1}Mpc^{-1}$ which are our only consistent results in that popular range, but favored by recent multi-parameter WMAP (6 free parameters)and BAO analyses (Komatsu et al., 2009). The *Simple-DE* model, which is most popular model currently, here presents values for H_0 of 67 to 75 using these four sets, so cannot really discriminate between the two controversial values towards each end of this range (Sandage et al., 2006).

A significant value for relativistic matter density was never found from any model with any of the the four data sets examined here; we do not present numerical estimates for Ω_r since these report much smaller than the calculated errors of Ω_{nr} for our two models. We do venture an upper bound for Ω_r of <0.001 from the present to near reionization. It seems likely this is because the portion of universal gravitation due to relativistic particles and photons is and has been $<<0.001$. The values of Ω_{nr} and Ω_m for the DE models, both presuming a flat Universe, differ widely. The more sophisticated Ω_r-DE model with a Ω_{nr} of about 0.05 or less suggests the Universe contains much less matter than previous estimates. On the other hand the *Simple*-DE model fit with an estimated Ω_m of 0.27±0.03 to the culled Hicken *et al.* data, which is similar to values published using log luminary distance data (Astier et al., 2006; Davis et al., 2007) or the official 6-parameter WMAP results(Komatsu et al., 2009) but is short of a total Ω_m of 0.32 found from reanalysis of WMAPLiu et al. (2009). A similar situation is found for the two models allowing spacetime curvature, *Analytic*-ST and Ω_r-ST, where a more "typical" value for Ω_m of $\approx$0.25 becomes $\approx$0.001 with the Ω_r-ST model. The two terms introduced to account for the different natures of matter and the Hubble flow in Eqs.(23,24) allow matter densities to vary with respect to Universe expansion; these more sophisticated models suggest a much lower average matter density in the current epoch. The differences in matter densities between the two DE and two ST models are consistent with the idea that a universe with more matter requires more energy to continue expanding.

We present two Figures (2 and 3) which are the results from a series of single free parameter fits, with data of the culled Hicken *et al.* set and modeling with the two *standard models*,

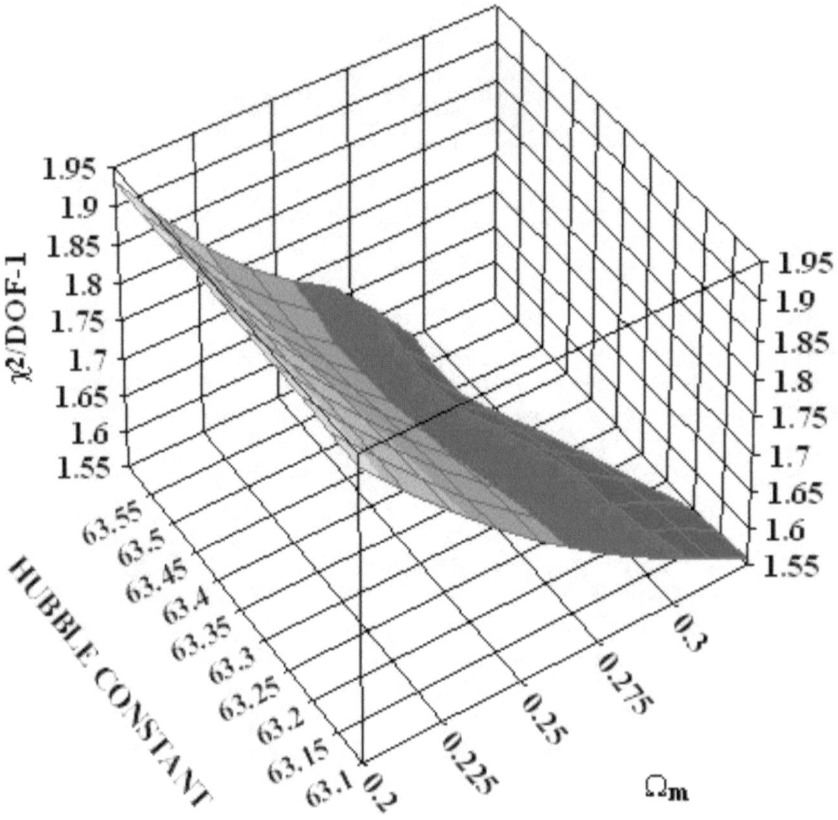

Fig. 2. Surface of Hubble constants and goodness of fits as functions of normalized matter densities with the *Simple*-DE model.

allowing Ω_m to be fixed but vary over the ranges close to values found here (Table 1) and solving for the Hubble constant and $\chi^2/(387)$ for the two *standard models*. Because Ω_m are tightly bound the fitting preference for any Ω_m at the minima are in the neighborhood but not exactly those determined *via* 2 and 3 free parameter modeling. These pseudo, three-dimensional figures display smooth, declining surfaces with increasing Ω_m for the *Simple*-DE model. On the other hand, the *Analytic*-ST model displays a fairly flat surface below Ω_m of 0.10. Notice the reduced $\chi^2/(387)$ for the *Analytic*-ST model at low values of Ω_m while the display of the *Simple*-DE fits suggests lower $\chi^2/(387)$ with increasing values of Ω_m. While the surfaces for the two models are quite different, the Hubble constants calculated for both are low and nearly invariant over the two ranges shown.

In Fig. 4 we present the results of fixing Ω_m at 0.27 and solving for the constant A_r from Eq. (24). We see the values found for this working constant are about 1/10 or less than those we derive from first principles. Unfortunately at this matter density, the goodness of fits are poor, but we can judge that the empirical value for K_r may be larger than those used by Chandrasekhar and this presentation. The figure also suggests a strong dependence of H_0 on the nature of matter density retarding Universe expansion. For instance, if Ω_m is really around 0.27, as used here, the effects of relativistic matter might be observed in SNe Ia signals.

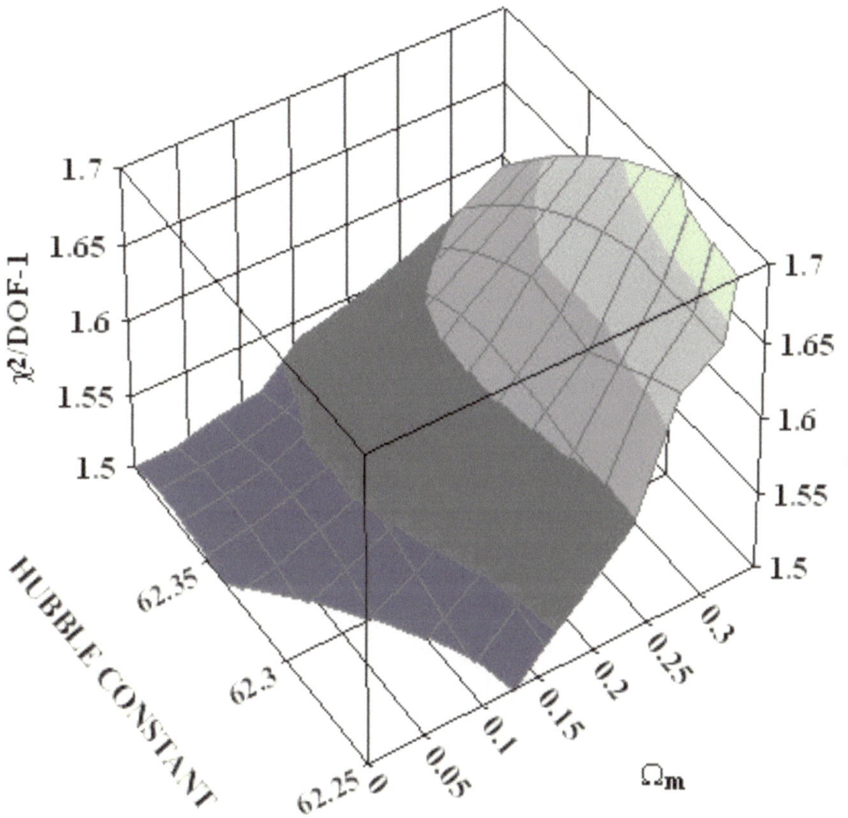

Fig. 3. Surface of Hubble constants and goodness of fits as functions of normalized matter densities with the *Analytic*-ST model

Using parameters from the Eisenstein (Eisenstein et al., 2005) investigation of cosmic BAO and presuming a flat Universe we have calculated several values for our coefficients B_{nr} and B_r of Eq. (28). We have selected three values for Ω_m from 0.273 preferred by Eisenstein to 0.32 preferred by Li and Liu Li & Liu (2009). Our results are reported in Table 5 as functions of Ω_m and x, where x is the ratio of Ω_r/Ω_m, with the sum of relativistic and non-relativistic matter being Ω_m.

In general, the evaluation errors are smallest for Ω_m of 0.273, where positive, though small, values were found for B_{nr} rather than 0. For this value of normalized matter density the relative ratios of Ω_{nr} to Ω_r might be considered of interest and perhaps even realistic. (Note the relative magnitudes of B_{nr} and B_r are not directly proportional to the weight fraction of these species in the Universe.) We also evaluate this routine at the much smaller values of Ω_m of 0.01 and 0.001 but the relative errors are about 2 orders of magnitude larger than for Ω_m of 0.273 and not worth reporting.

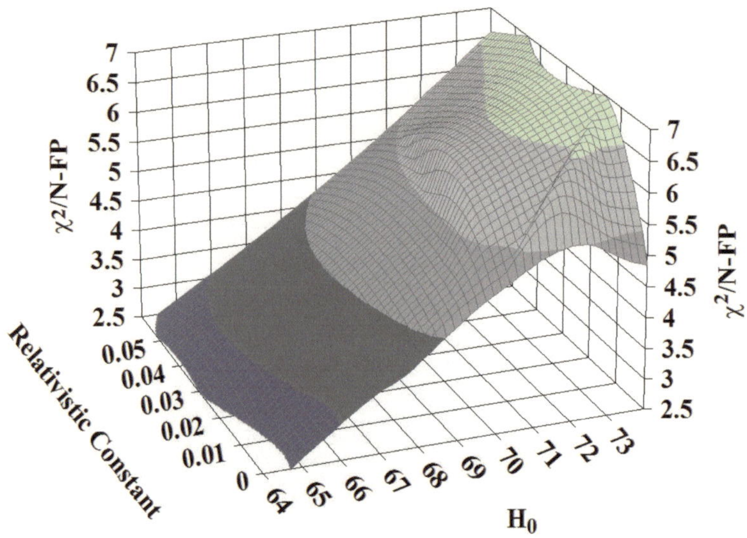

Fig. 4. Calculations of Relativistic Constant A_r from the Simple-DE model with Ω_m of 0.27.

Ω_m	0.75	0.5	0.25
0.32	$B_{nr} = 0$ $B_r = 0.326335$ $\delta = 3.3566$	$B_{nr} = 0$ $B_r = 0.980374$ $\delta = 0.1259$	$B_{nr} = 0$ $B_r = 2.26713$ $\delta = 1.7939$
0.3	$B_{nr} = 0$ $B_r = 0.294466$ $\delta = 1.2194$	$B_{nr} = 0$ $B_r = 0.916476$ $\delta = 1.5471$	$B_{nr} = 0$ $B_r = 2.144604$ $\delta = 3.0469$
0.273	$B_{nr} = 0.124236$ $B_r = 0.01$ $\delta = 0.3507$	$B_{nr} = 0.008885$ $B_r = 0.001121$ $\delta = 0.2151$	$B_{nr} = 0.004552$ $B_r = 0.003291$ $\delta = 0.2159$

δ is the relative calculation error as presented in Appendix 5.2.

Table 5. Calculation of the coefficients for Eq. 28 from BAO parameters

4. Conclusions

The data from the SDSS and SNe Ia collections are unique to science being by far the best ensembles of events stretching a large fraction of the Universe age. By using great times as a variable we can begin to answer questions which cannot be addressed by high energy experiments, perhaps even at the level of CERN. These data have been used to support several theories of spacetime and matter expansion including many models of DE (Davis et al., 2007; Sahni & Starobinsky, 2006) or the related gravitational/DE quintessence (Caresia et al., 2004; Ratra & Peebles, 2003) or even a decline of light emission frequency with local absolute time (Oztas et al., 2008). Here we begin to address the problem of the average densities of both non-relativistic and relativistic matter following the polytropic approach of Chandrasekhar applying this to variations of the FRW model using SNe Ia data and BOA results. We did attempt our own calculations of polytropic constants of a more universal nature than those of Chandrasekhar from first principles, but found these not as well suited for fitting real data as those presented by that exceptional individual and used for decades to describe SNe Ia explosions. Our results with these constants are successful to a first approximation and suggest this general approach may be useful for investigations into other, seemingly unrelated, fields.

Using the SNe Ia data from several sources and our sophisticated models we have found the preponderance of the Universe is either spacetime or dark energy. Our low values for Ω_{nr} from both regimes, might be equated with the combined densities of baryonic and CDM, though we prefer only equivalence with Ω_b for the following reasons. Our small values are consistent with results reported for another fit of SNe Ia data of a 5-dimensional model of the Universe that lumps all matter and energy into a single term but without resorting to CDM(Hartnett & Oliveira, 2007). A small Ω_{nr} is consistent with two recently published values for the baryonic density of about 0.05 for the Universe within seconds of singularity (Fields & Sakhar, 2009; Schramm , 2006) so the value will obviously decline towards our results with Universe expansion. Our small values for Ω_{nr} is also consistent with recent results from WMAP analysis where Ω_b was calculated as 0.046(Hinshaw et al., 2009). The difference in our findings and others (Burles et al., 2001a;b) from those of astronomers (Komatsu et al., 2009) for Ω_m might be thought the difference between baryonic and CDM. To solve this problem, a polytropic model which includes a term for CDM would have to be investigated, realistic polytropic constants discovered and the model fit to the astronomical data. Unfortunately for this investigation, CDM seems absent in our neighborhood of the Milky Way as attested in several un-refuted reports (Bahcall et al., 1995; Bienayme et al., 2006; Creze et al., 1998) which makes discovery and characterization extremely difficult.

In general, our individually determined estimated errors for the Hubble constant are smaller than other reports Komatsu et al. (2009). These smaller values for H_0, with smaller estimated errors, mean an older Universe which is helpful for those trying to adjust estimates of the minimum age from radioactive decay (Dauphas, 2005), globular cluster star composition (Formicola et al., 2004) and suffer the demands of very early galaxy formation (Primack, 2005). Most of the models investigated here, and with different data sets, return values for H_0 slightly larger than the estimate by A. Sandage of 62.3$\pm$6.3 km s^{-1}Mpc^{-1} from his accumulated works (Sandage et al., 2006) but also lower than the 70 to 73 km s^{-1}Mpc^{-1} currently fashionable. Since he and his coworkers have spent lifetimes evaluating the Hubble constant, and the present work, one cannot discard values in the low to mid-60s without very serious consideration. It has been well argued that values of H_0 derived from FRW modeling should only be used to estimate the lower bound of Universe age (Melia, 2009). Use of the polytropic constants require values for H_0 deep within the normalized matter parameters in the models presented here. This "constant" is therefore, of even greater importance than heretofore imagined for determination of matter densities, spacetime curvature and perhaps DE. Unfortunately, H_0 seems the least well known of any important constant today and deserves continued, intense investigation(Huchra, 2008). Unfortunate too, because many astronomers consider H_0 a nuisance parameter forgetting this determination an important reason for throwing the Hubble satellite into outer space.

The nagging and serious *cosmological coincidence* problem remains, where the expectation value for DE differs by more than the Planck constant from expectation (Carroll, 2008), with no resolution on the horizon. Serious flaws in the mathematics of the cosmic constant, which present as discontinuities, have been published (Oztas & Smith, 2006). In addition to this, it has also been shown by fundamental argument that the concept of dark energy as currently employed should not be estimated by the cosmic constant, where use of Ω_Λ fails as the origin of distant signals approach the gravitational horizon (Melia, 2009). Our analysis does not support the superiority of the *Simple*-DE model since this does not fit the distance-frequency data any better than other models. The concept of DE itself demands more and better SNe Ia data with more analyses, for resolution of all these problems does not seem at hand (Carroll, 2008).

We suggest our very small result for Ω_r of $< 10^{-3}$ is probably the upper bound for the abundance of relativistic matter in the Universe, during epochs between the present and reionization. Estimates of the normalized relativistic matter density have been made and seem to lie between this value and about 10^{-6}, as suggested from WMAP 3 year data (Goobar et al., 2006). So the data from SNe observations are useful to establish limits for not only non-relativistic but relativistic matter. If relativistic matter consists primarily of neutrinos this is the upper bound of the current, small gravitational contribution of these particles to our Universe.

Our results following Chandrasekhar's reasoning and with his constants are moderately successful to the first approximation when applied to both SNe Ia data and BAO parameters and suggest the polytropic model is of a very general nature and might be used by investigators from other, seemingly unrelated, fields. We demonstrate that data from SNe Ia observations are useful to estimate limits not only for non-relativistic but relativistic matter; much more astronomical data are needed to better define these values. Solutions from BAO investigations may provide a particularly good method for investigating a polytropic Universe; such has already been predicted (Hu et al., 1998).

We should point out our approach approximates changing matter densities with lookback time, something simpler, *standard models* ignore. The *standard models* presume a constant matter density, which is obviously not the case when fitting data back to z≈1.5. By incorporating the Hubble constant into the matter density terms, our model corrects for changing matter densities with expansion and better fit the data. Since Chandrasekhar's insight is confirmed by daily supernova explosions across our Universe, serious consideration should be given to his polytropic approach when dealing with gravity, density and pressure in cosmology.

5. Appendix

5.1 Polytropic constants

We quickly review the classic derivation of the polytropic constants of Chandrasekhar. We begin with the usual relationships describing the heat capacities at constant volume, V, and constant pressure, P, as

$$C_V = (\frac{dQ}{dT}) \tag{29}$$

and

$$C_P = (\frac{dQ}{dT}) + R \tag{30}$$

where R is the ideal gas constant, so that $C_P - C_V = R$ and $\frac{C_P}{C_P} = \gamma$. For an ideal gas, such as monatomic H and He at high temperature in the primitive Universe these heat capacities are straightforwardly related to the gas constant by

$$C_P = \frac{5}{2}R \tag{31}$$

and

$$C_V = \frac{3}{2}R \tag{32}$$

with the ratio of Eq. (31) over (32) to be

$$\gamma = \frac{C_P}{C_V} = \frac{5}{3}. \tag{33}$$

For the adiabatic situation of a system without heat exchange, $dQ = 0$ the heat capacity at constant volume may be cast in the form

$$C_V dT + \frac{RT}{V} dV = 0 \tag{34}$$

and substituting $C_P - C_V$ for R we get

$$C_V \frac{dT}{T} + (C_P - C_V)\frac{dV}{V} = 0 \tag{35}$$

With separation of variables we can integrate the equation above to get

$$C_V log(T) + (C_P - C_V) log(V) = constant \tag{36}$$

which rearranges to the simple relationship of

$$TV^{\gamma-1} = constant \tag{37}$$

after substitution of T with $\frac{PV}{R}$ we get

$$\frac{PV}{R} V^{\gamma-1} = constant \tag{38}$$

which is more simply

$$PV^{\gamma} = constant. \tag{39}$$

In the case where the specific heat remains constant with a changing temperature, $\frac{dQ}{dT} = constant = c$ we use a similar argument as previously with

$$\gamma\prime = \frac{C_P - c}{C_V - c} \tag{40}$$

leading to a similar relationship with Eq. (37) as

$$PV^{\gamma\prime} = constant. \tag{41}$$

5.2 Evaluation of B_i coefficients

We briefly review several relationships presented by Eisenstein *et al.* used in their evaluation of SDSS data. We use these to evaluate numerical candidates for our B_i species and we are especially interested in their dependence on the normalized matter density, Ω_m. There might appear to be a hidden dependency of our B_{nr} and B_r on the Hubble constant, because ρ_c includes H_0, but the ratio of numerical values (for instance, dependence on Ω_r/Ω_m) are actually Hubble flow independent.

The parameter A, used for BAO evaluation, is dependent on several common cosmological parameters

$$A = D_V(0.35)\frac{\sqrt{\Omega_m H_0^2}}{0.35c} \tag{42}$$

where $D_V(z)$ is the distance to recent redshifts

$$D_V(z) = \left[D_M(z)^2 \frac{cz}{H(z)}\right]^{1/3}$$

and $H(z)$ is typically defined in terms of $E(z)$

$$E(z) = \frac{H_z^2}{H_0^2} = \left(\frac{\dot{a}}{a}\right)^2.$$

$E(z)$ at a given redshift is used in the present study as the traditional version for a flat Universe

$$E(z) = \sqrt{\Omega_m(1+z)^3 + \Omega_r(1+z)^4}$$

which is the special case of our interest. The term $D_M(z)$ is a value reflecting the redshift distance at recombination depending on the constant $D_H = c/H_0 = 3000/h$ (in Mpc) and the value for the integral of $1/E(z)$

$$D_M(z) = D_H \int_0^z \frac{dz}{E(z)}.$$

The value for R is a ratio of redshift dependents, where 0.35 is a typical present redshift and 1089 is a popular value for the redshift at recombination during the primitive Universe

$$R_{0.35} = \frac{D_V(0.35)}{D_M(1089)}. \tag{43}$$

For evaluation of relative errors we use $R_{0.35}$ of 0.0979 and A of 0.469 from Table 1 of Eisenstein *et al.* 2005 as reference values. The coefficients B_{nr} and B_r were calculated by minimizing the relative error of our $E(z)$ with respect to these two reference values as

$$\delta = \frac{|A - 0.469|}{0.469} + \frac{|R_{0.35} - 0.0979|}{0.0979}.$$

6. References

Astier, P., et al., (2006) The Supernova Legacy Survey: Measurement of Ω_m, Ω_Λ and w from the First Year Data Set, *Astron. Astrophys.* 447, 31-48.

Bahcall, N.A., *et al.*, (1995). Where is the Dark Matter? *Astrophys. J. Lett.* 447, L81-L85.

Bienayme, O., *et al.*, (2006). Vertical Distribution of Galactic Disk Stars, *Astron. Astrophys.* 446, 933-942.

Burles, S.; Nollett, K.M. & Turner, M.S. (2001). What is the Big-Bang-Nucleosynthesis Prediction for the Baryon Density and How Reliable is It? *Phys. Rev. D Part. Fields* 63, 063512, 6 pages.

Burles, S.; Nollett, K.M. & Turner, M.S. (2001). Big Bang Nucleosynthesis Predictions for Precision Cosmology, *Astrophys. J. Lett.* 552, L1-L5.

Caresia, P.; Matarrese, S. & Moscardini, L. (2004). Constraints on Extended Quintessence from High-Redshift Supernovae, *Astrophys. J.* 605, 21-28.

Carroll, S.M.; Press., W.H. & Turner, E.L. (1992). The Cosmological Constant, *Ann. Rev. Astron. Astrophys.* 30, 499-542.

Carroll, S.M. (2008). *Living Reviews in Relativity* 4 www.livingreviews.org/Articles/Volume4/2001-1carroll/
Max Planck Institute for Gravitational Physics, Germany.

Chandrasekhar, S. (1964). The Dynamical Instability of Gaseous Masses Approaching the Schwarzschild Limit in General Relativity, *Astrophys. J.* 140, 417-433.

Chandrasekhar, S. (1983). Nobel Lecture http://nobelprize.org/nobelprizes/physics/laureates/1983/chandrasekhar- lecture.html

Chandrasekhar, S. & Trooper, R.F. (1964). The Dynamical Instability of the White-Dwarf Configurations Approaching the Limiting Mass, *Astrophys. J.* 139, 1396-1398.

Creze, M., *et al.*, (1998). The Distribution of Nearby Stars in Phase Space Mapped by Hipparcos. I. The Potential Well and Local Dynamical Mass, *Astron. Astrophys.* 329, 920-936.

Dauphas, N. (2005). The U/Th Production Ratio and the Age of the Milky Way from Meteorites and Galactic Halo Stars, *Nature* 435, 1203-1205.

Davis, T.M., *et al.*, (2007). Scrutinizing Exotic Cosmological Models Using ESSENCE Supernova Data Combined with Other Cosmological Probes, *Astrophys. J.* 666, 716-725.

Eisenstein, D.J., *et al.*, (2005). Detection of the Baryon Acoustic Peak in the Large-Scale Correlation Function of SDSS Luminous Red Galaxies, *Astrophys. J.* 633, 560-574.

Fields, B.D. & Sakhar, S. (2009). *Big Bang Nucleosynthesis* http://pdg.lbl.gov/2009/reviews/rpp2009-rev-bbang-nucleosynthesis.pdf. & http://arxiv.org/abs/astro-ph/0601514

Formicola, A., *et al.*,. (2004). Astrophysical S-factor of $^{14}N(p,\gamma)^{15}O$, *Physics Letters B* 591, 61-68.

Goobar, A., *et al.*, (2006). The Neutrino Mass Bound from WMAP 3 Year Data, the Baryon Acoustic Peak, the SNLS Supernovae and the Lyman-α Forest, *J. Cosmol. Astropart. Phys.* 606, 19.

Hartnett, J.G. & Oliveira, F.J. (2007). Luminosity Distance, Angular Size and Surface Brightness in Cosmological General Relativity, *Found. Phys.* 37, 446-454.

Hicken, M., *et al.*, (2009). Improved Dark Energy Constraints from $\approx$100 New CfA Supernova type Ia Light Curves, *Astrophys. J.* 700, 1097-1140.

Hinshaw, G., *et al.*, (2009). Five-Year Wilkinson Microwave Anisotropy Probe Observations, *Astrophys. J. Supp.* 180, 225-245.

Hu, W.; Eisenstein, D.J. & Tegmark, M. (1998). Weighing Neutrinos with Galaxy Surveys, *Phys. Rev. Lett.* 80, 5255-5258.

Huchra, J.P., https://www.cfa.harvard.edu/ dfabricant/huchra/hubble/

Kelly, P.L., *et al.*, (2009). Hubble Residuals of Nearby SN Ia Are Correlated with Host Galaxy Masses, arXiv:0912.0929v2.

Kessler, R., *et at.*, (2009). First-Year Sloan Digital Sky Survey-II Supernova Results: Hubble Diagram and Cosmological Parameters, *Astrophys. J. Suppl.* 185, 32.

Komatsu, E., *et al.*, (2009). Five-Year Wilkinson Microwave Anisotropy Probe Observations: Cosmological Interpretation, *Astrophys. J. Supp.* 180, 330.

Kowalski, M., *et al.*, (2008). Improved Cosmological Constraints from New, Old, and Combined Supernova Data Sets, *Astrophys. J.* 686, 749-778.

Leibundgut, B. (2008). Supernovae and Cosmology, *Gen. Relativ. Gravit.* 40, 221-248.

Linden, A. (2009). Cosmological Parameter Extraction and Biases from Type Ia Supernova Magnitude Evolution, *Astron. Astrophys.* 506, 1095-1105.

Liu, H. & Li, T.-P. (2009). Improved CMB Map from WMAP Data, arXiv:0907.2731

Li, T.-P., *et al.*, (2009). Observation Number Correlation in WMAP data, *M.N.R.A.S.* 398, 47.

Liu, H. & Li, T.-P. (2010). Inconsistency Between WMAP Data and Released Map, *Chinese Sci Bull* 55, 907-909.

Liu, H. & Li, T.-P. (2011). Pseudo-Dipole Signal Removal from WMAP Data, *Chinese Sci. Bull.* 56, 29-33.

Melai, F. (2009). Constraints on Dark Energy from the Observed Expansion of Our Cosmic Horizon, *Int. J. Mod. Phys. D* 18, 1113-1127.

Moss, A., Scott, D. & Sigurdson, K. (2010). Induced CMB Quadrapole from Pointing Offsets, arXiv:1004.3995v1

Oztas, A.M., Smith, M.L. & Paul, J. (2008). Spacetime Curvature is Important for Cosmology Constrained with Supernova Emissions, *Int. J. Theoret. Phys.* 47, 725-740.

Oztas, A.M. & Smith, M.L. (2006). Elliptical Solutions to the Standard Cosmology Model with Realistic Values of Matter Density, *Int. J. Theoret. Phys.* 45, 925-936.

Peebles, P.J.E. (1993). *Principles of Physical Cosmology* Princeton University Press, Princeton, New Jersey.

Primack, J.R. (2005). Precision Cosmology, *New Astron. Rev.* 49, 25-35.

Peebles, P.J.E. & Ratra, B. (2003). The Cosmological Constant and Dark Energy, *Rev. Mod. Phys.* 75, 559-606.

Roukema, B.F. (2010). On the Suspected Timing Error in WMAP Map-making, arXiv:1004.4506v2

Sahni, V. & Starobinsky, A. (2006). Reconstructing Dark Energy, *Int. J. Mod. Phys. D* 15, 2105-2132.

Sandage, A., *el. al.*, (2006). The Hubble Constant: A Summary of the Hubble Space Telescope Program for the Luminosity Calibration of Type Ia Supernovae by Means of Cepheids, *Astrophys. J.* 653, 843-860.

Schaefer, B.E. (2007). The Hubble Diagram to Redshift >6 from 69 Gamma-Ray Bursts, *Astrophys. J.* 660, 16-46.

Schramm, D.N. (2006). Summary of Recent Developments in Primordial Nucleosynthesis, *Ann. N. Y. Acad. Sci.* 688, 776-791.

Setare, M.R. (2009). Generalized Chaplygin Gas Model as a New Agegraphic Dark Energy in Non-flat Universe, *Int. J. Theoret. Phys.* 48, 3365-3371.

Sharaf, M.A. & Sendi, A.M. (2010). Computational Developments for Distance Determination of Stellar Groups, *J. Astrophys. Astron.* 31, 3-16.

Smith, M.L., *et at.*, (2010). *Constraints on Dark Energy and Dark Matter from Supernovae and Gamma Ray Burst Data* in Dark Energy: Developments and Implications, Nova Science, Hauppauge, New York.

Vikhlinin, A., *et al.*, (2008). Chandra Cluster Cosmology Project III: Cosmological Parameter Constraints, arXiv:0812.2720.

Vikhlinin, A., *et al.*, (2009). X-ray Cluster Cosmology, arXiv:0903.5320.

Wood-Vasey, W.M., *et al*, (2007). Observational Constraints on the Nature of Dark Energy: First Cosmological Results from the ESSENCE Supernova Survey, *Astrophys. J.* 666, 694-715.

16

Loop Quantum Cosmology: Effective Theory and Related Applications

Li-Fang Li, Kui Xiao and Jian-Yang Zhu
Department of Physics, Beijing Normal University, Beijing
China

1. Introduction

The effort toward making quantum mechanics and general relativity compatible (quantum gravity) has lasted more than a century. The string theory and loop quantum gravity both stand out as strong candidates and currently they both make popular research subjects in quantum gravity. As we known, Loop quantum gravity (LQG)(Ashtekar & Lewandowski, 2004; Rovelli, 1998; 2004; Thiemann, 2007) is a background independent and non-perturbative canonical quantum gravity theory. LQG has made many breakthroughs in recent years: the establishment of the quantum Einstein equations Ashtekar & Tate (1994); Ashtekar et al. (1995a); Corichi & Zapata (1997); Lewandowshi & Thiemann (1999); Rovelli & Smolin (1994); Thiemann (1996; 1998a;b;c; 2001), the proof that the Riemannian operators have discrete eigenvalues Ashtekar et al. (1995b); Lewandowski (1997); Loll (1995a;b; 1997a;b); Rovilli & Smolin (1995); Thiemann (1998d;e),results concerning the entropy of the black hole horizon and cosmological horizon entropy with statistical mechanics Ashtekar et al. (1998; 1999; 2000; 2001; 2002; 2003a;b); Berreira et al. (1996); Rovelli (1996a;b); Smolin (1995), and so on. As an application of loop quantum gravity to cosmology, loop quantum cosmology (LQC) Bojowald (2005a; 2008); Date (2002) also presents itself as a possible path toward answers to the cosmological and astrophysical riddles.

As a symmetry reduced model of LQG, LQC inherits the quantum schemes originated from LQG that dealt with the isotropic and homogeneous universe firstly and then extended to the inhomogeneous and anisotropic model Bojowald (2002a). It plays an important role in connecting the LQG theory and the measureable world. On one hand, it is used to test the full theory, which, in its own form, is extremely complex and difficult to directly apply.On the other hand, making connections to the real world sheds light on further improvement of the LQG theory. These reasons make LQC a promising and enlightening subject to study.

In LQC, the collapsing and expanding phases are connected by the cyclic or oscillatory models Lidsey et al. (2004), and the universe is automatically born with a small scale factor at the fixed point near the Planck phase Bojowald (2005); Mulryne et al. (2005a). Unlike in the emergent universe model, this fixed point here is stable and allows the universe to start in an initial phase of oscillation. Then an inflationary phase is entered, which is the relevant regime for structure formation. In LQC, there are many different inflationary scenarios Artymowski et al. (2009); Bojowald et al. (2004); Mulryne et al. (2005b); Zhang & Ling (2007), among which the one without inflation is the most attractive, mainly because it can explain the inflationary phase directly from LQG. But, unfortunately, it is difficult to study the structure formation.

To know how and why this is true requires a study of the mathematical structure and the physical meaning of LQC.
As is well known, LQC is based on the connection dynamics. So far, the successful quantization of the cosmological model is still confined to the homogeneous Bianchi A class because one can refer to the diagonal technique Bojowald (2003). With the Ashtekar's new variables, the Hamilton constraint can be written as a difference equation. One of the major successes in LQC is that the big bang singularity can be replaced by the big bounce Ashtekar et al. (2006a;b; 2008). (The robust demonstration of bounces in LQC is confined to the cases where quantum back reaction can be safely ignored.). For the general case, the quantization is more complex and the research is still going on. For the general inhomogeneous cosmology, there have been three different approaches. The first one is to introduce the inhomogeneous matter, but still based on the isotropic quantum geometry and its effective theory. The second approach starts with the full constraint, splits it into the homogeneous part and the inhomogeneous part, and then obtains the effective theory Bojowald et al. (2006; 2007; 2008; 2009) (it has achieved a series of successes so far). The third approach is to deal with the inhomogeneous symmetric model explicitly, and to shed light on the full theory. At the time of writing, we still do not know whether the general solution is a difference equation or not.
As we all know, the difference equation of state is difficult to analyze even in the homogeneous and isotropic model. Thus we need a new tool to extract physical information out of the theory. That is why the effective theory comes in. The effective theory shares the form of the classical theory, but contains correction terms from the quantum theory (LQC). The commonly considered ones include the inverse volume correction, the holonomy correction and the back reaction correction. The inverse volume correction Bojowald (2002b;c) is used to solve the quantization problem $p^{-\frac{3}{2}}$ in the matter Hamiltonian which cannot be quantized directly. Instead, we write the equivalent form of $p^{-\frac{3}{2}}$ in the classical form, which can be promoted to the well-defined operator in the quantum theory. This brings a correction term to the classical theory. And the matter Hamiltonian derived in this way will behave differently from the classical one on small scales. The matter Hamiltonian shows a repulsive behavior. The holonomy correction originates from the fact that there does not exist an operator in the quantum theory corresponding to the connection c Banerjee & Date (2005); Date & Hossain (2004). When quantized, c should be expressed as $\sin(\mu c)/\mu$. It is obvious that the classical expression recovers only at the small value of c. This provides a new correction to the classical equation. The back reaction correction is the main source of correction in genuine quantum systems Bojowald et al. (2007); Chiou (2008). If brought together, they may counteract each other. Therefore, it is important to bring all possible quantum corrections together in a consistent manner and study the corresponding physics. Based on these theories, many interesting cosmological riddles have been studied, such as the big bang nucleosynthesis, the already mentioned inflationary scenario, the anisotropy of CMB, and the gravitational wave and so on. At the same time, many issues still need to be clarified in LQC, such as the ambiguity problemBojowald et al. (2004) and the different quantum schemes Chiou & Li (2009a;b); Mielczarek & Szyd (2008); Yang et al. (2009).
Very recently, an integral formulation of loop quantum cosmology with the Feynman procedure has been discussed, which again shows that the loop quantum cosmology is different from the Wheeler-DeWitt theory Ashtekar (2010). The spin foam model of LQC has also been constructed (for recent progress see Ashtekar et al. (2009)). These theories add to the appeal of LQC from different perspectives.

In this Chapter, we will focus our attention on the effective theory from which we can easily extract the physical result, and analyzing the explicit model can in turn shed light on the full theory. The effective LQC theory is a semiclassical theory, and can be derived in different ways, such as the WKB approximation, the coherent state, and so on. And they are consistent with each other in the leading term. The effective theory is valid in the semiclassical region,in between the quantum one and the classical one. And this theory will go back to the classical theory in the classical region. So we can use effective loop quantum cosmology for both the semiclassical and the classical region. One the other hand, along with the development of modern space technology and high-precision measurement techniques, we have accumulated a large amount of experimental data,probably more than that can be explained well by the current theory, such as the Pioneer anomaly, dark matter, and the accelerating expansion of the universe,to name a few. Close connection between the theoretical results and real experiments is crucial, and motivates us to apply LQC to explicit physical models, and then to compare theoretical results with real or gedarken experiments. The works to be summarized in this chapter can be divided into the following three parts.

(I) We discuss the stability properties of an autonomous system in the effective LQC Xiao & Zhu (2010). The system is described by a self-interacting scalar field ϕ with positive potential V, coupled with a barotropic fluid in the Universe. With $\Gamma = VV''/V'^2$ considered as a function of $\lambda = V'/V$, the autonomous system is extended from three dimensions to four dimensions. We find that the dynamic behaviors of a subset, not all, of the fixed points are independent of the form of the potential. Considering the higher-order derivatives of the potential, we get an infinite-dimensional autonomous system which can describe the dynamical behavior of the scalar field with more general potential. We find that there is just one scalar-field-dominated scaling solution in the loop quantum cosmology scenario.

(II) We discuss the null energy condition in the effective LQC Li & Zhu (2009). Wormhole and time machine are objects of great interest in general relativity. However, it takes exotic matters which are impossible on the classical level to support them. But if we introduce the quantum effects of gravity into the stress-energy tensor, these peculiar objects can be constructed self-consistently. LQC, with the potential to bridge the classical theory and quantum gravity, provides a simple way to study quantum effect in the semiclassical case. We investigate the averaged null energy condition in LQC in the framework of effective Hamiltonian, and find out that LQC do violate the averaged null energy condition in the massless scalar field coupled model.

(III) We consider the covariant entropy bound conjecture in the effective LQC Li & Zhu (2010a). The covariant entropy bound conjecture is an important hint for the quantum gravity, with several versions available in the literature. For cosmology, Ashtekar and Wilson-Ewing showed the consistence between the loop gravity theory and one version of this conjecture. Recently, S. He and H. Zhang proposed a version for the dynamical horizon of the universe, which validates the entropy bound conjecture for the cosmology filled with perfect fluid in the classical scenario when the universe is far away from the big bang singularity. But their conjecture breaks down near the big bang region. We examine this conjecture in the context of LQC. With the example of photon gas, this conjecture is protected by the quantum geometry effects.

2. Effective Theory of Loop Quantum Cosmology

LQC is a symmetry-reduced sector of LQG. It is a direct application of the quantization technique that originated from LQG. This section serves two purposes: First, we show that one can indeed find semiclassical solution which is an approximation to the classical Einstein equations at late times. Second, we derive the effective equations incorporating the dominating quantum corrections within the framework of geometric quantum mechanics. The derivation of the effective equation utilizes two main tools: the geometric quantum mechanics and the "shadow state framework".

2.1 Classical framework

The classical phase space Γ is constructed by the Ashtekar variables (A^i_a, E^a_i), where A^i_a is an $SU(n)$ connection and E^a_i is the corresponding canonically conjugate variable. With the Ashtekar variables, the classical constraint of the gravitational part can be expressed as

$$C_{grav} = -\gamma^{-2} \int_{\mathcal{V}} d^3x \epsilon_{ijk} e^{-1} E^{ai} E^{bj} F^i_{ab}. \tag{1}$$

In the following, we will see that the Ashtekar variables can describe the classical theory very well, as the ordinary ADM variables do. Considering isotropic and homogeneous universe, the pair, (A^i_a, E^a_i), is equivalent to the following form:

$$A^i_a = c^o V^{-\frac{1}{3}} {}^o\omega^i_a, \tag{2}$$

$$E^a_i = p\sqrt{{}^oq} {}^oV^{-\frac{2}{3}} {}^oe^a_i, \tag{3}$$

where ${}^oe$ is a fiducial background triad, ${}^o\omega^i_a$ is the connection, and ${}^oV$ is the volume of the fiducial cell. From the above set of equations, we see that all the information about A^i_a and E^a_i are contained in the pair of new variables (c, p). The classical Hamiltonian constraint for a spatially flat FRW universe with a free massless scalar field is

$$C = -\frac{3}{\kappa\gamma^2} c^2 p^{\frac{1}{2}} + \frac{1}{2} \frac{p_\phi^2}{p^{\frac{3}{2}}} = 0. \tag{4}$$

For convenience, we replace the pair (c, p) with (β, V) through the following relationship

$$\beta = \frac{c}{\sqrt{p}}, \tag{5}$$

$$V = p^{\frac{3}{2}}. \tag{6}$$

Then the Hamiltonian constraint can be expressed as

$$C = -\frac{3}{\kappa\gamma^2} \beta^2 V + \frac{1}{2} \frac{p_\phi^2}{V} = 0, \tag{7}$$

with the symplectic structure $\Omega = \frac{2}{\kappa\gamma} d\beta \wedge dV + d\phi \wedge dp_\phi$. The phase space Γ consists of all possible points (β, V, ϕ, p_ϕ). The Poisson bracket on the phase space is given by

$$\{f, g\} = \frac{\kappa\gamma}{2} \left(\frac{\partial f}{\partial \beta}\frac{\partial g}{\partial V} - \frac{\partial g}{\partial \beta}\frac{\partial f}{\partial V}\right) + \frac{\partial f}{\partial \phi}\frac{\partial g}{\partial p_\phi} - \frac{\partial g}{\partial \phi}\frac{\partial f}{\partial p_\phi}. \tag{8}$$

The evolution of the canonical variable depends on the Poisson equations. Therefore, the classical equations of motion are given by

$$\dot{\beta} = \{\beta, C\} = -\frac{3}{2}\frac{\beta^2}{\gamma} - \frac{\kappa\gamma}{4}\frac{p_\phi^2}{V^2}, \tag{9}$$

$$\dot{V} = \{p, C\} = 3\frac{\beta}{\gamma}V, \tag{10}$$

$$\dot{\phi} = \{\phi, C\} = \frac{p_\phi}{V}, \tag{11}$$

$$\dot{p_\phi} = \{p_\phi, C\} = 0. \tag{12}$$

We can verify that the above equation set is equivalent to the Friedmann equation for a free scalar field when it is written in terms of the ordinary ADM variables. The new canonically conjugate variables are related to the old geometrodynamics variables via $\beta = \gamma\frac{\dot{a}}{a}$ and $V = a^3$. Then, with the definition of the density $\rho = \frac{1}{2}\frac{p_\phi^2}{p^3}$, we can get the classical Friedmann equation and the Raychauduri equation as follows

$$H^2 = \frac{\kappa}{3}\rho, \tag{13}$$

$$3\frac{\ddot{a}}{a} = -2\kappa\rho. \tag{14}$$

2.2 Quantum framework

LQC is the symmetry reduced model of LQG, and it inherits the quantization schemes of LQG. LQC is essentially different from the WDW theory. In LQC, the kinematical Hilbert space is in the "polymer representation" for p, instead of the standard Schrödinger representation. There is no operator corresponding to c. In addition, it is not densely defined if the inverse volume function $|p|^{-3/2}$ is naively quantized as the operator with eigenvalues equal to the inverse of the volume eigenvalues. Thus, to construct the Hamiltonian constraint operator, we have to express the classical constraint in terms of the triad variable p and the holonomy $h_k^{(\bar{\mu})}$, both of which have direct quantum analogs.

In the development of LQC, there exist two different quantum schemes: the μ_0 scheme and the $\bar{\mu}$ scheme. In the μ_0 scheme, μ_0 is a constant on the phase space, and the difference equation is in uniform step size. The greatest success of this scheme so far is that it can replace the big bang singularity with the big bounce, which reflects the nature of the quantum geometry effect. Unfortunately, it suffers from serious physical problems. For example, the critical value, ρ_{crit}, of the matter density at which the bounce occurs can be made arbitrarily small by increasing the initial momentum p_ϕ of the scalar field. In other words, large values of p_ϕ are permissible in the late universe, which leads to bounce at low matter density. This is a serious drawback because we do not expect the quantum effect to modify the evolution of the universe in the classical region. In the $\bar{\mu}$ scheme, $\bar{\mu}$ is a function on the phase space, unlike the constant μ_0 in the μ_0 scheme. This difference turns out to be enough to remove the major weakness of the μ_0 scheme, while keeping the desirable features of the original scheme. In this section, we only review the quantization procedure in the $\bar{\mu}$-scheme.

In the $\bar{\mu}$ scheme,one can shrink the loop $\Box_{ij}$ until the area of the loop approaches the area gap Δ, measured by the physical metric q_{ab}. The physical area of the elementary cell is $|p|$. Each side of $\Box_{ij}$ is $\bar{\mu}(p)$ times the edge of the elementary cell, which leads to

$$\bar{\mu}^2|p| = \Delta \equiv (2\sqrt{3}\pi\gamma)l_{pl}^2. \tag{15}$$

Therefore, $\bar{\mu}$ is a non-trivial function on the phase space and can be rewritten as

$$\bar{\mu} = \sqrt{\frac{\Delta}{p}}, \tag{16}$$

where γ is the Immirzi parameter[1].
Following Dirac, in the quantum theory, we should first construct a kinematical description. The Hilbert space H_{kin}^{grav} is the space $L^2(\mathbb{R}_{Bohr}, d\mu_{Bohr})$ consisting of square integrable functions on the Bohr compactification of the real line. To specify states concretely, we work with the representation of the operator $\widehat{p}$ in which the operator $\widehat{p}$ is diagonal. Eigenstates of $\widehat{p}$ are labeled by a real number μ and satisfy the orthonormality relation:

$$\langle \mu_1 | \mu_2 \rangle = \delta_{\mu_1, \mu_2}\,. \tag{17}$$

The right-hand side of the above equation is the Kronecker delta rather than the Dirac delta distribution. A general state in H_{kin}^{grav} can be expressed as a countable sum $\langle \Psi | = \sum_n c^{(n)} \langle \mu_n |$, where $c^{(n)}$ are complex coefficients and the inner product is given by

$$\langle \Psi_1 | \, \Psi_2 \rangle = \sum_n \bar{c}_1^{(n)} \, c_2^{(n)}.$$

The fundamental operators are $\hat{p}$ and $\exp(\widehat{i\bar{\mu}c/2})$. The action of the operator $\hat{p}$ on its eigenvalue is

$$\hat{p}\,|\mu\rangle \;=\; \frac{8\pi\gamma l p^2}{6}\,\mu\,|\mu\rangle \tag{18}$$

and the action of the operator $\exp(\widehat{i\bar{\mu}c/2})$ on $|\mu\rangle$ will be given later.
In order to achieve quantization, we should represent the Hamiltonian constraint operator $\widehat{C_{grav}}$ in terms of the above well-defined operators. Following the full theory, with the Thiemann trick, we can rewrite the term that involves the inverse triad e^{-1} as

$$e^{-1}\epsilon_i^{\,jk} E_j^a E_k^b = \sum_l \frac{{}^o q^{1/3}}{2\pi G\gamma\bar{\mu}L}\epsilon^{jklo} e_j^{ao} e_k^b \mathrm{Tr}(h_l^{(\bar{\mu})}\{(h_l^{(\bar{\mu})})^{-1}, V\}\tau_i), \tag{19}$$

where the holonomy

$$h_l^{(\bar{\mu})} = \mathcal{P}\exp\int_0^{\bar{\mu}L} \tau_i A_a^i dx^a = \exp\left(\bar{\mu}c\tau_l\right) \tag{20}$$

[1] The general form of $\bar{\mu} = (\frac{\Delta}{p})^x$, $0 < x < 1$. Here we choose $x = \frac{1}{2}$ according to the suggestions of Ashtekar. However, we still cannot determine it for the following two reasons. (1) The coordinate area is more natural than an invariant geometrical area when we consider the quantization of the single curvature components. (2)The quantization requires the area operator which is not well understood in the full theory.

is along the edge of coordinate length $\bar{\mu}L$ aligned with the direction of ${}^{o}e^{a}_{l}\partial_a$. τ_i is the generator of the $SU(2)$ and satisfies the relationship $[\tau_i, \tau_j] = \epsilon^{k}_{ij}\tau_k$, where $\tau_i = \frac{1}{2i}\sigma_i$ and σ_i are the Pauli matrices in the standard convention.
According to the techniques in the gauge theory, the field strength F^{i}_{ab} can be expressed as

$$F^{k}_{ab} = -2 \lim_{Ar_{\Box}\to 0} \mathrm{Tr}(\frac{h^{(\bar{\mu})}_{\Box_{ij}} - 1}{\bar{\mu}^2 V_0^{2/3}})\tau^{k\,o}\omega^{i}_{a}{}^{o}\omega^{j}_{b}, \tag{21}$$

where

$$h^{(\bar{\mu})}_{\Box_{jk}} := h^{(\bar{\mu})}_{j} h^{(\bar{\mu})}_{k} (h^{(\bar{\mu})}_{j})^{-1} (h^{(\bar{\mu})}_{k})^{-1} \tag{22}$$

is the holonomy along the the four edges of $\Box_{ij}$.
Combining Eq.(19) with Eq.(21), the classical Hamiltonian constraint for the gravitational sector can be rewritten as

$$\begin{aligned} C_{grav} &= -\frac{4sgn(p)}{8\pi\gamma^3\bar{\mu}^3}\sum_{ijk}\epsilon^{ijk}\mathrm{Tr}(h^{(\bar{\mu})}_{j}h^{(\bar{\mu})}_{k}(h^{(\bar{\mu})}_{j})^{-1}(h^{(\bar{\mu})}_{k})^{-1}h^{(\bar{\mu})}_{i}\{(h^{\bar{\mu}}_{i})^{-1}, V\}) \\ &= \sin(\bar{\mu}c)[-\frac{4}{8\pi\gamma^3}\frac{sgn(p)}{\mu^3}\sum_{k}\mathrm{Tr}\tau_k h^{(\bar{\mu})}_{i}\{(h^{(\bar{\mu})}_{i})^{-1}, V\}]\sin(\bar{\mu}c), \end{aligned} \tag{23}$$

where in the last step we have used a symmetric ordering of the three terms for later convenience.
Now, we consider the action of the operator $\widehat{e^{\frac{i\bar{\mu}c}{2}}}$ on the state $|\mu\rangle$. Although the geometrical meaning of this action of $\widehat{\exp i(\bar{\mu}c/2)}$ is simple, its expression in the $|\mu\rangle$ representation is complicated because μ is not an affine parameter along the integral curve of the vector $\bar{\mu}\frac{d}{d\mu}$. After calculation, we get

$$\widehat{e^{i\frac{\bar{\mu}c}{2}}}\tilde{\Psi}(\mu) = \tilde{\Psi}(sgn(\tilde{\mu})|\tilde{\mu}|^{\frac{2}{3}}), \text{ where } \tilde{\mu} = sgn(\mu)|\mu|^{\frac{3}{2}} + \frac{1}{K}. \tag{24}$$

Next, we will change to the v representation for simplicity. In the v representation, the action of $\widehat{e^{i(\bar{\mu}c/2)}}$ on $|v\rangle$ is extremely simple

$$\widehat{e^{\frac{i\bar{\mu}c}{2}}}\Psi(v) = \Psi(v+1), \tag{25}$$

and the action of the volume operator on it is

$$\hat{V}|v\rangle = (\frac{8\pi\gamma}{6})^{\frac{3}{2}}\frac{|v|}{K}|v\rangle, \tag{26}$$

where $v = Ksgn(\mu)|\mu|^{\frac{3}{2}}$ and $K = \frac{2\sqrt{2}}{3\sqrt{3\sqrt{3}}}$. Therefore, we will use the v representation in the following.

We can straightforwardly get the action of the operators $\widehat{\sin(\frac{\bar{\mu}c}{2})}$ and $\widehat{\cos(\frac{\bar{\mu}c}{2})}$ on v representation,

$$\widehat{\sin(\frac{\bar{\mu}c}{2})}|v\rangle = \frac{1}{2i}[|v+1\rangle - |v-1\rangle], \tag{27}$$

$$\widehat{\cos(\frac{\bar{\mu}c}{2})}|v\rangle = \frac{1}{2}[|v+1\rangle + |v-1\rangle]. \tag{28}$$

Then we promote the corresponding physical quantities $\sin(\frac{\bar{\mu}c}{2})$,$\cos(\frac{\bar{\mu}c}{2})$ and V to the operators. We have

$$\begin{aligned}\widehat{C_{grav}} &= \widehat{\sin(\frac{\bar{\mu}c}{2})}[\frac{24isgn(\mu)}{8\pi\gamma^3\bar{\mu}^3}(\widehat{\sin(\frac{\bar{\mu}c}{2})}\hat{V}\widehat{\cos(\frac{\bar{\mu}c}{2})} - \widehat{\cos(\frac{\bar{\mu}c}{2})}\hat{V}\widehat{\sin(\frac{\bar{\mu}c}{2})})]\widehat{\sin(\frac{\bar{\mu}c}{2})} \\ &=: \widehat{\sin(\bar{\mu}c)}\hat{A}\widehat{\sin(\bar{\mu}c)}.\end{aligned} \tag{29}$$

The action of the operator $\hat{A}$ on the state $\Psi(v)$ can be written as

$$\hat{A}\Psi(v) = -\frac{27K}{4}\sqrt{\frac{8\pi}{6\gamma^{3/2}}}|v||v-1| - |v+1||\Psi(v). \tag{30}$$

One needs to use caution when deriving this expression. $sgn(v)$ is unambiguously defined only on states other than the point $v = 0$. Since the right-hand side vanishes at $v = 0$, it is just the domain of $sgn(\mu)$, where $\hat{A}$ is well defined. Therefore, the operator on the right-hand side is densely defined.
The action of the gravitational constraint on $\Psi(v)$ is given by

$$\widehat{C_{grav}}\Psi(v) = f_+(v)\Psi(v+4) + f_0(v)\Psi(v) + f_-(v)\Psi(v-4) \tag{31}$$

with

$$f_+(v) = \frac{27}{16}\sqrt{\frac{8\pi}{6}\frac{Kl_{pl}}{\gamma^{3/2}}}|v+2||v+1| - |v+3|| \tag{32}$$

$$f_-(v) = f_+(v-4) \tag{33}$$

$$f_0(v) = -f_+(v) - f_-(v). \tag{34}$$

From the above, we can see that the gravitational constraint is again a difference operator. Compared with the μ_0-scheme, the new constraint involves steps which are constant in the eigenvalues of the volume operator $\hat{V}$, not in the eigenvalues of $\hat{p}$.
Next, we continue to quantize the matter part of the constraint

$$C_{matt} = 8\pi|p|^{-\frac{3}{2}}p_\phi^2. \tag{35}$$

It turns out that despite the existence of an inverse operator $\hat{p}^{-\frac{3}{2}}$, one can quantize it successfully. With the Thiemann trick, one can always write the inverse operator in an equivalent way, which contains the quantities that can be easily promoted to the operators. We express the inverse triad as

$$|p|^{-n} = (|\frac{3}{\gamma\kappa l\bar{\mu}j(j+1)(2j+1)}\sum_i \mathrm{Tr}({}^j\tau_i{}^j h_i^{\bar{\mu}}\{{}^j(h_i^{\bar{\mu}})^{-1}, V^{2l/3}\})|)^{\frac{n}{1-l}}$$
$$= (|\frac{9}{\gamma\kappa l\bar{\mu}j(j+1)(2j+1)}\mathrm{Tr}({}^j\tau_3{}^j h_3^{\bar{\mu}}\{{}^j(h_3^{\bar{\mu}})^{-1}, V^{2l/3}\})|)^{\frac{n}{1-l}}. \quad (36)$$

In this quantization, there are two ambiguities Bojowald (2002b; 2005), labeled by a half integer j and a real number l in the range $0 < l < 1$. Following the considerations in Perez (2006); Vandersloot (2005), we will set $j = 1/2$, and the general case for j can be found in Chiou & Li (2009a;b). For l, there is no universally accepted concept, and $l = 1/2$ and $l = 3/4$ have been used in the literature. Fortunately, the results do not change qualitatively with the exact choice. Here, we choose $j = 1/2$ and $l = 3/4$. Then

$$\widehat{|p|^{-\frac{3}{2}}}\Psi(v) = \left(\frac{6}{8\pi\gamma}\right)^{3/2} B(v)\,\Psi(v) \quad (37)$$

where

$$B(v) = \left(\frac{3}{2}\right)^3 K\,|v|\,\left||v+1|^{1/3} - |v-1|^{1/3}\right|^3. \quad (38)$$

Combining all the results above, we can write down the full constraint

$$\hat{C}\,\Psi(v) = \left(\hat{C}_{\mathrm{grav}} + \hat{C}_{\mathrm{matt}}\right)\,\Psi(v) = 0 \quad (39)$$

as follows:

$$p_\phi^2\Psi(v,\phi) = [B(v)]^{-1}\,\left(C^+(v)\,\Psi(v+4,\phi) + C^o(v)\,\Psi(v,\phi) + C^-(v)\,\Psi(v-4,\phi)\right)$$
$$=: -\Theta\,\Psi(v,\phi) \quad (40)$$

where the coefficients $C^{\pm}(v)$ and $C^o(v)$ are given by:

$$C^+(v) = \frac{3\pi KG}{8}\,|v+2|\,\left||v+1| - |v+3|\right|$$
$$C^-(v) = C^+(v-4)$$
$$C^o(v) = -C^+(v) - C^-(v). \quad (41)$$

2.3 Effective theory

The effective theory can be derived through the geometric quantum mechanics method. Because of the fiber bundle structure, any horizontal section can be identified with the classical phase space. If we can find such a section, then the quantum dynamics on it can be expressed in terms of effective Hamiltonian, which is simply the expectation value of the quantum Hamiltonian constraint operator. The expectation value yields the classical term as the leading term and has quantum correction in the subleading terms. This is the key idea for deriving the effective equation. Here we can look for a natural section that is approximately preserved by the flow of the Hamiltonian constraint in a precise sense.

In order to obtain the effective theory, we should choose a coherent state. Here we use a Gaussian coherent state mostly for the reason that it is the simplest for getting the effective equation with a late-time, large-volume approximation. We can also choose a more general form to do this. The Gaussian coherent state is as follows:

$$\begin{aligned}(\psi_{\beta',V';\phi',p'_\phi}| &= \int dp_\phi \sum_v e^{-\frac{1}{2}\epsilon^2(v-v')^2} e^{\frac{i}{2}\sqrt{\Delta}\beta'(v-v')} \\ &\quad \times e^{-\frac{1}{2}\epsilon_\phi^2(p_\phi-p'_\phi)^2} e^{-i\phi'(p_\phi-p'_\phi)}(v;p_\phi| \\ &=: \int dp_\phi \sum_v \overline{\psi_n}(p_\phi)(v;p_\phi|, \end{aligned} \tag{42}$$

where v and v' are defined as $V = (\frac{8\pi\gamma}{6})^{\frac{3}{2}} \frac{l_p^3}{K} v$ and $V' = (\frac{8\pi\gamma}{6})^{\frac{3}{2}} \frac{l_p^3}{K} v'$, here $K = \frac{2\sqrt{2}}{3\sqrt{3\sqrt{3}}}$. Additionally, we should put on it three constraints:

(1)$v' \gg 1$, $\sqrt{\delta}\beta' \ll 1$. This pair of conditions means that the scalar factor is much larger than the Planck length and demands that the rate of change of the scale factor is much smaller than the speed of light, which holds even in the early universe.

(2)$v'\epsilon \gg 1$ and $\epsilon \ll \sqrt{\delta}\beta'$. This pair means that the spreads of operator $\hat{v}$ and $\hat{\beta}$ must be small.

(3)$\phi \gg \epsilon_\phi$ and $p_\phi \epsilon_\phi \gg 1$. The last pair of restrictions on parameters demands that the spreads of ϕ and p_ϕ are small.

We need to show that the semiclassical state is sharply peaked at the classical point $(\beta', V', \phi', p'_\phi)$. Here, we face two difficulties. First, the operator corresponding to $\hat{p}$ in the Schrödinger representation is not defined in the polymer framework. So we need to define a fundamental operator in $\mathcal{H}_{poly}$ which is approximated by $\hat{p}$ of the Schrödinger representation. We define this operator as follows:

$$\widehat{\beta_\Delta} = \frac{1}{i\sqrt{\Delta}}(\widehat{e^{\frac{i}{2}\sqrt{\Delta}\beta}} - \widehat{e^{-\frac{i}{2}\sqrt{\Delta}\beta}}). \tag{43}$$

The operator $\widehat{\beta_\Delta}$ agrees approximately with the classical β in the regime $\sqrt{\Delta}\beta \ll 1$. And its action on the basis kets $|v; p_\phi\rangle$ is

$$\hat{\beta}|v; p_\phi\rangle = \frac{1}{i\sqrt{\Delta}}(|v+1; p_\phi\rangle - |v-1; p_\phi\rangle). \tag{44}$$

The second difficulty is that the coherent state defined in Eq.(42) lies on Cyl^*. But there is no inner product on the Cyl^*. In other words, the solutions to the constraints do not reside in the kinematical Hilbert space, but rather in its algebraic dual space, therefore the required expectation values cannot be defined on it. Fortunately, we can carry out calculation within the"shadow state framework". Each shadow captures only a part of the information contained in our state, but the collection of shadows can be used to determine the full properties of the state in Cyl^*. We can indeed prove this Gaussian coherent state is sharply peaked at some classical values $(\beta', V', \phi', p'_\phi)$. Here we do not give the detailed proof, and refer interested readers to Taveras (2008).

The required projection $\hat{P}_\gamma$ from Cyl^* to Cyl_γ can be defined as

$$(\Psi|\,\hat{P}_\gamma := \sum_{x_j \in \gamma} \Psi(x_j)\,|x_j\rangle \equiv |\Psi_\gamma^{\rm shad}\rangle. \tag{45}$$

The corresponding shadow state in our case is

$$|\Psi_{\gamma}^{shad}\rangle = \int dp_{\phi} \sum_{n} e^{-\frac{1}{2}\epsilon^2(n-N)^2} e^{-\frac{i}{2}\sqrt{\Delta}\beta'(n-N)} \times e^{-\frac{1}{2}\epsilon_{\phi}^2(p_{\phi}-p'_{\phi})^2} e^{i\phi'(p_{\phi}-p'_{\phi})} |n; p_{\phi}\rangle. \quad (46)$$

Then one can compute the expectation value of the constraint operator directly

$$\langle \hat{C} \rangle = -\frac{3}{16\pi G\gamma^2\bar{\mu}'^2} p^{\frac{1}{2}} \left[1 + e^{-4\epsilon^2}\left(2\sin^2(\sqrt{\Delta}\beta') - 1\right)\right] + \frac{1}{2}\left(p_{\phi}'^2 + \frac{1}{2\epsilon_{\phi}^2}\right)\left(\frac{6}{8\pi\gamma l_p^2}\right)^{\frac{3}{2}} K\left[\frac{1}{v'} + O(v'^{-3}, v'^{-3}\epsilon^{-2})\right]. \quad (47)$$

Furthermore, we want to know that the equations of motion for the other physical quantities O. We will use the commutator between O and the Hamiltonian. The corresponding results are as follows:

$$\begin{aligned}
\langle \dot{\beta} \rangle &\simeq -\frac{1}{16\pi}\frac{27}{16}\left(\frac{8\pi\gamma}{6}\right)^{\frac{1}{2}} \frac{K}{\gamma^2\sqrt{\Delta}}\left[4e^{-\frac{25}{4}\epsilon^2}\cos\left(\frac{5}{2}\sqrt{\Delta}\beta'\right) + 4e^{-\frac{9}{4}\epsilon^2}\cos\left(\frac{3}{2}\sqrt{\Delta}\beta'\right)\right. \\
&\quad \left. -8e^{-\frac{1}{4}\epsilon^2}\cos\left(\frac{1}{2}\sqrt{\Delta}\beta'\right)\right] - \left(p_{\phi}'^{\,2} + \frac{1}{2\epsilon_{\phi}^2}\right)\frac{1}{2V'^2}, \\
\langle \dot{V} \rangle &\simeq \frac{3V'}{\gamma} e^{-4\epsilon^2} \frac{\sin(2\sqrt{\Delta}\beta')}{2\sqrt{\Delta}}, \\
\langle \dot{\phi} \rangle &\simeq \frac{p'_{\phi}}{V'} + O\left(\frac{1}{V'^3}\right), \\
\langle \dot{p_{\phi}} \rangle &= 0.
\end{aligned}$$

2.4 Ordinary formalism

As mentioned above, in the geometric quantization picture, we take the expectation values as our basic observables, and try to obtain an effective description in terms of these variables. We denote the expectation of β, V, ϕ, p_{ϕ} as $\bar{\beta}$, $\bar{V}$, $\bar{\phi}$, $\bar{p}_{\phi}$, respectively. Then we obtain the effective equations of motion to the first order as follows

$$\bar{C} = -\frac{3}{\kappa\gamma^2}\bar{V}\bar{\beta}^2\left(1 - \frac{1}{4}\Delta\bar{\beta}^2\right) - \frac{6\epsilon^2}{\kappa\gamma^2}\frac{\bar{V}}{\Delta} + \frac{\bar{p}_{\phi}^2}{2\bar{V}}\left[1 + O(\bar{V}^{-2}, \bar{V}^{-2}\epsilon^{-2})\right], \quad (48)$$

$$\dot{\bar{\beta}} = \frac{3}{4\gamma}\sqrt{1 - \frac{1}{4}\Delta\bar{\beta}^2}\left[-2\bar{\beta}^2 + \Delta\bar{\beta}^4\right] - \frac{\kappa}{4}\sqrt{1 - \frac{1}{4}\Delta\bar{\beta}^2}\frac{\bar{p}_{\phi}^2}{V'^2}[1 + O(\bar{V}^{-2}, \bar{V}^{-2}\epsilon^{-2})], \quad (49)$$

$$\dot{\bar{V}} = 3\frac{\bar{\beta}}{\gamma}\bar{V}\sqrt{1 - \frac{\Delta\bar{\beta}^2}{4}}\left(1 - \frac{1}{2}\Delta\bar{\beta}^2\right), \quad (50)$$

$$\dot{\phi} = \frac{\bar{p}_\phi}{\bar{V}} + O(\bar{V}^{-3}), \tag{51}$$

$$\dot{\bar{p}}_\phi = 0. \tag{52}$$

Combining Eq.(48) with Eq.(49), we get the effective Friedmann equation

$$H^2 = \frac{\kappa\rho}{3}(1 - \frac{\rho}{\rho_{crit}}) + O(\epsilon^2), \tag{53}$$

where $H = \frac{\dot{a}}{a}$ and $\rho_{crit} = \frac{3}{\kappa\gamma^2\Delta}$ originated from the quantum effect.
From the effective theory, we can easily see that the semiclassical states follow the classical trajectory until the scalar field density is on the order of 1% of the Planck density where deviations from the classical trajectory emerge. Then there can be major deviations from the classical theory. The existence of the correction ρ_{crit} can allow $H = 0$, meaning a bounce is possible.
Also, from the equation of motion, we can get the conservation equation and the corrected Raychaudhuri equation for this effective theory. With the Poisson bracket, we can calculate $\dot{\rho} = \{\rho, \bar{C}\}$, and get the conservation equation

$$\dot{\rho} + 6\frac{\dot{a}}{a}\rho = 0, \tag{54}$$

which is the same as the classical conservation equation.
Similarly, we calculate $\ddot{V} = \{\dot{V}, \bar{C}\}$, and express it in terms of the scale factor a. Then we get

$$3\frac{\ddot{a}}{a} = -2\kappa\rho(1 - \frac{5}{2}\frac{\rho}{\rho_c}) + o(\epsilon^2). \tag{55}$$

Compared with the classical Raychaudhuri equation, it also gets a quantum correction term $\frac{\rho^2}{\rho_c}$. Classically, this equation is always negative, but in the effective framework, there is a bounce at the $\frac{\rho}{\rho_c} = 1$, which makes $\ddot{a}$ positive when $\rho > \frac{2}{5}\rho_c$.
In the end, there are still two points that need to be clarified. First, theoretically, Eq.(53) cannot describe the correct dynamics near the bounce point because the bounce point $\frac{\rho}{\rho_c} = 1$ is outside of the regime of our approximation. The effective framework is applicable only to a late-time, large-volume universe because $\epsilon \ll \sqrt{\Delta}\beta'$ is violated badly at the point of $\frac{\rho}{\rho_c} = 1$. However, numerical work Ashtekar et al. (2006a;b) has shown that the dynamics derived by the above describes the evolution of the universe very well even at the bounce, and hence the results obtained in the effective framework continue to be reliable even beyond their expected regime. Second, we omit the high-order correction term $o(\epsilon^2)$ in the modified Friedmann equation Eq.(53). We should note that near the bounce point, the term in parentheses in the modified Friedmann equation is approaching 0. It is not known if $O(\epsilon^2)$ should be omitted there because $\epsilon \ll \sqrt{\Delta}\beta'$ is violated. But the numerical results show that the modified Friedmann equation holds with negligible $O(\epsilon^2)$ corrections. Therefore, the effective theory remains valid beyond the domain for which it is constructed.

2.5 Phenomenological analysis

In the above subsection, we obtain the effective framework of LQC systematically. In this one, we analyze the holonomy correction and the inverse volume correction phenomenologically, which is easy to handle and can describe the evolution of the universe correctly. Meanwhile, it can lead us to new physics heuristically.

2.5.1 The holonomy correction

As mentioned above, the classical Hamiltonian with a free scalar field for the $k = 0$ FRW model is given by

$$H_{cl} = -\frac{3N}{8\pi G\gamma^2} c^2 \sqrt{|p|} + \frac{N p_\phi^2}{2\,|p|^{3/2}} \tag{56}$$

in terms of the Ashtekar variables c and p. p_ϕ is the conjugate momentum of ϕ. N is the lapse function and γ is the Barbero-Immirzi parameter. At the heuristic level, we can impose the loop quantum corrections of LQC phenomenologically. We simply replace c with

$$c \longrightarrow \frac{\sin(\bar{\mu}c)}{\bar{\mu}}. \tag{57}$$

Then we get the description of "holonomization". This effective dynamics is solved as if the dynamics was classical but governed by the new "holonomized" Hamiltonian, which reads as

$$H_{eff} = -\frac{3N}{8\pi G\gamma^2} \frac{\sin^2 \bar{\mu}c}{\bar{\mu}^2} \sqrt{|p|} + \frac{N p_\phi^2}{2\,|p|^{3/2}}. \tag{58}$$

As to be expected, the bouncing scenario can be easily obtained at the level of heuristic effective dynamics without invoking the sophisticated features of LQC. In particular, with the "improved" scheme imposed for $\bar{\mu}$, the modified Hamiltonian constraint $C_{\bar{\mu}} = 0$ immediately sets an upper bound for the matter density:

$$\rho_\phi := \frac{p_\phi^2}{2|p|^3} = \frac{3}{8\pi G\gamma^2\Delta} \sin^2 \bar{\mu}c \le 3\rho_{\rm Pl}, \tag{59}$$

where the Planckian density is defined as

$$\rho_{\rm Pl} := (8\pi G\gamma^2\Delta)^{-1}. \tag{60}$$

With this effective Hamiltonian, we have the canonical equation

$$\dot{p} = \left\{p, H_{eff}\right\} = -\frac{8\pi\gamma}{3}\frac{\partial H_{eff}}{\partial c}, \tag{61}$$

or,

$$\dot{a} = \frac{\sin(\bar{\mu}c)\cos(\bar{\mu}c)}{\gamma\bar{\mu}}. \tag{62}$$

Combining that with the constraint on Hamiltonian, $H_{eff} = 0$, we obtain the modified Friedmann equation

$$H^2 = \frac{8\pi}{3}\rho\left(1 - \frac{\rho}{\rho_c}\right), \tag{63}$$

where $H \equiv \frac{\dot{a}}{a}$ denotes the Hubble rate, and $\rho_c \equiv \frac{3}{8\pi\gamma^2\bar{\mu}^2 p}$ is the quantum critical density. Compared with the standard Friedmann equation, we can define the effective density

$$\rho_{eff} = \rho\left(1 - \frac{\rho}{\rho_c}\right). \tag{64}$$

Taking derivative of Eq.(63) and also using the conservation equation of matter, $\dot{\rho} + 3H(\rho + P) = 0$, we obtain the modified Raychaudhuri equation

$$\frac{\ddot{a}}{a} = \dot{H} + H^2 = -\frac{4\pi}{3}\left\{\rho(1 - \frac{\rho}{\rho_c}) + 3\left[P(1 - \frac{2\rho}{\rho_c}) - \frac{\rho^2}{\rho_c}\right]\right\}. \tag{65}$$

Comparing that with the standard Raychaudhuri equation, we can define the effective pressure,

$$P_{eff} = P\left(1 - \frac{2\rho}{\rho_c}\right) - \frac{\rho^2}{\rho_c}. \tag{66}$$

In terms of the effective density and the effective pressure, the modified Friedmann, Raychaudhuri and conservation equations take the following forms,

$$H^2 = \frac{8\pi}{3}\rho_{eff}, \tag{67}$$

$$\frac{\ddot{a}}{a} = \dot{H} + H^2 = -\frac{4\pi}{3}\left(\rho_{eff} + 3P_{eff}\right), \tag{68}$$

$$\dot{\rho}_{eff} + 3H\left(\rho_{eff} + P_{eff}\right) = 0. \tag{69}$$

Therefore, we can get the important features of LQC without going into the detailed construction of LQC at all. This can help us to extract the physics easily and provide insight into the full theory.

2.5.2 The inverse volume correction

The inverse volume $|p|^{-\frac{3}{2}}$ of the matter Hamiltonian can get a quantum correction and the matter Hamiltonian obtained in this manner will behave differently at small p. This is called the inverse volume correction, and can be interpreted as providing a natural curvature cut-off. For a scalar field, the modified matter Hamiltonian is

$$\mathcal{H}_{matt} = \frac{1}{2}d(a)p_\varphi^2 + a^3 V(\varphi), \tag{70}$$

where p_φ is the momentum canonically conjugate to φ, and $d(a)$, which is classically $1/a^3$, encodes the quantum corrections. In the semi-classical regime, where spacetime may be treated as continuous, it is given by

$$d(a) = \frac{D(q)}{a^3}, \; q = \frac{a^2}{a_*^2}, \; a_* = \sqrt{\frac{\gamma j}{3}}\,\ell_{\rm Pl}. \tag{71}$$

and

$$D(q) = q^{-3/2}\left\{\frac{3}{2l}\left(\frac{1}{l+2}\left[(q+1)^{l+2} - |q-1|^{l+2}\right]\right.\right.$$
$$\left.\left. - \frac{q}{1+l}\left[(q+1)^{l+1} - \mathrm{sgn}(q-1)|q-1|^{l+1}\right]\right)\right\}^{3/(2-2l)}. \tag{72}$$

Here γ is the Barbero-Immirzi parameter. Since the expression for a^{-1} is rather complicated, the final quantization contains quantization ambiguities of different types. Here, j (a half integer) is resulted from the use of arbitrary representations, and $0 < l < 1$ is another quantization parameter. The approximation to the eigenvalues becomes better for values of j larger than the minimal one, $1/2$.

The scale below which non-perturbative modifications become important is given by a_*. Typically, one chooses $j \gg 1$, so that $a_* \gg \ell_{\rm Pl}$. The Planck scale marks the onset of discrete spacetime effects. For $\ell_{\rm Pl} < a \ll a_*$, the universe is in the semiclassical regime. In this regime, $q \ll 1$. With the Taylor expansion, the geometrical density $d(a)$ behaves as

$$d(a) \sim \left[\left(\frac{3}{1+l}\right)^{3/(2-2l)} \left(\frac{a}{a_*}\right)^{3(2-l)/(1-l)}\right] \frac{1}{a^3} \,. \tag{73}$$

The Hamiltonian determines the dynamics completely. The equation of motion for the matter is

$$\dot{\varphi} = \{\varphi, \mathcal{H}\} = d(a) p_\varphi \,. \tag{74}$$

Combined with the Hamiltonian equation of motion for p_φ, the above equation can be cast into a second order equation for φ Bojowald & Vandersloot (2002); Singh & Toporensky (2004); Vereshchagin (2004); Tsujikawa et al. (2004),

$$\ddot{\varphi} + \left(3H - \frac{\dot{D}}{D}\right) \dot{\varphi} + D\, V_{,\varphi} = 0 \,. \tag{75}$$

For $\ell_{\rm Pl} < a \ll a_*$, we find that $\dot{D}/D > 3H$, which leads to the classical frictional term for an expanding universe. The case is the opposite if the universe is contracting.

The Friedman equation and the Raychaudhuri equation are as follows:

$$H^2 = \frac{8\pi G}{3}\left[\frac{\dot{\varphi}^2}{2D(a)} + V(\varphi)\right], \tag{76}$$

$$\frac{\ddot{a}}{a} = -\frac{8\pi G}{3}\left[\frac{\dot{\varphi}^2}{D}\left(1 - \frac{\dot{D}}{4HD}\right) - V(\varphi)\right]. \tag{77}$$

The Friedman equation implies a bounce in the scale factor, i.e., $\dot{a} = 0$ and $\ddot{a} > 0$, which requires a negative potential. (In a closed model, the curvature term allows for a bounce with positive potential Singh & Toporensky (2004); Vereshchagin (2004).) Vanishing Hubble parameter at the bounce implies

$$\dot{\varphi}^2 = -2D(a)V(\varphi) \,, \tag{78}$$

so that at the bounce,

$$\frac{\ddot{a}}{a} = \frac{4\pi G}{3}\left(6 - \frac{d \ln D}{d \ln a}\right) V \,. \tag{79}$$

Classically, i.e., for $D = 1$, a bounce for a negative $V(\varphi)$ is not allowed. With the modified $D(a)$, however, $d \ln D/d \ln a > 6$ will hold for sufficiently small a, so $\ddot{a} > 0$ is possible. Thus, the universe has to collapse sufficiently deep into the modified regime before it can bounce back.

So far, the two corrections appear independently in the discussion. We hope to find a consistent way to bring them together and study their impact on the universe. All the works discussed in the following are based on the effective theory.

3. Stability analysis of an autonomous system

Scalar field plays an important role in modern cosmology. Indeed, scalar field cosmological models have great importance in the study of the early universe, especially in the investigation of inflation. The dynamical properties of scalar fields also make an interesting research topic for modern cosmological studies Copeland et al. (2006); Coley (2003). The dynamical behavior of scalar field coupled with a barotropic fluid in spatially flat Friedmann-Robertson-Walker (FRW) universe has been studied by many authors (see Copeland et al. (2006); Coley (2003); Leon et al. (2010), and the first section of Copeland et al. (2009)).

The phase-plane analysis of the cosmological autonomous system is a useful method for studying the dynamical behavior of scalar field. One always considers the dynamical behavior of a scalar field with an exponential potential in the classical cosmology Copeland et al. (1998); Hao & Li (2003; 2004) or modified cosmology Li & Hao (2004); Samart & Gumjudpai (2007). And, if one considers the dynamical behavior of a scalar field coupled with a barotropic fluid, the exponential potential is also the first choice Billyard & Coley (2000); Ferreira & Joyce (1998); Hoogen et al. (1999); Yu & Wu (2008). The exponential potential V leads to the facts that the variables $\Gamma = VV''/V'^2$ equals 1 and that $\lambda = V'/V$ is also a constant. Then the autonomous system is always 2-dimensional in the classical cosmology Copeland et al. (1998), and 3-dimensional in LQC Samart & Gumjudpai (2007). Although one can also consider a more complex case with λ being a dynamically changing quantity Copeland et al. (2006); Macorra & Piccinelli (2000); Ng et al. (2001), the fixed point is not a real one, and this method is not exact. Recently, Zhou *et al* Fang et al. (2009); Zhou (2008) introduced a new method by which one can make Γ a general function of λ. Then the autonomous system is extended from 2-dimensional to 3-dimensional in the classical cosmology. They found that this method can help investigate many quintessence models with different potentials. One of our goals is to extend this method for studying the dynamical behavior of a scalar field with a general potential coupled with a barotropic fluid in LQC.

Based on the holonomy modification, the dynamical behavior of dark energy in LQC scenario has recently been investigated by many authors Fu et al. (2008); Lamon & Woehr (2010); Li & Ma (2010); Samart & Gumjudpai (2007); Wei & Zhang (2007); Xiao & Zhu (2010). The attractor behavior of scalar field in LQC has also been studied Copeland et al. (2008); Lidsey (2004). It was found that the dynamical properties of dark-energy models in LQC are significantly different from those in the classical cosmology. In this section, we examine the background dynamics of LQC dominated by a scalar field with a general positive potential coupled with a barotropic fluid. By considering Γ as a function of λ, we investigate scalar fields with different potentials. Since the Friedmann equation has been modified by the quantum effect, the dynamical system will be very different from the one in classical cosmology, e.g., the number of dimensions of autonomous system will change to four in LQC. It must be pointed out that this method cannot be used to describe the dynamical behavior of scalar field with arbitrary potential. To overcome this problem, therefore, we should consider an infinite-dimensional autonomous system.

This section is organized as follows. We present in Subsection 3.1 the basic equations and the 4-dimensional dynamical system, and discuss in Subsection 3.2 the properties of this system. In Subsection 3.3, we give more discussions on the autonomous system, and also on an infinite-dimensional autonomous system. We conclude the section in the last subsection.

3.1 Basic equations

We focus on the flat FRW cosmology. The modified Friedmann equation in the effective LQC with holonomy correction can be written as

$$H^2 = \frac{1}{3}\rho\left(1 - \frac{\rho}{\rho_c}\right), \tag{80}$$

in which ρ is the total energy density and the natural unit $\kappa = 8\pi G = 1$ is adopted for simplicity. We consider a self-interacting scalar field ϕ with a positive potential $V(\phi)$ coupled with a barotropic fluid. Then the total energy density can be written as $\rho = \rho_\phi + \rho_\gamma$, with the energy density of scalar field $\rho_\phi = \frac{1}{2}\dot{\phi}^2 + V(\phi)$ and the energy density of barotropic fluid ρ_γ. We consider that the energy momenta of this field to be covariant conserved. Then one has

$$\ddot{\phi} + 3H\dot{\phi} + V' = 0, \tag{81}$$

$$\dot{\rho}_\gamma + 3\gamma H\rho_\gamma = 0, \tag{82}$$

where γ is an adiabatic index and satisfies $p_\gamma = (\gamma - 1)\rho_\gamma$ with p_γ being the pressure of the barotropic fluid, and the prime denotes the differentiation w.r.t. the field ϕ. Differentiating Eq. (80) and using Eqs. (81) and (82), one can obtain

$$\dot{H} = -\frac{1}{2}\left(\dot{\phi}^2 + \gamma\rho_\gamma\right)\left[1 - \frac{2(\rho_\gamma + \rho_\phi)}{\rho_c}\right]. \tag{83}$$

Eqs. (80)-(82) or Eqs. (81)-(83) characterize a closed system which can determine the cosmic behavior. To analyze the dynamical behavior of the universe, one can further introduce the following variables Copeland et al. (1998); Samart & Gumjudpai (2007):

$$x \equiv \frac{\dot{\phi}}{\sqrt{6}H}, \quad y \equiv \frac{\sqrt{V}}{\sqrt{3}H}, \quad z \equiv \frac{\rho}{\rho_c}, \quad \lambda \equiv \frac{V'}{V}, \tag{84}$$

where the z term is a special variable in LQC (see Eq. (80)). In the LQC scenario, the total energy density ρ should be less or equal to the critical energy density ρ_c, and thus $0 \le z \le 1$. Notice that, in the classical region, $z = 0$ for $\rho \ll \rho_c$. Using these new variables, one can obtain

$$\frac{\rho_\gamma}{3H^2} = \frac{1}{1-z} - x^2 - y^2, \tag{85}$$

$$\frac{\dot{H}}{H^2} = -\left[3x^2 + \frac{3\gamma}{2}\left(\frac{1}{1-z} - x^2 - y^2\right)\right](1 - 2z \tag{86}$$

Using the new variables (84), and considering Eqs. (85) and (86), one can rewrite Eqs. (80)-(82) in the following forms,

$$\frac{dx}{dN} = -3x - \frac{\sqrt{6}}{2}\lambda y^2 + x\left[3x^2 + \frac{3\gamma}{2}\left(\frac{1}{1-z} - x^2 - y^2\right)\right] \times (1-2z), \tag{87}$$

$$\frac{dy}{dN} = \frac{\sqrt{6}}{2}\lambda xy + y\left[3x^2 + \frac{3\gamma}{2}\left(\frac{1}{1-z} - x^2 - y^2\right)\right] \times (1-2z), \tag{88}$$

$$\frac{dz}{dN} = -3\gamma z - 3z(1-z)\left(2x^2 - \gamma x^2 - \gamma y^2\right), \tag{89}$$

$$\frac{d\lambda}{dN} = \sqrt{6}\lambda^2 x(\Gamma - 1), \tag{90}$$

where $N = \ln a$ and

$$\Gamma \equiv \frac{VV''}{V'^2}. \tag{91}$$

Note that the potential $V(\phi)$ is positive in this section, but one can also discuss a negative potential. Just as Heard & Wands (2002) shown, the negative scalar potential could slow down the growth of the scale factor and cause the Universe to be in a collapsing phase. The dynamical behavior of scalar field with positive and negative potential in brane cosmology has been discussed by Copeland et al. (2009). In this section we are concerned only with an expanding universe, and both the Hubble parameter and the potential are positive.
Differentiating λ w.r.t. the scalar field ϕ, we obtain the relationship between λ and Γ,

$$\frac{d\lambda^{-1}}{d\phi} = 1 - \Gamma. \tag{92}$$

If we only consider a special case of the potential, like exponential potential Billyard & Coley (2000); Copeland et al. (1998); Ferreira & Joyce (1998); Hao & Li (2003; 2004); Hoogen et al. (1999); Li & Hao (2004); Samart & Gumjudpai (2007); Yu & Wu (2008), then λ and Γ are both constants. In this case, the 4-dimensional dynamical system, Eqs. (87)-(90), reduces to a 3-dimensional one, since λ is a constant. (In the classical dynamical system, the z term does not exist, and then the dynamical system is reduced from 3-dimensional to 2-dimensional.) The cost of this simplification is that the potential of the field is restricted. Recently, Zhou *et al* Fang et al. (2009); Zhou (2008) considered the potential parameter Γ as a function of another potential parameter λ, which enables one to study the fixed points for a large number of potentials. We will follow this method in this and the next subsections to discuss the dynamical behavior of scalar field in the LQC scenario, and we have

$$\Gamma(\lambda) = f(\lambda) + 1. \tag{93}$$

In this case, Eq. (93) can cover many scalar potentials.
For completeness' sake, we briefly review the discussion of Fang et al. (2009) in the following. From Eq. (92), one can obtain

$$\frac{d\lambda}{\lambda f(\lambda)} = \frac{dV}{V}. \tag{94}$$

Integrating out $\lambda = \lambda(V)$, one has the following differential equation of potential

$$\frac{dV}{V\lambda(V)} = d\phi. \tag{95}$$

Then, Eqs. (94) and (95) give a route for obtaining the potential $V = V(\phi)$. If we consider a concrete form of the potential (e.g., an exponential potential), the dynamical system is specialized (e.g., the dynamical system is reduced to 3-dimensional if one considers the exponential potential for $d\lambda/dN = 0$). These specialized dynamical systems are too special if one hopes to distinguish the fixed points that are the common properties of scalar field from those that are just related to the special potentials Fang et al. (2009). If we consider a more general λ, then we can get the more general stability properties of scalar field in the LQC scenario. We will continue the discussion of this topic in Subsection 3.3. In this case, Eq. (90) becomes

$$\frac{d\lambda}{dN} = \sqrt{6}\lambda^2 x f(\lambda). \tag{96}$$

Hereafter, Eqs. (87)-(89) along with Eq. (96) are definitely describing a dynamical system. We will discuss the stability of this system in the following subsection.

3.2 Properties of the autonomous system

Obviously, the terms on the right-hand side of Eqs. (87)-(89) and (96) only depend on x, y, z, λ, but not on N or other variables. Such a dynamical system is usually called an autonomous system. For simplicity, we define $\frac{dx}{dN} = F_1(x,y,z,\lambda) \equiv F_1$, $\frac{dy}{dN} = F_2(x,y,z,\lambda) \equiv F_2$, $\frac{dz}{dN} = F_3(x,y,z,\lambda) \equiv F_3$, and $\frac{d\lambda}{dN} = F_4(x,y,z,\lambda) \equiv F_4$. The fixed points $(x_c, y_c, z_c, \lambda_c)$ satisfy $F_i = 0, i = 1,2,3,4$. From Eq. (96), it is straightforward to see that $x = 0, \lambda = 0$ or $f(\lambda) = 0$ can make $F_4(x,y,z,\lambda) = 0$. Also, we must consider $\lambda^2 f(\lambda) = 0$. Just as Fang et al. (2009) argued, it is possible that $\lambda^2 f(\lambda) \neq 0$ and $\frac{d\lambda}{dN} \neq 0$ when $\lambda = 0$. Thus the necessary condition for the existence of the fixed points with $x \neq 0$ is $\lambda^2 f(\lambda) = 0$. Taking into account these factors, we can easily obtain all the fixed points of the autonomous system described by Eqs. (87)-(89) and (96), and they are shown in Tab. (1).

The properties of each fixed points are determined by the eigenvalues of the Jacobi matrix

$$\mathcal{M} = \left.\begin{pmatrix} \frac{\partial F_1}{\partial x} & \frac{\partial F_1}{\partial y} & \frac{\partial F_1}{\partial z} & \frac{\partial F_1}{\partial \lambda} \\ \frac{\partial F_2}{\partial x} & \frac{\partial F_2}{\partial y} & \frac{\partial F_2}{\partial z} & \frac{\partial F_2}{\partial \lambda} \\ \frac{\partial F_3}{\partial x} & \frac{\partial F_3}{\partial y} & \frac{\partial F_3}{\partial z} & \frac{\partial F_3}{\partial \lambda} \\ \frac{\partial F_4}{\partial x} & \frac{\partial F_4}{\partial y} & \frac{\partial F_4}{\partial z} & \frac{\partial F_4}{\partial \lambda} \end{pmatrix}\right|_{(x_c, y_c, z_c, \lambda_c)} . \tag{97}$$

According to Lyapunov's linearization method, the stability of a linearized system is determined by the eigenvalues of the matrix $\mathcal{M}$ (see Chapter 3 of Slotine & Li (1991)). If all of the eigenvalues are strictly in the left-half complex plane, then the autonomous system is stable. If at least one eigenvalue is strictly in the right-half complex plane, then the system is unstable. If all of the eigenvalues are in the left-half complex plane, but at least one of them is on the $i\omega$ axis, then one cannot conclude anything definite about the stability from the linear approximation. By examining the eigenvalues of the matrix $\mathcal{M}$ for each fixed point shown in Tab. (1), we find that points $P_{1,2,4-8,10}$ are unstable and point P_9 is stable only under some conditions. We cannot determine the stability properties of P_3 from the eigenvalues, and we have gave the full analysis of P_3 in the appendix of Xiao & Zhu (2010).

Some remarks on Tab.(1):

1. Apparently, points P_2 and P_6 have the same eigenvalues, and the difference between these two points is just on the value of λ. As noted in the caption of Tab. (1), λ_* means that λ can be any value, and λ_a is just the value that makes $f(\lambda) = 0$. Obviously, λ_a is just a special value of λ_*, and point P_6 is a special case of point P_2. P_6 is connected with $f(\lambda)$, but P_2 is not. From now on, we do not consider separately the special case of P_6 when we discuss the property of P_2. Hence the value of λ_a is contained in our discussion ofλ_*.
2. It is straightforward to check that, if $x_c = \lambda_c = 0$, y_c can be any value y_* when it is greater than or equal 1. But, if $y_* > 1$, then $z_c = 1 - 1/y_*^2 < 1$, and this means that the fixed

Fixed-points	x_c	y_c	z_c	λ_c	Eigenvalues	Stability
P_1	0	0	0	0	$\mathbf{M}^T = (0, -3\gamma, \frac{3}{2}\gamma, -3 + \frac{3}{2}\gamma)$	U, for all γ
P_2	0	0	0	λ_*	$\mathbf{M}^T = (0, \frac{3}{2}\gamma, -3\gamma, -3 + \frac{3}{2}\gamma)$	U, for all γ
P_3	0	1	0	0	$\mathbf{M}^T = (-3, -3\gamma, 0, 0)$	S, for $\gamma = 1, f_1(0) \geq 0$ U, for $\gamma = \frac{4}{3}$, if $f_1(0) \neq 0$ S, for $\gamma = \frac{4}{3}$, if $f_1(0) = 0$
P_4	1	0	0	0	$\mathbf{M}^T = (0, -6, 0, 6 - 3\gamma)$	U,for all γ
P_5	−1	0	0	0	$\mathbf{M}^T = (0, -6, 0, 6 - 3\gamma)$	U,for all γ
P_6	0	0	0	λ_a	$\mathbf{M}^T = (0, \frac{3}{2}\gamma, -3\gamma, -3 + \frac{3}{2}\gamma)$	U,for all γ
P_7	1	0	0	λ_a	$\mathbf{M}^T = \left(-6, 6 - 3\gamma, \frac{1}{2}\sqrt{6}\lambda_a + 3, \sqrt{6}\lambda_a A\right)$	U,for all γ
P_8	−1	0	0	λ_a	$\mathbf{M}^T = \left(-6, 6 - 3\gamma, -\frac{1}{2}\sqrt{6}\lambda_a + 3, -\sqrt{6}\lambda_a A\right)$	U,for all γ
P_9	$-\frac{\sqrt{6}}{6}\lambda_a$	$\sqrt{1 - \frac{\lambda_a^2}{6}}$	0	λ_a	$\mathbf{M}^T = \left(-\lambda_a^2, -3 + \frac{1}{2}\lambda_a^2, \lambda_a^2 - 3\gamma, -\lambda_a^3 - f_1(\lambda_a)\right)$	S, for $f_1(\lambda_a) > \lambda_a$, and $\lambda_a < 3\gamma$ U, for $f_1(\lambda_a) < \lambda_a$ and/or $\lambda_a > 3\gamma$
P_{10}	$-\sqrt{\frac{3}{2}\frac{\gamma}{\lambda_a}}$	$\sqrt{\frac{3}{2\lambda_a^2}\gamma(2-\gamma)}$	0	λ_a	See the Eq. (98)	U, for all γ

Table 1. The stability analysis of an autonomous system in LQC. The system is described by a self-interacting scalar field ϕ with positive potential V coupled with a barotropic fluid ρ_γ. Explanation of the symbols used in this table: P_i denotes the fixed points located in the 4-dimensional phase space, which are earmarked by the coordinates $(x_c, y_c, z_c, \lambda_c)$. λ_* means that λ can be any value. λ_a is the value that makes $f(\lambda) = 0$. $\mathbf{M}^T$ means the inverted matrix of the eigenvalues of the fixed points. $f_1(\Lambda) = \left.\frac{df(\lambda)}{d\lambda}\right|_{\lambda=\Lambda}$ with $\Lambda = 0, \lambda_a$. $A = \left[2f(\lambda_a) + \lambda_a\left(\left.\frac{df(\lambda)}{d\lambda}\right|_{\lambda_a}\right)\right]$. U stands for unstable, and S stands for stable.

point is located in the quantum-dominated regions. Although the stability of this point in the quantum regions may depend on $f(\lambda)$, it is not necessary to analyze its dynamical properties, since it does not have any physical meanings. The reason is the following. If the universe is stable, it will not evolve to today's pictures. If the Universe is unstable, it will always be unstable. We will just focus on point P_3 staying in the classical regions. Then $y_c = y_* = 1, z_c = 1 - 1/y_*^2 = 0$, i.e., for the classical cosmology region, $\rho \ll \rho_c$.

3. Since the adiabatic index γ satisfies $0 < \gamma < 2$ (in particular, for radiation $\gamma = \frac{4}{3}$ and for dust $\gamma = 1$), all the terms that contain γ should not change sign. A more general situation of γ is $0 \leq \gamma \leq 2$ Billyard et al. (1998). If $\gamma = 0$ or $\gamma = 2$, the eigenvalues corresponding to points $P_{1,2,4,5,9}$ will have some zero elements and some negative ones. To analyze the stability of these points, we need to resort to other more complex methods, just as we did in the appendix of Xiao & Zhu (2010) for the dynamical properties of point P_3. In this subsection, we just consider the barotropic fluid which includes radiation and dust, and $\gamma \neq 0, 2$. Notice that if one considers $\gamma = 0$, the barotropic fluid describes the dark energy. This is an interesting topic, but will not be considered here for the sake of simplicity.
4. $-\sqrt{6} < \lambda_a < \sqrt{6}, \lambda_a \neq 0$ should hold for point P_9, hence $-3 + \frac{1}{2}\lambda_a^2 < 0$.
5. $\lambda_a > 0$ should hold, since $y_c > 0$ for point P_{10}. The eigenvalue of this point is

$$\mathbf{M} = \begin{pmatrix} -3\gamma \\ -3\lambda_a \gamma f_1(\lambda_a) \\ -\frac{3}{2} + \frac{3}{4}\gamma + \frac{3}{4\lambda_a}\sqrt{(2-\gamma)(\lambda_a^2(2-\gamma) + 8\gamma + 24\gamma^2)} \\ -\frac{3}{2} + \frac{3}{4}\gamma - \frac{3}{4\lambda_a}\sqrt{(2-\gamma)(\lambda_a^2(2-\gamma) + 8\gamma + 24\gamma^2)} \end{pmatrix}. \tag{98}$$

Since we just consider $0 < \gamma < 2$ in this subsection, it is easy to check that $(2-\gamma)(\lambda_a^2(2-\gamma) + 8\gamma + 24\gamma^2) > 0$ is always satisfied. And this point is unstable with $f_1(\lambda_a) = \left.\frac{df(\lambda)}{d\lambda}\right|_{\lambda=\lambda_a}$ being either negative or positive, since $-\frac{3}{2} + \frac{3}{4}\gamma + \frac{3}{4\lambda_a}\sqrt{(2-\gamma)(\lambda_a^2(2-\gamma) + 8\gamma + 24\gamma^2)}$ is always positive.

Based on Tab. (1) and the related remarks above, we have the following conclusions.

1. Points $P_{1,2}$: The related critical values, eigenvalues and stability properties do not depend on the specific form of the potential, since $\lambda_c = 0$ or λ can be any value λ_*.
2. Point P_3: The related stability properties depend on $f_1(0) = \left.\frac{df(\lambda)}{d\lambda}\right|_{\lambda=0}$.
3. Points $P_{4,5}$: The related eigenvalues and stability properties do not depend on the form of the potential, but the critical values of these points should satisfy $\lambda^2 f(\lambda) = 0$ since $x_c \neq 0$.
4. Point P_6: It is a special case of P_2, but $f(\lambda_a) = 0$ should be satisfied.
5. Points $P_{7,8}$: Same as P_6, they would not exist if $f(\lambda_a) \neq 0$.
6. Point $P_{9,10}$: $f(\lambda_a) = 0$ should hold. The fixed values and the eigenvalues of these two points depend on $f_1(\lambda_a) = \left.\frac{df(\lambda)}{d\lambda}\right|_{\lambda=\lambda_a}$.

Thus, only points $P_{1,2}$ are independent of $f(\lambda)$.
Comparing the fixed points in LQC and the ones in the classical cosmology (see the Table **I** of Fang et al. (2009)), we can see that, even though the values of the coordinates (x_c, y_c, λ_c) are

the same, the stability properties are very different. This is reasonable, because the quantum modification is considered, and the autonomous system in the LQC scenario is very different from the one in the classical scenario, e.g., the autonomous system is 4-dimensional in LQC but 3-dimensional in the classical scenario. Notice that all of the fixed points lie in the classical regions, and therefore the coordinates of fixed points remain the same from classical to LQC, which we also pointed out in Xiao & Zhu (2010).
Now we focus on the later time attractors: point P_3 under the conditions of $\gamma = 1, f_1(0) \geq 0$ and $\gamma = 4/3, f_1(0) = 0$, and point P_9 under the conditions of $\lambda_a^2 < 6, f_1(\lambda_a) > \lambda_a, \lambda_a < 3\gamma$. Obviously, these points are scalar-field dominated, since $\rho_\gamma = H^2(1/(1-z_c) - x_c^2 - y_c^2) = 0$. For point P_3, the effective adiabatic index $\gamma_\phi = (\rho_\phi + p_\phi)/\rho_\phi = 0$, which means that the scalar field is an effective cosmological constant. For point P_9, $\gamma_\phi = \lambda_a^2/2$. This describes a scaling solution that, as the universe evolves, the kinetic energy and the potential energy of the scalar field scale together. And we can see that there is not any barotropic fluid coupled with the scalar field dominated scaling solution. This is different from the dynamical behavior of scalar field with exponential potential $V = V_0 \exp(-\lambda\kappa\phi)$ in the classical cosmology Billyard & Coley (2000); Copeland et al. (1998); Ferreira & Joyce (1998); Hao & Li (2003; 2004); Hoogen et al. (1999); Li & Hao (2004); Samart & Gumjudpai (2007); Yu & Wu (2008), and also is different from the properties of the scalar field in brane cosmology Copeland et al. (2009), in which $\lambda =$ const. (notice that the definition of λ in Copeland et al. (2009) is different from the one in this section) and Γ is a function of $L(\rho(a))$ and $|V|$. In these models, the universe may enter a stage dominated by scalar field coupled with fluid when λ, γ satisfy some conditions Copeland et al. (1998; 2009).
We discuss the dynamical behavior of the scalar field by considering Γ as a function of λ in this and the preceding subsections. But Γ can not always be treated as a function of λ. We need to consider a more general autonomous system, which we will introduce in the next subsection.

3.3 More discussions on the autonomous system

The dynamical behavior of scalar field has been discussed by many authors (e.g., see Billyard & Coley (2000); Copeland et al. (1998; 2006); Coley (2003); Ferreira & Joyce (1998); Hao & Li (2003; 2004); Hoogen et al. (1999); Li & Hao (2004); Samart & Gumjudpai (2007); Yu & Wu (2008)). If one wants to get the potentials that yield the cosmological scaling solutions beyond the exponential potential, one can add a $\frac{d\phi}{dN}$ term into the autonomous system Nunes & Mimoso (2000). All of these methods deal with special cases of the dynamical behavior of scalar fields in backgrounds of some specific forms. By considering Γ as a function of λ, one can treat potentials of more general forms and get the common fixed points of the general potential, as shown in Fang et al. (2009); Zhou (2008) and in the two preceding subsections. However, as is discussed in Fang et al. (2009), sometimes Γ is not a function of λ, and then the dynamical behaviors of scalar fields discussed above are still not general in the strict sense. For a more general discussion, we must consider the higher order derivatives of the potential. We define

$$
{}^{(1)}\Gamma = \frac{VV_3}{V'^2}, \quad {}^{(2)}\Gamma = \frac{VV_4}{V'^2}, \quad {}^{(3)}\Gamma = \frac{VV_5}{V'^2},
$$
$$
\ldots \quad {}^{(n)}\Gamma = \frac{VV_{n+2}}{V'^2}, \quad \ldots \tag{99}
$$

in which $V_n = \frac{d^nV}{d\phi^n}, n = 3, 4, 5, \cdots$. Then we can get

$$\frac{d\Gamma}{dN} = \sqrt{6}x\left[\Gamma\lambda + {}^{(1)}\Gamma - 2\lambda\Gamma^2\right], \tag{100}$$

$$\frac{d\left({}^{(1)}\Gamma\right)}{dN} = \sqrt{6}x\left[{}^{(1)}\Gamma\lambda + {}^{(2)}\Gamma - 2\lambda\Gamma\left({}^{(1)}\Gamma\right)\right], \tag{101}$$

$$\frac{d\left({}^{(2)}\Gamma\right)}{dN} = \sqrt{6}x\left[{}^{(2)}\Gamma\lambda + {}^{(3)}\Gamma - 2\lambda\Gamma\left({}^{(2)}\Gamma\right)\right], \tag{102}$$

$$\frac{d\left({}^{(3)}\Gamma\right)}{dN} = \sqrt{6}x\left[{}^{(3)}\Gamma\lambda + {}^{(4)}\Gamma - 2\lambda\Gamma\left({}^{(3)}\Gamma\right)\right], \tag{103}$$

$$\cdots\cdots$$

$$\frac{d\left({}^{(n)}\Gamma\right)}{dN} = \sqrt{6}x\left[{}^{(n)}\Gamma\lambda + {}^{(n+1)}\Gamma - 2\lambda\Gamma\left({}^{(n)}\Gamma\right)\right], \tag{104}$$

$$\cdots\cdots$$

To discuss the dynamical behavior of scalar field with more general potential, e.g., when neither λ nor Γ is constant, we need to consider a dynamical system described by Eqs. (87)-(90) coupled with Eqs. (100)-(104). It is easy to see that this dynamical system is also an autonomous one. We can discuss the values of the fixed points of this autonomous system. Considering Eq. (90), we can see that the values of fixed points should satisfy $x_c = 0$, $\lambda_c = 0$, or $\Gamma_c = 1$. Then, we can get the fixed points of this infinite-dimensional autonomous system.

1. If $x_c = 0$, considering Eqs. (87)-(89), one can get $(y_c, z_c, \lambda_c) = (0, 0, 0)$ or $(y_c, z_c, \lambda_c) = (0, 0, \lambda_*)$, and $\Gamma_c, {}^{(n)}\Gamma_c$ can be any values.
2. If $\lambda_c = 0$, considering Eqs. (87)-(89), one can see that the fixed points of (x, y, z) are $(x_c, y_c, z_c) = (0, y_*, 1 - 1/y_*^2)$and $(x_c, y_c, z_c) = (\pm 1, 0, 0)$. If $x_c = 0$, Γ_c and ${}^{(n)}\Gamma_c$ can be any values, and if $x_c = \pm 1$, ${}^{(n)}\Gamma_c = 0$.
3. If $\Gamma_c = 1$, considering Eqs. (87)-(89), one can get that the fixed points of (x, y, z, λ) are $(x_c, y_c, z_c, \lambda_c) = (0, 0, 0, \lambda_*)$ and $(x_c, y_c, z_c, \lambda_c) = (\pm 1, 0, 0, \lambda_*)$. And ${}^{(n)}\Gamma_c$ should satisfy ${}^{(n)}\Gamma_c = \lambda_*^n$. There are other fixed points, which will be discussed below.

Based on the above analysis and Tab. (1), one can find that points P_{1-10} are just special cases of the fixed points of an infinite-dimensional autonomous systems. Considering the definition of Γ (see Eq. (91)), the simplest potential is an exponential potential when $\Gamma_c = 1$. The properties of these fixed points have been discussed by many authors Billyard & Coley (2000); Copeland et al. (1998); Ferreira & Joyce (1998); Hao & Li (2003; 2004); Hoogen et al. (1999); Li & Hao (2004); Samart & Gumjudpai (2007); Yu & Wu (2008). If $x_c = 0$ and $y_c = 0$, this corresponds to a fluid-dominated universe, which we do not consider here. If $x_c = \pm 1$, $\Gamma_c = 0$ and ${}^{(n)}\Gamma_c = 0$, we do not need to consider the Γ and the ${}^{(n)}\Gamma$ terms. Then the stability properties of these points are the same as points $P_{4,5}$ in Tab. (1), and there are unstable points. The last case is $(x_c, y_c, z_c, \lambda_c) = (0, y_*, 1 - 1/y_*^2, 0)$ and $\Gamma, {}^{(n)}\Gamma$ can be any value. To analyze the dynamical properties of this autonomous system, we need to consider the ${}^{(n)}\Gamma_c$ terms. We will get an infinite series. In order to solve this infinite series, we must truncate it by setting a sufficiently high-order ${}^{(M)}\Gamma$ to be a constant, for a positive integer M, so that $d\left({}^{(M)}\Gamma\right)/dN = 0$. Thus we can get an $(M+4)$-dimensional autonomous

system. One example is the quadratic potential $V = \frac{1}{2}m^2\phi^2$ with some positive constant m that gives a 5-dimensional autonomous system, and another example is the Polynomial (concave) potential $V = M^{4-n}\phi^n$ Kallosh et al. (1991); Linde et al. (1991; 1994) that gives an $(n+3)$-dimensional autonomous system. Following the method we used in the two preceding subsections, we can get the dynamical behavior of such finite-dimensional systems.
In the rest of this subsection, we discuss whether this autonomous system has scaling solution. If $x_c = 0$, then $\Gamma_c \neq 0, {}^{(n)}\Gamma_c \neq 0$, and the stability of the fixed points may depend on the truncation. As an example, if we choose ${}^{(2)}\Gamma = 0$, then we can get a 6-dimensional autonomous system. The eigenvalues for the fixed point $(x_c, y_c, z_c, \lambda_c, \Gamma_c, {}^{(1)}\Gamma_c) = (0, 0, 0, \lambda_b, \Gamma_*, {}^{(1)}\Gamma_*)$, where $\lambda_b = 0$ or $\lambda_b = \lambda_*$, is

$$\mathbf{M}^T = (0, 0, 0, \frac{3}{2}\gamma, -3\gamma, -3 + \frac{3}{2}\gamma).$$

Obviously, this is an unstable point, and it has no scaling solution. The eigenvalues for the fixed point $(x_c, y_c, z_c, \lambda_c, \Gamma_c, {}^{(1)}\Gamma_c) = (0, 1, 0, 0, \Gamma_*, {}^{(1)}\Gamma_*)$ is

$$\mathbf{M}^T = (0, 0, 0, 0, -3\gamma, -3 - 3\gamma).$$

According to the center manifold theorem (see Chapter 8 of Khalil (1996), there are two non-zero eigenvalues, and we need to reduce the dynamical system to 2-dimensional to get the stability properties of the autonomous system. This point may have scaling solution, but we need more complex mathematical method. But it is easy to find that this point is scalar field dominated if it has a scaling solution.
We discuss the last case. If $\Gamma_c = 1$, we can consider an exponential potential. Then the autonomous system is reduced to 3-dimensional. It is easy to check that the values (x_{ec}, y_{ec}, z_{ec}) of the fixed points are just the values (x_c, y_c, z_c) of points P_{6-10} in Tab. (1). We focus on the two special fixed points:

$$F_1 : (x_{ec}, y_{ec}, z_{ec}) = (-\lambda/\sqrt{6}, \sqrt{1 - \lambda^2/6}, 0),$$
$$F_2 : (x_{ec}, y_{ec}, z_{ec}) = (-\sqrt{3/2}\gamma/\lambda, \sqrt{3\gamma(2-\gamma)/(2\lambda^2)}, 0).$$

Using Lyapunov's linearization method, we can find that F_2 is unstable and F_1 is stable if $\lambda < 3\gamma$. It is easy to check that $\rho_\gamma = H^2[1/(1 - z_{ec}) - x_{ec}^2 - y_{ec}^2] = 0$ when $(x_{ec}, y_{ec}, z_{ec}) = (-\lambda/\sqrt{6}, \sqrt{1 - \lambda^2/6}, 0)$. From the above analysis, we find that there is just the scalar-field-dominated scaling solution when we consider the autonomous system to be described by a self-interacting scalar field coupled with a barotropic fluid in the LQC scenario.

3.4 Conclusions

To discuss the dynamical properties of scalar field in the LQC scenario, we take Γ as a function of λ, and extend the autonomous system from 3-dimensional to 4-dimensional. We find this extended autonomous system has more fixed points than the 3-dimensional one does. And we find that for some fixed points, the function $f(\lambda)$ affects either their values, e.g., for points P_{4-10}, or their stability properties, e.g., for points $P_{3,9}$. In other words, the dynamical properties of these points depend on the specific form of the potential. But some other fixed points, e.g., points $P_{1,2}$,are independent of the potential. The properties of these fixed points are satisfied by all scalar fields. We also find that there are two later time attractors, but the universe is scalar-field-dominated since $\rho_\gamma = 0$ at these later time attractors.

The method developed by Fang et al. (2009); Zhou (2008) can describe the dynamical behavior of the scalar field with potential of a more general form than, for example, an exponential potential Billyard & Coley (2000); Copeland et al. (1998); Ferreira & Joyce (1998); Hao & Li (2003; 2004); Hoogen et al. (1999); Li & Hao (2004); Samart & Gumjudpai (2007); Yu & Wu (2008). But it is not all-encompassing. If one wants to discuss the dynamical properties of a scalar field with an arbitrary potential, one needs to consider the higher-order derivatives of the potential $V(\phi)$. Hence the dynamical system will extend from 4-dimensional to infinite-dimensional. This infinite-dimensional dynamical system is still autonomic, but it is impossible to get all of its dynamical behavior unless one considers $\Gamma_c = 1$ which just gives an exponential potential. If one wants to study as much as possible the dynamical properties of this infinite-dimensional autonomous system, one has to consider a truncation that sets ${}^{(M)}\Gamma = \text{Const.}$, with M above a certain positive integer. Then the infinite-dimensional system can be reduced to $(M+4)$-dimensional. And we find that there is just the scalar-field-dominated scaling solution for this autonomous system.
We only get the scalar-field-dominated scaling solutions, whether we consider Γ as a function of λ or consider the higher order derivatives of the potential. This conclusion is very different from the autonomous system which is just described by a scalar field with an exponential potential Samart & Gumjudpai (2007).

4. Averaged null energy condition

Wormholes and time machines are attractive objects in general relativity, always among top reasons that draw young minds to the study of this subject Morris & Thorn (1988), and they continue to be active research fields in general relativity Lobo (2007). The stress-energy tensor components of these exotic spacetime violate all known pointwise energy conditions, which is forbidden in classical general relativity. In contrast, the energy condition violation can be easily met in the semiclassical case because of quantum fluctuations Epstein et al. (1965); Klinkhanmmer (1991); Pitaevsky & Zeldovich (1971). For example, the Casimir vacuum for the electromagnetic field between two perfectly conducting plates has a negative local energy density Casimir (1948); squeezed states of light can result in negative energy densities. Based on semiclassical gravitational analysis, many self-consistent wormhole solutions have been found Barcelo & Visser (1999); Garattini (2005); Garattini & Lobo (2007); Hochberg et al. (1997); Khusnutdinov (2003); Sushkov (1992). On the other hand, the topological censorship theorem proved by Friedman, Schleich, and Witt Friedman et al. (1993) implies that the existence of macroscopic traversable wormholes requires the violation of the averaged null energy condition (ANEC). ANEC can be stated as

$$\int_\gamma T_{\mu\nu} k^\mu k^\nu dl \geq 0, \tag{105}$$

where the integral is along any complete, achronal null geodesic γ , k^μ denotes the geodesic tangent, and l is an affine parameter. Unfortunately the quantum effects in semiclassical gravitational analysis are always confined to an extremely thin band Roman (2004). So it seems impossible to find a macroscopic traversable wormhole based on semiclassical gravitational analysis.
As a quantum gravitational theory, loop quantum gravity (LQG) Ashtekar & Lewandowski (2004); Rovelli (1998); Smolin (2004); Thiemann (2007) is a non-perturbative and background-independent quantization of gravity. Physically, the Einstein equation in LQG is modified, while the stress-energy tensor $T_{\mu\nu}$ is unchanged. But mathematically, we can

move the terms modified by LQG to the side of the stress-energy tensor, and combine them with the stress-energy tensor to get an effective stress-energy tensor. This viewpoint allows to directly apply some previous analysis results in the considered situation. For instance, that proof of censorship theorem in Friedman et al. (1993) uses the Einstein equation extensively. But the proof places no restrictions on the form of matter. Instead, the geometric quantities are of paramount importance in the proof. Therefore with the concept of effective stress-energy tensor, the proof in Ref. Friedman et al. (1993) can be directly applied in the LQG-corrected spacetime.

The application of the techniques of LQG to the cosmological sector is known as loop quantum cosmology (LQC). Some of the main features of LQG such as discreteness of spatial geometry are inherited in LQC. A major success of LQC is that it resolves the problem of classical singularities both in an isotropic model Bojowald (2001) and in a less symmetric model Bojowald (2003), replacing the big bang spacetime singularity of cosmology with a big bounce. This bouncing scenario depends crucially on the discreteness of the theory. It has also been shown that non-perturbative modification of the matter Hamiltonian leads to a generic phase of inflation Bojowald (2002c); Date & Hossain (2005); Xiong & Zhu (2007a). These inflation models are built by taking only certain modification terms which affect the stress-energy tensor while ignoring the discretized geometry effect. But these modifications are also negligible for semiclassical gravitational theory. The effective stress-energy tensor in the inflation models has already been found to violate several kinds of energy conditions. For example, in loop quantum cosmology, non-perturbative modification to a scalar matter field at short scales induces a violation of the strong energy condition Xiong & Zhu (2007b). The ANEC is different from the other energy conditions such as the strong energy condition mentioned above in that the ANEC is an integral along any complete null-like geodesic, instead of being confined to the neighborhood of a certain point of the space-time. For a system without symmetry, it is a very complicated issue, making it almost impossible to calculate. But in the context of isotropic LQC, we can get an exact result, which can provide a hint for studying the wormholes in LQG and testing the validity of LQG. In this paper, we adopt effective method to study the quantum effect of effective stress-energy in loop quantum cosmology. From our calculation, we find that LQC does violate the averaged null energy condition in the massless scalar field coupled model.

This section is organized as follows. We introduce an exactly solvable model containing a massless scalar field in Subsec. 4.1. Then in Subsec. 4.2, we investigate the averaged null energy condition in this exactly solvable model. Finally, Subsec. 4.3 contains the discussion of our results and their implications. In this section we adopt $c = \hbar = G = 1$.

4.1 An exactly solvable model

The effective dynamics of LQC was formulated in Refs. Ashtekar et al. (2006a;b); Singh & Vandersloot (2005); Taveras (2008). Here we follow closely Mielczarek et al. (2008) to consider a universe containing a massless scalar field. Then the matter Hamiltonian in equation (58) can be written as

$$H_M(p,\phi) = \frac{1}{2}\frac{p_\phi^2}{p^{3/2}},$$

where p_ϕ is the conjugate momentum for the scalar field ϕ. The complete equations of motion for the universe containing a massless scalar field are

$$\begin{cases} \dot{c} = -\frac{1}{\gamma}\frac{\partial}{\partial p}\left(\sqrt{p}\left[\frac{\sin(\bar{\mu}c)}{\bar{\mu}}\right]^2\right) - \frac{\kappa\gamma}{4}\frac{p_\phi^2}{p^{5/2}}, \\ \dot{p} = \frac{2}{\gamma}\frac{\sqrt{p}}{\mu}\sin(\bar{\mu}c)\cos(\bar{\mu}c), \end{cases} \tag{106}$$

and

$$\begin{cases} \dot{\phi} = p^{-\frac{3}{2}}p_\phi, \\ \dot{p}_\phi = 0, \end{cases} \tag{107}$$

where $\kappa = 8\pi$. In addition, the Hamiltonian constraint $H_{eff} = 0$ becomes

$$\frac{3}{8\pi\gamma^2\bar{\mu}^2}\sqrt{p}\sin^2(\bar{\mu}c) = \frac{1}{2}\frac{p_\phi^2}{p^{3/2}}. \tag{108}$$

Combining equations (106) with (108), we obtain

$$\left(\frac{dp}{dt}\right)^2 = \Omega_1 p^{-1} - \Omega_3 p^{-4}, \tag{109}$$

with $\Omega_1 = \frac{2}{3}\kappa p_\phi^2$ and $\Omega_3 = \frac{1}{9}\kappa^2\gamma^2\xi^2 p_\phi^4$. Equation (107) implies that p_ϕ is a constant which characterizes the scalar field in the system. To solve equation (109) we introduce a new dependent variable u in the form

$$u = p^3. \tag{110}$$

With this new variable u, Eq.(109) becomes

$$\left(\frac{du}{dt}\right)^2 = 9\Omega_1 u - 9\Omega_3, \tag{111}$$

and has a solution

$$u = \frac{\Omega_3}{\Omega_1} + \frac{9}{4}\Omega_1 t^2 - \frac{9}{2}\Omega_1 C_1 t + \frac{9}{4}\Omega_1 C_1^2, \tag{112}$$

where C_1 is an integral constant. We can choose $C_1 = 0$ through coordinate freedom. Then the solution for p is

$$p = \left[\frac{\Omega_3}{\Omega_1} + \frac{9}{4}\Omega_1 t^2\right]^{1/3}. \tag{113}$$

4.2 The averaged null energy condition in LQC

Based on the above discussion, we calculate the averaged null energy condition in the context of LQC in this section. Because of the homogeneity of the universe, the null geodesic curves through different spatial points are the same. To investigate the ANEC, we only need to consider one of the null geodesic lines through any point in space. Due to the isotropy of the FRW metric, the null geodesic curves through the same point in different directions are also the same. Therefore, our problem is reduced to test any but one null geodesic line. Specifically, we consider a null geodesic line generated by vectors

$$\left(\frac{\partial}{\partial t}\right)^\mu + \frac{1}{a}\left(\frac{\partial}{\partial x}\right)^\mu.$$

According to the definition of affine parameter, $(\frac{\partial}{\partial l})^\mu \nabla_\mu (\frac{\partial}{\partial l})^\nu = 0$, we can reparameterize it with l to get

$$k^\mu = \left(\frac{\partial}{\partial l}\right)^\mu = \frac{1}{a}\left(\frac{\partial}{\partial t}\right)^\mu + \frac{1}{a^2}\left(\frac{\partial}{\partial x}\right)^\mu .$$

Then we can get the relationship between t and the affine parameter l,

$$t = \frac{l}{a}.$$

For the considered universe containing a massless scalar field, the energy density and the pressure of matter can be expressed as

$$\rho = \frac{1}{2}\dot{\phi}^2, \tag{114}$$

$$P = \frac{1}{2}\dot{\phi}^2, \tag{115}$$

according to the definition of the density and pressure. The effective energy density and pressure of matter take the forms

$$\rho_{eff} = \frac{1}{2}\dot{\phi}^2\left(1 - \frac{1}{2}\frac{\dot{\phi}^2}{\rho_c}\right), \tag{116}$$

$$P_{eff} = \frac{1}{2}\dot{\phi}^2\left(1 - \frac{\dot{\phi}^2}{\rho_c}\right) - \frac{1}{4}\frac{\dot{\phi}^4}{\rho_c}. \tag{117}$$

Since the effective stress-energy tensor takes an ideal fluid form,

$$\begin{aligned} T^{eff}_{\mu\nu} &= \rho_{eff}\,(dt)_\mu\,(dt)_\nu + a^2 P_{eff} \\ &\times \left[(dx)_\mu\,(dx)_\nu + (dy)_\mu\,(dy)_\nu + (dz)_\mu\,(dz)_\nu\right], \end{aligned} \tag{118}$$

the average null energy condition (105) for the effective stress-energy tensor becomes

$$\begin{aligned} \int_\gamma T^{eff}_{\mu\nu} k^\mu k^\nu dl &= \int_{-\infty}^{\infty} \frac{1}{a}\left(\rho_{eff} + P_{eff}\right) dt \\ &= \int_{-\infty}^{\infty} \frac{1}{a}\left(\dot{\phi}^2 - \frac{\dot{\phi}^4}{\rho_c}\right) dt \\ &= p_\phi^2 \int_{-\infty}^{\infty} (p^{-7/2} - p^{-13/2}\frac{p_\phi^2}{\rho_c})dt. \end{aligned}$$

In the last line we have used equation (107) and the relationship between p and a. Substituting the exact solution (113) into the above equation, we get

$$\int_\gamma T^{eff}_{\mu\nu} k^\mu k^\nu dl = -\frac{\Gamma(5/6)\Gamma(2/3)}{7\rho_c \Omega_I (\frac{\Omega_{III}}{\Omega_I})^{13/6}\sqrt{\frac{\pi}{\Omega_{III}}}} p_\phi^4, \tag{119}$$

where Γ is the gamma function. From the above result it is obvious that

$$\int_\gamma T^{eff}_{\mu\nu} k^\mu k^\nu dl < 0. \tag{120}$$

The above result shows that, in addition to the violation of some local energy conditions, the effective stress-energy tensor of loop quantum cosmology also violates the averaged null energy condition.

4.3 Conclusions and discussion

Given some kinds of local energy condition violation in loop quantum cosmology and motivated by the topological censorship theorem which rules out traversable wormholes in spacetime where the averaged null energy condition is satisfied, we investigate this kind of nonlocal energy condition in the context of LQC. Our analysis is based on a flat universe containing a massless scalar field. This model can be solved analytically. With the help of the analytical solution and taking advantage of the homogeneity and isotropy of the universe, we calculate the average of energy directly. Although the quantum correction is focused on the early universe around the Planck scale, the correction is so strong that it makes the universe violate the null averaged energy condition. Mathematically we have written the modified Einstein equation in LQC in the standard form but with effective stress-energy tensor. This form of equations allows to directly apply the original proof of Ref. Friedman et al. (1993) in the effective LQC. So the ANEC (for the original stress-energy tensor instead of the effective one) argument cannot forbid the existence of traversable wormhole once the Loop Quantum Gravity effects are taken into account. But we do not expect the existence of wormhole in LQC, because it is too symmetric to support wormholes. For less symmetric situations, the traversable wormholes might exist if quantum gravity effects make the effective stress-energy tensor violate the ANCE. On the other hand, LQC adopts the essence of LQG, so our result can shed some light on the ANEC of LQG. And we hope this result can give some hints on looking for wormhole solutions in the LQG theory. These interesting objects will provide some gedanken-experiments to test our quantum gravity theory.

5. Dynamical horizon entropy bound conjecture

The thermodynamical property of spacetime is an important hint for the quantization of gravity. Starting from Hawking's discovery of black hole's radiation Hawking (1975), a theory of thermodynamics of spacetime is being constructed gradually. Recently, the second law of this thermodynamics was generalized to the covariant entropy bound conjecture Bousso (1999). It states that the entropy flux S through any null hypersurface generated by geodesics with non-positive expansion, emanating orthogonally from a two-dimensional (2D) spacelike surface of area A, must satisfy

$$\frac{S}{A} \leq \frac{1}{4l_p^2}, \tag{121}$$

where $l_p = \sqrt{\hbar}$ is the Planck length. Here and in what follows, we adopt the units $c = G = k_B = 1$. Soon, Flanagan, Marolf and Wald Flanagan et al. (2000) proposed a new version of the entropy bound conjecture. If one allows the geodesics generating the null hypersurface from a 2D spacelike surface of area A to terminate at another 2D spacelike surface of area A' before coming to a caustic, boundary or singularity of spacetime, one can replace the above conjecture with

$$\frac{S}{A' - A} \leq \frac{1}{4l_p^2}. \tag{122}$$

More recently, He and Zhang related these conjectures to dynamical horizon and proposed a covariant entropy bound conjecture on the cosmological dynamical horizon He & Zhang (2007): Let $A(t)$ be the area of the cosmological dynamical horizon at cosmological time t, then the entropy flux S through the cosmological dynamical horizon between time t and t' ($t' > t$) must satisfy

$$\frac{S}{A(t') - A(t)} \leq \frac{1}{4l_p^2}, \tag{123}$$

if the dominant energy condition holds for matter.
Since it has been suggested that the holographic principle is a powerful hint and should be used as an essential building block for any quantum gravity theory Bousso (2002), it is important and tempting to investigate the covariant entropy bound conjecture in the framework of the LQC, which is a successful application of the non-perturbative quantum gravity scheme—the LQG. The authors of Ashtekar & Wilson-Ewing (2008) investigated the Bousso's covariant entropy bound Bousso (1999; 2002) with a cosmology filled with photon gas and found that the conjecture is violated near the big bang in the classical scenario. But they found the LQC can protect this conjecture even in the deep quantum region. In Zhang & Ling (2007), He and Zhang proposed a new version of the entropy bound conjecture for the dynamical horizon in cosmology and validated it through a cosmology filled with adiabatic perfect fluid, governed by the classical Einstein equation when the universe is far away from the big bang singularity. But when the universe approaches the big bang singularity, the strong quantum fluctuation does break down their conjecture. In analogy to Ashtekar and Wilson-Ewing's result Ashtekar & Wilson-Ewing (2008), one may wonder if He and Zhang's conjecture can also be protected by the quantum geometry effect of the LQG. Following Ashtekar & Wilson-Ewing (2008), we use photon gas as an example to investigate this problem. As expected, we find that the loop quantum effects can indeed protect the conjecture. Besides the result of Ashtekar & Wilson-Ewing (2008), our result presents one more evidence for the consistence between the loop gravity and the covariant entropy conjecture.
This section is organized as follows. In Subsec. 5.1, we describe the covariant entropy bound conjecture proposed by He and Zhang He & Zhang (2007). Then in Subsec. 5.2, we test this conjecture with cosmology filled with photon gas, and show that the LQC is able to protect the conjecture in all. We conclude the paper in Subsec. 5.3 and discuss the implications.

5.1 The covariant entropy bound conjecture

According to Zhang & Ling (2007), the cosmological dynamical horizon Bousso (2002) is defined geometrically as a three-dimensional hypersurface foliated by spheres, where at least one orthogonal null congruence with vanishing expansion exists. For a sphere characterized by any value of (t, r), there are two future directed null directions

$$k^a_\pm = \frac{1}{a}\left(\frac{\partial}{\partial t}\right)^a \pm \frac{1}{a^2}\left(\frac{\partial}{\partial r}\right)^a, \tag{124}$$

satisfying geodesic equation $k^b \nabla_b k^a = 0$. The expansion of these null directions is

$$\theta := \nabla_a k^a_\pm = \frac{2}{a^2}\left(\dot{a} \pm \frac{1}{r}\right), \tag{125}$$

where the dot denotes differential with respect to t, and the sign $+(-)$ represents the null direction pointing to larger (smaller) values of r. For an expanding universe, i.e. $\dot{a} > 0$, $\theta = 0$ determines the location of the dynamical horizon, $r_H = 1/\dot{a}$, by the definition of dynamical horizon given above. The LQC replaces the big bang with the big bounce, so the universe is symmetric with respect to the point of the bounce, expanding on one side of the bounce and contracting on the other side. The dynamical horizon in the contracting stage of the LQC corresponds to $r_H = -1/\dot{a}$, and all of the relations are similar to the ones given here. In this paper we only consider the expanding stage for the LQC, but note that the contracting stage is the same.

Since the area of the dynamical horizon is $A = 4\pi a^2 r_H^2 = 4\pi H^{-2}$, the covariant entropy bound conjecture in our question becomes

$$l_p^2 S \le \pi\left[H^{-2}(t') - H^{-2}(t)\right], \tag{126}$$

where S is the entropy flux through the dynamical horizon between cosmological time t and t' $(t' > t)$, and H is the Hubble parameter. Considering that the cosmology model discussed here is isotropic and homogeneous, we can write the entropy current vector as

$$s^a = \frac{s}{a^3}\left(\frac{\partial}{\partial t}\right)^a, \tag{127}$$

where s is the ordinary comoving entropy density, independent of space. If the entropy current of the perfect fluid is conserved, i.e., $\nabla_a s^a = 0$, s will be independent of t as well. For simplicity we restrict ourselves to this special case. The entropy flux through the dynamical horizon (shown in Fig.1) is given by

$$S = \int_{CDH} s^a \epsilon_{abcd} = \frac{4\pi s}{3}\left(r_H^3(t') - r_H^3(t)\right) \tag{128}$$

where $\epsilon_{abcd} = a^3 r^2 \sin\theta\,(dt \wedge dr \wedge d\theta \wedge d\phi)_{abcd}$ is the spacetime volume 4-form. So the conjecture is reduced to

$$H^{-2}(t') - \frac{4}{3} l_p^2 s \dot{a}^{-3}(t') \ge H^{-2}(t) - \frac{4}{3} l_p^2 s \dot{a}^{-3}(t), \;\; t' > t. \tag{129}$$

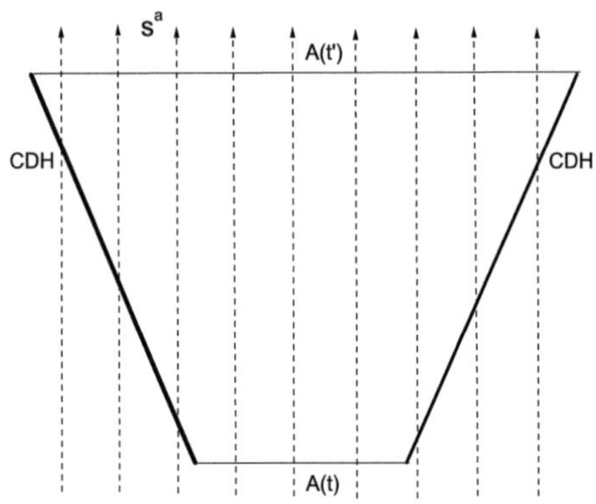

Fig. 1. A schematic of the entropy current flowing across the cosmological dynamical horizon. The thick solid line marked by "CDH" is the cosmological dynamical horizon. The thin solid line is the region enclosed by the CDH at time t and t' respectively. The dashed lines are the entropy current.

5.2 Conjecture test for a cosmology filled with perfect fluid

Given that the FRW universe is filled with photon gas, the energy momentum tensor can be expressed as

$$T_{ab} = \rho(t)(dt)_a(dt)_b + P(t)a^2(t)\left\{(dr)_a(dr)_b + r^2[(d\theta)_a(d\theta)_b + \sin^2\theta(d\phi)_a(d\phi)_b]\right\}. \qquad (130)$$

The pressure P and the energy density ρ satisfy a fixed equation of state

$$P = \omega\rho, \qquad (131)$$

where the constant $\omega = \frac{1}{3}$. From $\nabla^a T_{ab} = 0$, we have the conservation equation

$$\dot{\rho} + 3H(\rho + P) = 0. \qquad (132)$$

The comoving entropy density s is given by

$$s = a^3\frac{\rho + P}{T} = a^3(1+\omega)\frac{\rho}{T}, \qquad (133)$$

and ρ depends only on the temperature T,

$$\rho = K_o l_p^{-2-\frac{1+\omega}{\omega}} T^{\frac{1+\omega}{\omega}}, \qquad (134)$$

where K_o is a dimensionless constant depending on the density of energy state of the perfect fluid. For photon gas $K_o = \frac{\pi^2}{15}$. Plugging above thermodynamics relation into equation (133) we get $s = (1+\omega)K_o^{\frac{\omega}{1+\omega}} l_p^{-1-\frac{2\omega}{1+\omega}} \rho^{\frac{1}{1+\omega}} a^3$. Written the above conservation equation as

$$\dot{\rho} + 3(1+\omega)\rho\frac{\dot{a}}{a} = 0, \qquad (135)$$

we have an integration constant $C = \rho^{\frac{1}{1+\omega}} a^3$. Then $s = (1+\omega) K_o^{\frac{\omega}{1+\omega}} l_p^{-1-\frac{2\omega}{1+\omega}} C$. Combining our equation of state Eq. (131) with the above conservation equation, we get the relationship between ρ and the Hubble parameter,

$$H = -\frac{1}{3(1+\omega)} \frac{\dot{\rho}}{\rho}. \tag{136}$$

Substituting the above relation (136) into the modified Friedmann equation (67), we can get

$$\rho = \frac{1}{6\pi(t+C_1)^2(1+\omega)^2 + \frac{1}{\rho_c}} \tag{137}$$

where C_1 is an integration constant without direct physical significance, and we can always drop it by resetting the time coordinate. Setting $C_1 = 0$ gives

$$H = \frac{4\pi t\,(1+\omega)}{6\pi t^2\,(1+\omega)^2 + \frac{1}{\rho_c}}. \tag{138}$$

With the definition of the Hubble parameter, we can integrate once again to get

$$a(t) = C^{1/3}\left[6\pi t^2(1+\omega)^2 + \frac{1}{\rho_c}\right]^{\frac{1}{3(1+\omega)}}. \tag{139}$$

When ρ_c goes to infinity, all of the above solutions become the same as the classical ones [2] presented in Zhang & Ling (2007). In the classical scenario,

$$H^{-2} - \frac{4}{3} l_p^2 s \dot{a}^{-3} = \frac{9}{4} t^2 (1+\omega)^2 - \frac{9 K_o^{\frac{\omega}{1+\omega}} l_p^{1-\frac{2\omega}{1+\omega}}}{2(6\pi)^{1/(1+\omega)}} t^{3-\frac{2}{1+\omega}} (1+\omega)^{4-\frac{2}{1+\omega}}. \tag{140}$$

When $t \ll 1$, $H^{-2} - \frac{4}{3} l_p^2 s \dot{a}^{-3} \sim -t^{3-\frac{2}{1+\omega}} = -t^{3/2}$ which is a decreasing function of t, so the conjecture breaks down when the universe approaches the big bang singularity.
We introduce a new variable $\tau = \sqrt{2\pi\rho_c}(1+\omega)t$ for the LQC to simplify the above expressions to

$$H = \sqrt{2\pi\rho_c}\frac{2\tau}{3\tau^2+1}, \tag{141}$$

$$a = C^{1/3} \rho_c^{-\frac{1}{3(1+\omega)}} \left(3\tau^2+1\right)^{\frac{1}{3(1+\omega)}}, \tag{142}$$

$$\dot{a} = aH = 2\tau C^{1/3} \rho_c^{-\frac{1}{3(1+\omega)}} \sqrt{2\pi\rho_c} \left(3\tau^2+1\right)^{\frac{1}{3(1+\omega)}-1}. \tag{143}$$

[2] Note that the original result in Zhang & Ling (2007) used conformal time η, while we use universe time t in this paper. η can be negative which divides the discussion into two cases. t is always positive and makes the discussion simpler.

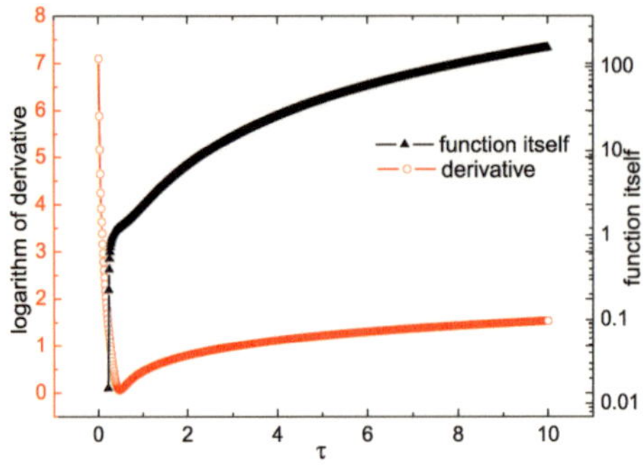

Fig. 2. Function $H^{-2} - \frac{4}{3}l_p^2 s\dot{a}^{-3}$ and its derivative respect to τ for photon gas.

Then

$$H^{-2} - \frac{4}{3}l_p^2 s\dot{a}^{-3}$$
$$= \frac{1}{2\pi\rho_c}\left[\left(\frac{3}{2}\tau + \frac{1}{2\tau}\right)^2 - \frac{(1+\omega)}{6\sqrt{2\pi}}K_o^{\frac{\omega}{1+\omega}}(\frac{\sqrt{3}}{16\pi^2\gamma^2})^{\frac{1}{1+\omega}-\frac{1}{2}}\frac{(3\tau^2+1)^{3-\frac{1}{(1+\omega)}}}{\tau^3}\right]. \tag{144}$$

It is obvious that the necessary and sufficient condition for meeting the covariant entropy bound conjecture is that the above expression increases with τ. In order to investigate the monotone property of above function, we plot $H^{-2} - \frac{4}{3}l_p^2 s\dot{a}^{-3}$ itself and its derivative respect to τ in Fig.2. The minimal value of the derivative is about $1.16 > 0$. The covariant entropy bound conjecture for dynamical horizon in cosmology is fully protected by loop quantum effect.

5.3 Conclusions and discussion

The covariant entropy bound conjecture comes from the holographic principle and is an important hint for the quantum gravity theory. In the recent years we have witnessed more and more success of the loop quantum gravity, especially for the problem of the big bang singularity in cosmology. The entropy bound conjecture usually breaks down in the strong gravity region of spacetime where the quantum fluctuation is strong, and one would expect the loop quantum correction to protect the conjecture from the quantum fluctuation. And Ashtekar and Wilson-Ewing do find a result in Ashtekar & Wilson-Ewing (2008) which is consistent with above expectation. In this paper, we generalized the covariant entropy conjecture for the cosmological dynamical horizon proposed in Zhang & Ling (2007) to the loop quantum cosmology scenario. We found that the quantum geometry effects of the loop quantum gravity can also protect the conjecture. Our result gives out one more evidence for the consistence of covariant entropy conjecture and loop quantum gravity theory. This adds one more encouraging result of loop quantum gravity theory besides previous ones.

6. Summary

Based on quantum geometry, the mathematical structure of LQC has been strictly defined. LQC inherits the non-perturbative and background-independent quantization schemes of LQG. In the framework of LQC, the evolution of the universe is described by the difference equation. In the past years, LQC achieved great successes. The most outstanding result is replacing the big bang space singularity of cosmology with a big bounce. LQC also gives a quantum suppression of classical chaotic behavior near singularities in the Bianchi-IX models. Furthermore, it has been shown that the non-perturbative modification of the matter Hamiltonian leads to a generic phase of inflation. LQC gives some possible answers to cosmological riddles due to the discreteness of the quantum geometry.

With the geometric quantum mechanics and the shadow state framework, we can get the effective theory of LQC. In this effective theory, the classical Hamiltonian gets a quantum correction, and the classical Einstein equation is replaced by the equations of motion induced by the effective Hamiltonian. Our works are based on this effective theory.

Due to the space limitations, this Chapter includes only our recently following three works: (1) We discussed the dynamical behavior of a scalar field with a general potential coupled with a barotropic fluid in LQC. (2) We found that the averaged null energy condition is violated in LQC which provides the possibility for the traversable time machine. (3) We found that the dynamical horizon entropy bound conjecture breaks down in classical general relativity near the big bang region but is protected by the quantum geometry effects in LQC.

It is undeniable that LQC is developing rapidly and showing its power in solving cosmological riddles. However, it still faces two major challenges:

- So far, we have not proven that LQC is truly derived from the full theory;
- At the time of writing, the predictions made by LQC are yet to be compared with cosmological observations.

Here involved is only a big branch of LQC. There are other issues worth studying, such as the perturbation theory, the cosmological spin foam theory, and so on. There is still a long way to go before the arrival of the victory.

Acknowledgments: This work was supported by the National Natural Science Foundation of China under Grant No. 10875012 and the Fundamental Research Funds for the Central Universities.

7. References

Artymowski, M., Lalak, Z. & Szulc, L. (2009). Loop Quantum Cosmology: holonomy corrections to inflationary models, Journal of Cosmology and Astroparticle Physics, Vol.2009(01): 04.

Ashtekar, A. & Tate, R. S. (1994). An algebraic extension of Dirac quantization: Examples, J. Math. Phys. Vol.35(12): 6434-6470.

Ashtekar, A., Lewandowski, J. & Marolf, D. (1995). Quantization of diffeomorphism invariant theories of connections with local degrees of freedom, J. Math. Phys. Vol.36(11): 6456-6493.

Ashtekar, A. & Lewandowski, J. (1995). Differential geometry on the space of connections via graphs and projective limits, J. Geom. Phys. Vol.17(3): 191-230.

Ashtekar A. & Lewandowski, J. (1997). Quantum theory of geometry: I. Area operator, Class. Quant. Grav. Vol.14(1997): A55-A81.

Ashtekar, A. & Lewandowski, J. (1998). Quantum Theory of Geometry II: Volume operators, Adv. Theor. Math. Phys. Vol.15(1998): 388-429.

Ashtekar, A., Baez, J., Corichi, A. & Krasnov, K. (1998). Quantum Geometry and Black Hole Entropy, Phys. Rev. Lett. Vol.80: 904-907.

Ashtekar, A., Corichi, A. & Krasnov, K. (1999). Isolated Horizons: the Classical Phase Space, Adv. Theor. Math. Phys. Vol.3(1999): 418-478.

Ashtekar, A., Baez, C. &Krasnov K., (2000). Quantum Geometry of Isolated Horizons and Black Hole Entropy, Adv. Theor. Math. Phys. Vol.4(1): 1-94.

Ashtekar, A., Beetle, C. & Lewandowski, J. (2001). Mechanics of rotating isolated horizons, Phys. Rev. D Vol.64: 044016.

Ashtekar, A., Beetle, C.&Lewandowski, J. (2002). Geometry of generic isolated horizons, Class. Quantum Grav. Vol.19(6): 1195-1225.

Ashtekar, A., Corichi, A. & Sudarski, D. (2003). Non-minimally coupled scalar fields and isolated horizon, Class. Quantum Grav. Vol.20(15): 3413-3425.

Ashtekar, A. & Corichi, A. (2003). Non-minimal couplings, quantum geometry and black-hole entropy, Class. Quantum Grav. Vol.20(20): 4473-4484.

Ashtekar, A. & Lewandowski, J. (2004). Background independent quantum gravity: a status report, Class. Quantum Grav. Vol.21(15): R53-R152.

Ashtekar, A., Pawlowski, T. & Singh, P. (2006). Quantum nature of the big bang: An analytical and numerical investigation, Phys. Rev. D Vol.73: 124038.

Ashtekar, A., Pawlowski, T. & Singh, P. (2006). Quantum nature of the big bang: Improved dynamics, Phys. Rev. D Vol.74: 084003.

Ashtekar, A., Corichi, A. & Singh, P. (2008). Robustness of key features of loop quantum cosmology, Phys. Rev. D Vol.77: 024046.

Ashtekar, A. & Wilson-Ewing, E. (2008). Covariant entropy bound and loop quantum cosmology, Phys. Rev. D Vol.78: 064047.

Ashtekar, A., Campiglia, M. & Henderson, A. (2009). Loop quantum cosmology and spin foams, Phys. Lett. B Vol.681(4),347-352.

Ashtekar, A., Campiglia, M. & Henderson, A. (2010). Path Integrals and the WKB approximation in Loop Quantum, Phys. Rev. D Vol.82: 124043.

Banerjee, K. & Date, G. (2005). Discreteness corrections to the effective Hamiltonian of isotropic loop quantum cosmology, Class. Quantum Grav., Vol.22(11): 2017-2033.

Barcelo, C. & Visser, M. (1999). Traversable wormholes from massless conformally coupled scalar fields, Phys. Lett. B Vol.466(1-4): 127-134.

Barreira, M., Carfora, M. &Rovelli, C. (1996). Physics with nonperturbative quantum gravity: Radiation from a quantum black hole, Gen. Rel. Grav. Vol.28(11): 1293-1299.

Billyard, A. P., Coley, A.A. & van den Hoogen, R.J. (1998). Stability of cosmological scaling solutions, Phys. Rev. D Vol.58: 123501.

Billyard, A. P. & Coley, A.A. (2000). Interactons in scalar field cosmology, Phys. Rev. D Vol.61: 083503.

Bojowald, M. (2001). Absence of a Singularity in Loop Quantum Cosmology, Phys. Rev. Lett. Vol.86: 5227-5230.

Bojowald, M. (2001). Inverse scale factor in isotropic quantum geometry, Phys. Rev. D Vol.64: 084018.

Bojowald, M. (2002). Isotropic loop quantum cosmology, Class. Quantum Grav., Vol.19(10): 2717-2742.

Bojowald, M. (2002). Quantization ambiguities in isotropic quantum geometry, Class. Quant. Grav., Vol.19(20): 5113-5130.
Bojowald, M. (2002). Inflation from Quantum Geometry, Phys. Rev. Lett. Vol.89: 261301.
Bojowald, M. & Vandersloot, K. (2003). Loop quantum cosmology, boundary proposals, and inflation, Phys. Rev. D Vol.67: 124023.
Bojowald, M. (2003). Homogeneous loop quantum cosmology, Class. Quantum Grav., Vol.20(13): 2595-2616.
Bojowald, M. & Date, G. (2004). Quantum Suppression of the Generic Chaotic Behavior Close to Cosmological Singularities, Phys. Rev. Lett. Vol.92: 071302 .
Bojowald, M. (2004). Loop Quantum Cosmology: Recent Progress, Pramana, Vol.63: 765-776.
Bojowald, M. (2004). The Bianchi IX model in loop quantum cosmology, Class. Quantum Grav. Vol.21(14): 3541-3570.
Bojowald, M., Lidsey, J. E., Mulryne, D. J., Singh, P. & Tavakol, R. (2004). Inflationary cosmology and quantization ambiguities in semiclassical loop quantum gravity, Phys. Rev. D Vol.70: 043530.
Bojowald, M. (2005). The early universe in loop quantum cosmology, J. Phys.: Conf. Ser., Vol.24(1): 77-86.
Bojowald, M. (2008). Loop Quantum Cosmology, Living Rev. Relativity, Vol.11: 4-131.
Bojowald, M. (2005). Original Questions, Nature, Vol.436: 920-921.
Bojowald, M., Kagan, M., Singh, P., Hernandez, H. & Skirzewski, A. (2006). Hamiltonian cosmological perturbation theory with loop quantum gravity corrections, Phys. Rev. D, Vol.74: 123512.
Bojowald, M. & Hossain, G. (2007). Cosmological vector modes and quantum gravity effect, Class. Quantum Grav., Vol.24(18): 4801-4816.
Bojowald, M. & Hossain, G. (2008). Loop quantum gravity corrections to gravitational wave dispersion, Phys. Rev. D Vol.77: 023508.
Bojowald, M., Hossain, G., Kagan, M., Mulryne, D., Nunes N. & Shankaranarayanan, S. (2009). Gauge invariant cosmological perturbation equations with corrections from loop quantum gravity, Phys. Rev. D Vol.79: 043505.
Bojowald, M. (2005). Loop Quantum Cosmology, Living Rev. Relativity, Vol.11: 4-131.
Bojowald, M. (2006). Loop quantum cosmology and inhomogeneities, Gen. Relativ. Gravit. Vol.38(11): 1771-1795.
Bojowald, M., Hector, H., Mikhail, K. & Aureliano, S. (2007). Effective constraints of loop quantum gravity, Phys. Rev. D Vol.75: 064022.
Bojowald, M. & Hossain, G. M. (2008). Loop quantum gravity corrections to gravitational wave dispersio, Phys. Rev. D Vol.77: 023508 .
Bousso, R. (1999). Holography in General Space-times, JHEP Vol.07(1999): 004.
Bousso, R. (2002). The holographic principle, Rev. Mod. Phys. Vol.74: 825-874.
Casimir, H. (1948). On the attraction between two perfectly conducting plates, Proc. K. Ned. Akad. Wet. B Vol.51: 793.
Chiou, Dah-Wei (2008). Effective equations of motion for constrained quantum systems: A study of the Bianchi I loop quantum cosmology, arXiv:0812.0921v1(gr-qc).
Chiou,Dah-Wei & Li, L.F. (2009). How loopy is the quantum bounce? A heuristic analysis of higher order holonomy corrections in loop quantum cosmology, Phys. Rev. D Vol.79: 063510.

Chiou,Dah-Wei & Li, L.F. (2009). Loop quantum cosmology with higher order holonomy corrections, Phys. Rev. D Vol.80: 043512.

Corichi, A. & Zapata, J. A. (1997). On diffeomorphism invariance for lattice theories, Nucl. Phys. B Vol.493(1-2): 475-490.

Copeland, E.J., Liddle, A.R., & Wands, D. (1998). Exponential potentials and cosmological scaling solutions, Phys. Rev. D Vol.57: 4686.

Copeland, E.J., Sami, M. & Tsujikawa, S. (2006). Dynamics of dark energy, Int. J. Mod. Phys. D Vol.15(11): 1753-1936.

Copeland, E.J., Mulryne, D.J., Nunes, N.J. & Shaeri, M. (2008). Superinflation in loop quantum cosmology, Phys. Rev. D Vol.77: 023510.

Copenland, E.J., Mizuno, S. & Shaeri, M. (2009). Dynamics of a scalar field in Robertson-Walker spacetimes, Phys. Rev. D Vol.79: 103515.

Coley, A.A. (2003). Dynamical systems and cosmology, Kluwer Academic Publishers, Dordrecht/Boston/London.

Date, G. (2002). Quantum Geometric Description of Cosmological Models, Mod. Phys. Lett. A, Vol.17(15-17): 967-976.

Date G. & Hossain, G. M. (2004). Effective Hamiltonian for isotropic loop quantum cosmology, Class. Quantum Grav., Vol.21(21): 4941-4954.

Date G. & Hossain, G. M. (2005). Genericness of Inflation in Isotropic Loop Quantum Cosmology, Phys. Rev. Lett. Vol.94: 011301.

Epstein, H., Glaser, V. & Jaffe, A. (1965). Nonpositivity of the energy density in quantized field theories, Nuovo Cimento Vol.36(3), 1016-1022.

Friedman, J., Schleich, K., & Witt, D. M. (1993). Topological censorship, Phys. Rev. Lett. Vol.71: 1486-1489.

Ferreira, P. G. & Joyce, M. (1998). Cosmology with a primordial scaling field, Phys. Rev. D Vol.58: 023503.

Flanagan, E., Marolf D. & Wald, R. (2000). Proof of classical versions of the Bousso entropy bound and of the generalized second law, Phys. Rev. D Vol.62: 084035.

Fu, X.Y., Yu, H.W. & Wu, P.X. (2008). Dynamics of interacting phantom scalar field dark energy in LQC, Phys. Rev. D Vol.78:063001.

Fang, W., Li, Y., Zhang, K. & Lu, H.Q. (2009). Exact analysis of scaling and dominant attractors beyond the exponential potential, Class. Quant. Grav. Vol.26(15): 155005.

Garattini, R. (2005). Self sustained traversable wormholes? Class. Quant. Grav. Vol.22(6): 1105-1117.

Garattini, R. & Lobo, F. (2007). Self-sustained phantom wormholes in semi-classical gravity, Class. Quant. Grav. Vol.24(9): 2401-2413.

Hao, J.G. & Li, X.Z. (2003). Attractor solution of phantom field, Phys. Rev. D Vol.67: 107303.

Hao, J.G. & Li, X.Z. (2004) Phantom cosmic dynamics: Tracking attractor and cosmic doomsday, Phys. Rev. D Vol.70: 043529.

Hawking, S. W. (1975). Particle creation by black holes, Commun. Math. Phys. Vol.43(3): 199-220.

He, S. & Zhang, H. (2007). The black hole dynamical horizon and generalized second law of thermodynamics, JHEP Vol.10(2007): 077.

Heard, I.P.C. & Wands, D. (2002). Cosmology with positive and negative exponential potentials, Classical Quantum Gravity, Vol.19(21): 5435-5447.

Hochberg, D., Popov, A. & Sushkov, S.V. (1997). Self-Consistent Wormhole Solutions of Semiclassical Gravity, Phys. Rev. Lett. Vol.78: 2050-2053.

van den Hoogen, R.J., Coley, A.A. & Wands, D. (1999). Scaling solutions in Robertson-Walker spacetimes, Class. Quantum Grav. Vol.16(6): 1843-1851.

Hossain, G. (2005). On energy conditions and stability in effective loop quantum cosmology, Classical Quantum Gravity, Vol.22(13): 2653-2670.

Hrycyna, O., Mielczarek, J. & Szydlowkski, M. (2009). Effects of the quantization ambiguities on the Big Bounce dynamics, Gen. Relativ. Gravit. Vol.41(5): 1025-1049.

Khalil, H.K. (1996). Nonlinear system, Englewood Cliffs, NJ: Prentice Hall.

Khusnutdinov, N. R. (2003). Semiclassical wormholes, Phys. Rev. D Vol.67: 124020.

Klinkhammer, G. (1991). Averaged energy conditions for free scalar fields in flat spacetime, Phys. Rev. D Vol.43: 2542-2548.

Lewandowski J. & Thiemann T. 1999. Diffeomorphism-invariant quantum field theories of connections in terms of webs, Class. Quantum Grav. Vol.16(7): 2299-2322.

Lewandowski, J. (1997). Volume and quantizations, Class. Quantum Grav. Vol.14(1): 71-76.

Leon, G., Silveira, P., & Fadragas, C.R. (2010). Phase-space of flat Friedmann-Robertson-Walker models with both a scalar field coupled to matter and radiation, arXiv: 1009.0689.

Li, X.Z. & Hao, J.G. (2004). Phantom field with O(N) symmetry in an exponential potential, Phys. Rev. D Vol.69: 107303.

Lidsey, J.E. (2004). Early universe dynamics in semi-classical loop quantum cosmology, J. Cosmol. Astropart.Phys. Vol.12(2004): 007.

Lidsey, J. E., Mulryne, D. J., Nunes, N. J. & Tavakol, R. (2004). Oscillatory universes in loop quantum cosmology and initial conditions for inflation, Phys. Rev. D, Vol.70: 063521.

Li, L.F.& Zhu, J.Y. (2009). Averaged null energy condition in loop quantum cosmology, Phys. Rev. D Vol.79: 044011.

Li, L.F.& Zhu, J.Y. (2010). Dynamical Horizon Entropy Bound Conjecture in Loop Quantum Cosmology, arXiv:1001.0288(gr-qc).

Lamon, R. & Woehr, A.J. (2010). Dynamical behavior of interacting dark energy in loop quantum cosmology, Phys.Rev.D Vol.81: 024026.

Li, S. & Ma, Y.G. (2010). Dark Energy Interacting with Dark Matter in Classical Einstein and Loop Quantum Cosmology, Eur. Phys. J. C Vol.68(1-2): 227-239.

Kallosh, R., Kratochvil, J., Linde, A., Linder E.V. & Shmakova, M.J. (2003). Observational bounds on cosmic doomsday, J. Cosmol. Astropart.Phys. Vol.10(2003): 015.

Linde, A.D. (1991). Axions in inflationary cosmology, Phys. Lett. B Vol.259(1-2): 38-47.

Linde, A.D. (1994). Hybrid inflation, Phys. Rev. D. Vol.49: 748-754.

F. Lobo, (2008). Exotic solutions in General Relativity: Traversable wormholes and 'warp drive' spacetimes, Classical and Quantum Gravity Research, PP.1-78, Nova Sci. Pub.

Loll, R. (1995). Volume Operator in Discretized Quantum Gravity, Phys. Rev. Lett. Vol.75: 3048-3051.

Loll, R. (1995). Spectrum of the volume operator in quantum gravity, Nucl. Phys. B Vol.460(1): 143-154.

Loll, R. (1997). Simplifying the spectral analysis of the volume operator, Nucl. Phys. B Vol.500(1-3): 405-420.

Loll, R. (1997). Further results on geometric operators in quantum gravity, Class. Quantum Grav. Vol.14(7): 1725-1741.

de la Macorra, A. & G. Piccinelli, (2000). Cosmological evolution of general scalar fields and quintessence, Phys. Rev. D Vol.61: 123503.
Mielczarek, J., Stachowiak, T. & Szydlowski, M. (2008). Exact solutions for a big bounce in loop quantum cosmology, Phys. Rev. D Vol.77: 123506.
Mielczarek J. & Szyd lowski, M. (2008). Emerging singularities in the bouncing loop cosmology, Phys. Rev. D Vol.77: 124008.
Morris, M. & Thorn, K. (1988). Wormholes in spacetime and their use for interstellar travel: A tool for teaching general relativity, Am. J. Phys. Vol.56(5): 395-412.
Mulryne, D.J., Tavakol, R., Lidsey, J. E. & Ellis, G. F. R. (2005). An emergent universe from a loop, Phys. Rev. D, Vol.71: 123512.
Mulryne, D.J., Nunes, N.J., Tavakol, R. & Lidsey, J. E. (2005). Inflationary cosmology and oscillating universes in loop quantum cosmology, Int. J. Mod. Phys. A Vol.20(11): 2347-2357.
Ng, S.C.C., Nunes, N.J. & Rosati, F. (2001). Applications of scalar attractor solutions to cosmology, Phys. Rev. D Vol.64: 083510.
Nunes, A. & Mimoso, J.P. (2000). On the potentials yielding cosmological scaling solutions, Phys. Lett. B Vol.488(3-7): 423-427.
Perez, A. (2006). Regularization ambiguities in loop quantum gravity, Phys. Rev. D, Vol.73: 044007.
Pitaevsky, L. & Zeldovich, Y. (1971). On the possibility of the creation of particles by a classical gravitational field, Commun. Math. Phys. Vol.23: 185-188.
Rovelli, C. & Smolin, L. (1994). The physical Hamiltonian in nonperturbative quantum gravity, Phys. Rev. Lett. Vol.72: 446-449.
Rovelli, C. & Smolin, L. (1995). Discreteness of area and volume in quantum gravity, Nucl. Phys. B Vol.442(3): 593-619.
Rovelli, C. (1996). Black Hole Entropy from Loop Quantum Gravity, Phys. Rev. Lett. Vol.77: 3288-3291.
Rovelli, C. (1996). Loop quantum gravity and black hole physics, Physics Helv. Phys. Acta. Vol.69: 582-611.
Rovelli, C. (1998). Loop Quantum Gravity, Living Rev. Relativity, Vol.1(1998): 1-75.
Rovelli, C. (2004). *Quantum Gravity*. Cambridge Universe Press, Cambridge.
Roman, T.A. (2004). Some Thoughts on Energy Conditions and Wormholes, arXiv:gr-qc/0409090.
Riess, A., et al., (2004). Type Ia Supernova Discoveries at $z > 1$ from the Hubble Space Telescope: Evidence for Past Deceleration and Constraints on Dark Energy Evolution, ApJ Vol.607(2):665-687.
Samart, D. & Gumjudpai, B. (2007). Phantom field dynamics in loop quantum cosmology, Phys. Rev. D Vol.76: 043514.
Smolin, L. (1995). Linking topological quantum field theory and nonperturbative quantum gravity, J. Math. Phys. Vol.36(11):6417-6455.
Smolin, L. (2004). An invitation to loop quantum gravity, arXiv:hep-th/0408048.
Singh, P. & Toporensky, A. (2004). Big crunch avoidance in k=1 semiclassical loop quantum cosmology, Phys. Rev. D Vol.69: 104008.
Vereshchagin, G.V. (2004).A qualitative approach to semi-classical loop quantum cosmology, JCAP Vol.07(2004): 13.

Singh, P. & Vandersloot, K. (2005). Semiclassical states, effective dynamics, and classical emergence in loop quantum cosmology, Phys. Rev. D Vol.72: 084004 .
Slotine, J.J.E. & Li ,W.P.(1991). Applied nonlinear control, Englewood Cliffs, NJ: Prentice-Hall.
Sushkov, S.V. (1992). A selfconsistent semiclassical solution with a throat in the theory of gravity, Phys. Lett. A Vol.164(1): 33-37.
Taveras, V. (2008). Corrections to the Friedmann equations from loop quantum gravity for a universe with a free scalar field, Phys. Rev. D, Vol.78: 064072.
Thiemann, T. (1996). Anomaly-free formulation of non-perturbative, four-dimensional Lorentzian quantum gravity, Phys. Lett. B Vol.380(3-4): 257-264.
Thiemann, T. (1998). Quantum spin dynamics (QSD), Class. Quantum Grav. Vol.15(4): 839-874.
Thiemann, T. (1998). Quantum spin dynamics (QSD): III. Quantum constraint algebra and physical scalar product in quantum general relativity, Class. Quantum Grav. Vol.15(5): 1207-1248.
Thiemann, T. (1998). Quantum spin dynamics (QSD): V. Quantum gravity as the natural regulator of the Hamiltonian constraint of matter quantum field theories, Class.Quantum Grav. Vol.15(5): 1281-1314.
Thiemann, T. (2001). Quantum spin dynamics (QSD): VII. Symplectic structures and continuum lattice formulations of gauge field theoriesčň Class. Quantum Grav. Vol.18(17): 3293-3338.
Thiemann, T. (1998). Closed formula for the matrix elements of the volume operator in canonical quantum gravity, J. Math. Phys. Vol.39(6):3347-3371.
Thiemann, T. (1998). A length operator for canonical quantum gravity, J. Math. Phys. Vol.39(6):3372-3392.
Thiemann, T. (2007). *Modern Canonical Quantum General Relativity*. Cambridge Universe Press, Cambridge.
Tsujikawa, S., Singh, P. & Maartens, R. (2004). Loop quantum gravity effects on inflation and the CMB, Class. Quantum Grav. Vol.21(24):5767-5776.
Vandersloot, K. (2005). Hamiltonian constraint of loop quantum cosmology, Phys. Rev. D, Vol.71: 103506.
Wei, H. & Zhang, S.N. (2007). Dynamics of quion and hessence energies in LQC, Phys. Rev. D Vol.76: 063005.
Xiao, K. & Zhu, J.Y. (2010). Dynamical behavior of interacting dark energy in loop quantum cosmology, Int. J. Mod. Phys. A Vol.25(26): 4993-5007.
Xiao, K. & Zhu, J.Y. (2010). Stability analysis of an autonomous system in loop quantum cosmology, Rhys. Rev. D Vol.83: 083501.
Xiong, H.H.& Zhu, J.Y. (2007). Tachyon field in loop quantum cosmology: Inflation and evolution picture, Phys. Rev. D Vol.75: 084023.
Xiong, H.H. & Zhu, J.Y. (2007). Violation of strong energy condition in effective loop quantum cosmology, Inter. J. Mod. Phys. A, Vol.22(18): 3137-3146.
Yang, J., Ding, Y. & Ma, Y. (2009). Alternative quantization of the Hamiltonian in loop quantum cosmology, Phys. Lett. B Vol.682(1):1-7.
Fu, X.Y., Yu, H.W. & Wu, P.X. (2008). Dynamics of interacting phantom scalar field dark energy in loop quantum cosmology, Phys. Rev. D Vol.78: 063001.
Zhang, X. & Ling, Y. (2007). Inflationary universe in loop quantum cosmology, JCAP Vol.8:(2007) 12.

Zhou, S.Y. (2008). A new approach to quintessence and a solution of multiple attractors, Phys. Lett. B Vol.660(1-2): 7-12.

17

Singularities and Thermodynamics of Geodesic Surface Congruences

Yong Seung Cho and Soon-Tae Hong
Ewha Womans University
Korea

1. Introduction

In this chapter, we study the singularity of the geodesic surface congruence for time-like and null strings using the expansion of the universe in the string theory. We have Raychaudhuri type equation for the expansion. Assuming the stringy strong energy condition and initial convergence, we induce the existence of a Hawking-Penrose type singularity and obtain the same inequality equation of the stringy strong energy condition for timelike and null stringy congruences.

We consider the variation of the surface spanned by closed strings in a space-time manifold. Using the Nambu-Goto string action, we induce the geodesic surface equation and the geodesic surface deviation equation which yields a Jacobi field, and we define the index form of a geodesic surface as in the case of point particles to discuss conjugate strings on the geodesic surface.

Using the Nambu-Goto string action in the space of the surfaces spanned by closed strings in a spacetime manifold, we investigated the geodesic surface equation in the space of surfaces joining two given strings and the geodesic surface deviation equation in geodesic surface congruence which yields a Jacobi field along a given geodesic surface, and singularities in geodesic surface congruences. Assuming that the singularity exists in geodesic surface congruences in a conformally symmetric manifold, we compute the Jacobi fields.

In the standard cosmology, the universe expanded from a single point, namely Big Bang, and then a radiation dominated phase occurred followed by a matter dominated one, together with a phase transition between these two phases. Stringy cosmology displays features that are different from standard cosmology. One may be surprised in that in the stringy scenario there is no phase transition between the radiation and matter dominated phases, and the universe repeats a cyclic pattern in its evolution similar to brane cyclic cosmology. The Large Hadron Collider at CERN is expected to be able to determine which cosmology among several scenarios including the stringy cosmology is viable through detecting the quark-gluon plasma state and its ensuing evolution of the tiny universe. This chapter is mainly based on the articles Cho & Hong (2007; 2008); Cho (2010); Cho & Hong (2010).

2. Morse theory in the space of surfaces

2.1 Stringy geodesic surfaces in Morse theory

In analogy of the relativistic action of a point particle, the Nambu-Goto action Goto (1971); Nambu (1970) for a closed string is proportional to the area of the surface spanned in spacetime manifold M by the evolution of the string. In order to define the action on the curved manifold, let (M, g_{ab}) be an n-dimensional manifold associated with the metric g_{ab}. Given g_{ab}, we can have a unique covariant derivative ∇_a satisfying Wald (1984) $\nabla_a g_{bc} = 0$, $\nabla_a \omega^b = \partial_a \omega^b + \Gamma^b{}_{ac}\, \omega^c$ and

$$(\nabla_a \nabla_b - \nabla_b \nabla_a)\omega_c = R_{abc}{}^d\, \omega_d. \tag{1}$$

We parameterize the surface $\gamma(\tau, \sigma)$ spanned by a closed string by two internal coordinates σ and τ, and then we have the corresponding vector fields $\xi^a = (\partial/\partial\tau)^a$ and $\zeta^a = (\partial/\partial\sigma)^a$. The Nambu-Goto string action is then given by Goto (1971); Nambu (1970)

$$S = -\int_0^T \int_0^{2\pi} d\tau d\sigma f(\tau, \sigma) \tag{2}$$

where

$$f(\tau, \sigma) = [(\xi \cdot \zeta)^2 - (\xi \cdot \xi)(\zeta \cdot \zeta)]^{1/2}. \tag{3}$$

We now perform an infinitesimal variation of the tubes $\gamma_\alpha(\tau, \sigma)$ traced by the closed string during its evolution in order to find the geodesic surface equation from the least action principle. Here we impose the restriction that the length of the string circumference is τ independent for simplicity. Let the vector field $\eta^a = (\partial/\partial\alpha)^a$ be the deviation vector which represents the displacement to an infinitesimally nearby tube, and let Σ denote the three-dimensional submanifold spanned by the tubes $\gamma_\alpha(\tau, \sigma)$. We then may choose τ, σ and α as coordinates of Σ to yield the commutator relations,

$$\pounds_\xi \eta^a = \xi^b \nabla_b \eta^a - \eta^b \nabla_b \xi^a = 0, \tag{4}$$

$$\pounds_\zeta \eta^a = \zeta^b \nabla_b \eta^a - \eta^b \nabla_b \zeta^a = 0, \tag{5}$$

$$\pounds_\xi \zeta^a = \xi^b \nabla_b \zeta^a - \zeta^b \nabla_b \xi^a = 0. \tag{6}$$

Now we find the first variation as follows

$$\begin{aligned} \frac{dS}{d\alpha} &= -\int\int d\tau d\sigma\ \eta^a \nabla_a f \\ &= -\int\int d\tau d\sigma\ (P^b_\tau \xi^a \nabla_a \eta_b + P^b_\sigma \zeta^a \nabla_a \eta_b) \\ &= \int\int d\tau d\sigma\ \eta_b (\xi^a \nabla_a P^b_\tau + \zeta^a \nabla_a P^b_\sigma) \\ &\quad - \int d\sigma\ P^b_\tau \eta_b |_{\tau=0}^{\tau=T} - \int d\tau\ P^b_\sigma \eta_b |_{\sigma=0}^{\sigma=2\pi}, \end{aligned} \tag{7}$$

where

$$\begin{aligned} P^a_\tau &= \frac{1}{f}[(\xi \cdot \zeta)\zeta^a - (\zeta \cdot \zeta)\xi^a], \\ P^a_\sigma &= \frac{1}{f}[(\xi \cdot \zeta)\xi^a - (\xi \cdot \xi)\zeta^a] \end{aligned} \tag{8}$$

are the energy-momentum currents. Using the endpoint conditions $\eta^a(0) = \eta^a(T) = 0$ and periodic condition $\eta^a(\sigma + 2\pi) = \eta^a(\sigma)$, we have the geodesic surface equation

$$\xi^a \nabla_a P^b_\tau + \zeta^a \nabla_a P^b_\sigma = 0, \tag{9}$$

and the constraint identities

$$\begin{aligned} P_\tau \cdot \zeta = 0,\ P_\tau \cdot P_\tau + \zeta \cdot \zeta = 0, \\ P_\sigma \cdot \xi = 0,\ P_\sigma \cdot P_\sigma + \xi \cdot \xi = 0. \end{aligned} \tag{10}$$

Let $\gamma_\alpha(\tau, \sigma)$ denote a smooth one-parameter family of geodesic surfaces: for each $\alpha \in \mathbf{R}$, the tube γ_α is a geodesic surface parameterized by affine parameters τ and σ. For an infinitesimally nearby geodesic surface in the family, we then have the following geodesic surface deviation equation

$$\xi^b \nabla_b(\eta^c \nabla_c P^a_\tau) + \zeta^b \nabla_b(\eta^c \nabla_c P^a_\sigma) + R_{bcd}{}^a(\xi^b P^d_\tau + \zeta^b P^d_\sigma)\eta^c \equiv (\Lambda\eta)^a = 0. \tag{11}$$

For an infinitesimal α, our goal is to investigate the variation vector field η^a by comparing $S(0)$ with $S(\alpha)$ with $S(0)$ along the time direction τ. The second variation $d^2S/d\alpha^2(0)$ is then needed only when $dS/d\alpha(0) = 0$. Explicitly, the second variation is given by

$$\begin{aligned} \frac{d^2S}{d\alpha^2}|_{\alpha=0} &= -\int\int d\tau d\sigma\ \eta^b \nabla_b(\eta^a \nabla_a f) \\ &= -\int\int d\tau d\sigma\ \Big[(\eta^c \nabla_c P^b_\tau)(\xi^a \nabla_a \eta_b) + (\eta^c \nabla_c P^b_\sigma)(\zeta^a \nabla_a \eta_b) \\ &\quad -R_{acb}{}^d(\xi^a P^b_\tau + \zeta^a P^b_\sigma)\eta^c \eta_d\Big] \\ &\quad -\int d\sigma\ P^b_\tau \eta^a \nabla_a \eta_b|^{\tau=T}_{\tau=0} - \int d\tau\ P^b_\sigma \eta^a \nabla_a \eta_b|^{\sigma=2\pi}_{\sigma=0}. \end{aligned} \tag{12}$$

Here the boundary terms vanish for the fixed endpoint and the periodic conditions, even though on the geodesic surface we have breaks which we will explain later. After some algebra using the geodesic surface deviation equation, we have

$$\frac{d^2S}{d\alpha^2}|_{\alpha=0} = \int\int d\tau d\sigma\ \eta_a(\Lambda\eta)^a. \tag{13}$$

2.2 Jacobi fields in orthonormal gauge

The string action and the corresponding equations of motion are invariant under reparameterization $\tilde{\sigma} = \tilde{\sigma}(\tau, \sigma)$ and $\tilde{\tau} = \tilde{\tau}(\tau, \sigma)$. We have then gauge degrees of freedom so that we can choose the orthonormal gauge as follows

$$\xi \cdot \zeta = 0, \quad \xi \cdot \xi + \zeta \cdot \zeta = 0, \tag{14}$$

where the plus sign in the second equation is due to the fact that $\xi \cdot \xi$ is timelike and $\zeta \cdot \zeta$ is spacelike. In this parameterization the energy-momentum currents (8) satisfying the constraints (10) are

$$P^a_\tau = -\xi^a, \quad P^a_\sigma = \zeta^a. \tag{15}$$

The geodesic surface equation and the geodesic surface deviation equation read

$$-\xi^a \nabla_a \xi^b + \zeta^a \nabla_a \zeta^b = 0,$$
$$-\xi^b \nabla_b (\xi^c \nabla_c \eta^a) + \zeta^b \nabla_b (\zeta^c \nabla_c \eta^a) - R_{bcd}{}^{a} (\xi^b \xi^d - \zeta^b \zeta^d) \eta^c = (\Lambda \eta)^a = 0. \quad (16)$$

We now restrict ourselves to strings on constant scalar curvature manifold such as S^n. We take an ansatz that on this manifold the string shape on the geodesic surface γ_0 is the same as that on a nearby geodesic surface γ_α at a given time τ. We can thus construct the variation vectors $\eta^a(\tau)$ as vectors associated with the centers of the string of the two nearby geodesic surfaces at the given time τ. We then introduce an orthonormal basis of spatial vectors e_i^a ($i = 1, 2, ..., n-2$) orthogonal to ξ^a and ζ^a and parallelly propagated along the geodesic surface. The geodesic surface deviation equation then yields for ($i, j = 1, 2, ..., n-2$)

$$\frac{d^2 \eta^i}{d\tau^2} + (R_{\tau j \tau}{}^{i} - R_{\sigma j \sigma}{}^{i}) \eta^j = 0. \quad (17)$$

The value of η^i at time τ must depend linearly on the initial data $\eta^i(0)$ and $\frac{d\eta^i}{d\tau}(0)$ at $\tau = 0$. Since by construction $\eta^i(0) = 0$ for the family of geodesic surfaces we must have

$$\eta^i(\tau) = A^i{}_j(\tau) \frac{d\eta^j}{d\tau}(0). \quad (18)$$

Inserting (18) into (17) we have the differential equation for $A^i{}_j(\tau)$

$$\frac{d^2 A^i{}_j}{d\tau^2} + (R_{\tau k \tau}{}^{i} - R_{\sigma k \sigma}{}^{i}) A^k{}_j = 0, \quad (19)$$

with the initial conditions

$$A^i{}_j(0) = 0, \quad \frac{dA^i{}_j}{d\tau}(0) = \delta^i{}_j. \quad (20)$$

Note that in (19) we have the last term from the contribution of string property. Next we consider the second variation equation (12) under the above restrictions

$$\frac{d^2 S}{d\alpha^2}|_{\alpha=0} = \int\int d\tau d\sigma \left(\frac{d\eta^i}{d\tau} \frac{\eta_i}{d\tau} - (R_{\tau j \tau}{}^{i} - R_{\sigma j \sigma}{}^{i}) \eta^j \eta_i \right). \quad (21)$$

Define the index form I_γ of a geodesic surface γ as the unique symmetric bilinear form $I_\gamma : T_\gamma \times T_\gamma \to \mathbf{R}$ such that

$$I_\gamma(V, V) = \frac{d^2 S}{d\alpha^2}|_{\alpha=0}(V, V) \quad (22)$$

for $V \in T_\gamma$. From (21) we can easily find

$$I_\gamma(V, W) = \int\int d\tau d\sigma \left(\frac{dW^m}{d\tau} \frac{dV_m}{d\tau} - (R_{\tau j \tau}{}^{m} - R_{\sigma j \sigma}{}^{m}) W^j V_m \right). \quad (23)$$

If we have breaks $0 = \tau_0 < \cdots < \tau_{k+1} = T$, and the restriction of γ to each set $[\tau_{i-1}, \tau_i]$ is smooth, then the tube γ is piecewise smooth. The variation vector field V of γ is always piecewise smooth. However $dV/d\tau$ will generally have a discontinuity at each break τ_i $(1 \leq i \leq k)$. This continuity is measured by

$$\Delta \frac{dV}{d\tau}(\tau_i) = \frac{dV}{d\tau}(\tau_i^+) - \frac{dV}{d\tau}(\tau_i^-), \tag{24}$$

where the first term is derived from the restrictions $\gamma|[\tau_i, \tau_{i+1}]$ and the second from $\gamma|[\tau_{i-1}, \tau_i]$. If γ and $V \in T_\gamma$ have the breaks $\tau_1 < \cdots < \tau_k$ then except at breaks, we have

$$\frac{dW^m}{d\tau}\frac{dV_m}{d\tau} = \frac{d}{d\tau}\left(V_m \frac{dW^m}{d\tau}\right) - V_m \frac{d^2W^m}{d\tau^2}, \tag{25}$$

and

$$\sum_{i=0}^{k} \int_{\tau_i}^{\tau_{i+1}} \frac{d}{d\tau}\left(V_m \frac{dW^m}{d\tau}\right) d\tau = \sum_{i=0}^{k} V_m \frac{dW^m}{d\tau}\Big|_{\tau_i}^{\tau_{i+1}} = -\sum_{i=0}^{k} V_m \Delta \frac{dW^m}{d\tau}(\tau_i), \tag{26}$$

to yield

$$I_\gamma(V, W) = -\int\int d\tau d\sigma\, V^m \left(\frac{d^2W^m}{d\tau^2} + (R_{\tau j \tau}{}^{m} - R_{\sigma j \sigma}{}^{m})W^j\right) - \sum_{i=0}^{k} \int d\sigma\, V_m \Delta \frac{dW^m}{d\tau}(\tau_i). \tag{27}$$

Here note that if we do not have the breaks, (21) yields

$$\frac{d^2S}{d\alpha^2}\Big|_{\alpha=0} = -\int\int d\tau d\sigma\, \eta_i \left(\frac{d^2\eta^i}{d\tau^2} + (R_{\tau j \tau}{}^{i} - R_{\sigma j \sigma}{}^{i})\eta^j\right). \tag{28}$$

A solution η^a of the geodesic surface deviation equation (17) is called a Jacobi field on the geodesic surface γ. A pair of points $p, q \in \gamma$ defined by the centers of the closed strings on the geodesic surface is then conjugate if there exists a Jacobi field η^a which is not identically zero but vanishes at both p and q. Roughly speaking, p and q are conjugate if an infinitesimally nearby geodesic surface intersects γ at both p and q. From (18), q will be conjugate to p if and only if there exists nontrivial initial data: $d\eta^i/d\tau(0) \neq 0$, for which $\eta^i = 0$ at q. This occurs if and only if $\det A^i{}_j = 0$ at q, and thus $\det A^i{}_j = 0$ is the necessary and sufficient condition for a conjugate point to p. Note that between conjugate points, we have $\det A^i{}_j \neq 0$ and thus the inverse of $A^i{}_j$ exists. Using (19) we can easily see that

$$\frac{d}{d\tau}\left(\frac{dA_{ij}}{d\tau}A^i{}_k - A_{ij}\frac{dA^i{}_k}{d\tau}\right) = 0. \tag{29}$$

In addition, the quantity in parenthesis of (29) vanishes at p, since $A^i{}_j(0) = 0$. Along a geodesic surface γ, we thus find

$$\frac{dA_{ij}}{d\tau}A^i{}_k - A_{ij}\frac{dA^i{}_k}{d\tau} = 0. \tag{30}$$

If γ is a geodesic surface with no point conjugate to p between p and q, then $A^i{}_j$ defined above will be nonsingular between p and q. We can then define $Y^i = (A^{-1})^i{}_j\eta^j$ or $\eta^i = A^i{}_jY^j$. From (28) and (30), we can easily verify

$$\frac{d^2S}{d\alpha^2}|_{\alpha=0} = \int\int d\tau d\sigma \left(A_{ij}\frac{dY^j}{d\tau}\right)^2 \geq 0. \tag{31}$$

Locally γ minimizes the surface of the string, if γ is a geodesic surface with no point conjugate to p between p and q.

On the other hand, if γ is a geodesic surface but has a conjugate point r between p and q, then we have the Jacobi field J^i along γ which vanishes at p and r. Extend J^i to q by putting it zero in $[r, q]$. Then $dJ^i/d\tau(r^-) \neq 0$, since J^i is nonzero. But $dJ^i/d\tau(r^+) = 0$ to yield

$$\Delta\frac{dJ^i}{d\tau}(r) = -\frac{dJ^i}{d\tau}(r^-) \neq 0. \tag{32}$$

Choose any $k^i \in T_\gamma$ such that

$$k_i\Delta\frac{dJ^i}{d\tau}(r) = c, \tag{33}$$

with positive constant c. Let η^i be $\eta^i = \epsilon k^i + \epsilon^{-1}J^i$ where ϵ is some constant, then we have

$$I_\gamma(\eta,\eta) = \epsilon^2 I_\gamma(k,k) + 2I_\gamma(k,J) + \epsilon^{-2}I_\gamma(J,J). \tag{34}$$

By taking ϵ small enough, the first term in (34) vanishes and the third term also vanishes due to the definition of the Jacobi field. Using (33) we have $I_\gamma(k,J) = -2\pi c$ and thus

$$\frac{d^2S}{d\alpha^2}|_{\alpha=0} = -4\pi c, \tag{35}$$

which is negative definite. From the above arguments, we conclude that given a smooth timelike curve γ connecting two points $p, q \in M$, the necessary and sufficient condition that γ locally minimizes the surface of the closed string between p and q over smooth one parameter variations is that γ is a geodesic surface with no point conjugate to p between p and q. It is also interesting to see that on S^n we have $n-1$ conjugate points as in the case of point particle. Moreover, on the manifold with the constant scalar curvature R, the geodesic surfaces have no conjugate points for $R < 0$ or $R = 0$, while conjugate points occur for $R > 0$.

3. Singularities in geodesic surface congruence for the time-like direction

3.1 Introduction

The Hawking-Penrose singularity Hawking & Penrose (1970) is assumed to exist at the beginning of the universe. In the standard inflationary cosmology based on the Hawking-Penrose singularity theorem and inflationary scenario, the universe is believed to expand from the Big Bang. Assuming that the early universe was filled with a perfect fluid consisting of massive particles and/or massless particles and using the strong energy condition which was used to show the Hawking-Penrose singularity theorem, one could find equations of state for each particle.

In the inflationary standard cosmology, it is believed that, after the Big Bang explosion, radiation dominated phase occurred followed by matter dominated one, even though there was a hot thermalization period of radiation and matter immediately after the Big Bang. Moreover, a phase transition exists between massive particle and massless particle phases in the universe. The equation of state of the massive particle is different from that of the massless particle, and thus the massive particle phase is not the same as the massless particle one.

Applying the string theory Green et al. (1987); Polchinski (1999) to cosmology, we might investigate the expansion of the universe in terms of the Hawking-Penrose singularity in geodesic surface congruences for the time-like and null strings Cho & Hong (2008; 2010). Taking an *ansatz* that the expansion of the stringy congruence is constant along the string coordinate direction, we derive the Raychaudhuri type equation, which is an evolution equation for the expansion, possessing correction terms associated with the stringy configurations. Assuming the stringy strong energy condition, we induce the Hawking-Penrose type inequality equation which produces the same inequality equation for both the time-like and null stringy congruences.

3.2 Congruence of strings

The action for a string is proportional to the area of the surface spanned in spacetime manifold M by the evolution of the string. In order to define the action on the curved manifold, we let (M, g_{ab}) be a D-dimensional manifold associated with the metric g_{ab}. Given g_{ab}, we can have a unique covariant derivative ∇_a satisfying (1).

Parameterize the surface generated by the evolution of a string by two world sheet coordinates τ and σ, and then we have the corresponding vector fields $\xi^a = (\partial/\partial\tau)^a$ and $\zeta^a = (\partial/\partial\sigma)^a$. Since we have gauge degrees of freedom, we can choose the orthonormal gauge as follows Scherk (1975)

$$\xi \cdot \zeta = 0, \quad \xi \cdot \xi + \zeta \cdot \zeta = 0, \tag{36}$$

where the plus sign in the second equation is due to the fact that $\xi \cdot \xi = -1$ is timelike and $\zeta \cdot \zeta = 1$ is spacelike. In the orthonormal gauge, we introduce tensor fields B_{ab} and $\bar{B}_{ab}$ defined as

$$B_{ab} = \nabla_b \xi_a, \quad \bar{B}_{ab} = \nabla_b \zeta_a, \tag{37}$$

which satisfy the following identities

$$\begin{aligned} B_{ab}\xi^a &= 0, \quad \bar{B}_{ab}\zeta^a = 0, \\ -B_{ab}\xi^b + \bar{B}_{ab}\zeta^b &= 0. \end{aligned} \tag{38}$$

Here in the last equation, we have used the geodesic surface equation

$$-\xi^a \nabla_a \xi^b + \zeta^a \nabla_a \zeta^b = 0. \tag{39}$$

In particular, the timelike curves of the strings are geodesic, then the geodesic surface equation holds.
Let the vector field $\eta^a = (\partial/\partial\alpha)^a$ be the deviation vector which represents the displacement to an infinitesimally nearby world sheet, and we let Σ denote the three-dimensional submanifold spanned by the world sheets $\gamma_\alpha(\tau,\sigma)$. We then may choose τ, σ and α as coordinates of Σ to yield the commutator relations (4)-(6). Using the above relations, we have

$$\xi^a \nabla_a \eta^b - \zeta^a \nabla_a \eta^b = (B^b{}_a - \bar{B}^b{}_a)\eta^a. \tag{40}$$

Next introduce the metrics h_{ab} and $\bar{h}_{ab}$,

$$h_{ab} = g_{ab} + \xi_a \xi_b, \quad \bar{h}_{ab} = g_{ab} - \zeta_a \zeta_b, \tag{41}$$

which satisfy

$$\begin{array}{lll} h_{ab}\xi^a = 0, & h_{ab}\xi^b = 0, & h_{ab}g^{bc}h_{cd} = h_{ad}, \\ \bar{h}_{ab}\zeta^a = 0, & \bar{h}_{ab}\zeta^b = 0, & \bar{h}_{ab}g^{bc}\bar{h}_{cd} = \bar{h}_{ad}, \\ h_{ab}h^{ab} = D-1, & \bar{h}_{ab}\bar{h}^{ab} = D-1, & h_{ab}\bar{h}^{ab} = D-2. \end{array} \tag{42}$$

Here note that h_{ab} and $\bar{h}_{ab}$ are the metrics on the hypersurfaces orthogonal to ξ^a and ζ^a, respectively. Moreover, we can define projection operators $h^a{}_b$ and $\bar{h}^a{}_b$ as follows

$$h^a{}_b = g^{ac}h_{cb}, \quad \bar{h}^a{}_b = g^{ac}\bar{h}_{cb}. \tag{43}$$

These operators fulfil

$$\begin{array}{ll} h^a{}_b h^b{}_c = h^{ab}h_{bc} = h^a{}_c, & \bar{h}^a{}_b \bar{h}^b{}_c = \bar{h}^{ab}\bar{h}_{bc} = \bar{h}^a{}_c, \\ h_{ab}h^{bc}h_{cd} = h_{ad}, & \bar{h}_{ab}\bar{h}^{bc}\bar{h}_{cd} = \bar{h}_{ad}. \end{array} \tag{44}$$

Now, decompose B_{ab} into three pieces

$$B_{ab} = \frac{1}{D-1}\theta h_{ab} + \sigma_{ab} + \omega_{ab}, \tag{45}$$

where the expansion θ, the shear σ_{ab} and the twist ω_{ab} of the stringy congruence are given by

$$\theta = B^{ab}h_{ab}, \quad \sigma_{ab} = B_{(ab)} - \frac{1}{D-1}\theta h_{ab}, \quad \omega_{ab} = B_{[ab]}. \tag{46}$$

Similarly, $\bar{B}_{ab}$ is also decomposed into three parts

$$\bar{B}_{ab} = \frac{1}{D-1}\bar{\theta}\bar{h}_{ab} + \bar{\sigma}_{ab} + \bar{\omega}_{ab}, \tag{47}$$

where

$$\bar{\theta} = \bar{B}^{ab}\bar{h}_{ab}, \quad \bar{\sigma}_{ab} = \bar{B}_{(ab)} - \frac{1}{D-1}\bar{\theta}\bar{h}_{ab}, \quad \bar{\omega}_{ab} = \bar{B}_{[ab]}. \tag{48}$$

We then find

$$\begin{aligned} \sigma_{ab}h^{ab} &= 0, & \omega_{ab}h^{ab} &= 0, \\ \bar{\sigma}_{ab}\bar{h}^{ab} &= 0, & \bar{\omega}_{ab}\bar{h}^{ab} &= 0, \\ -\sigma_{ab}\xi^{b} + \bar{\sigma}_{ab}\zeta^{b} &= 0, & -\omega_{ab}\xi^{b} + \bar{\omega}_{ab}\zeta^{b} &= 0, \end{aligned} \tag{49}$$

and

$$-\xi^{c}\nabla_{c}B_{ab} + \zeta^{c}\nabla_{c}\bar{B}_{ab} = B^{c}{}_{b}B_{ac} - \bar{B}^{c}{}_{b}\bar{B}_{ac} - R_{cbad}(\xi^{c}\xi^{d} - \zeta^{c}\zeta^{d}). \tag{50}$$

Exploiting (50) one arrives at

$$\begin{aligned} -\xi^{a}\nabla_{a}\theta + \zeta^{a}\nabla_{a}\bar{\theta} = &\frac{1}{D-1}(\theta^{2} - \bar{\theta}^{2}) + \sigma_{ab}\sigma^{ab} - \bar{\sigma}_{ab}\bar{\sigma}^{ab} - \omega_{ab}\omega^{ab} + \bar{\omega}_{ab}\bar{\omega}^{ab} \\ &+ R_{ab}(\xi^{a}\xi^{b} - \zeta^{a}\zeta^{b}). \end{aligned} \tag{51}$$

3.3 Expansion of timelike stringy congruence

The motion types of stringy congruence can be described in terms of expansion, twist and shear. In this section, we will pedagogically summarize the previous results Cho & Hong (2008; 2010) on the expansion rate of stringy congruence in the early universe for the sake of completeness. We will consider the twist and shear motions in the next section.

Taking an *ansatz* that the expansion $\bar{\theta}$ is constant along the σ-direction, from (51) one obtains a Raychaudhuri type equation

$$\frac{d\theta}{d\tau} = -\frac{1}{D-1}(\theta^{2} - \bar{\theta}^{2}) - \sigma_{ab}\sigma^{ab} + \bar{\sigma}_{ab}\bar{\sigma}^{ab} + \omega_{ab}\omega^{ab} - \bar{\omega}_{ab}\bar{\omega}^{ab} - R_{ab}(\xi^{a}\xi^{b} - \zeta^{a}\zeta^{b}). \tag{52}$$

Assume that $\omega_{ab} = \bar{\omega}_{ab}$, $\sigma_{ab} = \bar{\sigma}_{ab}$ and a stringy strong energy condition

$$R_{ab}(\xi^{a}\xi^{b} - \zeta^{a}\zeta^{b}) = 8\pi\left(T_{ab}(\xi^{a}\xi^{b} - \zeta^{a}\zeta^{b}) + \frac{2}{D-2}T\right) \geq 0, \tag{53}$$

where T_{ab} and T are the energy-momentum tensor and its trace, respectively. The Raychaudhuri type equation (52) then has a solution of the form

$$\frac{1}{\theta(\tau)} \geq \frac{1}{\theta(0)} + \frac{1}{D-1}\left(\tau - \int_{0}^{\tau} d\tau \left(\frac{\bar{\theta}}{\theta}\right)^{2}\right). \tag{54}$$

Assume that $\theta(0)$ is negative so that the congruence is initially converging as in the point particle case. The inequality (54) implies that $\theta(\tau)$ must pass through the singularity within a proper time

$$\tau \leq \frac{D-1}{|\theta(0)|} + \int_{0}^{\tau} d\tau \left(\frac{\bar{\theta}}{\theta}\right)^{2}. \tag{55}$$

For a perfect fluid, the energy-momentum tensor given by

$$T_{ab} = \rho\, u_{a}u_{a} + P\,(g_{ab} + u_{a}u_{b}) \tag{56}$$

where ρ and P are the mass-energy density and pressure of the fluid as measured in its rest frame, respectively, and u^{a} is the time-like D-velocity in its rest frame Misner et al. (1972);

Wald (1984), the stringy strong energy condition (53) yields only one inequality equation

$$\frac{D-4}{D-2}\rho+\frac{D}{D-2}P\geq 0. \tag{57}$$

Now, we consider the point particle limit of the timelike stringy congruence. If the fiber space F in the fibration $\pi : M \rightarrow N_4$ is a point, then the total space M is the same as the base spacetime four manifold N_4. In this case, the geodesic surfaces are geodesic in N_4, the congruence of time-like geodesic surfaces is a congruence of time-like geodesics, and so $\bar{B}_{ab} = \bar{\theta} = \bar{\sigma}_{ab} = \bar{\omega}_{ab} = 0$. If the congruence is hypersurface orthogonal, then we have $\omega_{ab} = 0$. Suppose that the strong energy condition $R_{ab}\xi^a\xi^b \geq 0$ is satisfied to yield two inequalities Carroll (2004); Hawking & Penrose (1970); Wald (1984)

$$\rho+3P\geq 0, \quad \rho+P\geq 0. \tag{58}$$

We then have the differential inequality equation

$$\frac{d\theta}{d\tau}+\frac{1}{3}\theta^2\leq 0 \tag{59}$$

which has a solution in the following form

$$\frac{1}{\theta(\tau)}\geq\frac{1}{\theta(0)}+\frac{1}{3}\tau. \tag{60}$$

If we assume that $\theta(0)$ is negative, the expansion $\theta(\tau)$ must go to the negative infinity along that geodesic within a proper time

$$\tau\leq\frac{3}{|\theta(0)|}, \tag{61}$$

whose consequence coincides with that of Hawking and Penrose Hawking & Penrose (1970).

4. Singularities in geodesic surface congruence for the null direction

4.1 Formalism of null stringy congruence

Next, we investigate the congruence of the null strings, where the tangent vector of a null curve is normal to itself. See Karlhede & Lindström (1986); Roshchupkin & Zheltukhin (1995; 1999); Schild (1977) for the proper definition and propagation of the classical null strings. We consider the evolution of vectors in a $(D-2)$-dimensional subspace of spatial vectors normal to the null tangent vector field $k^a = (\partial/\partial\lambda)^a$, where λ is the affine parameter, and to an auxiliary null vector l^a which points in the opposite spatial direction to k^a, normalized by $l^a k_a = -1$ Carroll (2004) and is parallel transported, namely, $k^a\nabla_a l^b = 0$. The spatial vectors in the $(D-2)$-dimensional subspace are then orthogonal to both k^a and l^a.

We now introduce the metrics n_{ab} defined below and $\bar{h}_{ab}$ defined in (41),

$$n_{ab} = g_{ab}+k_a l_b+l_a k_b. \tag{62}$$

Similarly to the time-like case, we introduce tensor fields $B_{ab} = \nabla_b k_a$ and $\bar{B}_{ab}$ in (37) satisfying the identities $B_{ab}k^a = \bar{B}_{ab}\zeta^a = 0$ and $-B_{ab}k^b + \bar{B}_{ab}\zeta^b = 0$. We also define the deviation vector $\eta^a = (\partial/\partial\alpha)^a$ representing the displacement to an infinitesimally nearby world sheet so that

we can choose λ, σ, and α as coordinates of the three-dimensional submanifold spanned by the world sheets. We then have the commutator relations $\pounds_k \eta^a = \pounds_\zeta \eta^a = \pounds_k \zeta^a = 0$ and $k^a \nabla_a \eta^b - \zeta^a \nabla_a \eta^b = (B^b{}_a - \bar{B}^b{}_a)\eta^a$.
We decompose B_{ab} into three pieces

$$B_{ab} = \frac{1}{D-2}\theta n_{ab} + \sigma_{ab} + \omega_{ab}, \tag{63}$$

where the expansion, shear, and twist of the stringy congruence along the affine direction are defined as $\theta = B^{ab} n_{ab}$, $\sigma_{ab} = B_{(ab)} - \frac{1}{D-2}\theta n_{ab}$ and $\omega_{ab} = B_{[ab]}$. It is noteworthy that even though we have the same notations for B_{ab}, θ, σ_{ab} and ω_{ab} in (45) and (63), the differences of these notations among the time-like sting cases and null string cases are understood in the context. Similarly, we decompose $\bar{B}_{ab}$ into three parts as in the time-like case.

4.2 Expansion of null stringy congruence

Taking the *ansatz* that the expansion $\bar{\theta}$ is constant along the σ-direction as in the time-like case, we have another Raychaudhuri type equation

$$\begin{aligned}\frac{d\theta}{d\lambda} &= -\frac{1}{D-2}\theta^2 + \frac{1}{D-1}\bar{\theta}^2 - \sigma_{ab}\sigma^{ab} + \bar{\sigma}_{ab}\bar{\sigma}^{ab} \\ &\quad + \omega_{ab}\omega^{ab} - \bar{\omega}_{ab}\bar{\omega}^{ab} - R_{ab}(k^a k^b - \zeta^a \zeta^b).\end{aligned} \tag{64}$$

Assuming $\omega_{ab} = \bar{\omega}_{ab}$, $\sigma_a b = \bar{\sigma}_a b$ and a stringy strong energy condition for null case $R_{ab}(k^a k^b - \zeta^a \zeta^b) \geq 0$ and exploiting the energy-momentum tensor of the perfect fluid, we reproduce the inequality (57) in the time-like congruence of strings. The Raychaudhuri type equation (64) for the null strings then has a solution in the following form:

$$\frac{1}{\theta} \geq \frac{1}{\theta_0} + \frac{1}{D-2}\left(\lambda - \frac{D-2}{D-1}\int_0^\lambda d\lambda \left(\frac{\bar{\theta}}{\theta}\right)^2\right), \tag{65}$$

where θ_0 is the initial value of θ at $\lambda = 0$. We assume again that θ_0 is negative. The inequality (65) then implies that θ must pass through the singularity within an affine length

$$\lambda \leq \frac{D-2}{|\theta_0|} + \frac{D-2}{D-1}\int_0^\lambda d\lambda \left(\frac{\bar{\theta}}{\theta}\right)^2. \tag{66}$$

In the point particle limit with $\bar{B}_{ab} = \bar{\theta} = \bar{\sigma}_{ab} = \bar{\omega}_{ab} = 0$ and $\omega_{ab} = 0$, we assume that the strong energy condition $R_{ab} k^a k^b \geq 0$ is satisfied to yield the second inequality of (58) Carroll (2004); Hawking & Penrose (1970); Wald (1984). If we assume that the initial value is negative, the expansion θ must go to the negative infinity along that geodesic within a finite affine length Hawking & Penrose (1970).

5. Jacobi fields in geodesic surface congruence

5.1 Geodesic surface deviation equation

In this subsection, we consider a fiber bundle $\pi : M \to X$ over a spacetime four-manifold X with a Calabi-Yau manifold as a fiber space. Let M be a n-dimensional manifold with a metric g_{ab}. The action for a string is proportional to the area of the surface spanned in the total space

M by the evolution along the time direction of the string in F. We parametrize the surface generated by a closed string by two world sheet coordinates τ and σ, and then we have the corresponding vector fields $\xi^a = (\frac{\partial}{\partial\tau})^a$ and $\zeta^a = (\frac{\partial}{\partial\sigma})^a$. The Nambu-Goto string action is given by Goto (1971); Nambu (1970)

$$S = -\int\int d\tau d\sigma f(\tau,\sigma) \tag{67}$$

where $0 \leq \tau \leq T$ and $0 \leq \sigma \leq 2\pi$, and $f(\tau,\sigma) = [(\xi,\xi)^2 - (\xi,\xi)(\zeta,\zeta)]^{1/2}$.
We perform a variation of the surfaces $\gamma_a(\tau,\sigma)$ traced by the closed string during its evaluation to find the geodesic surface equation by the least action principle. Let the vector field $\eta^a = \left(\frac{\partial}{\partial\alpha}\right)^a$ be the deviation vector field. Since we have gauge degrees of freedom, we can choose the orthonormal gauge $\xi \cdot \zeta = 0$ and $\xi \cdot \xi + \zeta \cdot \zeta = 0$. The first variation is

$$\frac{ds}{d\alpha} = \int\int d\tau d\sigma\, \eta_b(-\xi^a\nabla_a\xi^b + \zeta^a\nabla_a\zeta^b), \tag{68}$$

where we use the endpoint condition $\eta^a(0) = \eta^a(\tau) = 0$ and periodic condition $\eta^a(\sigma + 2\pi) = \eta^a(\sigma)$. In Cho & Hong (2007) we have the geodesic surface equation.

$$-\xi^a\nabla_a\xi^b + \zeta^a\nabla_a\zeta^b = 0 \tag{69}$$

When $\frac{ds}{d\alpha}|_{\alpha=0} = 0$, then the second variation is given by

$$\left.\frac{d^2S}{d\alpha^2}\right|_{\alpha=0} = -\int\int d\tau d\sigma[\frac{d^2\eta^i}{d\tau^2} + (R_{\tau j\tau}{}^i - R_{\sigma j\sigma}{}^i)]\eta_i. \tag{70}$$

We have the geodesic surface deviation equation in Cho & Hong (2007)

$$\frac{d^2\eta^i}{d\tau^2} + (R_{\tau j\tau}{}^i - R_{\sigma j\sigma}{}^i)\eta^i = 0. \tag{71}$$

In Cho & Hong (2007), locally γ minimizes the Nambu-Goto string action if and only if γ is a geodesic surface with no string conjugate to $p = \gamma(0,\sigma)$ between $\gamma(0,\sigma) = p$ and $q = \gamma(\tau,\sigma)$.

5.2 Singularity in geodesic surface congruence

In the orthogonal gauge, we introduce two tensor fields :

$$B_{ab} = \nabla_b\xi_a, \quad \bar{B}_{ab} = \nabla_b\zeta_a. \tag{72}$$

Define the metrics $h_{ab} = g_{ab} + \xi_a\xi_b$ and $\bar{h}_{ab} = g_{ab} - \zeta_a\zeta_b$. Let B_{ab} split B_{ab} into three pieces Hawking & Penrose (1970); Wald (1984)

$$B_{ab} = \frac{1}{n-1}\theta h_{ab} + \sigma_{ab} + \omega_{ab}, \tag{73}$$

where the expansion, shear, and twist of the stringy congruence along the time direction are defined as $\theta = B^{ab}h_{ab}$, $\sigma_{ab} = B_{(ab)} - \frac{1}{n-1}\theta h_{ab}$ and $\omega_{ab} = B_{[a,b]}$. Similarly, we split

$$\bar{B}_{ab} = \frac{1}{n-1}\bar{\theta}, \bar{h}_{ab} + \bar{\sigma}_{ab} + \bar{\omega}_{ab}, \tag{74}$$

where $\bar{\theta} = \bar{B}^{ab}\bar{h}_{ab}$, $\bar{\sigma}_{ab} = \bar{B}_{(ab)} - \frac{1}{n-1}\bar{\theta}\bar{h}_{ab}$ and $\bar{\omega}_{ab} = \bar{B}_{[a,b]}$.
Taking an *ansatz* that the expansion $\bar{\theta}$ is constant along the σ-direction, we have a Raychaudhuri-type equation

$$\frac{d\theta}{d\tau} = \frac{1}{n-1}(\theta^2 - \bar{\theta}^2) - \sigma_{ab}\sigma^{ab} + \bar{\sigma}_{ab}\bar{\sigma}^{ab} + \omega_{ab}\omega^{ab} - \bar{\omega}_{ab}\bar{\omega}^{ab} - R_{ab}(\xi^a\xi^b - \zeta^a\zeta^b). \tag{75}$$

In Cho & Hong (2008), assuming that $\omega_{ab} = \bar{\omega}_{ab}$, $\sigma_{ab} = \bar{\sigma}_{ab}$ and a stringy strong energy condition $R_{ab}(\xi^a\xi^b - \zeta^a\zeta^b) \geq 0$, we have

$$\frac{1}{\theta} \geq \frac{1}{\theta_0} + \frac{1}{n-1}\left(\tau - \int_0^\tau d\tau \left(\frac{\bar{\theta}}{\theta}\right)^2\right), \tag{76}$$

where $\theta_0 = \theta(0)$ is the initial value. If $\theta_0 < 0$ so that the congruence is initially converging, then θ must be pass through the singular point within a proper time

$$\tau \leq \frac{n-1}{|\theta_0|} + \int_0^\tau d\tau \left(\frac{\bar{\theta}}{\theta}\right)^2. \tag{77}$$

5.3 Jacobi fields along geodesic surfaces

An n-dimensional manifold M is said to be conformally symmetric if there is a smooth map $\lambda : M \to \mathbf{R}$ and for parallel vector fields U, V and W along a curve γ is M there is a parallel vector field X along γ such that the curvature tensor

$$R(U,V)W = \lambda X. \tag{78}$$

For instance, locally symmetric manifolds or symmetric manifolds are conformally symmetric Green et al. (1987); Milnor (1963). From now on we assume that M is conformally symmetric. Let $\gamma_\alpha(\tau,\sigma)$ be a variation of geodesic surfaces with the corresponding vector fields $\xi = \frac{\partial}{\partial\tau}$, $\zeta = \frac{\partial}{\partial\sigma}$ and the deviation vector field $\eta = \frac{\partial}{\partial\alpha}$. Then the geodesic surface deviation equation for the Nambu-Goto string action yields

$$\frac{d^2\eta}{d\tau^2} + [R(\xi,\eta)\xi - R(\zeta,\eta)\zeta] = 0 \tag{79}$$

where we can construct the variation vectors $\eta(\tau)$ as vectors associated with the centers of the string of the two nearby geodesic surfaces at the time τ. The solution η of the geodesic surface deviation equation is called a Jacobi field on the geodesic surface γ_0. Let ξ_0 be the time direction velocity vector at the string $P = \gamma_0(0,0)$, and ζ_0 be the string direction velocity vector at the string $\gamma_0(0,0)$. We have that the linear transformation

$$R(\xi,\cdot)\xi - R(\zeta,\cdot)\zeta : T_pM \to T_pM \tag{80}$$

is self-adjoint. Indeed, for any velocity vectors X and X' in T_pX

$$\langle R(\xi, X)\xi - R(\zeta, X)\zeta, X'\rangle = \langle X, R(\xi, X')\xi - R(\zeta, X')\zeta\rangle. \tag{81}$$

Let vectors $U_1, \cdots, U_n \in T_pM$ be an orthonormal basis such that

$$R(\xi, U_i)\xi - R(\zeta, U_i)\zeta = \lambda_i U_i \tag{82}$$

where $\lambda_1, \cdots, \lambda_n$ are the eigenvalues of the self-adjoint operator $R(\xi, \cdot)\xi - R(\zeta, \cdot)\zeta$. Extend the U_i to vector fields along γ_0 by parallel translation. Since M is conformally symmetric, there are eigenfunctions $\lambda_i : M \to \mathbf{R}$ such that

$$R(\xi, U_i)\xi - R(\zeta, U_i)\zeta = \lambda_i U_i \tag{83}$$

along the geodesic surface γ_0. The Jacobi field along the geodesic surface γ_0 may be expressed uniquely as

$$\eta(\tau) = \eta_1(\tau)U_1(\tau) + \cdots \eta_n(\tau)U_n(\tau). \tag{84}$$

Then the geodesic surface deviation equation takes the form

$$\sum_{i=1}^{n}\left[\frac{d^2\eta_i(\tau)}{d\tau^2} + \lambda_i(\tau)\eta_i(\tau)\right]U_i(\tau) = 0. \tag{85}$$

Since the U_i are every where linearly independent, the equation is equivalent to the system of n equations

$$\frac{d^2\eta_i(\tau)}{d\tau^2} + \lambda_i(\tau)\eta_i(\tau) = 0, \quad i = 1, \cdots, n. \tag{86}$$

Suppose that $\tau = 0$ is the singular point of the expansion θ so that the congruence of geodesic surfaces is converging, and that the expansion is constant along the σ-direction. Assume that $\lambda_i(\tau) = \frac{1}{\tau}$, and $\eta_i(\tau) := c_{i0} + c_{i1}\tau + c_{i2}\tau^2 + \cdots$. Then the system of n geodesic surface deviation equations

$$\frac{d^2\eta_i(\tau)}{d\tau^2} + \lambda_i(\tau)\eta_i(\tau) = 0. \tag{87}$$

By the initial condition $\eta(0) = 0$, $\eta_i(0) = 0 = c_{i0}$ for all i. By the differentiation of $\eta_i(\tau)$ with respect to τ, we have

$$\frac{d\eta_i(\tau)}{d\tau} = c_{i1} + 2c_{i2}\tau + 3c_{i3}\tau^2 + \cdots + nc_{in}\tau^{n-1} + \cdots. \tag{88}$$

$$\frac{d^2\eta_i(\tau)}{d\tau} = (2\cdot 1)c_{i2} + (3\cdot 2)c_{i3}\tau^1 + \cdots + (n\cdot(n-1))c_{in}\tau^{n-2} + \cdots. \tag{89}$$

By multiplying $\lambda_i(\tau)$ to $\eta_i(\tau)$,

$$\lambda_i(\tau)\eta_i(\tau) = c_{i1} + c_{i2}\tau^1 + \cdots + c_{i(n-1)}\tau^{n-2} + \cdots. \tag{90}$$

Thus,

$$(2\cdot 1)\, c_{i2} = -c_{i1}, \quad (3\cdot 2)\, c_{i3} = -c_{i2}, \quad \cdots, \quad (n\cdot(n-1))\, c_{in} = -c_{i(n-1)}. \tag{91}$$

Then by induction we have

$$c_{in} = -\frac{1}{n(n-1)}c_{i(n-1)} = (-1)^{n-1}\frac{1}{n!(n-1)!}c_{i1}. \tag{92}$$

Let $c_{i1} = c_i$. Then the solution

$$\eta_i(\tau) = c_i \sum_{n=1}^{\infty}(-1)^{n-1}\frac{1}{n!(n-1)!}\tau^n, \quad i = 1, \cdots, n, \tag{93}$$

where c_i is a constant, $i = 1, \cdots, n$.

Suppose that the singularity exists in geodesic surface congruences of the locally symmetric, spacetime manifold. Then the Jacobi field along geodesic surface γ_0 in the geodesic surface congruence is given by

$$\eta(\tau) = \sum_{i=1}^{n}\eta_i(\tau)U_i(\tau), \tag{94}$$

where $\eta_i(\tau) = c_i \sum_{k=1}^{\infty}(-1)^{k-1}\frac{\tau^k}{k!(k-1)!}$, and c_i is constant, $i = 1, \cdots, n$.

5.4 Observations

If M is a symmetric manifold and if ξ and ζ are parallel, the Jacobi equation becomes

$$\frac{d^2\eta_i(\tau)}{d\tau^2} + \lambda_i(\tau)\eta_i(\tau) = 0, \tag{95}$$

where the eigenfunctions $\lambda_i(\tau)$ of the operator $R(\xi,\cdot)\xi - R(\zeta,\cdot)\zeta$ are independent of τ.

$$\begin{aligned} &\text{If } \lambda_i > 0, \text{ then } \eta_i(\tau) = c_i \sin(\sqrt{\lambda_i}\tau). \\ &\text{If } \lambda_i = 0, \text{ then } \eta_i(\tau) = c_i\tau. \\ &\text{If } \lambda_i < 0, \text{ then } \eta_i(\tau) = c_i \sinh(\sqrt{|\lambda_i|}\tau). \end{aligned} \tag{96}$$

Thus if $\lambda_i \leq 0$, $\eta_i(\tau)$ does not vanish except $\tau = 0$.

In our assumption, the eigenfunction $\lambda_i(\tau) = \frac{1}{\tau}$ is a kind of curvature at time τ. If $\tau \to \infty$ then $\lambda_i(\tau) \to 0$, since the universe is expanding by observation. If $\tau \to 0$, then $\lambda_i(\tau) \to \infty$, since the point $\tau = 0$ is the singular point which the universe comes into being by Big Bang. The sum

$$\eta_i(\tau) = c_i \sum_{k=1}^{\infty}(-1)^{k-1}\frac{1}{k!(k-1)!}\tau^k \tag{97}$$

converges absolutely for all τ. Indeed, by the ratio test

$$\lim_{k\to\infty}\left|\frac{(-1)^k\frac{1}{(k+1)!k!}\tau^{k+1}}{(-1)^{k-1}\frac{1}{k!(k-1)!}\tau^k}\right| = \lim_{k\to\infty}\frac{|\tau|}{(k+1)k} = 0 < 1, \text{ for all } \varnothing. \tag{98}$$

We may think topologically our manifold $M \simeq [0,\tau] \times D^3(\tau) \times X \simeq D^4(\tau) \times X$ where τ is the time and X is the CY-manifold which is hidden. M is compact and we live in $\partial D^4(\tau) = S^3(\tau)$ which is our visible universe at time τ.

If $\lambda_i(\tau) = \frac{a_i}{\tau}$ for some constant a_i along U_i direction, then the solutions of the Jacobi equations are

$$\eta_i(\tau) = c_i \sum_{n \geq 1} (-1)^{n-1} \frac{{a_i}^{n-1}}{n!(n-1)!} \tau^n, \ i = 1, \cdots, n. \tag{99}$$

The series of the right hand have the radius of convergence ∞.

6. Conclusions

We consider phase transition which is assumed to have happened in the evolution of the universe around 75,000 years after the Big Bang. Assuming that the early universe was filled with a perfect fluid consisting of massive particles and/or massless particles and using the strong energy condition, one could find equations of state for each particle.

Specifically, the equation of state of the massive particle is different from that of the massless particle, which indicates that the massive particle phase is not the same as the massless particle one. In the standard cosmology based on the Hawking-Penrose singularity theorem and inflationary scenario, it is believed that, after the Big Bang explosion, a massless particle (or radiation) dominated phase occurred followed by a massive particle (or matter) dominated one, even though there was a hot thermalization period of radiation and matter immediately after the Big Bang. Moreover, as in the phase transition in water, a phase transition exists between massive particle and massless particle phases in the universe.

In the above explanations of the Hawking-Penrose singularity theorem and the ensuing standard cosmology, they assumed that massive and massless particles are point-like. Now let us discuss stringy cosmology which has been developed recently Cho & Hong (2007) and is based on the string theory Polchinski (1999). The string theory has given us a better understanding of the universe and has provided an analytical tool for studying the nature of the universe. It was proposed in the string theory that massive and massless particles in the universe can be described by vibrating strings instead of point particles. As in the standard (point particle) cosmology, using the strong energy condition modified by stringy corrections, one can find the stringy singularity theorem, which is similar to the Hawking-Penrose singularity theorem except for the fact that evolution to a singularity occurs at different spacetime Cho & Hong (2007).

Exploiting the stringy strong energy condition together with the assumption that the early universe was occupied with a perfect fluid of massive stringy particles and/or massless stringy particles, one obtains equations of state for each stringy particle. Surprisingly, the massive stringy particles and massless stringy particles produce the same equation of state. This implies that both massive stringy particles and massless stringy particles are in the same phase and thus there is no phase transition in the stringy universe Cho & Hong (2007). Here, we emphasize that the equations of state obtained from the standard and stringy cosmologies govern the evolution of the universe, since these equations originate from the initial conditions of the (stringy) singularity theorems. Immediately after the Big Bang explosion, in the stringy cosmology, the massless stringy particle dominated phase and the massive stringy particle dominated phase took place simultaneously. In the stringy universe, a massless-massive stringy particle mixture state without any phase transitions exists. These features in the stringy universe are drastically different from those in the standard point particle cosmology. In fact, these differences originate from the lone fact that the particles are composed of strings instead of points.

In the standard cosmology, after the radiation and matter dominated phases a dark energy dominated phase exists. In this inflationary scenario, a sequence of epoches occurred: Big Bang, radiation dominated phase, matter dominated phase, and dark energy dominated phase. In brane cyclic cosmology, which is also based on the string (or D brane) theory, the universe is assumed to be cyclic in evolution: Big Bang, radiation dominated phase, matter dominated phase, dark energy dominated phase, Big Crunch, and again Big Bang Steinhardt & Turok (2002). Moreover, both in the standard and stringy cosmologies exploiting the (stringy) singularity theorems, there exists an initial state with a negative expansion rate and the ensuing evolution to the singularity. One can thus claim that the stringy cosmology is also cyclic, similar to the brane cyclic cosmology, but modified: Big Bang, radiation-matter mixture phase, dark energy dominated phase, Big Crunch, and again Big Bang, which is consistent with the fact that both the stringy and brane cyclic cosmologies share the same characteristics that these scenarios use the string features.
We want to detect the so-called quark-gluon plasma state, which is assumed to exist in an extremely hot soup of quarks and gluons. Both in the standard and stringy cosmologies, this quark-gluon plasma state is supposed to occur immediately after the Big Bang of the tiny early universe manufactured in the Large Hadron Collider. However, there are drastically different ensuing processes in these two scenarios. Namely, in the standard cosmology, the quark-gluon plasma state can exist shortly and disappear eventually to enter the radiation dominated phase, while in the stringy cosmology the quark-gluon plasma state can develop into particles such as protons and neutrons and sustain the radiation and matter mixture phase. We recall that as far as radiation and matter are concerned, the mixture of these two coexists in the present universe. It is expected that the Large Hadron Collider will be able to detect the procedure of particle states along with the evolution of the tiny universe planned to occur at the Large Hadron Collider and it will be able to determine which cosmology is viable. More details are reported in Cho & Hong (2010).

7. References

Carroll, S.M. (2004). *An Introduction to General Relativity: Spacetime and Geometry*, Addition Wesley, New York.

Cho, Y.S. & Hong, S.T. (2007). Stringy Jacobi fields in Morse theory, *Phys. Rev. D* 75: 127902.

Cho, Y.S. & Hong, S.T. (2008). Singularities in geodesic surface congruence, *Phys. Rev. D* 78: 067301.

Cho, Y.S. (2010). Jacobi fields in geodesic surface congruences, *Int. J. Geom. Meth. Mod. Phys.* 7: 1407-1412.

Cho, Y.S. & Hong, S.T. (2010). String reveals secrets of universe, *arXiv:1010.3485*.

Goto, T. (1971). Relativistic quantum mechanics of one-dimensional mechanical continuum and subsidiary condition of dual resonance model, *Prog. Theor. Phys.* 46: 1560-1569.

Green, M.B., Schwarz, J.H. & Witten, E. (1987). *Superstring Theory*, Cambridge Univ. Press, Cambridge.

Hawking, S.W. & Penrose, R. (1970). The Singularities of gravitational collapse and cosmology, *Proc. Roy. Soc. Lond. A* 314: 529-548.

Karlhede, A. & Lindström, U. (1986). The classical bosonic string in the zero tension limit, *Class. Quantum Grav.* 3: L73-L75.

Milnor, J. (1963). *Morse Theory*. Princeton University Press, Princeton.

Misner, C.W., Thorne K.P. & Wheeler, J.A. (1972). *Gravitation*, Freeman, San Francisco.

Nambu, Y. (1970). *Lecture at the Copenhagen Symposium.*

Polchinski, J. (1999). *String Theory*, Cambridge Univ. Press, Cambridge.

Roshchupkin, S.N. & Zheltukhin, A. A. (1995). Friedmann universes and exact solutions in string cosmology, *Class. Quantum Grav.* 12: 2519-2524;

Roshchupkin, S.N. & Zheltukhin, A. A. (1999). Variational principle and a perturbative solution of nonlinear string equations in curved space, *Nucl. Phys. B* 543: 365-386.

Scherk, J. (1975). An introduction to the theory of dual models and strings, *Rev. Mod. Phys.* 47: 123-164.

Schild, A. (1977). Classical null strings, *Phys. Rev. D* 16: 1722-1726.

Steinhardt, P.J. & Turok, N. (2002). A cyclic model of the universe, *Science* 296: 1436-1439.

Wald, R.M. (1984). *General Relativity*, University of Chicago Press, Chicago.

18

Cosmology: The Noncommutative Quantum and Classical Cosmology

E. Mena, M. Sabido and M. Cano
Deparment of Physics, Cienega Universitary Center of the University of Guadalajara, Ocotlán, Jalisco, Mexico
Physics Deparment of the Division of Science and Engieneering of the University of Guanajuato, León, Guanajuato Mexico.

1. Introduction

During the early days of quantum mechanics and quantum field theory, continuous space-time and Lorentz symmetry was considered inappropriate to describe the small scale structure of the universe. It was also argued that one should introduce a fundamental length scale, limiting the precision of position measurements. The introduction of fundamental length is suggested to cure the ultraviolet divergencies occurring in quantum field theory. H. Snyder was the first to formulate these ideas mathematically (1), introducing noncommutative coordinates brings an uncertainty in the position. The success of the renormalisation made people forget about these ideas for some time. But when the quantization of gravity was considered thoroughly, it became clear that the usual concepts of space-time are inadequate and that spacetime has to be quantised or noncommutative, in some way. Quantum cosmology, is a simplified approach to the study of the very early universe, which means that the gravitational and matter variables have been reduced to a finite number of degrees of freedom (these models were extensively studied by means of Hamiltonian methods in the 1970's, (for reviews see (2; 3)); for homogenous cosmological models the metric depends only on time, this permits to integrate the space dependence and obtain a model with a finite dimensional configuration space, minisuperspace, whose variables are the 3-metric components. One way to extract useful dynamical information is through a WKB type method. The semiclassical or WKB approximations are usually discussed in text books on nonrelativistic quantum mechanics in the context of stationary states, i.e., determination of the energy eigenvalues and eigenfunctions (4). This approximation can also be used to obtain approximate and in some cases exact solutions of the dynamical problem, i.e., full Schroedinger equation, so the utility of the semiclassical approximation in obtaining exact solutions of the Schroedinger equation has not yet fully explored.
The same seems to be the case for the relativistic quantum mechanics. The importance of the semiclassical approximation in the relativistic case is probably best appreciated in quantum cosmology (5), specifically, in the analysis of the Wheeler-DeWitt equation, which is essentially a Klein-Gordon equation on the minisuperspace (6). In the last few years there have been several attempts to study the possible effects of noncommutativity in the classical cosmological scenario (7–9). The proposal of authors in Ref 9 introduces the effects

of noncommutativity at the quantum level, namely quantum cosmology, by deforming the minisuperspace through a Moyal deformation of the Wheeler-DeWitt (WDW) equation, similar to noncommutative quantum mechanics (10). The aim of this chapter is to introduce a deformation in the minisuperscpace variables through the Moyal product of the Wheeler DeWitt equation and apply a WKB type method to noncommutative quantum cosmology, and find the noncommutative classical solutions, avoiding in this way the difficult task to solve this cosmological models in the complicated framework of noncommutative gravity (11). We know how to introduce noncommutativity at a quantum level, by taking into account the changes that the Moyal product of functions induces on the quantum equation, and from there calculate the effects of noncommutativity at the classical level. This also has the advantage that for some noncommutative models for which the quantum solutions can not be found, the noncommutative classical solutions arise very easily from this formulation. This procedure is presented through many examples: first the Kanstowski-Sachs (KS) cosmological model is presented and the formalism developed in this model, is the applied to the Friedmann-Robertson-Walker (FRW) universe coupled to a scalar field and cosmological constant, besides we show the noncommutative proposal applied to a stringy model and the Bianchi I with Baratropic perfect fluid and Λ cosmological.

2. Cosmology

We start by reviewing the quantum cosmological models in which we are interested, and find the classical evolution through the WKB-type approximation.

2.1 Kantowski-Sachs (KS) Cosmology

The first example that we are interested because the simplest anisotropic, is the KS universe, part of the interest in this universe model is due to the wide set of analitycal solutions it admits, even if particular types of matter are coupled to gravity.
The Kanstowski-Sachs line element (12) is

$$ds^2 = -N^2 dt^2 + e^{2\sqrt{3}\beta} dr^2 + e^{-2\sqrt{3}\beta} e^{-2\sqrt{3}\Omega} \left(d\theta^2 + \sin^2\theta d\varphi^2 \right). \tag{1}$$

from the general relativity lagrangian we can construct the canonical momenta,

$$\begin{aligned} \Pi_\Omega &= -\frac{12}{N} e^{-\sqrt{3}\beta - 2\sqrt{3}\Omega} \dot{\Omega} \\ \Pi_\beta &= \frac{12}{N} e^{-\sqrt{3}\beta - 2\sqrt{3}\Omega} \dot{\beta}, \end{aligned} \tag{2}$$

and the corresponding Hamiltonian

$$H = \frac{N}{24} e^{-\sqrt{3}\beta - 2\sqrt{3}\Omega} \left[-\Pi_\Omega^2 + \Pi_\beta^2 + 48 e^{-2\sqrt{3}\Omega} \right], \tag{3}$$

for this model we can use canonical quantization and obtain the Wheeler-DeWitt (WDW) equation. Using the usual identifications $\Pi_\Omega = -i\frac{\partial}{\partial\Omega}$ and $\Pi_\beta = -i\frac{\partial}{\partial\beta}$ we get

$$\left[\frac{\partial^2}{\partial\Omega^2} - \frac{\partial^2}{\partial\beta^2} - 48 e^{-2\sqrt{3}\Omega} \right] \psi(\Omega, \beta) = 0, \tag{4}$$

in this parametrization the WDW equation has very simple form; the solutions to this equation are given by

$$\psi = e^{\pm i\nu\sqrt{3}\beta} K_{i\nu}\left(4e^{-\sqrt{3}\Omega}\right), \tag{5}$$

where ν is the separation constant and $K_{i\nu}$ are the modified Bessel functions. We now proceed to apply the WKB method. For this we propose the wave function

$$\Psi(\beta,\Omega) = e^{i(S_1(\beta)+S_2(\Omega))}, \tag{6}$$

the WKB approximation is reached in the limit

$$\left|\frac{\partial S_1^2(\beta)}{\partial\beta^2}\right| << \left(\frac{\partial S_1(\beta)}{\partial\beta}\right)^2 \qquad \left|\frac{\partial S_2^2(\Omega)}{\partial\Omega^2}\right| << \left(\frac{\partial S_2(\Omega)}{\partial\Omega}\right)^2 \tag{7}$$

this gives the Einstein-Hamilton-Jacobi (EHJ) equation

$$-\left(\frac{\partial S_2(\Omega)}{\partial\Omega}\right)^2 + \left(\frac{\partial S_1(\beta)}{\partial\beta}\right)^2 - 48e^{-2\sqrt{3}\Omega} = 0, \tag{8}$$

solving (equation 8) gives the functions S_1, S_2 and using the definition for the momenta

$$\Pi_\beta = \frac{dS_1(\beta)}{d\beta}, \qquad \Pi_\Omega = \frac{dS_2(\Omega)}{d\Omega}, \tag{9}$$

which combined with (equation 2) and fixing the value of $N(t) = 24e^{-\sqrt{3}\beta-2\sqrt{3}\Omega}$ we find

$$\begin{aligned} S_1(\beta) &= P_{\beta_0}\beta, \\ S_2(\omega) &= -\frac{1}{\sqrt{3}}\sqrt{P_{\beta_0}^2 - 48e^{-2\sqrt{3}\Omega}} + \frac{P_{\beta_0}}{\sqrt{3}}\operatorname{arctanh}\left[\frac{\sqrt{P_{\beta_0}^2 - 48e^{-2\sqrt{3}\Omega}}}{P_{\beta_0}}\right], \end{aligned} \tag{10}$$

from this solutions and using (equation 2) and (equation 9) we obtain the classical solutions

$$\begin{aligned} \Omega(t) &= \frac{1}{2\sqrt{3}}\ln\left[\frac{48}{P_{\beta_0}^2}\cosh^2\left(2\sqrt{3}P_{\beta_0}(t-t_0)\right)\right], \\ \beta(t) &= \beta_0 + 2P_{\beta_0}(t-t_0), \end{aligned} \tag{11}$$

this solutions are the same that solving the field equations of General Relativity.

2.2 Friedmann-Robertson-Walker (FRW) Cosmology with scalar field and Λ

The next set of examples correspond to homogeneous and isotropic Universes, the so called Friedmann-Robertson-Walker (FRW) universe coupled to a scalar field and cosmological constant. The FRW metric is given by

$$ds^2 = -N^2(t)dt^2 + e^{2\alpha(t)}\left[\frac{dr^2}{1-kr^2} + r^2(d\vartheta^2 + sin^2\vartheta d\varphi^2)\right], \tag{12}$$

where $a(t) = e^{\alpha(t)}$ is the scale factor, $N(t)$ is the lapse function, and k is the curvature constant that takes the values $0, +1, -1$, which correspond to a flat, closed and open universes, respectively. The Lagrangian we are to work on, is composed by the gravity sector and the matter sector, which for the FRW universe endowed with a scalar field and cosmological constant Λ is

$$\mathcal{L}_{tot} = \mathcal{L}_g + \mathcal{L}_\phi = e^{3\alpha}\left[6\frac{\dot{\alpha}^2}{N} - \frac{1}{2}\frac{\dot{\phi}^2}{N} - N\left(2\Lambda + 6ke^{-2\alpha}\right)\right], \tag{13}$$

the corresponding canonical momenta are

$$\Pi_\alpha = \frac{\partial \mathcal{L}}{\partial \dot{\alpha}} = 12e^{3\alpha}\frac{\dot{\alpha}}{N}, \qquad \Pi_\phi = \frac{\partial \mathcal{L}}{\partial \dot{\phi}} - e^{3\alpha}\frac{\dot{\phi}}{N}, \tag{14}$$

proceeding as before the WDW equation is obtained from the classical Hamiltonian. By the variation of (equation 13) with respect to N, $\partial\mathcal{L}/\partial N = 0$, implies the well-known result $\mathcal{H} = 0$.

$$e^{-3\alpha}N\left[-\frac{1}{24}\frac{\partial^2}{\partial\alpha^2} + \frac{1}{2}\frac{\partial^2}{\partial\phi^2} + e^{6\alpha}\left(2\Lambda + 6ke^{-2\alpha}\right)\right]\Psi(\alpha,\phi) = 0. \tag{15}$$

Now that we have the complete framework and the corresponding WDW equation, we can proceed to study different cases.

In table 1 we can see the different cases that we solved[1], all of them are calculated by using the WKB type procedure, the classical solutions are the same we would get by solving Einstein's field equations. We can expect that this approximation includes all the gravitational degrees of freedom of the particular cosmological model under study. This almost trivial observation is central to the ideas we are presenting in the next section.

2.3 Stringy Quantum Cosmology

In the case of strings, this example is related to the graceful exit of pre-big bang cosmology (13), this model is based on the gravi-dilaton effective action in 3+1 dimensions

$$S = -\frac{\lambda_s}{2}\int d^4x\sqrt{-g}e^{-\phi}(R + \partial_\mu\phi\partial^\nu\phi + V), \tag{16}$$

in this expression λ_s is the fundamental string length, ϕ is the dilaton field with V the possible dilaton potential. Working with an isotropic background, and setting $a(t) = e^{\beta(t)/\sqrt{3}}$, after integrating by parts, we get

$$S = -\frac{\lambda_s}{2}\int d\tau\left(\bar{\phi}'^2 - \beta'^2 + Ve^{-2\bar{\phi}}\right), \tag{17}$$

we have used the time parametrization[2] $dt = e^{-\bar{\phi}}d\tau$, the gauge $g_{00} = 1$, and defined $\bar{\phi} = \phi - \ln\int\left(\frac{d^3x}{\lambda_s^3}\right) - \sqrt{3}\beta$. From this action we calculate the canonical momenta, $\Pi_\beta = \lambda_s\beta'$ and $\Pi_{\bar{\phi}} = -\lambda_s\bar{\phi}'$. From the classical hamiltonian we find the WDW equation

$$\left[\frac{\partial^2}{\partial\bar{\phi}^2} - \frac{\partial^2}{\partial\beta^2} + \lambda_s^2 V(\bar{\phi},\beta)e^{-2\bar{\phi}}\right]\Psi(\bar{\phi},\beta) = 0, \tag{18}$$

[1] The case $k \neq 0$, $\Lambda \neq 0$ does not have a closed analytical solution to the WDW equation.

[2] The prime denotes differentiation respect to τ.

case	Quantum Solution	Classical Solution
k=0, $\Lambda \neq 0$	$\psi = e^{\pm i\nu\frac{\sqrt{3}}{2}\phi} K_{i\nu}\left(4\sqrt{\frac{\Lambda}{3}}e^{3\alpha}\right)$ and J_ν for $\Lambda < 0$	$\phi(t) = \phi_0 - P_{\phi_0} t$, $\alpha(t) = \frac{1}{6}\ln\left(\frac{P^2_{\phi_0}}{4\Lambda}\right)$ $+\frac{1}{3}\ln\left(\text{sech}\left[\frac{\sqrt{3}}{2}P_{\phi_0}(t-t_0)\right]\right)$,
$k \neq 0$, $\Lambda = 0$	$\psi^{(1)} = e^{\pm i\frac{\nu}{\sqrt{3}}\phi} K_{i\nu}\left(6e^{2\alpha}\right)$ for $k = 1$, $\psi^{(2)} = e^{\pm i\frac{\nu}{\sqrt{3}}\phi} J_\nu\left(6e^{2\alpha}\right)$ for $k = -1$	$\phi(t) = \phi_0 - P_{\phi_0}(t - t_0)$, $\alpha(t) = \frac{1}{4}\ln\left[\frac{P^2_{\phi_0}}{12k}\right]$ $+\frac{1}{2}\ln\left(\text{sech}\left[\frac{1}{\sqrt{3}}P_{\phi_0}(t-t_0)\right]\right)$,
$k \neq 0$, $\Lambda \neq 0$	Unknown	$\phi(t) = \phi_0 - P_{\phi_0}(t - t_0)$, $\int \frac{d\alpha(t)}{\sqrt{P_{\phi_0} - 2e^{6\alpha}(2\Lambda + 6ke^{-2\alpha})}} = \frac{1}{\sqrt{12}}(t - t_0)$,

Table 1. Classical and quantum solutions for the FRW universe coupled to a scalar field ϕ. For the case $\Lambda \neq 0$ $k \neq 0$, the classical solution for the scale factor is given in an implicit expression. We have fixed the lapse function to $N(t) = e^{3\alpha}$.

in particular for a potential of the form $V(\bar{\phi}) = -V_0 e^{m\bar{\phi}}$, the quantum solution is

$$\Psi(\bar{\phi}, \beta) = e^{\pm - i\frac{m-2}{2}\nu\beta} K_{i\nu}\left[\frac{2\lambda_s\sqrt{V_0}}{m-2} e^{\left(\frac{m-2}{2}\right)\bar{\phi}}\right]. \tag{19}$$

The classical solutions for the scale factor and the dilaton are

$$\bar{\phi}(\tau) = \frac{1}{m-2}\ln\left[\frac{P^2_{\beta_0}}{V_0\lambda_s^2}\text{sech}^2\left(\frac{P_{\beta_0}}{2\lambda_s}(m-2)(\tau-\tau_0)\right)\right],$$
$$\beta(\tau) = \beta_0 + \frac{P_\beta}{\lambda_s}(\tau - \tau_0), \tag{20}$$

for $m = 0$ and $m = 4$, the solutions have been obtained in (13), and are used in connection to the graceful exit from pre-big bang cosmology in quantum string.

2.4 Isotropization in Bianchi I with barotropic perfect fluid and Λ Cosmological

In our final example let us begin by recalling canonical formulation of the ADM formalism to the diagonal Bianchi Class A cosmological models. The metrics have the form

$$ds^2 = -(N^2 - N^j N_j)dt^2 + e^{2\Omega(t)} e^{2\beta_{ij}(t)}\, \omega^i \omega^j, \tag{21}$$

where N and N_i are the lapse and shift functions, respectively, $\Omega(t)$ is a scalar and $\beta_{ij}(t)$ a 3x3 diagonal matrix, $\beta_{ij} = diag(\beta_+ + \sqrt{3}\beta_-, \beta_+ - \sqrt{3}\beta_-, -2\beta_+)$, ω^i are one-forms that characterize each cosmological Bianchi type model, and that obey $d\omega^i = \frac{1}{2}C^i_{jk}\omega^j \wedge \omega^k$, C^i_{jk} the structure constants of the corresponding invariance group (8). The metric for the Bianchi type I, takes the form

$$ds_I^2 = -N^2 dt^2 + e^{2\Omega} e^{2\beta_+ + 2\sqrt{3}\beta_-} dx^2 + e^{2\Omega} e^{2\beta_+ - 2\sqrt{3}\beta_-} dy^2 + e^{2\Omega} e^{-4\beta_+} dz^2, \tag{22}$$

The corresponding lagrangian density is

$$L_{Total} = \sqrt{-g}\,(R - 2\Lambda) + L_{matter}, \tag{23}$$

and using (equation 22), this have the following form

$$L = 6e^{3\Omega} \left[-\frac{\dot{\Omega}^2}{N} + \frac{\dot{\beta}_+^2}{N} + \frac{\dot{\beta}_-^2}{N} - \frac{\Lambda}{3} N + \frac{8}{3}\pi G N \rho \right]. \tag{24}$$

where the overdot denotes time derivatives. The canonical momentas to coordinate fields are defined in the usual way

$$P_\Omega = \frac{\partial L}{\partial \dot{\Omega}} = -12 e^{3\Omega} \frac{\dot{\Omega}}{N}, \quad P_+ = \frac{\partial L}{\partial \dot{\beta}_+} = 12 e^{3\Omega} \frac{\dot{\beta}_+}{N}, \quad P_- = \frac{\partial L}{\partial \dot{\beta}_-} = 12 e^{3\Omega} \frac{\dot{\beta}_-}{N}, \tag{25}$$

and the correspondent Hamiltonian function is

$$H = \frac{Ne^{-3\Omega}}{24} \left[-P_\Omega^2 + P_+^2 + P_-^2 - 48\Lambda e^{6\Omega} + 384\pi G M_\gamma e^{-3(\gamma-1)\Omega} \right] = 0, \tag{26}$$

together with barotropic state equation $p = \gamma \rho$, the Hamilton-Jacobi equation is obtained when we substitute $P_{q^\mu} \to \frac{dS_i}{dq^\mu}$ into (equation 26). In what follows, we should consider the gauge $N = 1$.

2.4.1 Classical Solutions á la WKB

The quantum Wheeler-DeWitt (WDW) equation for these models is obtained by making the canonical quantization P_{q^μ} by $-i\partial_{q^\mu}$ in (equation 26) with $q^\mu = (\Omega, \beta_+, \beta_-)$ is

$$\frac{e^{-3\Omega}}{24} \left[\frac{\partial^2}{\partial \Omega^2} - \frac{\partial^2}{\partial \beta_+^2} - \frac{\partial^2}{\partial \beta_-^2} - \lambda e^{6\Omega} + b_\gamma e^{-3(\gamma-1)\Omega} \right] \Psi = 0. \tag{27}$$

where $\lambda = 48\Lambda$, $b_\gamma = 384\pi G M_\gamma$. We now proceed to apply the WKB semiclassical approximation using the ansatz

$$\Psi(\Omega, \beta_\pm) = e^{i[S_1(\Omega) + S_2(\beta_+) + S_3(\beta_-)]}, \tag{28}$$

into (equation 27), and without any loss of generality, one can consider the condition $\frac{d^2 S_i}{dq_i^2}$ be small i.e.,

$$\left(\frac{dS_1}{d\Omega}\right)^2 >> \frac{d^2 S_1}{d\Omega^2}, \qquad \left(\frac{dS_2}{d\beta_+^2}\right)^2 >> \frac{d^2 S_2}{d\beta_+^2}, \qquad \left(\frac{dS_2}{d\beta_-^2}\right)^2 >> \frac{d^2 S_2}{d\beta_-^2}, \tag{29}$$

to get the classical Einstein-Hamilton-Jacobi equation

$$-\left(\frac{dS_1}{d\Omega}\right)^2 + \left(\frac{dS_2}{d\beta_+}\right)^2 + \left(\frac{dS_3}{d\beta_-}\right)^2 - \lambda e^{6\Omega} + b e^{-3(\gamma-1)\Omega} = 0, \tag{30}$$

which can be separate in a set of differential equations

$$-\left(\frac{dS_1}{d\Omega}\right)^2 + a_1^2 - \lambda e^{6\Omega} + be^{-3(\gamma-1)\Omega} = 0, \tag{31}$$

$$\left(\frac{dS_2}{d\beta_+}\right)^2 = n_1^2, \tag{32}$$

$$\left(\frac{dS_3}{d\beta_-}\right)^2 = p_1^2, \tag{33}$$

where a_1^2, n_1^2 and p_1^2 are the separation constants and their relations is $a_1^2 = n_1^2 + p_1^2$. Therefore using the relations between (equation 25), (equation 31), (equation 32) and (equation 33) we have the following equations of motion

$$\pm\sqrt{a_1^2 - \lambda e^{6\Omega} + b_\gamma e^{-3(\gamma-1)\Omega}} \equiv -12e^{3\Omega}\frac{\dot{\Omega}}{N}, \tag{34}$$

$$\pm n_1 \equiv 12e^{3\Omega}\frac{\dot{\beta}_+}{N}, \tag{35}$$

$$\pm p_1 \equiv 12e^{3\Omega}\frac{\dot{\beta}_-}{N}. \tag{36}$$

The main master equation to solved in the gauge $N = 1$, is

$$\frac{dt}{12} = \frac{d\Omega}{\sqrt{a_1^2 e^{-6\Omega} + b_\gamma e^{-3(\gamma+1)\Omega} - \lambda}}, \tag{37}$$

the other two equations (equation 35) and (equation 36) are trivially integrable. For particular stadium of the universe evolution, given by the γ parameter, we present these classical solutions in table 2.

2.4.2 Classical solutions via Hamiltonian formalism

In order to find the commutative equation of motion, we use the classical phase space variables $(\Omega, \beta_\pm)$, where the Poisson algebra for these minisuperspace variables are

$$\{\Omega, \beta_\pm\} = \{\beta_+, \beta_-\} = \{P_\Omega, P_\pm\} = \{P_+, P_-\} = 0, \qquad \{q^\mu, P_{q^\mu}\} = 1, \tag{38}$$

and recalling the Hamiltonian (equation 26), we obtain the classical solutions with the following procedure.
The classical equations of motion for the phase variables Ω, $\beta_\pm$, $P_\pm$, and P_Ω are

$$\dot{\Omega} = \{\Omega, H\} = -\frac{1}{12}e^{-3\Omega}P_\Omega, \tag{39}$$

$$\dot{\beta}_- = \{\beta_-, H\} = \frac{1}{12}e^{-3\Omega}P_-, \tag{40}$$

$$\dot{\beta}_+ = \{\beta_+, H\} = \frac{1}{12}e^{-3\Omega}P_+, \tag{41}$$

$$\dot{P}_\Omega = \{P_\Omega, H\} = \frac{1}{8}e^{-3\Omega}\left[-P_\Omega^2 + P_-^2 + P_+^2 + \lambda e^{6\Omega} + \gamma b_\gamma e^{-3(\gamma-1)\Omega}\right], \tag{42}$$

$$\dot{P}_- = \{P_-, H\} = 0, \qquad \rightarrow \qquad P_- = \pm p_1 = const. \tag{43}$$

$$\dot{P}_+ = \{P_+, H\} = 0, \qquad \rightarrow \qquad P_+ = \pm n_1 = const. \tag{44}$$

Case	Commutative solutions	
$\gamma=-1,\ \Lambda\neq 0,\ \rho_{-1}=M_{-1}$	$\Omega=\frac{1}{3}\,Ln\left[\frac{e^{2qt}-4a_1^2}{16qe^{qt}}\right],$	$q^2=24\pi GM_{-1}-3\Lambda,$
	$\beta_+=\pm\frac{2}{3}\frac{n_1}{a_1}\,arctanh\left[\frac{e^{qt}}{2a_1}\right],$	$a_1^2=n_1^2+p_1^2,$
	$\beta_-=\pm\frac{2}{3}\frac{p_1}{a_1}\,arctanh\left[\frac{e^{qt}}{2a_1}\right].$	
$\gamma=1,\ \Lambda<0,\ \rho_1=M_1e^{-6\Omega}$	$\Omega=\frac{1}{3}\,Ln\left[\frac{e^{2qt}-4a_1^2}{16qe^{qt}}\right],$	$q=\sqrt{3\lvert\Lambda\rvert},$
	$\beta_+=\pm\frac{2}{3}\frac{n_1}{a_1}\,arctanh\left[\frac{e^{qt}}{2a_1}\right],$	$a_1^2=n_1^2+p_1^2+384\pi GM_1,$
	$\beta_-=\pm\frac{2}{3}\frac{p_1}{a_1}\,arctanh\left[\frac{e^{qt}}{2a_1}\right].$	
$\gamma=1,\ \Lambda=0,\ \rho_1=M_1e^{-6\Omega}$	$\Omega=\frac{1}{3}Ln\left[\frac{a_1}{4}t\right],$	$a_1^2=n_1^2+p_1^2+384\pi GM_1,$
	$\beta_+=\pm Ln\left[t^{-\frac{n_1}{3a_1}}\right],$	
	$\beta_-=\pm Ln\left[t^{-\frac{p_1}{3a_1}}\right].$	
$\gamma=0,\ \Lambda=0,\ \rho_0=M_0e^{-3\Omega}$	$\Omega=\frac{1}{3}Ln\left[\frac{b_0t^2}{64}+\frac{a_1t}{4}\right],$	$b_0=384\pi GM_0,$
	$\beta_+=\pm\frac{n_1}{3a_1}\,Ln\left[\frac{16a_1+b_0t}{t}\right],$	$a_1^2=n_1^2+p_1^2,$
	$\beta_-=\pm\frac{p_1}{3a_1}\,Ln\left[\frac{16a_1+b_0t}{t}\right].$	

Table 2. Classical Solutions for $\gamma=-1,1,0,$ and constraints q, a_1 and b_0.

Introducing (equation 26) into (equation 42), we have

$$8e^{-3\Omega}\dot{P}_\Omega=2\lambda+(\gamma-1)b_\gamma e^{-3(\gamma+1)\Omega}, \tag{45}$$

which can be integrate to obtain the relation for P_Ω

$$P_\Omega=\pm\sqrt{a_1^2-\lambda e^{6\Omega}+b_\gamma e^{-3(\gamma-1)\Omega}}, \tag{46}$$

where $a_1^2=n_1^2+p_1^2$.

The set of equations (equation 39), (equation 40) and (equation 41) are equivalents to the set of equations (equation 34), (equation 35) and (equation 36), equations used to obtain the classical solutions.

Just to remark, the solutions obtained with the Hamiltonian formalism and the WKB-like procedure are equivalent to solving GR field equations.

3. Noncommutative Quantum Cosmology

There is a huge interest to noncommutative theories to explain the appropriate modification of Classical General Relativity, and hence of spacetime symmetries at short-distance scales, that implies modifications at large scales. General Quantum Mechanics arguments indicate that, it is not possible to measure a classical background spacetime at the Planck scale, due to the effects of gravitational backreaction (14). It is therefore tempting to incorporate the dynamical features of spacetime at deeper kinematical level using the standard techniques of noncommutative classical field theory based in the so called Moyal product in which for all calculations purposes (differentiation, integration, etc.) the space time coordinates are treated as ordinary (commutative) variables and noncommutativity enters into play in the way in which fields are multiplied (16). Using a deformation in the minisuperspace of this space variables in the Hamilton approach, as we are trying with the idea of noncommutative space time, we propose that the minisuperspace variables do not commute, for that purpose we will modified the Poisson structure, this approach does not modify the hamiltonian structure in the noncommutative fields.

Finding the classical cosmological solutions for any cosmological model in noncommutative gravity (5) is a very difficult task, this is a consequence of the highly nonlinear character of the theory. To avoid these difficulties, we will follow the original proposals of noncommutative quantum cosmology that was developed in (12). We start by presenting, in quite a general form the construction of noncommutative quantum cosmology and the WKB type method to calculate the classical evolution.

Let us start with a generic form for the commutative WDW equation, this is defined in the minisuperspace variables x, y. As mentioned in (12) a noncommutative deformation of the minisuperspace variables is assumed

$$[x, y] = i\theta, \tag{47}$$

this noncommutativity can be formulated in terms of noncommutative minisuperspace functions with the Moyal product of functions

$$f(x,y) \star g(x,y) = f(x,y) e^{i\frac{\theta}{2}\left(\overleftarrow{\partial_x}\overrightarrow{\partial_y} - \overleftarrow{\partial_y}\overrightarrow{\partial_x}\right)} g(x,y). \tag{48}$$

Then the noncommutative WDW equation can be written as

$$\left(-\Pi_x^2 + \Pi_y^2 - V(x,y)\right) \star \Psi(x,y) = 0, \tag{49}$$

we know from noncommutative quantum mechanics, that the symplectic structure is modified changing the commutator algebra. It is possible to return to the original commutative variables and usual commutation relations if we introduce the following change of variables

$$x \to x + \frac{\theta}{2}\Pi_y \qquad \text{and} \qquad y \to y - \frac{\theta}{2}\Pi_x, \tag{50}$$

the efects of the Moyal star product are reflected in the WDW equation, only through the potential

$$V(x,y) \star \Psi(x,y) = V(x + \frac{\theta}{2}\Pi_y, y - \frac{\theta}{2}\Pi_x), \tag{51}$$

taking this into account and using the usual substitutions $\Pi_{q^\mu} = -i\partial_{q^\mu}$ we arrive to

$$\left[\frac{\partial^2}{\partial x^2} - \frac{\partial^2}{\partial y^2} - V\left(x - i\frac{\theta}{2}\frac{\partial}{\partial y}, y + i\frac{\theta}{2}\frac{\partial}{\partial x}\right)\right]\Psi(x,y) = 0, \tag{52}$$

this is the Noncommutative WDW equation (NCWDW) and its solutions give the quantum description of the noncommutative universe. We can use the NCWDW to find the temporal evolution of our noncommutative cosmology by a WKB type procedure.

3.1 Noncommutative Kantowski-Sachs Cosmology

Using the method outlined and using (equation 4) we find the NCWDW equation

$$\left[\frac{\partial^2}{\partial \Omega^2} - \frac{\partial^2}{\partial \beta^2} - 48e^{-2\sqrt{3}\left(\Omega - i\frac{\theta}{2}\frac{\partial}{\partial \beta}\right)}\right]\psi(\Omega, \beta) = 0, \tag{53}$$

assuming that we can write $\Psi(\Omega,\beta) = e^{\sqrt{3}\nu\beta}X(\Omega)$ the equation for $X(\Omega)$ is

$$\left[-\frac{d^2}{d\Omega^2} + 48e^{-3i\nu\theta}e^{-2\sqrt{3}\Omega} + 3\nu^2\right] X(\Omega) = 0, \tag{54}$$

then the solutions of the NCWDW equation are

$$\Psi(\Omega,\beta) = e^{\pm i\sqrt{3}\nu\beta} K_{i\nu}\left(4e^{-\sqrt{3}\pm\frac{3}{2}\nu\theta}\right), \tag{55}$$

as already mentioned the NCWDW equation has the same problems as its commutative counterpart, it has no time dependence and unfortunately it can not be normalized. Usually the next step is to construct a " Gaussian" wave packet that can be normalized and do the physics with the new wave function. This is not needed for our purposes, as we will be applying the WKB method as in the previous section. Using equations (equation 6) and (equation 7) we arrive at

$$\begin{aligned} S_1(\beta) &= P_{\beta_0}\beta, \\ S_2(\beta) &= -\frac{1}{\sqrt{3}}\sqrt{P_{\beta_0}^2 - 48e^{-\sqrt{3}\theta P_{\beta_0}}e^{-2\sqrt{3}\Omega}} \\ &+ \frac{P_{\beta_0}}{\sqrt{3}}\text{arctanh}\left[\frac{\sqrt{P_{\beta_0}^2 - 48e^{-\sqrt{3}\theta P_{\beta_0}}e^{-2\sqrt{3}\Omega}}}{P_{\beta_0}}\right], \end{aligned} \tag{56}$$

using (equation 50) we get

$$\dot{\beta_C} = \frac{1}{2}\beta - \frac{1}{2}\dot{P_\Omega}, \qquad \dot{\Omega_C} = \frac{1}{2}\Omega + \frac{1}{2}\dot{P_\beta} \tag{57}$$

and the fact that the momenta are not modified we arrive to

$$\begin{aligned} \Omega(t) &= \frac{1}{2\sqrt{3}}\ln\left[\frac{48}{P_{\beta_0}^2}\cosh^2\left(2\sqrt{3}P_{\beta_0}(t-t_0)\right)\right] - \frac{1}{2}\theta P_{\beta_0}, \\ \beta(t) &= \beta_0 + 2P_{\beta_0}(t-t_0) \\ &- \frac{\theta}{2}P_{\beta_0}\tanh^2\left[2\sqrt{3}P_{\beta_0}(t-t_0)\right], \end{aligned} \tag{58}$$

this solutions have already been obtained in (15), in that paper they do a deformation of the simplectic structure at a classical level, modifying the Poisson brackets, to include noncommutativity.

3.2 Noncommutative FRW Cosmology with scalar field and Λ

We can use the NCWKB type method to FRW universe coupled to a scalar field. Proceeding as before the corresponding NCWDW equation is

$$\left[-\frac{1}{24}\frac{\partial^2}{\partial\alpha^2} + \frac{1}{2}\frac{\partial^2}{\partial\phi^2} + e^{6(\alpha - i\frac{\theta}{2}\frac{\partial}{\partial\phi})}\left(2\Lambda + 6ke^{-2(\alpha - i\frac{\theta}{2}\frac{\partial}{\partial\phi})}\right)\right]\Psi = 0. \tag{59}$$

From the NCWDW equation, we use the method developed in the previous sections and calculate the classical evolution by appliying the NCWKB type method. These results are presented in the next table

case	NC Quantum Solution	NC Classical Solution
k=0, $\Lambda \neq 0$	$\psi = e^{\pm i\nu \frac{\sqrt{3}}{2}\phi} K_{i\nu}\left[4\sqrt{\frac{\Lambda}{3}}\, e^{3\left(\alpha - \frac{3}{2}\nu\theta\right)}\right]$ and J_ν for $\Lambda < 0$	$\phi(t) = \phi_0 - P_{\phi_0} t$ $-\sqrt{3}\theta P_{\phi_0} \tanh\left(\frac{\sqrt{3}}{2} P_{\phi_0}(t - t_0)\right)$, $\alpha(t) = \frac{\theta}{2} P_{\phi_0} + \frac{1}{6}\ln\left(\frac{P^2_{\phi_0}}{4\Lambda}\right)$ $+\frac{1}{3}\ln\left(\text{sech}\left[\frac{\sqrt{3}}{2} P_{\phi_0}(t - t_0)\right]\right)$,
k $\neq 0$, $\Lambda = 0$	$\psi^{(1)} = e^{\pm i\frac{\nu}{\sqrt{3}}\phi} K_{i\nu}\left[6e^{2\left(\alpha - \frac{\theta}{2}\nu\right)}\right]$ for k = 1 $\psi^{(2)} = e^{\pm i\nu/\sqrt{3}\phi} J_\nu\left[6e^{2\left(\alpha - \frac{\theta}{2}\nu\right)}\right]$, for k = −1	$\phi(t) = \phi_0 - P_{\phi_0}(t - t_0)$ $-\sqrt{3}\theta P_{\phi_0} \tanh\left(\frac{P_{\phi_0}}{\sqrt{3}}(t - t_0)\right)$, $\alpha(t) = \frac{\theta}{2} P_{\phi_0} + \frac{1}{4}\ln\left[\frac{P^2_{\phi_0}}{12k}\right]$ $+\frac{1}{2}\ln\left(\text{sech}\left[\frac{1}{\sqrt{3}} P_{\phi_0}(t - t_0)\right]\right)$
k $\neq 0$, $\Lambda \neq 0$	Unknown	$\phi(t) = \phi_0 - P_{\phi_0}(t - t_0)$ $+6\theta \int e^{6\alpha}\left(\Lambda + 2e^{-2\alpha}\right) dt$, $\int \frac{d\alpha(t)}{\sqrt{P_{\phi_0} - 2e^{6\alpha + 3\theta P_{\phi_0}}\left(2\Lambda + 6ke^{-2\alpha - \theta P_{\phi_0}}\right)}}$ $= \frac{1}{\sqrt{12}}(t - t_0)$,

Table 3. Classical and quantum solutions for noncommutative FRW universe coupled to a scalar field. For these models noncommutativity is introduced in the gravitational and matter sectors. As in the commutative scenario, for $\Lambda \neq 0$ and k $\neq 0$ the noncommutative classical solution is given in an implicit form, and there is not a closed analytical quantum solutions. As in the commutative case we have fixed the value of the lapse function $N(t) = e^{3\alpha}$.

3.3 Stringy Noncommutative Quantum Cosmology

As in the previous examples we introduce the noncommutative relation $[\bar{\phi}, \beta] = i\theta$, and from the classical hamiltonian we find the NCWDW equation

$$\left[\frac{\partial^2}{\partial \bar{\phi}^2} - \frac{\partial^2}{\partial \beta^2} - \lambda_s^2 V(\bar{\phi}, \beta) e^{(m-2)\left(\bar{\phi} - i\frac{\theta}{2}\frac{\partial}{\partial \beta}\right)}\right] \Psi(\bar{\phi}, \beta) = 0. \tag{60}$$

The noncommutative wave function is

$$\Psi(\bar{\phi}, \beta) = e^{\pm - i\frac{m-2}{2}\nu\beta} K_{i\nu}\left[\frac{2\lambda_s\sqrt{V_0}}{m-2} e^{(m-2)\left(\bar{\phi} \mp \frac{m-2}{4}\theta\nu\right)}\right], \tag{61}$$

using the NCWKB type method the classical solutions for the noncommutative stringy cosmology are

$$\begin{aligned}
\bar{\phi}(\tau) &= \frac{1}{m-2}\ln\left[\frac{P^2_{\beta_0}}{V_0\lambda_s^2}\text{sech}^2\left(\frac{P_{\beta_0}}{2\lambda_s}(m-2)(\tau - \tau_0)\right)\right] \\
&\quad - \frac{\theta}{2} P_{\beta_0}, \\
\beta(\tau) &= \beta_0 + \frac{P_\beta}{\lambda_s}(\tau - \tau_0) \\
&\quad + \theta\frac{P_{\beta_0}}{2}\tanh\left[\frac{P_{\beta_0}}{2\lambda_s}(m-2)(\tau - \tau_0)\right],
\end{aligned} \tag{62}$$

the classical evolution for string cosmology can be calculated for $m = 0$ and $m = 4$. An interesting issue concerns the B field that is turned off in the string cosmology model (13) and does not contribute to the effective action. In open string theory, however noncommutativity arises precisely in the low energy limit of string theory in the presence of a constant B field. The θ parameter we have introduced in the minisuperspace could then be understood as a kind of B-field related with the Neveu-Schwarz B-field.

3.4 Noncommutative solutions of the isotropization in Bianchi I with barotropic perfect fluid and Λ cosmological

Let us begin introducing the noncommutative deformation of the minisuperspace in the WDW equation, this time, between all the variables of the minisuperspace, assuming that Ω_{nc} and $\beta_{\pm nc}$ obey the commutation relation

$$[\Omega_{\rm nc}, \beta_{-\rm nc}] = i\theta_1, \qquad [\Omega_{\rm nc}, \beta_{+\rm nc}] = i\theta_2, \qquad [\beta_{-\rm nc}, \beta_{+\rm nc}] = i\theta_3. \tag{63}$$

Instead of working directly with the physical variables Ω and $\beta_\pm$ we may achieve all the above solutions by making use of the auxiliary canonical variables $\Omega_{\rm nc}$ and $\beta_{\pm nc}$ defined as

$$\Omega_{nc} \equiv \Omega - \frac{\theta_1}{2}P_- - \frac{\theta_2}{2}P_+, \tag{64}$$

$$\beta_{-nc} \equiv \beta_- + \frac{\theta_1}{2}P_\Omega - \frac{\theta_3}{2}P_+, \tag{65}$$

$$\beta_{+nc} \equiv \beta_+ + \frac{\theta_2}{2}P_\Omega + \frac{\theta_3}{2}P_-. \tag{66}$$

maintaining the usual commutation relations between the fields, i.e., $[q^\mu, q^\nu] = 0$ and the identifications $P_\Omega = P_{\Omega\rm nc}$ and $P_\pm = P_{\pm\rm nc}$. With this shift and the usual canonical quantization $P_{q^\mu} \to -i\partial_{q^\mu}$, we arrive to the noncommutative WDW equation

$$\left[\frac{\partial^2}{\partial\Omega^2_{nc}} - \frac{\partial^2}{\partial\beta^2_{+nc}} - \frac{\partial^2}{\partial\beta^2_{-nc}} - \lambda e^{6\Omega_{nc}} + b_\gamma e^{-3(\gamma-1)\Omega_{nc}}\right]\Psi(\Omega,\beta_\pm) = 0, \tag{67}$$

where $\lambda = 48\Lambda$, $b_\gamma = 384\pi G M_\gamma$. At this point we have a noncommutative WDW equation and noncommutative hamiltonian. In what follows, we shall consider a wave function and apply the WKB procedure to obtain classical solutions.

3.4.1 Noncommutative classical solutions á la WKB

In order to find noncommutative classical solutions through the WKB approximation, we use the fact that $e^{i\theta\frac{\partial}{\partial x}}e^{\eta x} \equiv e^{i\eta\theta}e^{\eta x}$, and the ansatz for the wavefunction $\Psi(\Omega_{nc}, \beta_{\pm nc}) = e^{i[S_1(\Omega_{nc})\pm n_1\beta_{+nc}\pm p_1\beta_{-nc}]}$, where we use explicitly $S_2(\beta_{+nc}) = \pm n_1\beta_{+nc}$ and $S_3(\beta_{-nc}) = \pm p_1\beta_{-nc}$ to get the classical noncommutative Einstein-Hamilton-Jacobi (EHJ) equation

$$-\left(\frac{dS_1}{d\Omega_{nc}}\right)^2 + \left(\frac{dS_2}{d\beta_{+nc}}\right)^2 + \left(\frac{dS_3}{d\beta_{-nc}}\right)^2 - \lambda e^{6\Omega_{nc}} + be^{-3(\gamma-1)\Omega_{nc}} = 0, \tag{68}$$

which can be separate in a set of differential equations with $m_1^2 = n_1^2 + p_1^2$. We have the

Case	Noncommutative Solutions
$\gamma=-1,\ \Lambda\neq 0,\ \rho_{-1}=M_{-1}$	$\Omega_{nc}=\frac{1}{3}Ln\left[\frac{e^{2qt}-4a_1^2}{16qe^{qt}}\right]-\frac{\theta_1}{2}p_1-\frac{\theta_2}{2}n_1,$
$a_1^2=n_1^2+p_1^2,$	$\beta_{+nc}=\pm\frac{2}{3}\frac{n_1}{a_1}\,arctanh\left[\frac{e^{qt}}{2a_1}\right]+\frac{\theta_2}{8}\left(\frac{e^{qt}}{4}+a_1^2e^{-qt}\right)-\frac{\theta_3}{2}p_1,$
$q^2=24\pi GM_{-1}-3\Lambda,$	$\beta_{-nc}=\pm\frac{2}{3}\frac{p_1}{a_1}\,arctanh\left[\frac{e^{qt}}{2a_1}\right]+\frac{\theta_1}{8}\left(\frac{e^{qt}}{4}+a_1^2e^{-qt}\right)+\frac{\theta_3}{2}n_1,$
$\gamma=1,\ \Lambda<0,\ \rho_1=M_1e^{-6\Omega}$	$\Omega_{nc}=\frac{1}{3}Ln\left[\frac{e^{2qt}-4a_1^2}{16qe^{qt}}\right]-\frac{\theta_1}{2}p_1-\frac{\theta_2}{2}n_1,\quad q=\sqrt{3\lvert\Lambda\rvert},$
$a_1^2=n_1^2+p_1^2+384\pi GM_1,$	$\beta_{+nc}=\pm\frac{2}{3}\frac{n_1}{a_1}\,arctanh\left[\frac{e^{qt}}{2a_1}\right]+\frac{\theta_2}{8}\left(\frac{e^{qt}}{4}+a_1^2e^{-qt}\right)-\frac{\theta_3}{2}p_1,$
	$\beta_{-nc}=\pm\frac{2}{3}\frac{p_1}{a_1}\,arctanh\left[\frac{e^{qt}}{2a_1}\right]+\frac{\theta_1}{8}\left(\frac{e^{qt}}{4}+a_1^2e^{-qt}\right)+\frac{\theta_3}{2}n_1,$
$\gamma=1,\ \Lambda=0,\ \rho_1=M_1e^{-6\Omega}$	$\Omega_{nc}=\frac{1}{3}Ln\left[\frac{a_1}{4}t\right]-\frac{\theta_1}{2}p_1-\frac{\theta_2}{2}n_1,$
$a_1^2=n_1^2+p_1^2+384\pi GM_1,$	$\beta_{+nc}=\pm Ln\left[t^{-\frac{n_1}{3a_1}}\right]+\frac{\theta_2}{2}a_1-\frac{\theta_3}{2}p_1,$
	$\beta_{-nc}=\pm Ln\left[t^{-\frac{p_1}{3a_1}}\right]+\frac{\theta_1}{2}a_1+\frac{\theta_3}{2}n_1,$
$\gamma=0,\ \Lambda=0,\ \rho_0=M_0e^{-3\Omega}$	$\Omega_{nc}=\frac{1}{3}Ln\left[\frac{b_0t^2}{64}+\frac{a_1t}{4}\right]-\frac{\theta_1}{2}p_1-\frac{\theta_2}{2}n_1,$
$b_0=384\pi GM_0,$	$\beta_{+nc}=\pm\frac{n_1}{3a_1}\,Ln\left[\frac{16a_1+b_0t}{t}\right]+\frac{\theta_2}{2}\sqrt{a_1^2+\frac{b_0t^2}{64}+\frac{a_1t}{4}}-\frac{\theta_3}{2}p_1,$
$a_1^2=n_1^2+p_1^2,$	$\beta_{-nc}=\pm\frac{p_1}{3a_1}\,Ln\left[\frac{16a_1+b_0t}{t}\right]+\frac{\theta_1}{2}\sqrt{a_1^2+\frac{b_0t^2}{64}+\frac{a_1t}{4}}+\frac{\theta_3}{2}n_1.$

Table 4. Noncommutative solutions for, $\gamma=-1,1,0$, and constraints q, a_1 and b_0.

following noncommutative equations of motion

$$\pm\sqrt{a_1^2-\lambda e^{6\Omega_{nc}}+b_\gamma e^{-3(\gamma-1)\Omega_{nc}}}\equiv-12e^{3\Omega_{nc}}\frac{\dot{\Omega}_{nc}}{N}, \tag{69}$$

$$\pm n_1\equiv 12e^{3\Omega_{nc}}\frac{\dot{\beta}_{+nc}}{N}, \tag{70}$$

$$\pm p_1\equiv 12e^{3\Omega_{nc}}\frac{\dot{\beta}_{-nc}}{N}. \tag{71}$$

One just need to be careful in (equation 69), (equation 70) and (equation 71), and apply the chain rule to the variables (equation 64), (equation 65) and (equation 66), in order to get the right solution, $\dot{\beta}_{-nc}=\frac{\partial\beta_-}{\partial t}+\frac{\partial\beta_{-nc}}{\partial P_\Omega}\frac{\partial p_\Omega}{\partial t}+\frac{\partial\beta_{-nc}}{\partial P_+}\frac{\partial p_+}{\partial t}+\frac{\partial\beta_{-nc}}{\partial P_-}\frac{\partial p_-}{\partial t}=\dot{\beta}_-+\frac{\theta_1}{2}\dot{P}_\Omega$. In this sense, all solutions to find in the commutative case, remain for the noncommutative case with the corresponding shift, as we show in the table 4.

3.4.2 Noncommutative classical solutions á la Hamilton

In the commutative model we know that the solutions to hamiltons equations are the same as in General Relativity. Now the natural extension is to consider the noncommutative version of our model, with the idea of noncommutative between the three variables $(\Omega_{nc},\beta_{\pm nc})$, so we apply a deformation of the Poisson algebra. For this we start with the usual hamiltonian (equation 26), but the symplectic structure is modify as follow

$$\{P_\Omega,P_\pm\}_\star=\{P_+,P_-\}_\star=0,\qquad \{q^\mu,P_{q^\mu}\}_\star=1, \tag{72}$$

$$\{\Omega,\beta_-\}_\star=\theta_1,\quad \{\Omega,\beta_+\}_\star=\theta_2,\quad \{\beta_-,\beta_+\}_\star=\theta_3. \tag{73}$$

where the $\star$ is the Moyal product. In the second case, the hamiltonian is modify by the shift (equation 64),(equation 65) and (equation 66) resulting

$$H_{nc} = \frac{Ne^{-3\Omega_{nc}}}{24}\left[-P_\Omega^2 + P_+^2 + P_-^2 - \lambda e^{6\Omega_{nc}} + b_\gamma e^{-3(\gamma-1)\Omega_{nc}}\right] = 0, \tag{74}$$

but the symplectic structure is the one that we know, the commutative one (equation 38). The noncommutative equations of motion, for the first formalism that we exposed have the original variables, but with the variables modified,

$$\begin{aligned} \dot{q}^{\mu}_{nc} &= \{q^\mu, H\}_\star, \\ \dot{P}^{\mu}_{nc} &= \{P^\mu, H\}_\star, \end{aligned} \tag{75}$$

and for the second formalism we use the shifted variables but with the original (commutative) symplectic structure

$$\begin{aligned} \dot{q}^{\mu}_{nc} &= \{q^\mu_{nc}, H_{nc}\}, \\ \dot{P}^{\mu}_{nc} &= \{P^\mu_{nc}, H_{nc}\}, \end{aligned} \tag{76}$$

in both approaches we have the same result. Therefore the equations of motion take the form

$$\dot{\Omega}_{nc} = \{\Omega, H\}_\star = \{\Omega_{nc}, H_{nc}\} = -\frac{e^{-3\Omega_{nc}}}{12}P_\Omega, \tag{77}$$

$$\dot{\beta}_{-nc} = \{\beta_-, H\}_\star = \{\beta_{-nc}, H_{nc}\} = \frac{e^{-3\Omega_{nc}}}{12}P_- + \frac{\theta_1}{2}\dot{P}_\Omega, \tag{78}$$

$$\dot{\beta}_{+nc} = \{\beta_+, H\}_\star = \{\beta_{+nc}, H_{nc}\} = \frac{e^{-3\Omega_{nc}}}{12}P_+ + \frac{\theta_2}{2}\dot{P}_\Omega, \tag{79}$$

$$\dot{P}_\Omega = \{P_\Omega, H\}_\star = \{P_\Omega, H_{nc}\} = \frac{e^{-3\Omega_{nc}}}{8}\left[6\lambda e^{6\Omega_{nc}} + 3(\gamma-1)b_\gamma e^{-3(\gamma-1)\Omega_{nc}}\right], \tag{80}$$

$$\dot{P}_- = \{P_-, H\}_\star = \{P_-, H_{nc}\} = 0, \quad \to \quad P_- = p_1, \tag{81}$$

$$\dot{P}_+ = \{P_+, H\}_\star = \{P_+, H_{nc}\} = 0, \quad \to \quad P_+ = n_1. \tag{82}$$

if we proceed as in the commutative case we get the solutions showed in the table IVA.

4. Conclusions and outlook

In this chapter we have presented the NCWKB type method for noncommutative quantum cosmology and with this procedure, found the noncommutative classical solutions for several noncommutative quantum cosmological models.

Noncommutativity is incorporated in the minisuperspace variables, in a similar manner as it is a proposal that originally emerged at the quantum level, by this reason we considered it as in standard quantum mechanics. By means of the WKB approximation on the corresponding NCWDW equation, one gets the noncommutative generalized Einstein-Hamilton-Jacobi equation (NCEHJ), from which the classical evolution of the noncommutative model is obtained. The examples we studied were the Kantowski-Sachs cosmological model, the FRW universe with cosmological constant and coupled to a scalar field, a string quantum cosmological model and Bianchi I with Barotropic perfect fluid and Λ Cosmological.

In the commutative scenario, that the classical solutions found from the WKB-type method are solutions to the corresponding Einsteins field equations. Due to the complexity of the noncommutative theories of gravity(5), classical solutions to the noncommutative field equations are almost impossible to find, but in the approach of noncommutative quantum cosmology and by means of the WKB-type procedure, they can be easily constructed. Also the quantum evolution of the system is not needed to find the classical behavior, from table 2 we can see that for the case $\Lambda \neq 0$ and $k \neq 0$ the wave function can not be analitacally calculated, but still the noncommutative effects can be incorporated and the classical evolution is found implicitly. This procedure gives a straightforward algorithm to incorporate noncommutative effects to cosmological models. In this approach the effects of noncommutativity are encoded in the potential through the Moyal product of functions (equation 51). We only need the NCWDW equation and the approximations (equation 7), to get the NCEHJ and from it, the noncommutative classical behavior can easily be constructed. As already mentioned, in(11) the effects of noncommutativity were studied in connection with inflation, but the noncommutative deformation was only done in the matter sector neglecting the gravity sector. For completeness to the section 2.4 and 3.4 we present the solutions in the gauge $N = 24e^{3\Omega}$ (see appendix A and B), one of the advantages of this gauge is that the solutions are very9 simple, this is something to take into account when we introduce a more complex form of matter, where in the gauge $N(t) = 1$ analytical solutions can not be found. The procedure developed here has the advantage that we can implement noncommutativity in both sectors in a straightforward way and find the classical solutions (i.e. inflationary models). The study on deformed phase space for all of this cosmologies should be constructed with this noncommutative proposal. These ideas are being explored and will be reported elsewhere.

5. Acknowledgments

We will like to thank M. P. Ryan for enlightening discussions on quantum cosmology. This work was partially supported by PROMEP 103.5/10/6209.

6. Appendix

A Commutative classical solutions in the Gauge $N = 24e^{3\Omega}$

In this appendix we present the classical solutions in the gauge $N = 24e^{3\Omega}$, the equations are much more simpler to solve in this guage.

A.1 Commutative Classical Solutions á la WKB

The master equation becomes

$$2dt = \frac{d\Omega}{\sqrt{a_1^2 - \lambda e^{6\Omega} + b_\gamma e^{-3(\gamma-1)\Omega}}}, \tag{83}$$

and the other two equations are immediately integrable. Again for particular cases in the γ parameter, we present the classical solutions, table 5

A.2 Classical solutions via Hamiltonian formalism

With the gauge fixed to $N = 24e^{3\Omega}$ we can see that the hamiltonian takes the form

$$H = -P_\Omega^2 + P_+^2 + P_-^2 - \lambda e^{6\Omega} + b_\gamma e^{-3(\gamma-1)\Omega} = 0. \tag{84}$$

Case	Commutative solutions
$\gamma=-1,\ \Lambda\neq 0,\ \rho_{-1}=M_{-1}$	$\Omega=\frac{1}{6}Ln\left[-\frac{a_1^2}{384\pi GM_{-1}-48\Lambda}\,Sech^2\left(6a_1t\right)\right],\quad a_1^2=n_1^2+p_1^2,$ $\beta_+=\pm 2n_1t,$ $\beta_-=\pm 2p_1t.$
$\gamma=1,\ \Lambda\neq 0,\ \rho_1=M_1e^{-6\Omega}$	$\Omega=\frac{1}{6}\,Ln\left[\frac{a_1^2}{48\Lambda}\,Sech^2(6a_1t)\right],\quad a_1^2=n_1^2+p_1^2+384\pi GM_1,$ $\beta_+=\pm 2n_1t,$ $\beta_-=\pm 2p_1t.$
$\gamma=1,\ \Lambda=0,\ \rho_1=M_1e^{-6\Omega}$	$\Omega=2\sqrt{a_1^2+b_1t},\qquad a_1^2=n_1^2+p_1^2+384\pi GM_1,$ $\beta_+=-2n_1t,$ $\beta_-=-2p_1t.$
$\gamma=0,\ \Lambda=0,\ \rho_0=M_0e^{-3\Omega}$	$\Omega=\frac{1}{3}Ln\left[-\frac{a_1^2}{b_0}\,sech^2(3a_1t)\right],\quad b_0=384\pi GM_0,$ $\beta_+=\pm 2n_1t,\quad a_1^2=n_1^2+p_1^2,$ $\beta_-=\pm 2p_1t.$
$\gamma=\frac{1}{3},\ \Lambda=0,\ \rho_0=M_{\frac{1}{3}}e^{-4\Omega}$	$\Omega=\frac{1}{2}Ln\left[-\frac{a_1^2}{b_{\frac{1}{3}}}\,sech^2(2a_1t)\right],\quad a_1^2=n_1^2+p_1^2,$ $\beta_+=\pm 2n_1t,\qquad b_{\frac{1}{3}}=384\pi GM_{\frac{1}{3}},$ $\beta_-=\pm 2p_1t.$

Table 5. Classical Solutions for $\gamma=-1,\frac{1}{3},1,0,$ and constraints a_1, b_0 and b_1.

The Poisson brackets structure yields to equations of motion

$$\dot{\Omega}=\{\Omega,H\}=-2P_\Omega, \tag{85}$$
$$\dot{\beta_-}=\{\beta_-,H\}=2P_-,\qquad\rightarrow\qquad \beta_-=\pm 2p_1t, \tag{86}$$
$$\dot{\beta_+}=\{\beta_+,H\}=2P_+,\qquad\rightarrow\qquad \beta_+=\pm 2n_1t, \tag{87}$$
$$\dot{P}_\Omega=\{P_\Omega,H\}=\left[+6\lambda e^{6\Omega}+3(\gamma-1)b_\gamma e^{-3(\gamma-1)\Omega}\right], \tag{88}$$
$$\dot{P}_-=\{P_-,H\}=0,\qquad\rightarrow\qquad P_-=\pm p_1=const. \tag{89}$$
$$\dot{P}_+=\{P_+,H\}=0,\qquad\rightarrow\qquad P_+=\pm n_1=const. \tag{90}$$

Using (equation 84), introducing (equation 89) and (equation 90), we obtain the expression for P_Ω

$$P_\Omega=\sqrt{m_1^2-\lambda e^{6\Omega}+b_\gamma e^{-3(\gamma-1)\Omega}}, \tag{91}$$

being self-consistent with equation (equation 88), where $a_1^2=n_1^2+p_1^2$. Introducing this equation into (equation 85) we get the master equation found to solve the Einstein field equation in this gauge, where the classical solutions are presented in table IIIA.

B Noncommutative classical solutions

B.1 Noncommutative classical solutions in the Gauge $N=24e^{3\Omega}$ á la WKB and via Hamiltonian formalism

The noncommutative solutions in the space q^μ become

Case	Noncommutative Solutions
$\gamma=-1,\ \Lambda\neq 0,\ \rho_{-1}=M_{-1}$ $a_1^2=n_1^2+p_1^2,$	$\Omega_{nc}=\frac{1}{6}Ln\left[-\frac{a_1^2}{384\pi GM_{-1}-48\Lambda}\,Sech^2\left(6a_1t\right)\right]-\frac{\theta_1}{2}p_1-\frac{\theta_2}{2}n_1,$ $\beta_{+nc}=\pm 2n_1t+\frac{\theta_2 a_1}{2}\,tanh(6a_1t)-\frac{\theta_3}{2}p_1,$ $\beta_{-nc}=\pm 2p_1t+\frac{\theta_1 a_1}{2}\,tanh(6a_1t)+\frac{\theta_3}{2}n_1,$
$\gamma=1,\ \Lambda\neq 0,\rho_1=M_1e^{-6\Omega}$ $a_1^2=n_1^2+p_1^2+384\pi GM_1,$	$\Omega_{nc}=\frac{1}{6}\,Ln\left[\frac{a_1^2}{48\Lambda}\,Sech^2(6a_1t)\right]-\frac{\theta_1}{2}p_1-\frac{\theta_2}{2}n_1,$ $\beta_{+nc}=\pm 2n_1t+\frac{\theta_2 a_1}{2}\,tanh(6a_1t)-\frac{\theta_3}{2}p_1,$ $\beta_{-nc}=\pm 2p_1t+\frac{\theta_1 a_1}{2}\,tanh(6a_1t)+\frac{\theta_3}{2}n_1,$
$\gamma=1,\ \Lambda=0,\ \rho_1=M_1e^{-6\Omega}$ $a_1^2=n_1^2+p_1^2+384\pi GM_1,$	$\Omega_{nc}=2a_1t-\frac{\theta_1}{2}p_1-\frac{\theta_2}{2}n_1,$ $\beta_{+nc}=-2n_1t+\frac{\theta_2}{2}a_1-\frac{\theta_3}{2}p_1,$ $\beta_{-nc}=-2p_1t+\frac{\theta_1}{2}a_1+\frac{\theta_3}{2}n_1,$
$\gamma=0,\ \Lambda=0,\ \rho_0=M_0e^{-3\Omega}$ $b_0=384\pi GM_0,$ $a_1^2=n_1^2+p_1^2,$	$\Omega_{nc}=\frac{1}{3}Ln\left[-\frac{a_1^2}{b_0}\,sech^2(3a_1t)\right]-\frac{\theta_1}{2}p_1-\frac{\theta_2}{2}n_1,$ $\beta_{+nc}=\pm 2n_1t+\frac{\theta_2 a_1}{2}\,tanh(3a_1t)-\frac{\theta_3}{2}p_1,$ $\beta_{nc}-=\pm 2p_1t+\frac{\theta_1 a_1}{2}\,tanh(3a_1t)+\frac{\theta_3}{2}n_1.$
$\gamma=\frac{1}{3},\ \Lambda=0,\ \rho_0=M_{\frac{1}{3}}e^{-4\Omega}$ $a_1^2=n_1^2+p_1^2,$	$\Omega_{nc}=\frac{1}{2}Ln\left[-\frac{a_1^2}{b_{\frac{1}{3}}}\,sech^2(2a_1t)\right]-\frac{\theta_1}{2}p_1-\frac{\theta_2}{2}n_1,$ $\beta_{+nc}=\pm 2n_1t+\frac{\theta_2 a_1}{2}\,tanh(2a_1t)-\frac{\theta_3}{2}p_1$ $\beta_{-nc}=\pm 2p_1t+\frac{\theta_1 a_1}{2}\,tanh(2a_1t)+\frac{\theta_3}{2}n_1.$

Table 6. Noncommutative solutions for $\gamma=-1,\frac{1}{3},1,0$, and constraints a_1, b_0 and b_1.

7. References

[1] H. Snyder, Phys. Rev. 71, 38 (1947).

[2] A. Connes, M. R. Douglas, and A. Schwarz, JHEP 9802:003 (1998).

[3] N. Seiberg and E. Witten, JHEP 9909:032 (1999).

[4] B. Jurco, L. Moller, S. Schraml, P. Schupp and J. Wess, Eur. Phys. J. C 21, 383 (2001),hep-th/0104153.

[5] H. García-Compeán, O. Obregón, C. Ramírez and M. Sabido, Phys. Rev. D 68 (2003) 044015; H. García-Compeán, O. Obregón, C. Ramírez and M. Sabido, Phys. Rev. D 68 (2003) 045010; M. Bañados, O. Chandia, N. Grandi, F.A. Schaposnik and G.A. Silva, Phys. Rev. D 64 (2001) 084012; H. Nishino and S. Rajpoot, Phys. Lett. B 532 (2002) 334; V.P. Nair, "Gravitational Fields on a Noncommutative Space", hep-th/0112114; S. Cacciatori, D. Klemm, L. Martucci and D. Zanon, Phys. Lett. B 536 (2002) 101; S. Cacciatori, A.H. Chamseddine, D. Klemm, L. Martucci, W.A. Sabra and D. Zanon, "Noncommutative Gravity in Two Dimensions", hep-th/0203038; Y. Abe and V.P. Nair,"Noncommutative Gravity: Fuzzy Sphere and Others", hep-th/0212270; M.A. Cardella and Daniela Zanon, "Noncommutative Deformation of Four-dimensional Einstein Gravity", hep-th/0212071; A.H. Chamseddine, "Invariant Actions for Noncommutative Gravity", hep-th/0202137; J.W. Moffat, Phys. Lett. B 491 (2000) 345; Phys. Lett. B 493 (2000) 142; V. O. Rivelles, Phys. Lett. B 558 (2003) 19; E. Harikumar and V. O. Rivelles, arXiv:hep-th/0607115.

[6] S. Minwalla, M. Van Raamsdonk and N. Seiberg, JHEP 0002, 020 (2000).

[7] M. R. Douglas and N. A. Nekrasov, Rev. Mod. Phys. 73, 977 (2001).

[8] M.P. Ryan and L.C. Shepley, in: *Homogeneous Relativistic Cosmologies*, Princeton University Press, Princeton, New Jersey (1975).

[9] M. MacCallum, in: *General Relativity: An Einstein Centenary Survey*, edited by S. Hawking and W. Israel (Cambridge University Press, Cambridge, England, 1979).
[10] J. M. Romero and J. A. Santiago, Mod. Phys. Lett. A 20 (2005) 78.
[11] R. Brandenberger and P. M. Ho, Phys. Rev. D 66 (2002) 023517; Q. G. Huang and M. Li, Nucl. Phys. B 713 (2005) 219; Q. G. Huang and M. Li, JHEP 0306 (2003) 014; H. s. Kim, G. S. Lee, H. W. Lee and Y. S. Myung, Phys. Rev. D 70 (2004) 043521; H. s. Kim, G. S. Lee and Y. S. Myung, Mod. Phys. Lett. A 20 (2005) 271; D. J. Liu and X. Z. Li, Phys. Rev. D 70 (2004) 123504.
[12] H. García-Compeán, O. Obregón and C. Ramírez, Phys. Rev. Lett. 88, 161301 (2002).
[13] M. Gasperini, J. Maharana and G. Veneziano, Nucl. Phys. B 472 (1996) 349.
[14] R.J. Szabo, Int. J. Mod. Phys. A 19, 1837 (2004), [arXiv:physics/0401142].
[15] G.D. Barbosa, N. Pinto-Neto, Phys. Rev. D 70, 103512 (2004), [arXiv:hep-th/0407111].
[16] R.J. Szabo, Phys. Rep. 378, 207 (2003), [arXiv:hep-th/0109162].
[17] J.M. Carmona, J.L. Cortes, J. Gamboa, and F. Mendez, JHEP 0303, 058 (2003). [arXiv:hep-th/0301248].
[18] B. Vakili, N. Khosravi, H.R. Sepangi, Class. Quantum Grav. 24, 931 (2007). [arXiv:gr-qc/0701075].

Part 6

A Finite Lifetime Universe

19

Small-Bang versus Big-Bang Cosmology

Antonio Alfonso-Faus
EUIT Aeronáutica, Universidad Politécnica de Madrid/ Escuela de Ingeniería Aeroespacial
Spain

1. Introduction

During the 1920´s the U.S. astronomer Hubble observed that the red shifts, from distant galaxies, were increasing with distance. The similarity with the well known Doppler Effect gave the way to a rapidly spreading idea: that the universe was expanding. Galaxies were thought to be receding from us at a speed proportional to their distance. Considering the universe as a "gas" of galaxies, each galaxy similar to one molecule in a gas, the expansion clearly implied that the universe was getting cooler and thinner with age. We know today that this cooling and thinning is correct: the universe is very old and its known temperature and density for today are very low: 2.7ºK (the cosmic microwave background radiation) of temperature and about $\sim 10^{-29}$ grams/cc average density.

Now, if we imagine a thought experiment and reverse the time, going backwards, we get the idea of a very hot and very dense universe at its initial stages. Going to the limit, getting closer and closer to a theoretically zero time, we have a mathematical singularity: infinite temperature and infinite density. As a result of this initial picture, we can imagine that these infinites were the result of a very big and sudden explosion: and that it expanded rapidly to a lower and lower temperature and density. Today we observe a cool temperature and a thin "gas". The British astronomer and cosmologist Fred Hoyle ironically called this a "Big-Bang". But if we take the imagination of a Big-Bang explosion as a fact, as many people have done, we are entitled to take as a fact too that a gas after an explosion decelerates to a lower and lower speed of expansion. Today we should observe a decelerating universe. And this is not the case.

Initial expansion (according to a hypothetical Big-Bang), and present acceleration of the universe, as observed thanks to the astrophysics of the supernovae Type Ia, are two very different things. While the expansion is very well based on observation, following the Hubble´s red shift findings, an initial explosion at a space-time point, the Big-Bang, is the result of a mathematical extrapolation, and therefore so far it is only speculation. On the other hand, the accelerated expansion of the universe is based on observation [1]. It is the result of the successful application of the scientific method, like the case of the expansion of the universe. Accelerated expansion is a very well based observation on scientific grounds, and in a direct way. This is not so for the assumed Big-Bang initial explosion.

It is very interesting to note that the cosmological model of a Big-Bang, as a frame of work, has been and still is the underpinning of the majority of the research work done in cosmology. It had, and still has, many drawbacks. One of them, a very serious one, was related to the fact that it could not explain the present size of the universe. Following the initial developments of this model the present size of the universe would be very small: may be of the order of meters. Obviously this is not the case, and one had to look for an explanation. Instead of looking for an alternative model, something that the many drawbacks of the model has demanded many times, the main stream of the scientific community in cosmology has always decided to add more and more "ad hoc" explanations to keep this frame of work alive. And it appears that this is going to go on for a long time. There is so much work, interests, beliefs, efforts, etc, behind the Big-Bang idea that the overall worldwide inertia created by this cosmological model is very big indeed.

At any rate, some of the "ad hoc" explanations to sustain the main stream ideas could be good ones. I mean good ones when one considers them isolated, independent of the reason that made them come into existence. For example: INFLATION. A very fast exponential expansion at the very early stages of the universe would bring it close to a reasonable size to avoid discrepancies. It has some predictions, flat universe, critical density, cosmic microwave background radiation properties (CMBR) etc. that have been observed. Then, it seems to be a good idea, a good scientific approach supported by the confirmation of some of its predictions.

Again, if one accepts INFLATION as a beginning of the universe, a fast exponential expansion during a very short time, one immediately imagines that after inflation a period of deceleration should follow. This is the case, but there is more to it. As mentioned above, during the last half of the age of the universe there is an observed accelerated expansion. And of course this must have a reason, a physical push to expand the universe. This physical mechanism must be of universal significance, because it has been accelerating the whole universe during the last half of its age. And during the first half it counteracted the inertial deceleration after inflation due to the gravitational universal attraction. And it reversed the deceleration giving the accelerated expansion we observe today. About half way in time the deceleration-acceleration transition implied zero acceleration. We see no need of an initial point like explosion. Inflation does the job.

History has already gone through this state of affairs. Almost one hundred years ago, when Einstein developed his cosmological equations, the general belief was to imagine the universe in a static state. Since gravitation was well known, as an attractive force, soon it was realized that a collapse was inevitable due to the pull of gravity. But no collapsing universe was observed. Then a pushing mechanism should be balancing gravitation to get a static universe. And Einstein introduced his well known cosmological constant, the lambda constant. Today we observe the universe in an accelerated expansion during the last half of its age. Then a kind of pushing mechanism is again required to explain this observation. And it could be a question of strength: the pushing force due to the Λ "constant" seems to be growing as the universe expands. On the contrary, the overall gravitational force is constant. This is enough to explain by itself the present acceleration of the expansion of the universe. And it may arrive at a disaggregation of everything in a finite time: expansion to infinity. Since the lambda constant is a very well known physical construct, the attention of most cosmologists is again in favor of such solution. The point is that such a Λ constant implies energy, and the immediate and easy way out is to imagine the existence of a kind of dark energy to explain Λ, so dark that no one has seen it yet. We do not know of any interaction between this postulated dark energy and any other well known energy we are used to observe and identify. So far, the dark energy is just a theoretical construct. But we have more choices to explain this pushing force. Aside from believing in dark energy one can believe in an equivalent mechanism to explain the push: creation of matter, *as we will see* [2]. Then the sequence of events to explain the dynamics of the universe would be: fast exponential inflation, and then a slow deceleration followed by a slow acceleration as of today. And our prediction is that this late acceleration is increasing and that it will disperse the whole universe to infinity in a finite time. Like a kind of second inflation at the end of the time of the universe as we know it. We may be now at about one half of the total age of our universe. The creation pressure [2] is always present, growing, and its effects are permanently present till the final stage.

Following the arguments given above, we can make now a straight forward proposal: there was no big bang at all. Instead we can say that we are the result of an initial small bang, just after inflation of an initial fluctuation, an initial quantum black hole whose inflation a little later decelerated. But this deceleration was overcome by the push of the creation pressure, the continuous creation of matter [2], [3], [4] and [5]. As we will see, most physicals properties of the universe are subject to this increase with time.

The above considerations are in agreement with the idea that the universe is a kind of black hole [6]. Black holes have a characteristic mass-size relation. Taking the gravitational constant G and the speed of light c as units, G = c = 1, the black hole mass M is equal to his size L, within a factor of 2. Then, dividing the size L by the speed of light c one gets a characteristic time t for the black hole. In these units 1 second equals 3×10^{10} cm, and this equals $\sim 10^{40}$ grams. We then have:

$$(\sim 2)\ M = L = t \tag{1}$$

For the universe $M = L = t \sim 10^{56}$ grams $\sim 10^{28}$ cm $\sim 10^{10}$ years. For the Planck scale, a quantum black hole, one has to divide (1) by $\sim 10^{61}$ to get the Planck´s units $m = l = t \sim 10^{-5}$ grams $\sim 10^{-33}$ cms $\sim 10^{-44}$ seconds. Possibly this may be the first quantum of everything in our universe. All the basic physical properties at the Planck scale (the so called natural units) differ by the factor $\sim 10^{-61}$ from the scale of the universe.

2. Scale cosmology

It looks like the universe can be considered to be structured in different scales. Each scale is a quantum black hole, as we will see, and is in itself a universe too. A black hole has its mass M and its size L connected by the simple relation

$$(\sim 2)\ GM/c^2 = L \tag{2}$$

On the other hand, a *quantum* black hole is characterized by its size L being equal to its the de Broglie wavelength (with a generalized Planck´s constant **H)**

$$L = \mathbf{H}/Mc \tag{3}$$

Now, combining (2) and (3) we get (for a general quantum black hole) the mass M, length L and time t as follows

$$M = (\mathbf{H}c/G)^{1/2}$$

$$L = (G\mathbf{H}/c^3)^{1/2} \tag{4}$$

$$t = (G\mathbf{H}/c^5)^{1/2}$$

If we use the natural Planck´s constant ħ in (4) we get the Planck´s units

$$m_p = (\hbar c/G)^{1/2} \approx 2.177\ 10^{-5}\ \text{grams}$$

$$l_p = (G\hbar/c^3)^{1/2} \approx 1.616\ 10^{-33}\ \text{cms} \qquad (5)$$

$$t_p = (G\hbar/c^5)^{1/2} \approx 5.39\ 10^{-44}\ \text{sec}$$

The scale of our universe is found to be the Planck´s scale (5) multiplied by the factor ~ 10^{61}or, equivalently, by using a universal Planck´s constant $\mathbf{H} \approx 10^{122}\,\hbar$ giving

$$M_u = (\mathbf{H}c/G)^{1/2} \approx 10^{56}\ \text{grams}$$

$$L_u = (G\mathbf{H}/c^3)^{1/2} \approx 10^{28}\ \text{cms} \qquad (6)$$

$$t_u = (G\mathbf{H}/c^5)^{1/2} \approx 5\ 10^{17}\ \text{sec}$$

There is a new scale that can be defined below the Planck´s scale. The point is that the quantum of gravity [7] has a mass m_g given by

$$m_g = \hbar/c^2 t_u \approx 2\ 10^{-66}\ \text{grams} \qquad (7)$$

and it defines a scale like Planck´s scale multiplied by, once again, the factor 10^{-61} . This is equivalent to obtain this new scale by using an equivalent generalized Planck´s constant $\mathbf{H} \approx 10^{-122}\,\hbar$ giving the sub-Planck scale

$$m_{sp} = 10^{-61}\,(\hbar c/G)^{1/2} \approx 2\ 10^{-66}\ \text{grams}$$

$$l_{sp} = 10^{-61}\,(G\hbar/c^3)^{1/2} \approx 10^{-94}\ \text{cms} \qquad (8)$$

$$t_{sp} = 10^{-61}\,(G\hbar/c^5)^{1/2} \approx 10^{-104}\ \text{sec}$$

The physical meaning of the sub-Planckian scale (8) is not yet very well known, except for the concept of the quantum of gravity m_g that we have introduced [7] in the past. It may also have a meaning related to information [8]: in a parallel way it can be given a sense as the unit of information, the bit, with the physical properties in (8). We can also give some sense for a quantum of time, defined as the minimum interval of time obtained using the mass of the universe

$$t_{sp} = \hbar/M_u c^2 \approx 10^{-104}\ \text{sec} \qquad (9)$$

This is a very suggestive relation: it means that the sub-Planckian scale (8) gives us the minimum quantum of mass, length and time. The three quantum black hole scales, (5), (6) and (8) are then the minimum scale (8), sub-Planckian, the "natural" scale (5), Planckian, and the scale of the universe (6) where we live.
There is a new physical parallel that gives a meaning to the sub-Planckian "quantum". It may be regarded as the unit of information, the bit [8]

3. Gravity as an emerging entropic force

Verlinde [9] has introduced the concept of the force of gravity as due to a gradient of entropy S, i.e. gravity as an emergent entropic force. Though the change in entropy S may be due to internal redistribution of masses in the system, it may also be due to a cosmological increase of mass with time, as we will see here. The basic idea can be expressed as the relation between temperature T, entropy S and energy Mc^2, according to the thermodynamic relation

$$T\,\Delta S = \Delta M c^2 = \Delta t/2 \tag{10}$$

We have used the Machian black hole relation $2GM/c^2 = ct$ to obtain the last term in (10) Dealing with a "quantum" black hole universe with $H \approx 10^{122}\ \hbar$, we have equations (1), and (2), and using the Hawking [10] and Bekenstein [11] black hole relation for the entropy S

$$S = 4\pi\, k/\hbar c\, GM^2 \tag{11}$$

we get from (11), with $G = c = \hbar = k = 1$

$$\Delta S = 4\pi\, 2M\, \Delta M \tag{12}$$

And using (10) and (12) we have

$$T\,\Delta S = 2\pi\, Tt\, \Delta t = \Delta t/2 \tag{13}$$

i.e.

$$4\pi\, Tt = k/\hbar = 1 \tag{14}$$

Then temperature varies inversely proportional to cosmological time. This is a well known relation in our universe. But here we have a surprising possibility: since the temperature T is a statistical parameter, then the time t may have this character too.
The mass of the universe must be time varying [14], so that the gradient of M in (12) is responsible for the increase in entropy ΔS, and therefore for the force of gravity. Verlinde´s ideas [9] may be extended to a distribution of mass in the whole universe varying with cosmological time.

4. The cosmological constant versus the pressure of creation

The cosmological constant Λ has been related to the vacuum energy, and therefore to a negative pressure, to explain the accelerated expansion of the universe. Recently we have an interesting suggestion [1]: it implies that there is no cosmological constant. Its theoretical need can also be fulfilled by a creation pressure p_c. At any rate, either Λ or a creation pressure implies (with c = 1), from Einstein cosmological equations:

$$\Lambda \approx 1/t^2 \tag{15}$$

And from (1) and (11) we get

$$S \approx t^2 \quad \text{i.e.} \qquad \Lambda S \approx 1 \tag{16}$$

The creation pressure p_c [1] has been presented as equivalent to the effect of a cosmological constant Λ. A creation pressure expressed as Ω_{cp}, a dimensionless parameter i.e.

$$\Omega_{cp} = (8\pi/3)\, Gp_c\,/\,(c^2H^2) \tag{17}$$

as usually done in cosmology, is equivalent to the effect of a cosmological constant Λ, with omega parameter Ω_Λ, if and only if the following relation holds:

$$-\,\Omega_{cp} \equiv 3\Omega_{\Lambda} \tag{18}$$

This follows from the first of the cosmological equations of Einstein, i.e.

$$1 - 2q + \Omega_p + \Omega_k = 3\,\Omega_\Lambda \tag{19}$$

Here q is the deceleration parameter and Ω_k the curvature. If we consider a creation pressure instead of a cosmological constant, usually taken as the dark energy constituent of the universe, we get from (18) and (19)

$$1 - 2q + \Omega_k = -\,\Omega_{cp} \tag{20}$$

And using the present observations that give $\Omega_k \ll 1$ we finally get for the creation pressure, instead of a cosmological constant

$$\Omega_{cp} = 2q - 1 \tag{21}$$

The present estimates [15] of the numerical values of the deceleration parameter q are: for very high red shift, close to the initial stages of the universe, $q \approx 0.5$ which implies $\Omega_{cp} \approx 0$. The initial creation pressure is very small, corresponding to a small dark energy component, if any. At this stage we should expect a small acceleration of the initial expansion that balances the gravitational attraction (may be after inflation has finished in a very short time). At the present time [15] we have the approximate value $q \approx -0.5$, which implies $\Omega_{cp} \approx -2$. The present creation pressure is then pretty high. From (18) it would correspond to a value of $\Omega_\Lambda \approx 2/3$, in complete agreement with the very well known value of this parameter for today. There is no known reason for this negative increase in the creation pressure (positive increase in Λ and therefore in accelerating the expansion of the universe) to stop in the near future. We can extrapolate and consider the rather strong possibility that the universe will spread to infinity, in a finite time, due to an ever increasing accelerated expansion [8].
The creation pressure is related to the creation rate Γ of the mass M [1] by the following expression

$$\Gamma = \dot{\rho} / \rho + 3\,\dot{R}/R = d\,(\ln \rho\ R^3)/dt = d\,(\ln M)/dt = \dot{M}/M \tag{22}$$

The creation pressure p_{cp} is defined in terms of the creation rate Γ and other physical quantities [1] and is

$$P_{cp} = -\,\rho\, c^2\, (\Gamma/3H) \tag{23}$$

If we consider the universe as a black hole [6] then we have

$$2\,GM/c^2 = R \qquad \text{i.e.} \quad \dot{M}/M = \dot{R}/R = H = \Gamma \tag{24}$$

where H is the Hubble parameter. The creation pressure in (23) becomes

$$p_{cp} = -\,(1/3)\,\rho\, c^2 \tag{25}$$

5. The cosmological constant versus the energy of the information

We can think of our universe as a kind of "quantum" black hole [6] and apply the Hawking-Bekenstein [10] and [11] formulation for its entropy S. Using the black hole relation (2) between its mass M and its size *a(t)*

$$2GM/c^2 = a(t) \tag{26}$$

and combining (11) and (26) we get (with the linear relation *a(t) ≈ ct)*

$$S = 4\pi\, k/\hbar c\; G(c^3t/2G)^2 = \pi\, k/\hbar\, (c^5/G)\, t^2 \tag{27}$$

And in natural units $G = c = \hbar = k = 1$ we finally get

$$S \approx t^2 \tag{28}$$

Going on using natural units, in Planck´s units of time we have then from (28)

$$S \approx 10^{122} \tag{29}$$

The entropy of the universe increases with time and will arrive at a maximum at $x = 2$, its lifetime, and has a value of the order of $\sim 10^{122}$.
The quantum of gravity with mass m_g has been presented [7] as

$$m_g = \hbar/(c^2t) \approx 10^{-65} \text{ grams} \tag{7}$$

Since the mass of the universe M is about 10^{56} grams, one has the number of gravity quanta N_g in the universe as

$$N_g \approx M/m_g \approx 10^{122} \tag{30}$$

The two very large numbers in (29) and (30), being of the same order of magnitude, give us a very strong reason to believe that the entropy S of the universe is the number of gravity quanta, as proposed 10 years ago [7], and this is the number of bits I that it contains:

$$I \approx S \approx N_g \approx M/m_g \approx 10^{122} \tag{31}$$

Then, the unit of information, the bit, can be interpreted as having a mass m_g and an energy $m_g c^2 \approx \hbar/t$, i.e.the quantum of gravitational energy $\hbar\omega \approx 10^{-45}$ ergs.
Now we can check the holographic principle, [12] and [13], for the universe: the amount of information (31) inside the whole universe is equal to the area of the event horizon in Planck´s units (28) and (29).

6. The accelerated expansion of the universe

The deceleration parameter q was defined in terms of the scale factor a(t) and its derivatives as follows:

$$q = -a''a/(a')^2 \tag{32}$$

We see that $\ddot{a}$ being a deceleration one has $\ddot{a} < 0$ and then the parameter q should be $q > 0$ for deceleration and $q<0$ for acceleration. We can take into account the definition of the Hubble parameter H

$$H = \dot{a}/a \tag{33}$$

So that equation (32) transforms to

$$H' + (1 + q)H2 = 0 \tag{34}$$

The importance of this equation cannot be overestimated. It means that given the measured values of q [15] one can approach its time variation by the linear relationship:

$$q(x) = - x + ½ \quad (35)$$

where we have defined $x = t/t_0$ the ratio of any age of the universe t to the present age of the universe $t_0 \approx 1.37\ 10^{10}$ years. Then close to the beginning of the universe we have $q \approx ½$ (i.e. $x = \varepsilon << 1$) and today $q \approx - ½$ ($x_0 \approx 1$). Rearranging equation (34) with the change $dt = t_0\, dx$ we get

$$\dot{H} / H^2 = - d (1/H)/ (t0\, dx) = - [1 + q(x)] \quad (36)$$

And integrating we have

$$1/H\, t_0 = \int [1 + q(x)]\, dx \;\; + \text{constant} \quad (37)$$

Using (35) we get

$$1/H\, t_0 = 1.5\, x - 0.5\, x^2 + \text{constant} \quad (38)$$

Choosing the limits of integration from 0 to x and taking into account that the present value of H is $H_0 \approx 1/t_0$, for x = 1, the constant in (38) has the value zero. With (33) equation (38) is then equivalent to

$$t_0\, \dot{a}/a = [1.5\, x - 0.5\, x2]\text{-}1 \; = d \ln a \,/\, dx \quad (39)$$

Integrating once more we get

$$\ln a = \int [1.5\, x - 0.5\, x2]\text{-}1\, dx \;\; + \ln a_0 \quad (40)$$

where a_0 is the present value of the cosmological scale parameter $a(t_0)$

$$a/a_0 = \exp \{ \int [1.5\, x - 0.5\, x2]\text{-}1\, dx \; \} \quad (41)$$

integrating (41) we have

$$a/a_0 = \exp\{(2/3) \ln [2x/(3\text{-}x)]\} = [2x/(3\text{-}x)]^{2/3} \quad (42)$$

The plot of this expression is shown below in Fig. 1.

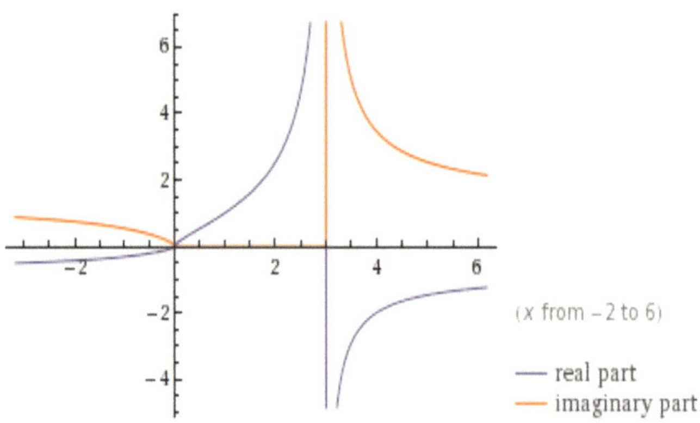

Fig. 1. In this figure 1 we have the plot of the scale factor of the universe (vertical axis), relative to its present value a_0. in terms of time t (horizontal axis), relative to the present age of the universe t_0. An infinite expansion appears at $t = 3t_0$.

7. The final inflation

Having used only the relations (32), (33), (34) and (35), without using the field equations of general relativity (only the observed values of the deceleration parameter q), the predicted final "inflation" at $t_f = 3t_0$ is a result of an extrapolation towards the future. The present day observations of q cover 1/3 of this time interval and strongly support the final expansion, the finite lifetime of the universe in a surprisingly rather short time from now (only 2 aeons).

We speculate that the initial inflation may have started from the Planck´s quantum black hole, bringing the universe close to its present size. After that, an almost linear expansion goes on due to the creation pressure, thus bringing the universe to its present size. The final inflation follows at about $4x10^{10}$ years of age, giving a finite lifetime for our universe. This is clearly an unexpected result that comes from the present observations of the values of the deceleration parameter q.

8. Conclusions

The generalization of the concept of a quantum black hole (giving the sub-Planckian scale, the Planckian scale and the scale of our universe) shows that there is a numerical factor, $\sim10^{61}$, that is equivalent to the total age of the universe in Planck´s units. It looks like this is the characteristic lifetime of a universe, in terms of the successive factors for the different scales, 10^{-61} , 1, 10^{61} , (or in terms of the generalized Planck´s constant, 10^{-122} , 1, 10^{122} . The age of a universe is intimately related to the choice of the unit of time interval. For the sub-Planckian scale we have 10^{-104} seconds, for the Planck scale 5 10^{-44} seconds and for our universe about 5 10^{17} seconds.

The picture that arises for the evolution of the universe is: no big-bang, an initial inflation (an exponential expansion) of a quantum black hole, Planck´s type, a slow deceleration followed by a slow acceleration. Then we have an almost linear expansion at the present time. And a final disaggregation to infinity at about 4 10^{10} years of age, the lifetime of our universe.

The cosmological constant Λ can be substituted by a creation pressure. This is in line with the idea of gravitation being an emerging entropic force. For the existence of this force an increase in mass with time (a Mass-Boom, [14) is necessary, giving a positive gradient of entropy for the universe and therefore the emergent gravitation.

9. Appendix

We are going to calculate now the following important cosmological parameters, in terms of the dimensionless age, $x = t/t_0$, and relative to the present size of the universe $a_0 = 1$:

1. The speed of expansion of the universe $a'(t)$
2. The Hubble parameter $a'(t)/a(t)$
3. The acceleration of the expansion $a''(t)$
4. The deceleration parameter $q= - a''(t)\, a(t)/ a'(t)^2$
    1. The speed of expansion of the universe $a'(t)$. We find the derivative of the scale factor $a(t)$ in (42) as

$$a`(t)/a(1) = 4\,(2x)^{-1/3}\,(3-x)^{-5/3} \tag{43}$$

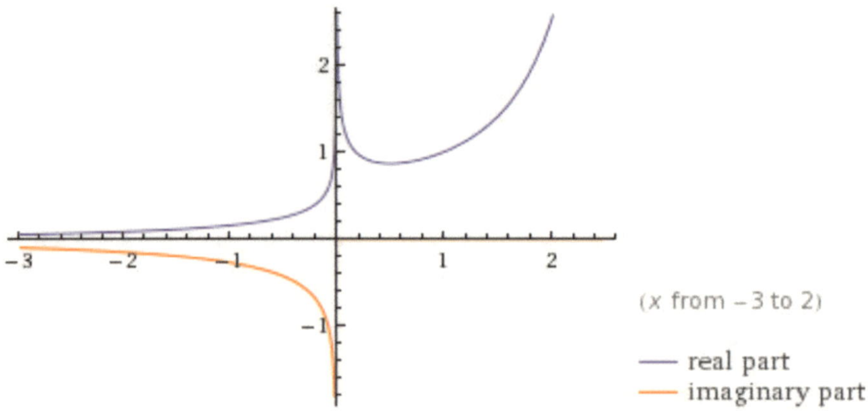

Fig. 2. The speed of expansion of the universe as in (43). There are two vertical asymptotes at x = 0 and at x = 3. They imply the initial inflation (x = 0) and the final disaggregation to infinity (x = 3) at about 4 10^{10} years.

2. The Hubble parameter H = *a´(t)/a(t)*. If we divide the expression (43) by the expression (42) we get for H

$$H = 2/x(3-x) \tag{44}$$

The following figure 3 gives the graph of this expression:

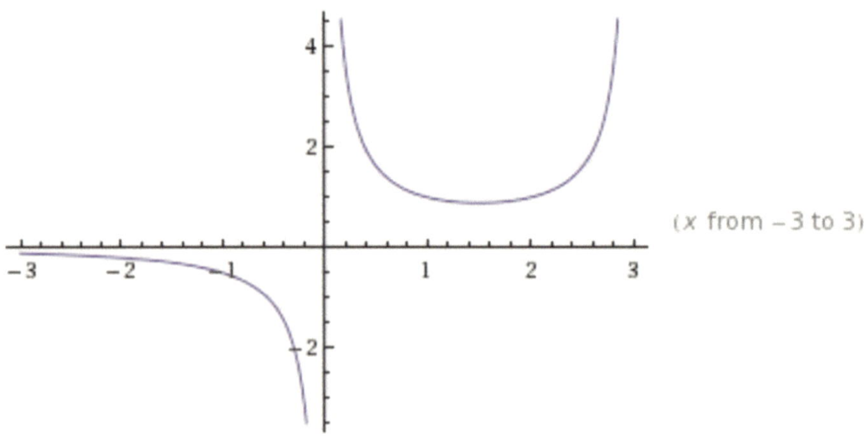

Fig. 3. The Hubble parameter H in terms of age x. We see again the initial inflation (x=0) and the final (x=3) disaggregation given by the two vertical asymptotes.

3. The acceleration of the expansion *a´´(t)*. Differentiating once more the expression (43) we get for the acceleration of the expansion of the universe

$$a'' = -8/3\,(1/2x)^{4/3}\,(1/(3-x))^{5/3} + 20/3\,(1/2x)^{1/3}\,(1/(3-x))^{8/3} \tag{45}$$

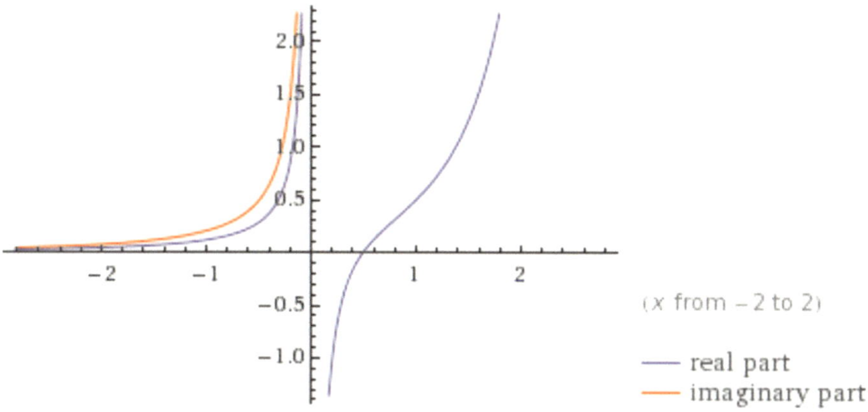

Fig. 4. The acceleration of the expansion of the universe is seen here again with two vertical asymptotes. Close to the origin the negative acceleration suggests the action of gravitation balancing the inflation phase. After half of the present age of the universe we see a positive acceleration, growing, and due to the pushing force that grows with the increasing size of the universe.

4. The deceleration parameter

$$q = - a''(t)\, a(t) / a'(t)^2 = 0.5 - x \quad (46)$$

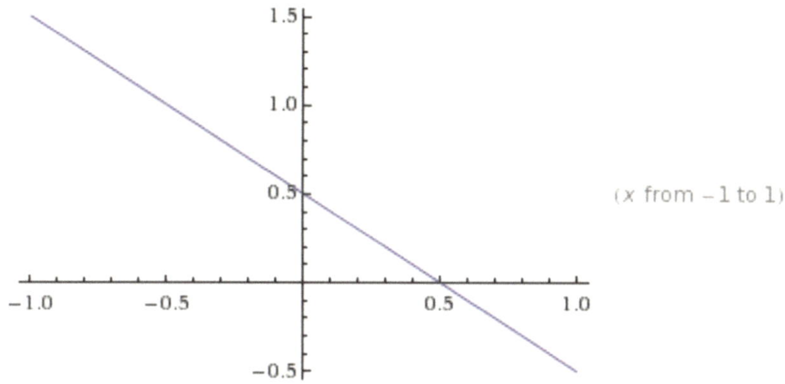

$$q = 0.5 - x \quad (46)$$

Fig. 5. The deceleration parameter. Using the expressions (42, (43) and (45) that define q gives back the original function assumed for q in (35).

9. Acknowledgement

I am thanking the owners of the Wolfram Mathematica Online Integrator that I have used to obtain the Fig. 1, 2, 3, 4 and 5 of this work.

10. References

[1] Hiterer, D., "The Accelerating Universe" arXiv: 1010.1162v1 astro-ph.CO, 6 Oct. 2010
[2] Lima J.A. et al. "CDM Accelerating Cosmology as an Alternative to ΛCDM model", arXiv: 0911.5727v1, astro-ph.CO, 30 Nov. 2009
[3] Alain de Roany and J.A. de Freitas Pacheco, "Continuous matter creation and the acceleration of the universe: the growth of density fluctuations" arXiv: 1007.4546v1, gr-qc, 26 Jul. 2010
[4] Lima J.A. et al. "Note on Continuous matter creation and the acceleration of the universe: the growth of density fluctuations" arXix; 1012.5069v1, astro-ph.CO, 22 Dec. 2010
[5] Alain de Roany and J.A. de Freitas Pacheco, "Continuous matter creation and the acceleration of the universe: a replay" arXiv: 1012.5781v1 astro-ph.CO, 28 Dec. 2010
[6] Alfonso-Faus, A., "The case for the Universe to be a quantum black hole" arXiv: 0912.1048 and Journal of Astrophys. Space Sci. 325:113-117, 2010
[7] Alfonso-Faus, A., "Gravity Quanta, Entropy and Black Holes" arXiv: gr-qc/0002065 and Journal: Phys. Essays 12 (1999) 673-677
[8] Alfonso-Faus, A., arXiv:1102.0443 "Cosmology, Holography, the Brain and the Quantum Vacuum" Also presented as "Plenary Lecture" WSEAS Aiked 11 Cambridge, 21-22 February 2011.
[9] Verlinde, Erik, "On the Origin of Gravity and the Laws of Newton", arXiv: 1001.0785, Jan. 2010
[10] Hawking, S.W., (1974), "Black hole explosions?" Nature 248, 30
[11] J. D. Bekenstein, (1973), Phys. Rev. D 7, 2333
[12] 't Hooft, Gerard (1993). "Dimensional Reduction in Quantum Gravity". *Preprint*. arXiv:gr-qc/9310026.
[13] Susskind, L., (1995) "The World as a hologram," J. Math. Phys. 36, 6377 and arXiv:hep-th/9409089.
[14] Alfonso-Faus, A., "Entropy, Gravity and the Mass-Boom" arXiv: 1009.2254, September 2010
[15] Santos, B. et al., "Current constraints on the epoch of cosmic acceleration" arXiv: 1009.2733v1, astro-ph.CO, 14 Sep. 2010